AF528606

Ornamental Fish in Hobby and Harvest

THE AUTHOR

Dr. Sagarika Chaudhuri completed her graduation from Miranda House, Delhi, obtained her M.Sc. and a Ph.D. degree from Delhi University and the University of Calcutta, respectively. She was a research fellow in Zoological Survey of India, Calcutta, and subsequently a Research Associate (CSIR)at the University of Calcutta. Later she joined teaching in Sree Chaitanya College, Habra, West Bengal. She has experience of twenty years of teaching at undergraduate along with a few years in Post-graduate level also. She completed three Minor Research Project (UGC) and one Major Research Project (DST). Dr. Chaudhuri has several research papers in reputed science journals. She authored an Honours level text-book entitled '*Economic Zoology*' in 2017.

Ornamental Fish in Hobby and Harvest

Sagarika Chaudhuri
M.Sc., Ph.D.
Former Associate Professor & Head
Department of Zoology
Sree Chaitanya College, Habra
Affiliated to West Bengal State University
West Bengal, India

2022
Daya Publishing House®
A Division of
Astral International Pvt. Ltd.
New Delhi – 110 002

ISBN: 978-93-90435-60-9

Published by : **Daya Publishing House®**
A Division of
Astral International Pvt. Ltd.
– ISO 9001:2015 Certified Company –
4736/23, Ansari Road, Darya Ganj
New Delhi-110 002
Ph. 011-43549197, 8130496929
E-mail: info@astralint.com
Website: www.astralint.com

Digitally Printed at : **Replika Press Pvt. Ltd.**

Dedicated

To the Memories of

my

Father—Murari Prosad Guha

Mother—Dipali Guha

Foreword

Ornamental fish keeping and its propagation have been an interesting activity for many, which provide not only aesthetic pleasure but also financial openings. About 2500 ornamental fish species have been reported worldwide and Indian waters possess a rich diversity of ornamental fish, with biodiversity hotspots like northeastern states, Western Ghats, Andaman & Nicobar Islands, and Lakshadweep. These fishes are fascinating organisms, abundantly distributed in freshwater, brackish water, and inshore marine ecosystems around us. They are often called 'living jewel' due to their colour, shape, behavior, and origin.

In recent years, ornamental fishes are receiving increased attention due to the local and global demand and the consequent growth of the export market and trade. Despite having immense potential, India's share of the global ornamental fish trade is less than one percent but, still she is projected as a "sleeping giant" because of yet untapped potential resources. If the resources are managed properly, India can be one of the leaders of the world's ornamental trade in the coming years.

More than 100 varieties of indigenous freshwater ornamental fish species are known in Indian waters. Though some of the species fetch a high price in the world market and support trade outside the country, it is regrettable that the endemic ornamental species were until recently neither promoted nor their potential realized.

In India, the ornamental fish sector is facing many problems including technical and economic aspects. The major constraints are lack of technical know-how, scientific information regarding breeding, types of feeds and feeding, and health management. Most of the entrepreneurs are not aware that this upcoming sector/ industry is far more lucrative than the shrimp, involving much less capital and little risk. India could not make any big headway in the export of marine ornamental fishes so far due to the non-availability of infrastructure. Most of the indigenous ornamental fishes of India are low-priced and used as food fish in rural areas/ small towns due to lack of awareness. The Indian ornamental fish trade is not well organized in many aspects. To conserve these species and to get economic benefit sustainably, development and due attention in this sector are required. More intensified research and development in both fresh, brackish, and marine water fishes can lead to improved culture technologies for many species of demand in the aquarium trade in the globe. More initiatives by the government like providing an incentive to established ornamental fish production units can attract considerable private investment to this industry,which would generate additional employment opportunities and improvement of livelihood of such fish producing communities.

The scope of our scientific understanding and technical skills in ornamental fish culture and trade have widened significantly over the years. The title of this book as "Ornamental Fish in Hobby and Harvest" written by Dr. Sagarika Chaudhuri is very appropriate since the discipline of knowledge concerns the sustainable development of a process linked with human wellbeing. Serving as a definitive guide this distinguished book is indeed a bounty for the students, researchers, hobbyists, and entrepreneurs, who would prefer to enter into the world of ornamental fishes. In this endeavour the author has contemplated evaluating the importance of these marvellous fishes and sketch out an account of varied aspects on the subject to develop a deeper understanding.

I believe that the people associated with ornamental fish keeping in various ways will find the present effort by the author interesting, useful, and worthy.

Dr. Sumit Homechaudhuri,
Professor, Department of Zoology,
University of Calcutta,
35, Ballygunge Circular Road,
Kolkata-19.

Preface

Attractive colourful fishes kept in enclosed tanks and containers have often been referred to as aquarium fishes or ornamental fishes. These are sometimes referred to as living jewels also. Keeping such fishes is considered the second-largest hobby in the world next to photography. At a global level, ornamental fish farming is fast emerging as a major branch of aquaculture, as these fishes have tremendous revenue earning potential. At present more than 125 countries are involved in this trade and there are about 2,500 species of ornamental fishes in the market. The growing importance of such fishes has also opened up huge economic opportunities and prospects for aquarium accessories, aquatic plant industry,with entrepreneurial development gaining popularity. Nowadays more and more people are becoming engaged in this business of cultivating and breeding ornamental fishes through farming.

India contributes only 0.4% of the total world export of ornamental fish and is considered a sleeping giant with yet untapped potential. So, for the development of such fisheries in India, initiatives, strategies, and plans of action along with substantial financial investments have been promulgated by the Ministry of Fisheries, Animal Husbandry & Dairy, Government of India along with State Fisheries. The increasing importance of ornamental fish keeping is being reflected in the course curriculum in the Fisheries colleges, Fisheries, and Agricultural Universities, National Research Institutes all over India. Despite its importance and inclusion in the curriculum, there is still an acute dearth of publications on the subject in an absolute-approach manner. And,so there is an essential need for a publication that is following the syllabus and is also learner-friendly.

Because of these, the endeavour here is to survey the fabulous disposition of the world of alluring fishes and provide a combined account of observations, contemplations, and rendition on the subject. The challenge is to act as a communicator who allows reading,ponder to wonder at the many mysteries, and develop a deeper understanding of it. This compact book on the modern field of a worthwhile biological inquest is intended for newcomers embarking on this fascinating study, to facilitate the hobbyists, and fish keepers in all spheres, as the subject is challenging both in teaching and learning.

It also aims to provide substantiated information on almost all the imperative areas of ornamental fish keeping. This book serves as a definitive guide towards identifying freshwater, brackish water, and marine fishes because the golden rule is always to recognize your work first. It further deals with the general characteristics, evolution, and taxonomic position of fishes in the animal kingdom.

Freshwater, brackish water, and marine ornamental fishes, both indigenous and exotic, have been listed and described.

Other aspects of ornamental fish keeping, such as history, aquatic setups, fabrication, aquarium accessories, water quality, water plants, food and feeding, breeding, other ornamentals in the trade, transportation, and diseases are considered in detail. Developments in ornamental fisheries, genetics and biotechnology, management practices, the establishment of ornamental fish farms, economics, employment opportunities, and ornamental fish trade have been contemplated to serve as a guiding verse, through the basic principles of aquarium management.

Author

Acknowledgements

I am very much thankful to Mr. Deepak Nopany of Asian Exports, for kindly permitting us the visits to his ornamental fish farm for the collection of some valuable data and photographs of fishes and aquariums, etc. Moreover, he provided a few photographs of indigenous ornamental fishes of India from his collection.

I convey my sincere thanks to the Director, Astral International Pvt Ltd for his support behind the publication of this book. I thankfully acknowledge Mr. Gopal Singh (co-ordinator, Astral International Pvt Ltd.), Mr. Kanav Kansra, and other staff members for their help in bringing the manuscript to the final form.

I sincerely thank Prof. (Dr.) Sumit Homechaudhuri for contributing the Foreword of the book, and for encouragement.

Artist Mr. Niranjan Giri is thankfully acknowledged for some of the drawings.

I am glad to have gathered much information from various aquariums inside and outside India for this book. I earnestly thank the people involved in selling ornamental fishes in huts (weekly local markets) and rearing and culturing these fishes on small and large scales. They were always a great source of information at various steps of the present project.

The role of those very near and dear to me in the family deserves special mention. The book was initiated and completed with the constant support and help from my husband Prof. (Dr.) Sibendu Chaudhuri. He took most of the photographs and drew some of the illustrations. He also helped me in undertaking various field works and tours, within the country and abroad,for the preparation of this book.

Inspiration to assume this project came from Dr. Swetaprovo Chaudhuri, my elder son, and Dr. Amrita Chaudhuri, my daughter-in-law. Mr. Chandramauli Chaudhuri, my younger son, besides being a source of encouragement provided a few photographs of some marine animals for the book.

I remain grateful to my former colleagues, friends,and students who were the source of inspiration to me.

Author

Contents

Foreword		*vii*
Preface		*ix*
Acknowledgment		*xi*
1.	**Recognising a Fish**	**1-6**
1.1	Introduction	1
1.2	Identifying Characters of a Fish	1
1.3.	Evolution of Fishes	1
1.4	Classification of Fishes	2
1.5	References	6
2.	**Ornamental Fish Diversity**	**7-390**
2.1	Introduction	7
	DIVISION 1: FRESHWATER ORNAMENTAL FISHES	
2.2	Freshwater Ornamental Fishes of India	7
2.3	An Explicit Endeavour of some Indian Freshwater Ornamentals	52
2.4	Freshwater Ornamental Fishes in the World Trade	103
2.5	Detail Endeavour of Popular Ornamental Freshwater Fishes of the World Trade	146
	DIVISION 2: BRACKISH WATER ORNAMENTAL FISHES	
2.6	Brackish Water Ornamental Fishes of India	222
2.7	Brackish Water Fishes in the Global Ornamental Fish Trade	231
2.8	Explicit Endeavour of some Brackish Water Ornamental Fishes	234
	DIVISION 3: MARINE ORNAMENTAL FISHES	
2.9	Marine Ornamental Fishes of India	245
2.10	Popular Marine Ornamental Fishes in the Trade	248
2.11	An Explicit Endeavour of some Marine Ornamental Fishes	304
2.12	References	388
3.	**Anecdote of Ornamental Fish Keeping**	**391-409**
3.1	History of Keeping Ornamental Fishes as Pets	391
3.2	The Ornamental Fish Abode—Aquatic Set-ups	392

	3.3 Ideas and Implementations Related to Aquariums	394
	3.4 References	409
4.	**Fabrication and Setting Up of Aquarium**	**410-418**
	4.1 Fabrication of an Aquarium	410
	4.2 Setting Up of Aquarium	414
	4.3 List of Accessories and Equipment Used	416
	4.4 References	418
5.	**Equipment, Accessories and Decors**	**419-442**
	5.1 Algae Scrubber	419
	5.2 Berlin Method	420
	5.3 Deep Sand Bed	421
	5.4 Live Sand	421
	5.5 Live Rock	421
	5.6 Substrate	422
	5.7 Bog-Wood	428
	5.8 Drift Wood	428
	5.9 Fish Cam	428
	5.10 Sump	429
	5.11 Air Stone	429
	5.12 Aquarium Heater	429
	5.13 Thermometer	430
	5.14 Aquarium Lighting	430
	5.15 Aquarium Fish Feeder	431
	5.16 Power Head(Pump)	431
	5.17 Aeration	431
	5.18 Aquarium Filters	434
	5.19 Systems in Marine Aquariums	440
	5.20 Calcium Reactor	441
	5.21 Refugium	441
	5.22 References	442
6.	**Water Quality Management**	**443-448**
	6.1 Water Chemistry	443
	6.2 Aquarium Care	447
	6.3 References	448

7. Aquarium Plants **449-498**

7.1 Introduction 449

7.2 Classification and Types of Aquatic Plants 449

7.3 Importance of Aquarium Plants 450

7.4 List of Freshwater Aquarium Plant Species 450

7.5 Aspects of some Popular Freshwater Plants chosen as Ornamentals 452

7.6 Brackish Water Ornamental Plant Species 483

7.7 Aspects of some Brackish Water Aquarium Plants 484

7.8 Marine Aquarium Plant Species 486

7.9 Aspects of some Marine Aquarium Plants 486

7.10 Propagation of Aquarium Plants 489

7.11 References 497

8. Ornamental Invertebrates and Vertebrates **499-540**

8.1 Introduction 499

8.2 Freshwater Aquarium Invertebrates and Vertebrates 499

8.3 Account of Freshwater Aquarium Species 500

8.4 Brackish Water Aquarium Invertebrates and Vertebrates 504

8.5 Detail of Brackish Water Ornamental Vertebrate Species 504

8.6 Marine Ornamental Invertebrates and Vertebrates 506

8.7 Detailed Account of some Marine Invertebrates and Vertebrates 510

8.8 References 540

9. Nutriment and Feeding of Ornamental Fishes **541-566**

9.1 Nutritional Requirements 541

9.2 Feeding Practices Followed for Ornamental Fishes 543

9.3 Feeding Rate and Frequency 543

9.4 Feeds and Feed Management 544

9.5 Marine Ornamental Fish Diets 552

9.6 Live Fish Food Organisms 553

9.7 Some Live/Natural Fish Food Organisms 554

9.8 Colour Enhancers in Ornamental Fish Feed 562

9.9 Recommendations 564

9.10 References 565

10. Reproductive Strategies in Ornamental Fishes **567-590**

10.1 Introduction 567

10.2 Breeding of Egg Layers 567

10.3 Captive Breeding Protocols 568
10.4 Stimulating Factors for Spawning 571
10.5 Breeding of some Egg Laying Ornamental Fishes in Captivity 572
10.6 Breeding of Livebearers 579
10.7 Success in Captive Breeding of Indigenous Ornamental Fish 588
10.8 Marine Ornamental Fish Culture and Breeding 589
10.9 References 590

11. Transport of Ornamental Fishes 591-598

11.1 Introduction 591
11.2 Basic Principles for Transport of Ornamental Fish 591
11.3 Pre-Conditioning Preparations—Step 1 592
11.4 Conditioning of Fish before Packing—Step 2 592
11.5 Packing—Step 3 593
11.6 Transporting and Transferring—Step 4 596
11.7 Conclusion 597
11.8 References 597

12. Ornamental Fish Diseases and Their Control 599-626

12.1 Introduction 599
12.2 How to Prevent Fish Diseases 599
12.3 Stress Related Diseases in Ornamental Fishes 600
12.4 Non-Infectious Diseases 601
12.5 Infectious Diseases 601
12.6 Protozoan Diseases 601
12.7 Internal Protozoan Parasites Causing Diseases 606
12.8 Bacterial Diseases 608
12.9 Fish Viral Disease 612
12.10 Fungal Diseases in Ornamental Fishes 615
12.11 Helminth Diseases 616
12.12 Acanthocephalan Diseases 619
12.13 Nematode Infection 619
12.14 Leech Infection 619
12.15 Crustacean Diseases 620
12.16 Antibiotics used in Ornamental Fish Culture 622
12.17 Quarantine and Health Certification Measures to Secure Health Status of Fish 623

12.18 Immunodiagnostic Methods in Ornamental Fish Disease Detection 623
12.19 Immunoprophylaxis 625
12.20 Future Strategies for Health Management 626
12.21 References 626

13. Application of Genetics and Biotechnology in Ornamental Fish Keeping 627-632

13.1 Introduction 627
13.2 Cytogenetic Studies in Fishes 627
13.3 Application of Biotechnology 630
13.4 Genetic Engineering and Transgenic Fishes 630
13.5 References 632

14. Application of Best Management Practices in Ornamental Fish Farming 633-635

14.1 Introduction 633
14.2 Adoption of BMPat the Ornamental Fish Farm Level 633
14.3 Appraisal of Adopting BMP 634
14.4 References 635

15. Entrepreneurship Development and Economics of Fish Breeding and Culture Units 636-644

15.1 Introduction 636
15.2 Essential Approaches for Successful Breeding and Culture of Ornamental Fishes 636
15.3 Culturable Ornamental Fish Species 638
15.4 Production Facilities 638
15.5 Types of Production Units in Ornamental Fish Farming Business 640
15.6 Use of Rain Water Harvesting System and Water-Recirculation Facility 640
15.7 Eco-Friendly Management of the Farm 641
15.8 Profit Analysis 641
15.9 References 644

16. The Potential Scope of Ornamental Fish Industry and the Trade 645-656

16.1 Introduction 645
16.2 Present Panorama 645
16.3 Ornamental Fishes in the Global Fish Industry 646
16.4 Recent Supply Trends of Ornamental Fishes 646
16.5 Constraints 647

16.6 Green Certification and Ecolabelling 647

16.7 Ornamental Fish Trade in India 648

16.8 Ornamental Fish Trade and Employment Opportunities in India 649

16.9 Recent Initiatives taken for the Development of Ornamental Fisheries in India 653

16.10 Strategies Adopted by the PMMSY,Government of India 654

16.11Action Plan for Development of Ornamental Fisheries in India: Aquarainbow Vision—2030 654

16.12 Financial Resources 655

16.13 Recommendations for Further Improvement in Ornamental Fisheries 655

16.14 References 655

Index **657-669**

1

Recognising a Fish

1.1 Introduction

Fishes, the aquatic vertebrates, satisfy human needs in various ways. They are found as food on one hand and as an object of hobby on the other, comprising both trade and livelihoods. While beginning with the entire endeavour on ornamental fishes which may be simply said as an object of hobby and trade, it is fine to understand a fish first, by viewing its general characteristics, evolution and its taxonomical position in the Animal Kingdom.

1.2 Identifying Characters of a Fish

Fishes are cold-blooded aquatic vertebrates that breathe using pharyngeal gills, propelling and balancing themselves by means of unpaired and paired fins. These vertebrates are identified primarily on the basis of some external morphological features and anatomical features. Among fishes, some (**cartilaginous fishes**) are with cartilaginous skeleton consisting of denticles (placoid scales). The skull lacks sutures, teeth are placed serially and usually not fused to jaws, fin rays are soft, swim bladder and lung absent. Fertilization is internal. On the other hand, in **bony fishes**, internal skeleton is highly ossified, skull is with sutures, lung or swim bladder are present. Homocercal tail is present, circular scales are either cycloid or ctenoid, and skull is with complex jaw mobility (Young, 1981; Pough et al., 1999; Kardong, 2007).

1.3 Evolution of Fishes

The earliest known vertebrates are a group of jawless vertebrates, whose teeth and dermal bones (armoured skin plates) appear in marine deposits of the late Cambrian period and whose fossils extend into the late Devonian (Strickberger, 2000). These jawless groups called "Ostracoderms" (bony skins), had a large muscular pharynx, and, they lasted till the Devonian period. New jawed and finned vertebrates arose during the mid-Devonian using hard bony tissue for structure and defence. Cartilage also appeared as the major structural tissue. The first jawed vertebrates appeared during the Silurian period and are divided into acanthodians and placoderms. The presence of jaws and the further development of paired fins were significant features in both acanthodians and placoderms. Entirely new forms, evolving from a group of jawed fish, began to appear during the Devonian period.

The earliest jawed groups, acanthodians and placoderms, died out and more efficient swimmers, cartilaginous fishes, and bony fishes with swim bladders replaced them (Strickberger, 2000). These two groups possess elaborate and highly evolved fin patterns including a caudal fin, dorsal and ventral fins for stability, and paired pectorals and pelvic fins for control. The early bony fishes are grouped into Actinopterygii (ray-finned fish) and Sarcopterygii (flesh-finned). Living descendants of the former group are sturgeon, garpike and teleost. Among these, teleost is the most successful of bony fishes as there are changes in skull, fin and scale structures along with the feeding and locomotory adaptations.

1.4 Classification of Fishes

About 32,000 species of fishes are present in the world. These vertebrates arose and began to radiate more than 500 million years ago and exhibit now and, in the past, incomparable diversity in their morphology, in the habitats, they occupy, in their physiology, and in their behaviour (Nelson et al., 2016). Freshwater habitats comprise only a small population of the Earth's surface water, but contain a disproportionately large number of the world's fish species (Nelson et al., 2016).

Classification of living fishes:

Nelson et al., (2016) has been followed for the classification of all extant fishes.

PHYLUM: CHORDATA

Subphylum	UROCHORDATA (Tunicata: the tunicates)
Subphylum	CEPHALOCHORDATA (Acrania, in part)
Subphylum	CRANIATA
Infraphylum	MYXINOMORPHI
Infraphylum	VERTEBRATA (vertebrates)
Superclass	PETROMYZONTOMORPHI
Superclass	GNATHOSTOMATA (jawed vertebrates)
Grade	CHONDRICHTHYOMORPHI
Class	CHONDRICHTHYES----cartilaginous fishes
Subclass	HOLOCEPHALI
Subclass	EUSELACHII -- -**shark** (Figure 1.1) **ray** (Figure 1.2) and related fossils
Order	HETERODONTIFORMES—bullhead sharks
Order	ORECTOLOBIFORMES—carpet sharks
Order	LAMNIFORMES—mackerel sharks
Order	CARCHARHINIFORMES---ground sharks
Order	HEXANCHIFORMES----six-gill sharks
Order	SQUALIFORMES----dogfish sharks
Order	ECHINORHINIFORMES – bramble sharks
Order	SQUATINIFORMES – angel sharks

Order	PRISTIOPHORIFORMES-----saw sharks
Order	TORPEDINIFORMES----electric rays
Order	RAJIFORMES---skates
Order	PRISTIFORMES---guitarfishes and saw fishes
Order	MYLIOBATIFORMES---stingrays
Grade	TELEOSTOMI
Class	OSTEICHTHYES----bony fishes and tetrapods
Subclass	SARCOPTERYGII----lobe-finned fishes and tetrapods
Order	COELACANTHIFORMES----coelacanths
Order	CERATODONTIFORMES-----living lungfishes and their fossil relatives
Subclass	ACTINOPTERYGII------ray-finned fishes
Order	POLYPTERIFORMES----bichirs
Order	ACIPENSERIFORMES----paddlefishes and sturgeons
Order	LEPISOSTEIFORMES-----gars
Order	AMIIFORMES-------bowfins
Cohort	ELOPOMORPHA
Order	ELOPIFORMES------tenpounders
Order	ALBULIFORMES-----bonefishes
Order	NOTACANTHIFORMES---halosaurs and deep-sea spiny eels
Order	ANGUILLIFORMES----eels
Cohort	OSTEOGLOSSOMORPHA
Order	HIODONTIFORMES----mooneyes
Order	OSTEOGLOSSIFORMES-----bonytongues
Cohort	OTOCEPHALA
Order	CLUPEIFORMES----herrings
Order	ALEPOCEPHALIFORMES----slickheads and tubeshoulders
Order	GONORHYNCHIFORMES----milkfishes
Order	CYPRINIFORMES--**carp** (Figure 1.3) loaches, minnows and relatives
Order	CHARACIFORMES----characins
Order	SILURIFORMES-------**catfish** (Figure 1.4)
Order	GYMNOTIFORMES---Neotropical knife fishes
Cohort	EUTELEOSTEI
Order	LEPIDOGALAXIIFORMES----salamander fishes

Order	SALMONIFORMES-----trout, salmon and whitefish
Order	ESOCIFORMES----pikes and mudminnows
Order	ARGENTINIFORMES----marine smelts
Order	GALAXIIFORMES-----galaxiiformes
Order	OSMERIFORMES----freshwater smelts
Order	STOMIIFORMES------dragonfishes
Order	ATELEOPODIFORMES----jellynose fishes
Order	AULOPIFORMES-----lizardfishes
Order	MYCTOPHIFORMES----lanternfishes
Order	LAMPRIFORMES----opahs
Order	POLYMIXIIFORMES---beardfishes
Order	PERCOPSIFORMES----trout-perches
Order	ZEIFORMES----dories
Order	STYLEPHORIFORMES----tube-eyes or thread-tails
Order	GADIFORMES----cods and hakes
Order	HOLOCENTRIFORMES----squirrelfishes
Order	TRACHICHTHYIFORMES---roughies
Order	BERYCIFORMES----beryciforms
Order	OPHIDIIFORMES----cusk-eels
Order	BATRACHOIDIFORMES---toadfishes
Order	KURTIFORMES----nurseryfishes and cardinal fishes
Order	GOBIIFORMES----gobies
Order	MUGILIFORMES---mullets
Order	CICHLIFORMES----cichlids and convict blennies
Order	BLENNIFORMES---blennies
Order	GOBIESOCIFORMES----clingfishes
Order	ATHERINIFORMES----silversides
Order	BELONIFORMES-----needlefishes
Order	CYPRINODONTIFORMES----killifishes
Order	SYNBRANCHIFORMES----swamp eels
Order	CARANGIFORMES-----jacks
order	ISTIOPHORIFORMES----barracudas and billfishes
Order	ANABANTIFORMES----labyrinth fishes
Order	PLEURONECTIFORMES----flatfishes

Order	SYNGNATHIFORMES----pipefishes and seahorses
Order	ICOSTEIFORMES----ragfishes
Order	CALLIONYMIFORMES---dragonets
Order	SCOMBROLABRACIFORMES----longfin escolars
Order	SCOMBRIFORMES----mackerels
Order	TRACHINIFORMES----swallowers etc.
Order	LABRIFORMES----wrasses and relatives
Order	PERCIFORMES----perches
Order	SCORPAENIFORMES----mail-cheeked fishes
Order	MORONIFORMES----temperate basses
Order:	ACANTHURIFORMES----surgeonfishes and relatives
Order:	SPARIFORMES----breams and porgies
Order:	CAPROIFORMES----boarfishes
Order:	LOPHIIFORMES-----angler fishes
Order:	TETRADONTIFORMES----plectognaths (puffers, boxfishes, triggerfishes, filefishes, triplespines, porcupinefishes).

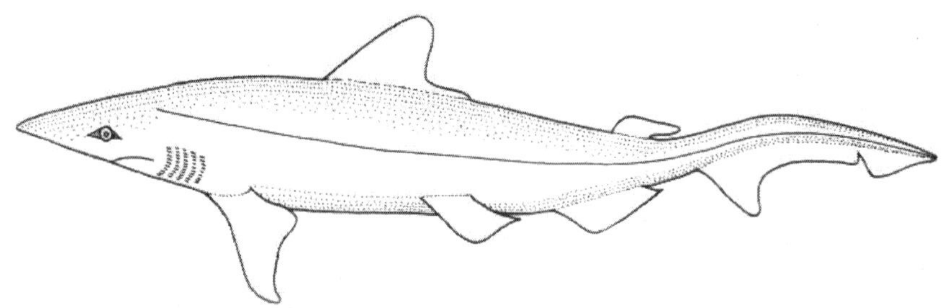

Figure 1.1 Shark

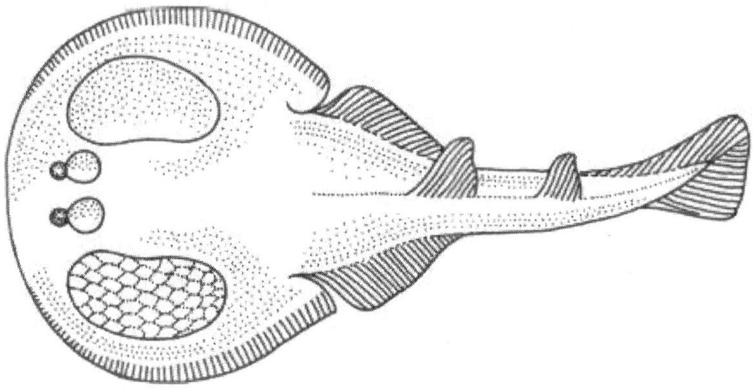

Figure 1.2 Ray

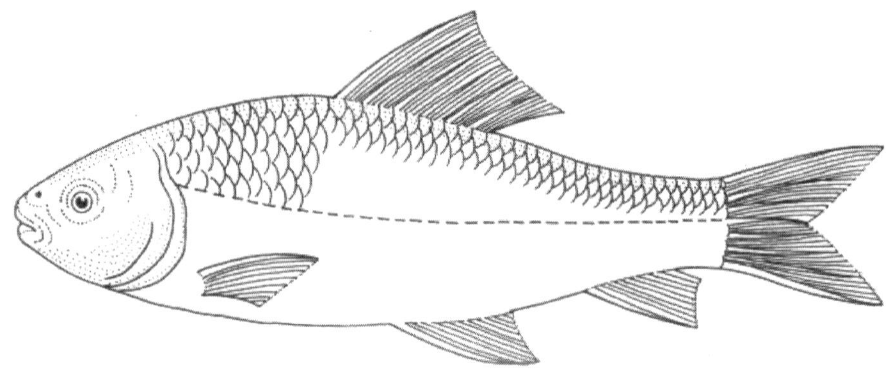

Figure 1.3 Carp

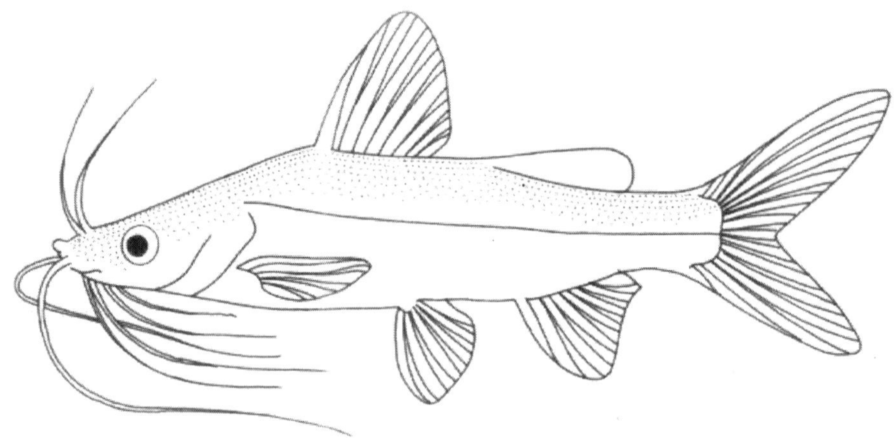

Figure 1.4 Catfish

1.5 References

Kardong K.V. *Vertebrates: Comparative anatomy, function, evolution*. 4th edition. Tata Mcgraw-Hill Publishing Company Limited. 2007, pp. 1-782.

Nelson J.S., Grande T.C., Wilson M.V.H. *Fishes of the world*. John Wiley & Sons, Inc. 2016, pp. 1-707.

Pough F.H., Heiser J.B., McFarland W.N. *Vertebrate Life*. Prentice-Hall, Inc. 1999, pp. 1-798.

Strickberger M.W. *Evolution*.3rd edition. Jones and Barlett Publishers. 2000, pp. 1-722.

Young J.Z. *The Life of Vertebrates*. 3rd edition. Oxford University Press. 1981, pp. 1-645.

2

Ornamental Fish Diversity

2.1 Introduction

More than 125 countries, all over the world, are involved in the **ornamental fish trade**, and there are about 2500 species of ornamental fishes available in the market. These ornamental fishes are generally marked either as **indigenous** (**native**) or **exotic** on the basis of their place of origin. And, are further categorized as **freshwater**, **brackish water,** and **marine**, based on their habitat and distribution. Popular ornamental fishes are also, placed under four categories. These are **tropical freshwater**, **tropical marine**, **brackish water** and **cold water**. Tropical freshwater fishes are the most popular ones as they are easily available and easy to keep. Depending on the breeding habits, ornamental fishes are also designated as egg layers and live bearers.

DIVISION 1: FRESHWATER ORNAMENTAL FISHES

2.2 Freshwater Ornamental Fishes of India

Freshwater systems *viz*., lakes, reservoirs and rivers though account for a small amount of about 2% of total water on the Earth (Alfred and Nandi, 2001), but contain a large number of the world's fish species.

Out of a rich biodiversity of 2,319 finfishes, distributed in different ecosystems of Indian waters (Swain et al., 2011), a large number of fishes are documented as inland fishes (Jayaram,1981; Talwar and Jhingran, 1991; Kapoor et al., 2002) and may be more than 806 species are inhabitants of freshwaters (Menon, 1999; Swain et al., 2010) and about 374 are indigenous freshwater fish species (Silas et al., 2011). North Eastern Region of India has a rich store of fish species, and over 50% of the 350, or, so, listed, are **endemics**. This **Hot Spot** region is rich with 250 ornamental fish species, and contributes 85% of the total trade, with export of 58 indigenous fishes. Western Ghats is with a resource of 327 fish species, of which 68% are endemics to the region and 155 are ornamental fish species (Karmakar and Das, 2005; Mercy et al., 2007; Sinha, 2008; Silas et al., 2011). The ornamental fish diversity is scattered throughout the country, out of these, about 33% are found in the North Eastern region, 24% in the southern region,23% in the eastern region, 6% in the western region, 3% in the northern region, 2% in central region and 9% throughout India (Swain et al., 2010). There is a great scope of more potential ornamental fishes being added (Sinha,2008; Chaudhuri, 2017) to the list (Table 2.1). Considering the abounding ornamental fish diversity throughout the country, there is an urgent need to build proper strategy and infrastructure to claim protection for as many biodiversity derived freshwater ornamental fishes as possible and a list of native freshwater ornamental fish species of India has been presented for **GI** (Silas, et al., 2011).

Table 2.1: Freshwater, indigenous egg-layer ornamental fishes

Grade: TELEOSTOMI; Class: OSTEICHTHYES (bony fishes); Subclass: ACTINOPTERYGII

Order: CYPRINIFORMES, Family: CYPRINIDAE, Subfamily: BARBINAE

Common Name	Scientific Name	Total length (TL) Standard length (SL) in cm	Distribution
Rosy barb	*Pethia conchonius* (Hamilton, 1822)	14.0 TL male/ unsexed	In the Ganga, Brahmaputra, Mahanadi and Cauvery river systems. In Assam, north-eastern Bengal, Bihar, Uttar Pradesh, Punjab and Maharashtra.
Golden dwarf barb	*Pethia gelius* (Hamilton, 1822)	5.1 TL male/unsexed	In the Ganga and Brahmaputra in northern India. Occurs in West Bengal, Bihar, Assam, Orissa and Madhya Pradesh.
Glass barb	*Pethia guganio* (Hamilton, 1822)	8.0 TL male/unsexed	Northern India, throughout the Ganga and Brahmaputra river systems with most of their major tributaries and in Assam, Orissa, Andhra Pradesh and Tamil Nadu.
Spotted sail barb, Dwarf barb, Phutuni barb	*Pethia phutunio* (Hamilton, 1822)	3.9 TL male/unsexed	Lower Ganges and Brahmaputra river basins in West Bengal, Bihar, Assam, Orissa, Tripura and Goa.
	Pethia atra (Linthoingambi & Viswanath, 2007)	5.8 SL male	Native/ Iril and Imphal rivers, Manipur ponds and lakes in Imphal valley.

Common Name	Scientific Name	Total length (TL) Standard length (SL) in cm	Distribution
	Pethia aurea Knight, 2013	2.36 SL male	Native/ ponds in South 24 parganas, West Bengal.
	Pethia canius (Hamilton, 1822)	3.0 SL male	Native/ Cooch Behar and Nadia in West Bengal.
	Pethia expletifois Dishma & Viswanath,2013	4.66 SL male	Native/Tuisi river in Mizoram.
	Pethia khugae Linthoingambi & Viswanath, 2007)	4.7 SL male	Manipur, Khuga River at Churachandpur district.
Long-tailed pethia	*Pethia longicauda* Katwate, Paingankar, Raghavan& Dahanukar, 2014	3.77 SL male	Native/ river in Kothapur district in Maharashtra.
Citron barb	*Pethia lutea* Katwate, Katwate, Paingankar & Dahanukar, 2014	3.9 SL male	Native/ Western Ghats.
	Pethia manipurensis (Menon, Rema Devi & Viswanath,2000)	6.2 SL male	Lake Loktak, Manipur.
	Pethia meingangbii (Arunkumar & Tombi Singh, 2003)	?	Native/ Yu river drainage in Manipur.
	Pethia muvattupuzhaensis Jameela Beevi & Ramachandran, 2005	4.5 SL male	Native /Kerala.
Narayan barb	*Pethia narayani* (Hora, 1937)	7.75 TL male	Endemic/Western Ghats.
	Pethia nigripinna (Knight, Rema Devi, Indra & Arunachalam, 2012)	4.5 SL male	Native/Western Ghats.
	Pethia ornatus (Viswanath& Laisaram, 2004)	4.23 TL male 4.34 TL female	Native/ Manipur and Assam.

Common Name	Scientific Name	Total length (TL) Standard length (SL) in cm	Distribution
	Pethia pookodensis (Mercy & Jacob, 2007)	4.3 SL male	Endemic/Kerala.
	Pethia punctata (Day, 1865)	7.5 TL male	Native/ Tamil Nadu and Western Ghats
	Pethia rutila Lalramliana, Knight & Laltlan hina , 2014	4.53 SL male	Native/ Mizoram, northeast India.
Sanjay's black -tip pethia	*Pethia sanjaymoluri* Katewate, Jadhav, Kumkar, Raghavan & Dahanukar, 2016	3.32 SL male	Native/Northern Western Ghats-Maharashtra.
	Pethia striata Atkora, Knight, Rema Devi & Krishnaswamy, 2015	3.31 SL male	Native/ Karnataka in Western Ghats.
Two-spot barb	*Pethia ticto* (Hamilton,1822)	10 unsexed	Native/ throughout India except Kerala and south Tamil Nadu.
	Pethia yuensis (Arunkumar& Tombi Singh, 2003)	5.5 SL male	Native/ Manipur.
	Puntius ambassis (Day, 1869)	7.5 TL male/unsexed	Native/ Andhra Pradesh and Tamil Nadu.
Scarlet -banded barb	*Puntius amphibius* (Valenciennes, 1842)	20.0 TL male/unsexed	Native/ in the Peninsular India.
Arenatus barb	*Puntius arenatus* (Day, 1878)	10.0 SL male/unsexed	Endemic/ Western Ghats.
Two-spot barb	*Puntius bimaculatus* (Bleeker, 1863)	7.0 male/unsexed	Native/ Peninsular India.
	Puntius burmanicus (Day, 1878)	10.0 TL male/ unsexed	Minicoy, Kerala.
Cauvery barb	*Puntius cauveriensis* (Hora, 1937)	7.4 male/unsexed	Native/ Cauvery river in Karnataka.
Chola barb	*Puntius chola* (Hamilton,1822)	15.0 TL male/unsexed	Native/ throughout India.
	Puntius crescentus Yazdani & Singh,1994		Native/ Karnataka.

Common Name	Scientific Name	Total length (TL) Standard length (SL) in cm	Distribution
Deccan barb	*Puntius deccanensis* Yazdani & Babu Rao,1976	3.7 SL male/unsexed	Endemic/ Western Ghats.
Long-fin Kerala barb	*Puntius dolichopterus* Plamoottil, 2015	6.8 SL male/unsexed	Native/ Kerala.
Long-snouted barb	*Puntius dorsalis* (Jerdon, 1849)	25.0 TL male/ unsexed	Native/ Western Ghats.
Dharna barb	*Puntius fraseri* (Hora & Mishra,1938)	4.60 TL male	Endemic/Western Ghats.
	Puntius khohi Dobriyal, Singh, Uniyal, Joshi, Phurailatpam & Bisht, 2004	?	Native/Uttaranchal.
	Puntius madhu-soodani Krish-nakumar,Benno Pereira & Rad-hakrishnan,2012	9.143 SL male	Native/Kerala.
Wynaad barb	*Puntius mahecola* (Valenciennes, 1844)	8.90 SL male/ unsexed	Native/Kerala.
	Puntius melanostigma (Day, 1878)	10.0 TL male	Native/ Kerala and Tamil Nadu.
	Puntius morehensis Arunkumar & Tombi Singh, 1998	12.0 SL male	Native/ Manipur.
	Puntius nangalensis Jayaram, 1990	?	Native/ Tamil Nadu.
Travancore yellow barb	*Puntius nelsoni* Plamoottil, 2014	9.10 SL male	Native/Travancore.
Malabar black-backed barb	*Puntius nigronotus* Plamoottil,2015	10.51 TL male	Native/ Kerala.
Parrah barb	*Puntius parrah* Day, 1865	15.0 TL male/ unsexed	Endemic/ Western Ghats.
	Puntius pinnauratus (Day, 1865)	9.05 SL male	Native/Kerala.
Puntio barb	*Puntius puntio* (Hamilton, 1822)	7.5 TL male	Native/ West Bengal.
	Puntius stigma (Valenciennes, 1844)	8.6 SL Male	Native/Kerala.

Common Name	Scientific Name	Total length (TL) Standard length (SL) in cm	Distribution
Spot-fin green barb	*Puntius viridis* Plamoottil & Abraham, 2014	8.10 SL male	Native/Kerala.
Khavli barb, Indian Maharaja barb	*Puntius sahyadriensis* Silas, 1953	7.0	Endemic/Western Ghats.
Kooli barb	*Puntius vittatus* Day, 1865	5.0 TL male	Native/ ponds, streams and lakes.
Spotfin swamp barb, Pool barb	*Puntius sophore* (Hamilton, 1822)	18.0 SL male	Native/ throughout India.
Onespot barb, Teri barb	*Puntius terio* (Hamilton,1822)	10.0 TL male	Native/ rivers, canals, ponds and ditches in northern and north-east India.
Denison barb	*Sahyadria denisonii* (Day, 1865)	15.0 TL male	Endemic/India-only from Western Ghats.
Aruli barb, Long-fin barb	*Dawkinsia arulius* (Jerdon, 1849)	12.0 unsexed	Endemic/ western Ghats, Cauvery river basin of south east India.
	Dawkinsia rohani (Rema Devi, Indra & Knight, 2010)	9.5 SL male	Native/hill streams of Tamil Nadu and inland waters.
	Dawkinsia assimilis (Jerdon, 1849)	11.15 SL male	Native/ Kerala.
	Dawkinsia rubrotinctus (Jerdon, 1849)	8.62 SL male	Native/Widespread in Cauvery River and its tributaries in Karnataka, Kerala and Tamil Nadu.
	Dawkinsia tambraparniei (Silas, 1954)	9.2 SL unsexed	Endemic/ Tambraparni river basin in Western Ghats.
Filament barb	*Dawkinsia filamentosa* (Valencinnes, 1844)	17.5 unsexed	Native/ freshwater streams in Goa, Karnataka, Kerala, Tamil Nadu and Andhra Pradesh.
	Dawkinsia exclamatio (Pethiyagoda & Kottelat, 2005)	8.34 SL male	Native/Kallada River in Western Ghats.

Common Name	Scientific Name	Total length (TL) Standard length (SL) in cm	Distribution
Melon barb	*Haludaria fasciata* (Jerdon, 1849)	7.0 SL male	Endemic/ Western Ghats and south India.
	Haludaria kannikattiensis (Arunachalam & Johnson, 1849)	11.15 SL male	Native/ Kerala.
	Haludaria melanampyx (Day, 1865)	7.5 TL male	Native/ Western Ghats.
Stedman barb	*Systomus clavatus* (McClelland, 1845)	24.0 TL male/unsexed	Native/ Assam, Manipur and West Bengal.
	Systomus jayarami (Vishwanath & Tombi Singh, 1986)	9.27 SL male	Native/Manipur.
Red finned kuruva	*Systomus rufus* Plamoottil, 2014	10.6 SL male	Native/Kerala.
Penninsular Olive barb	*Systomus sarana* (Hamilton, 1822)	42.0 TL male/unsexed	Native/ throughout India.
	Oreichthys coorgensis (Jayaram, 1982)	2.73 SL male	Native/ headwaters of Cauvery river in Karnataka.
	Oreichthys cosuatis (Hamilton, 1822)	8.0 TL male/unsexed	Native/ Assam, West Bengal, Orissa, Madhya Pradesh, Maharashtra, Karnataka and Kerala.
Neon high fin barb	*Oreichthys crenuchoides* Schafer, 2009	3.05 SL male	Native/River Jorai in West Bengal.
	Oreichthys duospilus Keith & Kumar, 2015	3.11 SL male	Native/ Karnataka.
	Oreichthys incognito Keith & Kumar, 2015	3.14 SL male	Native/ Kerala.
Bowany barb	*Neolissochilus bevanicus* (Day, 1877)	40 TL male	Endemic/ Western Ghats.
	Neolissochilus dukai (Day, 1878)	15.8 SL male	Native/ Teesta River in North Bengal.

Common Name	Scientific Name	Total length (TL) Standard length (SL) in cm	Distribution
Katli	*Neolissochilus hexagonolepis* (McClelland, 1839)	120 TL male/ unsexed	Native/ northern India.
	Neolissochilus hexastichus (McClelland, 1839)	?	Native/ Himalayan foothills.
	Neolissochilus spinulosus (McClelland, 1845)	?	Endemic/ Sikkim and Ganges.
	Osteobrama alfredianus (Valenciennes, 1844)	?	Native/ Type locality Mysore.
Malabar osteobrama	*Osteobrama bakeri* (Day, 1873)	11.0 TL male	Endemic to Western Ghats, also in Kerala.
Manipur osteobrama	*Osteobrama belangeri* (Valenciennes, 1844)	38.0 unsexed	Native/ Manipur, also in Godavari river.
	Osteobrama bhimensis Singh & Yazdani, 1992	?	Endemic to Western Ghats, also in Maharashtra.
Cotio	*Osteobrama cotio* (Hamilton, 1822)	10.0 unsexed	Native/ northern India, Madhya Pradesh and Western Ghats.
Cunma osteobrama	*Osteobrama cunma* (Day, 1888)	15.0 TL male	Native/ Manipur valley, Western Ghats.
Nilgiri osteobrama	*Osteobrama nielli* (Day, 1873)	12.0 unsexed	Endemic to Western Ghats, also in South India.
Peninsular osteobrama	*Osteobrama peninsularis* Silas, 1952	15.0 TL male	Endemic/ Peninsular India in Maharashtra, Orissa and Andhra Pradesh, and Western Ghats.
Bheema osteobrama	*Osteobrama vigorsii* (Sykes, 1839)	30.0 TL male	Endemic/ Orissa, Maharashtra and Krishna and Godavari river systems and Madhya Pradesh. South India and Western Ghats.

Common Name	Scientific Name	Total length (TL) Standard length (SL) in cm	Distribution
Nukta, double-mouth Tor barb	*Schismatorhynchos nukta* (Sykes, 1839)	30.0	Endemic/ rapid streams and rivers in Karnataka, Maharashtra, Tamil Nadu, and Western Ghats.
Dark mahseer	*Tor chelynoides* (McClelland, 1839)	41.0 TL	Native/ fast flowing mountain streams.
Mosal mahseer	*Tor mosal* (Hamilton, 1822)	24.0 SL male	Rivers of the Himalaya.
Jungha mahseer	*Tor progeneius* (McClelland, 1839)	64.0 SL	Large rapid streams with rocky bottoms Assam, Manipur, Meghalaya, Nagaland, and Uttaranchal.
Golden mahseer	*Tor putitora* (Hamilton, 1822)	180.0 TL unsexed	Native/ montane and submontane region, in streams and rivers.
Tor mahseer	*Tor tor* (Hamilton, 1822)	120.0 TL unsexed	Native/ rivers with rocky bottoms.
	Tor barakae Arunkumar& Basudha, 2003	13.6 SL male	Native/ Barak river system in Manipur.
Deccan mahseer	*Tor khudree* (Sykes, 1839)	100.0 TL	Native/ entire Peninsular India, Godavari river, Tamil Nadu, Madhya Pradesh and Western Ghats.
Dwarf mahseer	*Tor kulkarnii* Menon, 1992	20.8 SL male	Native/ Maharashtra.
	Tor macrolepis (Heckel, 1838)	26.0 TL male	Native/ Indus drainage system.
Malabar mahseer	*Tor malabaricus* (Jerdon, 1849)	40.0 TL male	Native/ Western Ghats, Tamil Nadu and Kerala.
	Tor remadevii Madhsoodana Kurup & Radhakrishnan, 2011	?	Native/ Kerala.

Common Name	Scientific Name	Total length (TL) Standard length (SL) in cm	Distribution
Chaguni	*Chagunius chagunio* (Hamilton, 1822)	50.0 TL male	Native/ large rivers with rocky bottoms-north India.
Ngara (Manipuri)	*Chagunius nicholsi* (Myers,1924)	30.0 TL male	Native/ lower course of river system-Manipur.
Alghad snowtrout	*Schizopyge niger* (Heckel,1838)	32.6 SL	Native/ Kashmir Valley.
Kinnaur snowtrout	*Schizopygopsis stoliczkai* Steindachner, 1866	34.0 TL male	Native/ upper reaches of the Sutlej river (Himachal Pradesh) and Indus river (Ladakh).
Dapeghat snowtrout	*Schizothorax curvifrons* Heckel, 1838	37.0 SL	Native/ Kashmir valley and Ladakh.
Chirruh snowtrout	*Schizothorax esocinus* Heckel, 1838	50.8 SL	Native/ Indus river and its tributaries in Ladakh and Kashmir Valley.
	Schizothorax huegelii Heckel, 1838	30.0 TL male	Native/ Kashmir Valley.
Kumaon snowtrout	*Schizothorax kumaonensis* Menon, 1971	18.5 TL male	Endemic/ Kumaon Himalaya in Uttar Pradesh.
Kunar snowtrout	*Schizothorax labiatus* (McClelland, 1842)	42.5 SL	Native/ Kashmir Valley and Ladakh.
Dongu snowtrout	*Schizothorax nasus* Heckel, 1838	30 TL male	Native/ Kashmir Valley.
Hill trout	*Schizothorax plagiostomus* Heckel, 1838	48.8 SL	Native/ Himalayan foot hills, Sikkim, Kashmir Valley and Nainital Lake.
Dinnawah snowtrout	*Schizothorax progastus* (McClelland, 1839)	30.0 unsexed	Native/ Jammu and Kashmir Valley, Ganges, Bagmati, Gandak rivers in Uttar Pradesh, Brahmaputra river in Assam and Kedar Valley.
Trout	*Schizothorax richardsonii* (Gray, 1832)	30.0 unsexed	Native/ along the Himalaya, Kedar Valley and Siwalik.

Common Name	Scientific Name	Total length (TL) Standard length (SL) in cm	Distribution
Family: CYPRINIDAE			
Curmuca barb	*Hypselobarbus curmuca* (Hamilton,1822)	120.0 TL male	Endemic/ Western Ghats.
Jerdon's carp	*Hypselobarbus jerdoni* (Day,1870)	46.0 TL male	Endemic/Western Ghats.
	Hypselobarbus keralaensis Arunachalam, Chinnaraja& Mayden, 2016	19.23 SL male	Native/Kerala.
Kolus	*Hypselobarbus kolus* (Sykes,1839)	60.0 TL male	Endemic/Rivers in Penninsular India, Madhya Pradesh and Western Ghats.
	Hypselobarbus kurali Menon& Rema Devi, 1995	35.0 TL male	Endemic/Kerala and Karnataka.
Canara barb	*Hypselobarbus lithopidos* (Day, 1874)	60.0 TL unsexed	Endemic/Western Ghats and Karnataka.
Korhi barb	*Hypselobarbus micropogon* (Valenciennes, 1842)	90.0 TL male	Endemic/Karnataka, Kerala and Western Ghats.
	Hypselobarbus pulchellus (Day, 1870)	44 TL male	Endemic/Karnataka.
Subfamily: DANIONINAE			
	Danio assamila Kullander, 2015	6.83 SL male	Native/tributaries of Brahmaputra River.
Zebra danio	*Danio rerio* (Hamilton, 1822)	3.8 SL male/unsexed	Native/Arunachal Pradesh, Assam, Karnataka, Meghalaya, Orissa, Uttar Pradesh, and West Bengal.
Olive danio/Dangila danio	*Danio dangila* (Hamilton, 1822)	5.0 NG unsexed	Native/ Ganges and Brahmaputra River, north and north-east India and Madhya Pradesh.

Common Name	Scientific Name	Total length (TL) Standard length (SL) in cm	Distribution
Queen danio	*Danio regina* (Fowler, 1934)	3.0 SL	Native/ throughout India.
	Danio jaintainensis (Sen, 2007)	?	Native/ Meghalaya.
	Danio meghalayensis Sen & Dey, 1985	5.05 SL male	Native/ Brahmaputra River drainage.
	Devario deruptotalea Ramachandra & Viswanath, 2014	6.34 SL male	Native/ Manipur.
Fraser danio	*Devario fraseri* (Hora, 1935)	10.0 unsexed	Endemic/ Maharashtra.
	Devario horai (Barman, 1983)	2.8 SL male	Native/ Arunachal Pradesh.
	Devario assamensis (Barman, 1984)	9.5 SL male	Native/ Darrang District in Assam.
Giant danio	*Devario aequipinnatus* (McClelland, 1839)	15.0 unsexed	Native/ throughout India.
Bengal danio	*Devario devario* (Hamilton, 1822)	10.0 unsexed	Native/ northern India and Krishna-Godavari river system.
Malabar danio	*Devario malabaricus* (Jerdon, 1849)	12.0 TL male	Native/ Maharashtra to Kerala, and Western Ghats.
	Devario manipurensis (Barman, 1987)	3.8 SL male	Native/ Manipur.
Naga danio	*Devario naganensis* (Chaudhuri, 1912)	4.5 TL male	Endemic/ Manipur and Nagaland.
Queen danio, Fowler danio	*Devario regina* (Fowler, 1934)	10.1 TL male	Native/ Assam.
Peninsular danio	*Devario neilgherriensis* (Day, 1867)	10.0 TL male	Endemic/ South India.
	Devario yuensis (Arunkumar & Tombi Singh, 1998)	8.8 TL male	Native/Manipur.

Common Name	Scientific Name	Total length (TL) Standard length (SL) in cm	Distribution
Manipur danio	*Devario acuticephala* (Hora, 1921)	4.0 SL male	Endemic/ Brahmaputra drainage system in Manipur and Naga hills to Arunachal Pradesh.
Mola carplet	*Amblypharyngodon mola* (Hamilton, 1822)	20.0 TL male/unsexed	Native/throughout India.
Burmese carplet	*Amblypharyngodon atkinsonii* (Blyth, 1860)	15.0 TL male/unsexed	Native/ northern India.
Attentive carplet	*Amblypharyngodon melettinus* (Valennciennes, 1844)	8.0 TL male/unsexed	Native/ Tamil Nadu, Kerala and Western Ghats.
Indian carplet	*Amblypharyngodon microlepis* (Bleeker, 1853)	10.0 TL male/unsexed	Native/ northern and southern parts of India.
Jaya	*Aspidoparia jaya* (Hamilton, 1822)	8.0 unsexed	Native/ northern India and northeast India.
	Aspidoparia ukhrulensis Selim & Vishwanath, 2001	?	Native/ Manipur.
Morari, Aspidoparia	*Cabdio morar* (Hamilton, 1822)	20.0 NG	Native/ throughout India and Western Ghats.
Silver hatchet chela	*Chela cachius* (Hamilton, 1822)	6.0 unsexed	Native/ throughout India.
	Chela macrolepis Knight & Rema Devi, 2014	3.8 SL male	Native/ Chembarampakkam Lake in Tamil Nadu.
	Chela khujairokinesis Arunkumar, 2000	?	Native/ Yu River, Manipur.
Dadio	*Laubuca dadiburjori* Menon, 1952	2.52 SL male; 3.0 SL female	Endemic/ Cochin in Kerala, Tamil Nadu to Goa.
Winged danio, Indian glass barb	*Laubuka laubuka* (Hamilton, 1822)	17.0 TL unsexed	Native/ throughout India.
Malabar hatchet chela	*Laubuca fasciata* (Silas, 1958)	6.0 TL male	Endemic/ Kerala and Western Ghats.
Gora chela	*Securicula gora* (Hamilton, 1822)	24.5 TL male	Native/ northern and eastern India.

Common Name	Scientific Name	Total length (TL) Standard length (SL) in cm	Distribution
Large razor belly minnow	*Salmostoma bacaila* (Hamilton, 1822)	18.0 TL male	Native/ northern India.
Bloch razor belly minnow	*Salmostoma balookee* (Sykes, 1839)	15.0 unsexed	Native/Tamil Nadu, Kerala, Chhattisgarh, Andhra Pradesh, Madhya Pradesh, Maharashtra, Karnataka and Gujrat.
Fine scale razor belly minnow	*Salmostoma phulo* (Hamilton, 1822)	12.0 unsexed	Native/ Maharashtra, West Bengal and Assam.
Silver razor belly minnow	*Salmostoma acinaces* (Valenciennes, 1844)	15.0 unsexed	Native/ Hooghly river in West Bengal, Maharashtra, Madras, south Karnataka and Western Ghats.
Boopis razor belly minnow	*Salmostoma boopis* (Day,1874)	12.0 unsexed	Endemic/ Western Ghats, Eastern Ghats, south and central India.
Hora razor belly minnow	*Salmostoma horai* (Silas, 1951)	10.0 unsexed	Endemic/ Coorg in Karnataka and Andhra Pradesh.
Novacula razor belly minnow	*Salmostoma novacula* (Valenciennes, 1840)	12.5 unsexed	Endemic/ South India, Central India and the Western Ghats.
Punjab razor belly minnow	*Salmostoma punjabense* (Day, 1872)	8 TL male	Native/ river systems in Punjab.
Sardinella razor belly minnow	*Salmostoma sardinella* (Valenciennes, 1844)	15 NG unsexed	Native/ Ganga, Brahmaputra and Mahanadi river systems.
Mahanadi razor belly minnow	*Salmostoma untrahi* (Day, 1869)	20 TL male	Endemic/ throughout India.
South Indian flying barb	*Esomus barbatus* (Jerdon, 1849)	12 TL male	Endemic/ South India, Tamil Nadu and Karnataka.

Common Name	Scientific Name	Total length (TL) Standard length (SL) in cm	Distribution
	Esomus thermoicos (Valenciennes, 1842)	12.7 TL male	Peninsular India, Kerala and Tamil Nadu.
Indian flying barb	*Esomus danrica* (Hamilton, 1822)	13.0 TL male/unsexed	Native/ throughout northern India.
	Esomus malabaricus Day, 1867	12.5 TL male	Native/ Kerala, Maharashtra and Tamil Nadu.
	Esomus manipurensis Tilak &Jain, 1990	5.8 TL male	Native/ Manipur.
Striped flying barb	*Esomus lineatus* Ahl, 1923	5.0 TL male, 6.0 TL female	Native/mouth of the Ganges river.
Blackline rasbora	*Rasbora daniconius* (Hamilton, 1822)	15.0 unsexed	Native/ throughout India.
Slender rasbora	*Rasbora labiosa* Mukerji, 1935	8.5 unsexed	Endemic/ Maharashtra.
Gangetic scissor tail rasbora	*Rasbora rasbora* (Hamilton, 1822)	13.0 TL male/unsexed	Native/ throughout India.
Cauvery rasbora	*Rasbora caverii* (Jerdon, 1849)	10 TL male	Native/ peninsular India in Karnataka and Tamil Nadu.
	Rasbora dandia (Valenciennes, 1844)	9.85 SL male	Native/ Karnataka, Kerala and Tamil Nadu.
	Rasbora microcephalus (Jerdon, 1849)	7.72 SL male	Native/ Coromandel coast of Tamil Nadu.
Malabar baril	*Barilius bakeri* Day, 1865	15.0 unsexed	Endemic/ Western Ghats of Kerala.
Barred baril	*Barilius barila* (Hamilton, 1822)	12.5 SL unsexed	Native/ North India and northeast India.
Barna baril	*Barilius barna* (Hamilton, 1822)	15.0 unsexed	Ganges, Brahmaputra and Mahanadi river systems.
Vagra baril	*Barilius vagra* (Hamilton, 1822)	12.83 TL unsexed	Native/ base of the Himalaya, Indus, Ganges, Brahmaputra river, Bhimtal, Naukuchiatal lakes and Western Ghats.

Common Name	Scientific Name	Total length (TL) Standard length (SL) in cm	Distribution
Hamilton's baril	*Barilius bendelisis* (Hamilton, 1822)	18.12 TL unsexed	Native/throughout India.
Indian trout	*Raiamas bola* (Hamilton, 1822)	35.0 unsexed	Native/Haryana, Himachal Pradesh, Uttar Pradesh, Bihar, Assam, Arunachal Pradesh, West Bengal and Orissa.
River carp baril	*Barilius gatensis* (Valenciennes, 1844)	15.0 unsexed	Endemic/ Western Ghats, Maharashtra, Karnataka, Kerala and Tamil Nadu.
Gunther's baril	*Barilius radiolatus* Gunther, 1868	10.0 TL male	Endemic/ Madhya Pradesh and Tripura.
Shacra baril	*Barilius shacra* (Hamilton, 1822)	14.0 TL male	Native/ Ganges, Yamuna and Brahmaputra.
Tileo baril	*Opsarius tileo* (Hamilton, 1822)	30.5 SL male/unsexed	Native/ Arunachal Pradesh, Assam, Bihar, West Bengal, Tripura, Uttaranchal, Jharkhand, Nagaland, Sikkim, Manipur, Meghalaya, Mizoram.
	Barilius ardens Knight, Rai, d'Souza & Vijaykrishnan, 2015	8.72 SL male	Native/ Western Ghats of Karnataka.
Jerdon's baril	*Barilius canarensis* (Jerdon, 1849)	15.0 unsexed	Endemic/ Karnataka and Kerala of Western Ghats.
Malabar baril	*Barilius malabaricus* (Jerdon, 1849)	9.1 SL male	Native/ Payaswini and Vallapattanam rivers in Kerala and southern Karnataka.
	Barilius chatricensis Selim & Vishwanath, 2002	8.64 SL male	Native/ Manipur.

Common Name	Scientific Name	Total length (TL) Standard length (SL) in cm	Distribution
	Barilius dimorphicus Tilak & Hussain,1990	18.5 TL male	Endemic/ Uttar Pradesh.
Manipur baril	*Barilius dogarsinghi* Hora, 1921	8.5 SL male	Endemic/ Manipur.
Day's baril	*Barilius evezardi*	11 TL male	Endemic/ Maharashtra.
	Barilius lairokensis Arunkumar & Tombi Singh, 2000	10.5 SL male	Native/ Manipur.
Indus baril	*Barilius modestus* Day, 1872	12.5 TL male	Native/ Jammu, Himachal Pradesh and Punjab.
	Barilius ngawa Vishwanath & Manojkumar, 2002	9.65 SL male	Native/ Manipur.
	Barilius profundus Dishma & Vishwanath, 2012	7.11 SL male	Native/ Mizoram.
Subfamily: LABEONINAE			
Angra labeo	*Labeo angra* (Hamilton,1822)	22 TL male	Native/ northern India.
Bata labeo	*Labeo bata* (Hamilton,1822)	61.0 unsexed	Native/ throughout India.
Boga labeo	*Labeo boga* (Hamilton,1822)	30.0 unsexed	Native/ throughout India except western face of Western Ghats.
Boggut labeo	*Labeo boggut* (Sykes, 1839)	29.0 TL male	Native/ northern India and up to Cauvery river system.
Brahmaputra labeo	*Labeo dyocheilus* (McClelland, 1839)	50 NG unsexed	Native/ all along the Himalaya, Assam. Mahanadi, Damoder rivers and Nainital Lake.
Kuria labeo	*Labeo gonius* (Hamilton, 1822)	100.0 unsexed	Native/ North India, Assam, Orissa. Krishna and Jayanti rivers.

Common Name	Scientific Name	Total length (TL) Standard length (SL) in cm	Distribution
Nandi labeo	*Labeo nandina* (Hamilton, 1822)	81.0 SL male	Native/ West Bengal, Assam and Tripura.
Pangusia labeo	*Labeo pangusia* (Hamilton, 1822)	60.0 unsexed	Native/ all along the Himalayas.
Karnataka labeo	*Labeo calbasu* (Hamilton, 1822)	100.0 NG unsexed	Native/ throughout India.
	Labeo dussumieri (Valenciennes, 1842)	55.0 TL unsexed	Native/ Western Ghats up to north Canara.
Fringed-lipped peninsular carp	*Labeo fimbriatus* (Bloch, 1795)	91.0 unsexed	Native/ throughout India except west face of Western Ghats.
Deccan labeo	*Labeo kawrus* (Sykes, 1839)	60 TL unsexed	Endemic/ Western Ghats up to the Deccan, and Maharashtra.
Pig-mouth carp	*Labeo kontius* (Jerdon, 1849)	61.0 unsexed	Endemic/ Western Ghats, Tamil Nadu and Karnataka.
	Labeo microphthalmus Day, 1877	25.0 TL male	Native/ Western Himalaya.
Bombay labeo	*Labeo porcellus* (Heckel, 1844)	30.0 TL unsexed	Native/ Western Ghats of the Bombay region, Deccan and Maharashtra.
Deccan labeo	*Labeo potail* (Sykes, 1839)	40.0 TL unsexed	Endemic/ Western Ghats-Kerala, Karnataka.
Roho labeo	*Labeo rohita* (Hamilton,1822)	96.0 TL unsexed	Native/ North and Central India, Maharashtra, Kerala, Tripura and Western Ghats.
Gangetic latia	*Crossocheilus latius* (Hamilton, 1822)	16.2 SL	Native/ Ganges, Brahmaputra and Mahanadi drainages, Western Ghats, Nainital and Chilka Lakes.

Common Name	Scientific Name	Total length (TL) Standard length (SL) in cm	Distribution
	Crossocheilus burmanicus Hora, 1936	15.0 NG unsexed	Native/ Assam.
Kashmir latia	*Crossocheilus diplocheilus* (Heckel, 1838)	17.3 TL male	Native/ Himachal Pradesh and Kashmir.
	Crossocheilus periyarensis Menon & Jacob, 1996	35 TL unsexed	Endemic/ Periyar River in Western Ghats, Kerala.
Annandale garra	*Garra annandalei* Hora, 1921	23 TL male	Native/ Himalaya, Darjeeling, Assam, West Bengal and Bihar.
Gotyla, sucker head	*Garra gotyla* (Gray, 1830)	15.0 unsexed	Native/ all along the Himalayas, Western Ghats, Indian peninsula, and Kedar valley.
Burmese garra	*Garra gravelyi* (Annandale, 1919)	12.0 unsexed	Native/ Manipur and Mizoram.
Kemp garra	*Garra kempi* Hora, 1921	11.1 SL unsexed	Native/ Arunachal Pradesh, Meghalaya, Mizoram and Nagaland.
Lamta garra	*Garra lamta* (Hamilton, 1822)	20.4 TL male	Native/ Assam, Sikkim, Western Ghats, Himalaya and Nainital Lake.
Khasi garra	*Garra lissorhynchus* (McClelland, 1842)	9.2 SL male	Endemic/ Brahmaputra River system, Meghalaya and Nagaland.
Cauvery garra	*Garra mcclellandi* (Jerdon, 1849)	17.4 SL male	Endemic/ Cauvery drainage, Western Ghats, Kerala, Karnataka and Periyar Lake.
Naga garra	*Garra naganensis* Hora, 1921	10.9 SL male	Endemic/ Nagaland and Arunachal Pradesh.

Common Name	Scientific Name	Total length (TL) Standard length (SL) in cm	Distribution
Khasi garra	*Garra nasuta* (McClelland, 1838)	20 NG male	Native/ Arunachal Pradesh, Manipur and Meghalaya.
Mishmi garra	*Garra rupecula* (McClelland, 1839)	6.5 SL male	Endemic/ Mishmi hills in Arunachal Pradesh and Manipur.
Mullya garra	*Garra mullya* (Sykes, 1839)	17.0 Unsexed	Native/throughout India except Assam and the Himalaya.
	Garra namyaensis Shangningam & Vishwanath, 2012	7.25 SL male	Native/Manipur.
	Garra nigricauda Arunachalam, Nandagopal & Mayden, 2013	14.28 SL male	Native/ Arunachal Pradesh.
	Garra alticaputus Arunachalam, Nandagopal & Mayden, 2013	7.16 SL male	Native/ Arunachal Pradesh.
	Garra elongata Vishwanath & Kosygin, 2000	11.25 SL male	Native/ Manipur.
	Garra emarginata Madhusoodana Kurup & Radhakrishnan, 2011	8.57 SL male	Native/ river Periyar in Kerala.
Cardamon garra	*Garra hughi* Silas, 1955	15.5 SL male	Endemic/ Western Ghats and Kerala.
	Garra jerdoni Day, 1867	16 SL male	Native/ Tamil Nadu.
	Garra kalpangi Nebeshwar, Bagra & Das, 2012	7.24 SL male	Native/ Arunachal Pradesh and Brahmaputra River system.
	Garra palaruvica Arunachalam, Raja, Nandagopal & Mayden, 2013	4.44 SL male	Native/ Kerala.

Common Name	Scientific Name	Total length (TL) Standard length (SL) in cm	Distribution
Nilgiris garra	*Garra stenorhynchus* (Jerdon, 1849)	15.5 SL male	Endemic/ Western Ghats and Nilgiris.
	Garra surendranathanii Shaji, Arun & Easa, 1996	14.7 SL Male/unsexed	Native/ Endemic in Kerala, Western Ghats in Kerala.
	Garra tyao Arunachalam, Nandagopal &Mayden, 2014	7.6 SL male	Native/ Mizoram.
Family: BOTIIDAE (Botiid loaches)			
Yoyo, Almora loach	*Botia almorhae* Gray, 1831	15.5 SL male	Native/ Almorah Hills, western Himalaya, eastern Himalaya, Yamuna, Chambal drainage, and Kosi River.
Birdi loach	*Botia birdi* Chaudhuri, 1909	18.6 TL male	Native/ Indus drainage, Punjab, Kashmir Valley and Himachal Pradesh.
Bengal loach, necktie loach	*Botia dario* (Hamilton, 1822)	15.1 TL male	Native/Assam, West Bengal, South Bihar, Uttar Pradesh, Ganges and Brahmaputra drainages, Western Ghats of Maharashtra.
Hora loach	*Botia dayi* Hora, 1932	12 TL male	Native/ all along the base of Himalaya, Brahmaputra drainage and Kedar Valley.
Burmese loach	*Botia histrionica* Blyth, 1860	10 NG unsexed	Native/Manipur, Assam and Uttar Pradesh.
Y loach	*Botia lohachata* Chaudhuri, 1912	15.4 TL Male/unsexed	Native/ Ganges drainage, river Gandak.
Gangetic loach	*Botia rostrata* Gunther, 1868	25 NG unsexed	Native/ Assam, West Bengal and Meghalaya.

Common Name	Scientific Name	Total length (TL) Standard length (SL) in cm	Distribution
Tiger loach, Zebra loach	*Botia striata* Narayan Rao, 1920	7.8 SL male	Endemic/ Western Ghats and Maharashtra.
Family: COBITIDAE (loaches)			
Gongota loach	*Canthophrys gongota* (Hamilton, 1822)	13.0 TL male	Native/ Assam, Meghalaya to Uttar Pradesh, North Bengal and Tripura.
Annandale loach	*Lepidocephalichthys annandalei* Chaudhuri, 1912	4.3 SL male	Native/ Assam, North Bengal, Meghalaya and Uttar Pradesh.
Berdmore's loach	*Lepidocephalichthys berdmorei* Blyth, 1860	8 NG unsexed	Native/ Manipur river drainage system.
Loktak loach	*Lepidocephalichthys irrorata* Hora, 1921	34 TL unsexed	Assam, Manipur and Meghalaya.
Malabar loach	*Lepidocephalichthys thermalis* (Valenciennes,1846)	8.0 Unsexed	Native/ Karnataka, Kerala, Maharashtra, Tamil Nadu and Western Ghats.
Guntea loach	*Lepidocephalichthys guntea* (Hamilton, 1822)	10.0 Unsexed	Native/throughout India except Karnataka, Kerala and North India.
	Lepidocephalichthys arunachalensis (Datta & Barman, 1984)	5.0 unsexed	Native/ Arunachal Pradesh.
	Lepidocephalichthys coromandelensis (Menon, 1992)	5.7 SL male	Native/Araku Valley in Andhra Pradesh.
	Lepidocephalichthys goalparensis Pillai & Yazdani, 1976	7.01 TL female	Native/ northern India and Assam.

Common Name	Scientific Name	Total length (TL) Standard length (SL) in cm	Distribution
Yo loach	*Lepidocephalichthys manipurensis* Arunkumar, 2000	56 TL male	Native/ Manipur.
Pangia coolie-loach	*Pangio pangia* (Hamilton, 1822)	6.5 TL male	Native/ North Bengal, Assam, Manipur.
	Pangio apoda Britz & Maclaine, 2007	3.82 SL male	Endemic/Tista drainage in West Bengal.
	Pangio oblonga (Valenciennes, 1846)	8 TL unsexed	Native/ north-eastern Bengal and Manipur.
Goalpara loach	*Neoeucirrhichthys maydelli* Banarescu & Nalbant, 1968	3.6 SL male	Native/ Brahmaputra drainage in Assam.
Banana fish	*Acantopsis dialuzona* Van Hasselt, 1823	4.5 SL unsexed	Native/ Assam.
	Acantopsis multistigmatus Viswanath & Laisram, 2005	?	Native/ a tributary of the Yu river in Manipur.
Family: NEMACHEILIDAE (stone loaches)			
	Nemacheilus anguilla Annandale, 1919	5.0 unsexed	Endemic/ Western Ghats in Maharastra and Karnataka.
Dotted -finned loach	*Nemacheilus semiarmatus* Day, 1867	5.6 SL unsexed	Native/ Peninsular India.
	Nemacheilus doonensis Tilak &Husain, 1977	3.2 SL male	Endemic/ Dehradun in Uttar Pradesh.
Spotted loach	*Nemacheilus corica* (Hamilton, 1822)	4.2 SL male	Native/ in Himalayas, Sutlej basin and Nainital Lake.
Guenther's loach	*Nemacheilus guentheri* Day, 1867	5.6 unsexed	Endemic/ Western Ghats, Peninsular India.
Kerala loach	*Nemacheilus keralensis* (Rita, Banarescu & Nalbant, 1978)	5.3 SL male	Endemic/ Western Ghats, Kerala.

Common Name	Scientific Name	Total length (TL) Standard length (SL) in cm	Distribution
	Nemacheilus pulchellus Day, 1873	4.6 unsexed	Endemic/ Western Ghats, Cauvery river system in South India, Kerala and Tamil Nadu.
	Nemacheilus rueppelli (Sykes, 1839)	7.4 unsexed	Endemic/ Western Ghats of Maharashtra and Karnataka.
	Nemacheilus singhi Menon, 1987	3.6 SL male	Endemic/ Nagaland and Kiphire.
	Nemacheilus subfusca (McClelland, 1839)	7.4 SL unsexed	Native/ Brahmaputra River.
Zodiac loach	*Nemacheilus triangularis* Day, 1865	7.1 TL unsexed	Endemic/ Karnataka, Kerala and Tamil Nadu.
	Neonoemacheilus assamensis (Menon, 1987)	4.3 SL male	Native/ Assam and Manipur.
	Neonoemacheilus labeosus (Kottelat, 1982)	10 NG unsexed	Native/ Brahmaputra drainage in Assam.
	Neonoemacheilus peguensis (Hora, 1929)	5.5 SL unsexed	Native/ Chindwin drainage, Manipur.
Red-tailed, striped sand loach	*Aborichthys elongates* (Hora, 1921)	7.4 SL male	Endemic/ Darjeeling, West Bengal.
Stone loach	*Aborichthys garoensis* Hora, 1925	8.95 SL male	Endemic/ Garo Hills of Meghalaya.
Abori loach	*Aborichthys kempi* Chaudhuri, 1913	8.1 SL male	Native/ Aor and Garo Hills Meghalaya and Arunachal Pradesh.
Tikaderi abor	*Aborichthys tikaderi* Barman, 1985	10.9 SL male	Endemic/ Namdapha in Arunachal Pradesh.
	Aborichthys cataracta Arunachalam, Raja, Malaiammal & Mayden, 2014	9.26 SL male	Native/ in a pool of Arunachal Pradesh.

Common Name	Scientific Name	Total length (TL) Standard length (SL) in cm	Distribution
	Aborichthys verticauda Arunachalam, Raja, Malaiammal & Mayden, 2014	7.75 SL male	Native/ stream in Arunachal Pradesh.
	Aborichthys waikhomi Kosygin, 2012	6.8 SL male	Native/ Namdapha in Arunachal Pradesh.
	Acanthocobitis botia (Hamilton, 1822)	11 TL male	Native/ Western Ghats.
	Acanthocobitis mooreh (Sykes, 1839)	4.4 unsexed	Endemic/ Peninsular India.
	Acanthocobitis rubidipinnis (Blyth, 1860)	10 NG unsexed	Native/India.
Beavani loach	*Schistura beavani* (Gunther, 1868)	8 TL male	Native/ Garo Hills (Meghalaya), North Bengal, Uttar Pradesh, Kedar Valley, Nainital, Bhimtal and Naukuchia Tal.
	Schistura cincticauda (Blyth, 1860)	5 NG unsexed	Native/ Arunachal Pradesh.
Murangi	*Schistura denisonii* (Day, 1867)	5.0 unsexed	Endemic/ Peninsular India and Bihar.
	Schistura devdevi (Hora, 1935)	4.0 SL male	Endemic/ Eastern Himalaya and Teesta Valley.
	Schistura kangjupkhulensis (Hora, 1921)	7.2 TL male	Native/ Manipur Valley and Naga Hills.
	Schistura manipurensis (Chaudhuri, 1912)	8 TL male	Endemic/ Manipur, Nagaland, Brahmaputra and Chindwin basins.
	Schistura multifasciata (Day, 1878)	9.8 SL male	Native/ Eastern Himalaya.
	Schistura nagaensis (Menon, 1987)	5.9 SL male	Endemic/ Brahmaputra basin in Nagaland.

Common Name	Scientific Name	Total length (TL) Standard length (SL) in cm	Distribution
	Schistura prashadi (Hora, 1921)	6.5 SL male	Endemic/ Manipur Valley.
	Schistura rendahli (Banarescu & Nalbant, 1968	?	Native/ Southern India.
Not known	*Schistura reticulofasciata* (Singh & Banarescu, 1982)	4.8 SL unsexed	Native/ Meghalaya.
	Schistura reticulata Vishwanath, Nebeshwar Sharma, 2004	7.70 SL male	Native/Manipur.
Stone loach	*Schistura rupicola* McClelland, 1838	6.7 SL male	Native/ Shimla, Kashmir, Doon Valley, Poonch Valley, Himachal Pradesh, Kedar Valley and Western Himalaya.
Half banded loach/ bicolor loach	*Schistura savona* (Hamilton, 1822)	4.0 SL male	Native/ Eastern Himalaya, Uttar Pradesh.
Grizzled loach	*Schistura sikmaiensis* (Hora, 1921)	7.8 SL male	Native/ Meghalaya.
	Schistura singhi (Menon, 1987)	3.6 cm SL unsexed	Native/ Nagaland.
Green striped loach	*Schistura scaturigina* McClelland, 1839	10 TL male	Native/ Eastern Himalaya, Meghalaya and Uttar Pradesh.
	Schistura vinciguerrae (Hora, 1935)	8 NG male	Native/ Manipur.
	Schistura sharvathiensis Sreekantha, Gururaja, Rema Devi, Indra & Ramchandra, 2006	2.9 SL male	Native/ Central Western Ghats, Karnataka.
	Schistura sijuensis (Menon, 1987)	5.1 SL male	Endemic/ Meghalaya.

Common Name	Scientific Name	Total length (TL) Standard length (SL) in cm	Distribution
	Schistura striata (Day, 1987)	5.0 unsexed	Endemic/ Kerala, Karnataka, Maharashtra, Western Ghats Rivers.
	Schistura tigrina Viswanath & Nebeshwar Sharma, 2005	9.35 SL male	Native/ Manipur.
	Schistura tirapensis Kottelat, 1990	5.6 SL male	Native/ Arunachal Pradesh.
	Schistura chindwinica (Tilak & Husain, 1990)	6.5 SL male	Endemic/ Manipur.
Gracilis triplophysa loach	*Triplophysa gracilis* (Day, 1877)	11.0 SL male	Native/ Kashmir Valley and Ladakh.
	Triplophysa kashmirensis (Hora, 1922)	10.6 SL male	Native/ Kashmir.
	Triplophysa ladacensis (Gunther, 1868)	10.5 SL male	Native/ Ladakh and Ghani in Jammu and Kashmir.
Kashmir triplophysa loach	*Triplophysa marmorata* (Heckel, 1838)	8 SL male	Endemic/ Kashmir lakes and Indus River.
Leh triplophysa-loach	*Triplophysa microps* (Steindachner, 1866)	6.8 TL male	Native/ Ladakh in Kashmir and Lahul-Spiti in Himachal Pradesh.
Tilak triplophysa loach	*Triplophysa shahensis* Tilak, 1987	13.6 SL male	Endemic/ Ladakh in Jammu and Kashmir.
	Triplophysa stewartia (Hora, 1922)	20.8 SL unsexed.	Native/ Mugleb, Ladakh, Man lagoon and Sikkim.
Stoliczka triplophysa loach	*Triplophysa stoliczkai* (Steindachner, 1866)	15 TL male	Native/ Ladakh in Jammu and Kashmir.
Tibetan triplophysa loach	*Triplophysa tenuicauda* (Steindachner, 1866)	9.9 SL unsexed	Native/ Ladakh and Nurba Valley.

Common Name	Scientific Name	Total length (TL) Standard length (SL) in cm	Distribution
Yasin triplophysa-loach	*Triplophysa yasinensis* (Alcock, 1898)	8.3 SL male	Native/ Ladakh and Kashmir Valley.
Family: PSILORHYNCHIDAE Mountain carps			
Balitora minnow	*Psilorhynchus balitora* (Hamilton, 1822)	7.0 TL male	Native/ North Bengal, Assam, Jamuna in Himachal Pradesh and Gomati in Uttar Pradesh.
Rainbow minnow	*Psilorhynchus gracilis* (Rainboth, 1983)	5.5	Native/Assam, Mizoram, West Bengal.
Homaloptera minnow	*Psilorhynchus homaloptera* (Hora& Mukherjee, 1935)	9.4 SL unsexed	Native/ Manipur, Nagaland and Assam.
River stone carp	*Psilorhynchus sucatio* (Hamilton, 1822)	8.2 TL male	Native/ Eastern Himalaya, Assam, Meghalaya, Ganges in Uttar Pradesh and Bihar.
	Psilorhynchus amplicephalus Arunachalam, Muralidharan & Sivakumar, 2007	5.68 SL male	Native/ Assam.
	Psilorhynchus arunachalensis (Nebeshwar, Bagra & Das, 2007)	?	Native/ Brahmaputra basin in Arunachal Pradesh.
	Psilorhynchus breviminor Conway & Mayden, 2008	4.87 SL male	Native/ Manipur.
	Psilorhynchus chakpiensis Shangningam & Viswanath, 2013	5.31 SL male	Native/ Manipur.
	Psilorhynchus hamiltoni Conway, Dittmer, Zezisek& Ng, 2013	3.06 SL male	Native/ Tista River in West Bengal.

Common Name	Scientific Name	Total length (TL) Standard length (SL) in cm	Distribution
	Psilorhynchus kaladanansis Lairamiiana, Lainuntiuanga & Laironunga, 2015	5.22 SL male	Native/ Kaladan River drainage, Mizorum.
	Psilorhynchus maculatus Shangningam & Vishwanath, 2013	6.12 SL male	Native/ Manipur.
Rainbow minnow	*Psilorhynchus nudithoracicus* Tilak & Husain, 1980	8.44 SL male	Native/ Ganges River drainage in West Bengal, Uttar Pradesh, Brahmaputra River drainage in Assam and West Bengal.
Family: BALITORIDAE hill stream and river loaches.			
Western Ghats loach	*Bhabania australis* (jerdon, 1849)	9.0 unsexed	Endemic/ Western Ghats in Karnataka, Tamil Nadu and Kerala.
Gray's stone loach	*Balitora brucei* Gray, 1830	10.5 SL male	Native/ Assam, Meghalaya and Darjeeling in West Bengal.
	Balitora chipkali Kumkar, Katwate, Raghavan & Dakanukar, 2016	5.06 SL male	Native/ Western Ghats of Northern Karnataka.
Silent Valley stone loach	*Balitora jaipalli* Raghavan, Tharian, Ali, Jadhav & Dahanukar, 2013	6.17 SL male	Native/ Silent Valley National Park, Kerala.
Palmas stone loach	*Balitora laticauda* Bhoite, Jadhav & Dahanukar, 2012	8.44 SL male	Native/ Maharashtra.
Slender stone loach	*Balitora mysorensis* Hora, 1941	5.0 unsexed	Endemic/Western Ghats.
Qrder: SILURIFORMES Family: BAGRIDAE—bagrid catfishes			
Kerala mystus	*Mystus armatus* (Day, 1865)	14.5 SL male	Native/ Western Ghats, Nagaland and Kerala.

Common Name	Scientific Name	Total length (TL) Standard length (SL) in cm	Distribution
Day's mystus	*Mystus bleekeri* (Day, 1877)	12.5 unsexed	Native/northern India, West Bengal, Kerala, Maharashtra and Tamil Nadu.
	Mystus canarensis Grant, 1999	10.1 SL male	Native/ Kerala.
	Mystus carico (Hamilton, 1822)	5 SL male	Native/ northern Bengal.
	Mystus catapogon Plamoottil, 2016	7.0 SL male	Native/ Kerala.
Gangetic mystus	*Mystus cavasius* (Hamilton, 1822)	30 unsexed	Native/ Uttar Pradesh, Bihar, West Bengal, Assam, Orissa, Maharastra, Tamil Nadu and Western Ghats.
	Mystus dibrugarensis (Chaudhuri, 1913)	?	Native/ Ganga-Brahmaputra basin in north-eastern India.
Elongated Kerala mystus	*Mystus heoki* Plamoottil & Abraham, 2013	13.7 SL male	Native/Kerala.
Central Travancore catfish	*Mystus indicus* Plamoottil & Abraham, 2013	10.7 SL male	Native/ Kerala.
Keletius mystus	*Mystus keletius* (Valenciennes, 1840)	18TL male	Native/ Tamil Nadu, Western Ghats at Kerala and Karnataka.
Long whiskered Kerala catfish	*Mystus keralai* Plamoottil & Abraham, 2014	5.9 SL male	Native/ Kerala.
Jerdon's mystus	*Mystus malabaricus* (Jerdon, 1849)	15 unsexed	Endemic/ Western Ghats of Kerala, Karnataka & Maharashtra.
Idukki mystus	*Mystus manoni* Plamoottil & Abraham, 2013	12.1 SL male	Native/ Kerala.
Wynaad mystus	*Mystus montanus* (Jerdon, 1849)	15 SL unsexed	Endemic/ Kerala, Karnataka, Assam, and Madhya Pradesh.

Common Name	Scientific Name	Total length (TL) Standard length (SL) in cm	Distribution
	Mystus ngasep Darshan, Vishwanath, Mahanta & Barat, 2011	10.87 SL male	Native/ Manipur.
	Mystus seengtee (Sykes, 1839)	15.65 SL male	Native/ Krishna & Cauvery river basins.
Tengara mystus	*Mystus tengara* (Hamilton, 1822)	18.0 TL male	Native/ northern India.
Striped dwarf catfish	*Mystus vittatus* (Bloch, 1794)	16.0 TL unsexed	Native/ throughout India, Western Ghats.
Menoda catfish	*Hemibagrus menoda* (Hamilton, 1822)	45.0 TL male	Native/ Assam, West Bengal, Bihar, Orissa, Maharashtra and Ganges in Uttar Pradesh.
Nilgiri mystus	*Hemibagrus punctatus* (Jerdon, 1849)	50.0 TL unsexed	Endemic/ Western Ghats in Kerala, Karnataka, Tamil Nadu; Madhya Pradesh and Assam.
Krishna mystus	*Hemibagrus maydelii* (Rossel, 1964)	116.0 TL unsexed	Endemic/ Krishna River drainage in southern India.
Irrawaddy mystus	*Hemibagrus microphthalmus* (Day, 1877)	150.0 NG unsexed	Native/ Manipur drainage.
	Olyra astrifera Arunachalam, Raja, Mayden & Chandran, 2013	3.3 SL male	Native/ southern Western Ghats.
	Olyra burmanica Day, 1872	11.0 TL male	Northeast region.
Hora olyra	*Olyra horae* (Prashad & Mukerji, 1929)	6.84 SL male	Native/ Meghalaya.
	Olyra kempi Chaudhuri, 1912	14.5 TL male	Native/ Mangaldai and Assam.
Himalayan olyra	*Olyra longicaudata* McClelland, 1842	11.0 SL male	Native/ Eastern Himalaya.
	Olyra saginata Ng, Lalramliana & Lalthanzara, 2014	8.58 SL male	Native/ Mizoram, north eastern India.

Common Name	Scientific Name	Total length (TL) Standard length (SL) in cm	Distribution
	Batasio affinis Blyth, 1860	7.44 SL male	Native/ Manipur.
Tista batasio	*Batasio batasio* (Hamilton, 1822)	12.5 unsexed	Native/ Teesta river system in North Bengal, Assam and Tripura.
	Batasio convexirostrum Darshan, Anganthoibi & Vishwanath, 2011		Native/ Mizoram.
	Batasio fasiolatus Ng, 2006	7.06 SL male	Native/ Tista River drainage.
	Batasio marianiensis (Chaudhuri, 1913)	6.57 SL male	Native/Assam.
	Batasio sharavatiensis Bhatt & Jayaram, 2004	10.7 SL male	Native/ Sharavati River in Karnataka.
	Batasio spilurus Ng, 2006	4.2 SL male	Native/ Brahmaputra River drainage in Assam.
Assamese batasio	*Batasio tengana* (Hamilton, 1822)	9.0 TL male	Native/ Punjab, North Bengal, Assam, Manipur.
Malabar batasio	*Batasio travancoria* Hora & Law, 1941	10.0 TL male	Endemic/ Western Ghats of Kerala.
Assam cory	*Chadramara chandramara* (Hamilton, 1822)	5.0 SL male	Native/ North Bengal, Assam, Meghalaya and Uttar Pradesh.
Long-whiskered catfish	*Sperata aor* (Hamilton, 1822)	74.0 TL unsexed	Native/ northern India up to the Krishna river system in the south.
Giant river catfish	*Sperata seenghala* (Sykes, 1839)	150.0 TL unsexed	Native/ Large rivers in India.
Family: AILIIDAE			
Bastar garua	*Clupisoma bastari* Datta & Karmakar, 1980	18.0 TL Male	Endemic/ Indravati river in Madhya Pradesh.

Common Name	Scientific Name	Total length (TL) Standard length (SL) in cm	Distribution
Guarchcha	*Clupisoma garua* (Hamilton, 1822)	60.9 unsexed	Native/ Jammu & Kashmir, eastern Punjab, U.P., Bihar, W.B., Assam, Orissa, M.P., Gangetic plains, Maharashtra.
Kocha garua	*Clupisoma montana* Hora, 1937	29.0 TL male	Native/ West Bengal, Uttar Pradesh, Bihar, Tripura, and Eastern Himalaya.
Burmese garua	*Clupisoma prateri* Hora, 1937	25.0 NG unsexed	Teesta river in North Bengal, Bihar, and Uttar Pradesh.
	Eutropiichthyes cetosus Ng, Lalramliana, Lalronunga & Lalnuntluanga, 2014	12.77 SL male	Native/ southern Mizoram.
	Eutropiichthys goongwaree (Sykes, 1839)	64.0 unsexed	Endemic/ Maharashtra, Andhra Pradesh.
Murius vacha	*Eutropiichthys murius* (Hamilton, 1822)	28.0 TL male	Native/ Yamuna, the rivers of West Bengal, Assam, Bihar, Orissa, Gangetic river systems, Gumti and Manu rivers, Tripura.
Batchwa vacha	*Eutropiichthys vacha* (Hamilton, 1822)	40.2 SL unsexed	Native/ Northern and north eastern India up to Orissa, Chilka lake, Tripura. Endangered in Western Ghats.
Gangetic ailia	*Ailia coila* (Hamilton, 1822)	18.0 unsexed	Native/ Krishna, Yamuna, Ganga, Brahmaputra.
Jamuna ailia	*Ailiichthys punctata* Day, 1872	10.0 TL male	Native/ Ganga Jamuna drainage systems.

Common Name	Scientific Name	Total length (TL) Standard length (SL) in cm	Distribution
Family: CHACIDAE Indian frog-mouth catfish	*Chaca chaca* (Hamilton, 1822)	17.5 TL unsexed	Native/ Ganges, Brahmaputra drainages, Bihar, West Bengal and Assam.
Family: PANGASIIDAE- shark catfish Pangas catfish	*Pangasius pangasius* (Hamilton, 1822)	300.0 SL unsexed	Large rivers, streams, lakes, coastal waters and estuaries.
Family: AMBLYCIPITIDAE- torrent catfish Indian torrent catfish	*Amblyceps mangois* (Hamilton, 1822)	12.5 cm SL unsexed	Native/foothills from Kanara valley to Assam and Krishna stream.
Family: ERETHISTIDAE- erethistid catfishes Hara loach	*Erethistes pusillus* Muller & Troschel, 1849	4.24 SL male	Native/ Bihar, Assam, Delhi, and West Bengal.
	Erethistes nareshi (Mahapatra & Kar, 2015)	2.5 SL male	Native/ Assam.
	Erethistoides infuscatus Ng, 2006	4.5 SL male	Native/ Northeast India.
	Erethistoides montana Hora, 1950	4.8 TL, male	Endemic/ Tripura, Assam, and Darjeeling in West Bengal.
	Erethistoides pipri Hora, 1950	3.05 SL, male	Endemic/ Uttar Pradesh and Bihar.
	Erethistoides senkhiensis Tamang, Chaudhry & Choudhury, 2008	4.34 SL male	Native/ Arunachal Pradesh.
	Erethistoides sicula Ng, 2005	3.9 SL male	Native/ northern West Bengal.
Kosi hara	*Hara hara* (Hamilton, 1822)	13.0 TL male	Native/ Uttar Pradesh, Bihar, North Bengal, Assam and Orissa, Manipur and Imphal.
Terai hara	*Hara horai* Misra, 1976	8.0 TL male	Endemic/ Terai and Duars in North Bengal.

Common Name	Scientific Name	Total length (TL) Standard length (SL) in cm	Distribution
Sylhet hara, stone catfish	*Hara jerdoni* Day, 1870	4.0 TL male	Native/ Bihar, Teesta, slow-moving hill streams, river in North Bengal, Assam, and Uttar Pradesh.
Conta catfish	*Conta conta* (Hamilton, 1822)	7.8 TL male	Native/ base of Himalaya in North Bengal, Assam, Meghalaya and Terai in Uttar Pradesh and Mahanadi River.
	Conta pectinate Ng, 2005	4.91 SL male	Native/ Brahmaputra River in Assam, Arunachal Pradesh.
Family: SILURIDAE--Sheatfishes			
Pabo catfish	*Ompok pabo (Hamilton, 1822)*	25.0 NG unsexed	Native/northeast India.
Pabda catfish	*Ompok pabda* (Hamilton, 1822)	30.0 TL male	Native/ throughout northern India.
Indian butter catfish	*Ompok bimaculatus* (Bloch, 1794)	40.0 NG unsexed	Native/ throughout India, sub montane regions.
Goan catfish	*Ompok malabaricus* (Valenciennes, 1840)	51.0 TL unsexed	Endemic/ Goa along the West face of Western Ghats to Kerala.
Boal, freshwater shark	*Wallago attu* (Bloch & Schneider, 1801)	240.0 TL	Native/ major rivers of India.
Family: SISORIDAE			
Clown catfish, Indian gagata	*Gagata cenia* (Hamilton, 1822)	12.5 unsexed	Native/ Ganges, Brahmaputra river systems, Punjab, Delhi, U.P., Bihar, W.B., Assam, Orissa.
Gangetic gagata	*Gagata gagata* (Hamilton, 1822)	31.0 TL male	Native/ U.P., W.B., in Ganga and Brahmaputra river systems.
Deccan nangra	*Gagata itchkeea* (Sykes, 1839)	7.6 TL male	Native/ Maharashtra, and Karnataka.

Common Name	Scientific Name	Total length (TL) Standard length (SL) in cm	Distribution
Koel gagata	*Gagata sexualis* Tilak, 1970	6.0 TL male	Endemic/ Yamuna in Delhi, upper reaches of Ganges. .
	Gagata youssoufi Ataur Rahman, 1976	5.0 SL unsexed	Native/ Bihar, Assam, Orissa, North Bengal, Manipur, and Imphal.
	Glyptothorax alaknandi Tilak, 1969	9.0 TL male	Native/Alaknanda river in Uttar Pradesh.
	Glyptothorax anamalaiensis Silas, 1952		Native/ Anamalai Hills in Kerala, Western Ghats.
Patharchatta	*Glyptothorax annandalei* Hora, 1923	17.1 SL unsexed	Native/ Western Ghats and the Vindhyas.
	Glyptothorax ater Anganthoibi & Viswanath, 2011	12.65 SL male	Native/ Koladyne basin in Mizoram.
	Glyptothorax botius (Hamilton, 1822)	7.91 SL male	Native/ Ganges River drainages.
Not known	*Glyptothorax brevipinnis* (Hora, 1923)	9.7 SL male	Native/ Kedar valley.
	Glyptothorax caudimaculatus Anganthoibi & Viswanath, 2011	5.98 SL male	Native/ Koladyne basin in Mizoram.
Not Known	*Glyptothorax cavia* (Hamilton, 1822)	27.0 cm SL unsexed	Native/ North Bengal, Yamuna river, and northern Assam.
Not known	*Glyptothorax saisii* (Jenkins, 1910)	90.0 SL male/ unsexed	Native/Bihar, West Bengal.
Not known	*Glyptothorax conirostris* (Steindachner, 1867)	14.0 cm TL	Native/ Himachal Pradesh.
	Glyptothorax housi Herre, 1942	10.0 TL male	Endemic/ Western Ghats.

Common Name	Scientific Name	Total length (TL) Standard length (SL) in cm	Distribution
Not known	*Glyptothorax indicus* Talwar, 1991	11.8 cm SL	Native/Kosi River, Rihand River and rivers of Vindhya mountain range, Punjab.
	Glyptothorax jayarami Rameshori & Vishwanath, 2012	10.45 SL male	Native/ Mizoram.
	Glyptothorax maculatum (Regan, 1905)	32.0 TL male	Native/ Ganges, Hooghly river system, West Bengal coast, Chilka lake and Pondicherry coast.
	Glyptothorax malabarensis Gopi, 2010	5.52 SL male	Native/ Western Ghats, Kerala, and Kannur.
Not known	*Glyptothorax manipurensis* Menon, 1955	15.11 cm SL	Native/ Assam, Manipur.
Nayid or River cat	*Glyptothorax pectinopterus* (McClelland, 1842)	17.8 cm SL	Native/ Kanara valley, Shimla Hills, Garhwal, Kumaon Himalayas.
Not known	*Glyptothorax platypogonoides* (Bleeker, 1855)	12.0 cm TL male	Nagaland, Manipur, Assam.
	Glyptothorax reticulatum McClelland, 1842	24.0 SL male	Native/ Ladakh and Kashmir Valley.
Not known	*Glyptothorax striatus* (McClelland, 1842)	21.5 cm	Endemic/ Khasi and Garo Hills, Meghalaya, Sikkim, Brahmaputra river systems, Surma-Meghna River system.
Telichatta	*Glyptothorax telchitta* (Hamilton, 1822)	12.5 cm TL unsexed.	Native/ North Bengal.
Three-lined catfish	*Glyptothorax trilineatus* Blyth, 1860	15 unsexed	Native/ Manipur.

Common Name	Scientific Name	Total length (TL) Standard length (SL) in cm	Distribution
	Sisor chennuah Ng Lakhar, 2003	9.8 SL male/unsexed	Native/ Brahmaputra River drainage in India.
Sisor catfish	*Sisor rabdophorus* Hamilton, 1822	15.0 SL unsexed	Native/Northern rivers of Bengal, Bihar and Delhi.
	Sisor rheophilus Ng, 2003	15.5 TL male	Native/ Middle and upper reaches of the Ganges River basin in Bihar and Uttar Pradesh.
	Sisor torsus Ng, 2003	9.5 SL male	Native/ Middle reaches of the Ganges River basin in Bihar and Delhi.
Sulcatus catfish	*Pseudecheneis sulcate* (McClelland, 1842)	20.0 TL male	Native/ Doon Valley in Uttar Pradesh, north Bengal, Khasi Hills in Meghalaya, Kedar Valley.
	Pseudecheneis suppaetula Ng, 2006	8.41 SL male	Native/ Ganges River drainage in Himachal Pradesh.
	Pseudacheneis ukhrulensis Vishwanath & Darshan, 2007		Native/streams of the Chindwin basin in Ukhrul district, Manipur.
Indian koskimonni	*Creteuchiloglanis kamengensis* (Jayaram, 1966)	19.8 SL unsexed	Native/Arunachal Pradesh.
	Creteuchiloglanis payjab Darshan, Dutta, Kachari, Gogoi, Aran & Das, 2014	15.6 SL male	Native/Arunachal Pradesh.
	Exostoma barakensis Vishwanath & Joyshree, 2007	9.8 SL male	Native/ Manipur.
	Exostoma labiatum (McClelland, 1842)	10.3 SL unsexed	Native/ North Bengal, Meghalaya and Mishmee Hills in Arunachal Pradesh.

Common Name	Scientific Name	Total length (TL) Standard length (SL) in cm	Distribution
	Nangra assamensis Sen & Biswas, 1984	10.5 TL male	Native/ Ganges and Brahmaputra Rivers.
Kosi nangra	*Nangra nangra* (Hamilton, 1822)	5.5 SL unsexed	Native/ Ganges, Kosi and Hooghly River systems.
Gangetic goonch	*Bagarius bagarius* (Hamilton, 1822)	230.0 unsexed	Native/throughout India.
Goonch	*Bagarius yarrelli* (Sykes, 1839)	200.0 NG unsexed	Native/ Western Ghats.
Family: HORABAGRIDAE			
Indian potasi	*Pachypterus atherinoides* (Bloch, 1794)	12.5 unsexed	Native/ throughout India except Kerala.
Khavalchor catfish	*Pachypterus khavalchor* (Kulkarni, 1952)	15.0 unsexed	Endemic/ Krishna river in Maharashtra and Pennar river in Andhra Pradesh.
Family: CLARIIDAE			
Magur	*Clarias batrachus* (Linnaeus, 1758)	40.0 SL unsexed	Freshwater plains of India.
Malabar clariid	*Clarias dayi* Hora, 1936	17.5 TL unsexed	Endemic/ Wynaad hills, Western Ghats of Kerala, Tamil Nadu.
Valenciennes clariid	*Clarias dussumieri* Valenciennes, 1840	50.0 TL unsexed	Endemic/ Goa, Karnataka, Kerala, Pondicherry, and Tamil Nadu.
Family: HETEROPNEUSTIDAE			
Stinging catfish	*Heteropneustes fossilis* (Bloch, 1794)	30.0 NG unsexed	Native/ throughout Indian plains and the Andamans.
	Heteropneustes microps (Gunther, 1864)	15.0 TL male	Native/ throughout India.
Order: ANGUILLIFORMES Family: ANGUILLIDAE			
Indian longfin eel	*Anguilla bengalensis* (Gray, 1831)	1500.0 NG Unsexed	Native/ throughout India.

Common Name	Scientific Name	Total length (TL) Standard length (SL) in cm	Distribution
Shortfin eel	*Anguilla bicolor* McClelland, 1844	120.0 SL female	Native/ Western Ghats.
Order: OSTEOGLOSSIFORMES Family: NOTOPTERIDAE			
Featherback	*Chitala chitala* (Hamilton, 1822)	122.0 SL male	Native/ Indus, Ganges, Brahmaputra, Mahanadi, Buxa, Adma, Jayanti rivers, and Western Ghats.
Featherback	*Notopterus notopterus* (Pallas, 1769)	40.0 SL unsexed	Native/ Indus, Ganges, Brahmaputra, rivers of southern India, rivers of West Bengal.
Order: CLUPEIFORMES Family: CLUPEIDAE			
Indian river shad	*Gudusia chapra* (Hamilton, 1822)	20.0 cm TL	Middle and upper reaches of rivers in India.
	Tenualosa ilisha (Hamilton, 1822)	60.0 TL, unsexed	Native/Chilka Lake, Ganga, Yamuna, Narmada, Tapi, Krishna, Godavary, Cauvery and Pennar rivers.
Family: ENGRAULIDAE			
Short hairfin anchovy	*Setipinna brevifilis* (Valenciennes, 1848)	26.0 SL male	Endemic/ Ganges river system and Jumma river.
Gangetic hairfin anchovy	*Setipinna phasa* (Hamilton, 1822)	40.0 TL male	Native/ Ganga river system, Orissa rivers and estuaries.
Order: SYNGNATHIFORMES Family: SYNGNATHIDAE			
Short-tailed pipefish	*Microphis brachyurus* (Bleeker, 1854)	22.0 SL male	Native/ Rivers of Kerala; Indian coasts.
Crocodile tooth pipefish	*Microphis cuncalus* (Hamilton, 1822)	17.5 SL male	Native/ throughout India.

Common Name	Scientific Name	Total length (TL) Standard length (SL) in cm	Distribution
Deocata pipefish	*Microphis deocata* (Hamilton, 1822)	15.0 SL male	Native/ North Bengal, Bihar, and Assam.
Order TETRADONTIFORMES Family: TETRADONTIDAE			
Ocellated pufferfish	*Leidon cutcutia* (Hamilton, 1822)	6.0 NG unsexed	Native/ Orissa, West Bengal, Bihar, Assam, Chilka lake, and West Tripura.
Order: SYNBRANCHIFORMES Family: SYNBRANCHIDAE			
Rice swamp eel	*Monopterus albus* Zuiew, 1793	90.0 SL unsexed	Native/ Meghalaya, Arunachal Pradesh, Assam, north eastern region.
Cuchla	*Monopterus cuchia* (Hamilton, 1822)	60.9 unsexed	Native/ northern, north eastern India, Maharashtra, and Tripura.
Family: MASTACEMBELIDAE			
Lesser spiny eel	*Macrognthus aculeatus* (Bloch, 1786)	25.0 unsexed	Native/ Buxa, Adma, Jayanti rivers, Tripura.
Spotless spiny eel	*Macrognathus albus* Plamoottil & Abraham, 2014	21.15 SL male	Native/ Manimala river in Kerala.
One-stripe spiny eel	*Macrognathus aral* (Bloch & Schneider, 1801)	40.0 NG unsexed	Native/ West Bengal, Assam and Tamil Nadu, Brahmaputra drainage, Chilka lake.
Yellow banded spiny eel	*Macrognathus fasciatus* Plamoottil & Abraham, 2014	30.65 SL male	Native/ Manimala river in Kerala.
Malabar spiny eel	*Macrognathus guentheri* (Day, 1865)	29.9 SL male	Endemic/ Kerala, Assam, and Karnataka.
	Macrognathus lineatomaculatus Britz, 2010	18.43 SL male	Native/ Jorai river, West Bengal.

Common Name	Scientific Name	Total length (TL) Standard length (SL) in cm	Distribution
	Macrognathus malabaricus (Jerdon, 1849)	26.0 TL male	Native/ Kerala.
	Macrognathus morehensis Arunkumar & Tambi Singh, 2000		Native/ Manipur.
Striped spiny eel	*Macrognathus pancalus* Hamilton, 1822	25.0 unsexed	Native/ Peninsular India, North Bengal, Tripura, Maharashtra, Chilka lake, Western Ghats.
Tire-track spiny eel	*Mastacembelus armatus* (Lacepede, 1800)	80.0 SL unsexed	Native/ Chilka lake, rivers in Maharashtra, Tamil Nadu, Kerala, West Bengal, Jammu, and Western Ghats.
Series: PERCOMORPHA Subseries: OVALENTARIA Family: AMBASSIDAE			
Elongate glass perchlet	*Chanda nama* (Hamilton, 1822)	11.0 cm TL unsexed	Native/ throughout India.
Malabar glassy perchlet	*Ambassis dussumieri* Cuvier, 1828	10.0 TL male	Native/southern India and Andaman Islands.
Day's glassfish	*Parambassis dayi* (Bleeker, 1874)	17.5 TL male	Endemic/ Western Ghats.
High fin glassy perchlet	*Parambassis lala* (Hamilton, 1822)	3.0 NG unsexed	Native/ West Bengal, Orissa, ponds, ditches, pools, lowlands of Ganges, Brahmaputra and Mahanadi basin.
Indian glassy fish	*Parambassis ranga* (Hamilton, 1822)	10.0 unsexed	Native/ throughout India.
Western Ghat glassy perchlet	*Parambassis thomassi* (Day, 1870)	12.0 SL male	Native/ Western Ghats of Kerala and Karnataka.
	Parambassis walkhomi Geetakumari & Basudha, 2012	3.98 SL male	Native/ Lohtak Lake, Manipur.

Common Name	Scientific Name	Total length (TL) Standard length (SL) in cm	Distribution
Order: ANABANTIFORMES Family: BADIDAE			
Badis	*Badis badis* (Hamilton, 1822)	10.0 unsexed	Native/ Yamuna, Ganges, Brahmaputra, Mahanadi River Systems.
	Badis triocellus Khynriam &Sen, 2013	4.874 TL male	Native/ Arunachal Pradesh, Assam, Meghalaya.
	Badis tuivaiei Vishwanath & Shanta, 2004	5.59 SL male	Native/ Manipur.
Family: NANDIDAE			
	Nandus andrewi Ng & Jafar, 2008	12.5 SL male	Native/ Ichamati River drainage in north eastern India.
Mottled nandus	*Nandus nandus* (Hamilton,1822)	12.5 unsexed	Native/ throughout India.
Family: CHANNIDAE			
Borna snakehead	*Channa amphibeus* (McClelland, 1845)	90 TL male	Native/ North Bengal.
	Channa andrao Britz, 2013		Native/ West Bengal.
	Channa aurantimaculata Musikasinthorn, 2000	19.08 SL male	Endemic/ Brahmaputra River basin in Assam and Arunachal Pradesh, north eastern India.
Barca snakehead	*Channa barca* (Hamilton, 1822)	105.0 cm TL male	Native/ Assam and West Bengal.
	Channa gachua (Hamilton, 1822)	15.6 SL unsexed	Native/ Kerala and Maharashtra, Bhimtal and Naukuchiatal Lakes.
Bulleye snakehead	*Channa marulius* (Hamilton, 1822)	90.0 unsexed	Native/ Maharashtra, Tamil Nadu, Andhra Pradesh, West Bengal, Tripura, Bhimtal, Naukuchiatal lakes, Western Ghats.

Common Name	Scientific Name	Total length (TL) Standard length (SL) in cm	Distribution
Walking snakehead, Asiatic snakehead	*Channa orientalis* Bloch & Schneider, 1801	18.0 unsexed	Native/ throughout India.
Meghalaya leopard	*Channa pardalis* Knight, 2016	14.11 SL male	Native/ Meghalaya.
Spotted snakehead	*Channa punctatus* (Bloch, 1793)	31.0 cm TL unsexed	Native/ common throughout the plains.
Assamese snakehead, golden snakehead	*Channa stewarti* (Playfair, 1867)	17.5 unsexed	Native/ Meghalaya, Bihar, West Bengal, Arunachal Pradesh, Eastern Himalayas, Brahmaputra, Kosi, Bagmati, Gandak.
Stripped or banded snakehead	*Channa striata* (Bloch, 1793)	100.0 cm SL	Native/ throughout India.
Family: ANABANTIDAE			
Gangetic koi	*Anabas cobojius* (Hamilton, 1822)	30.0 TL male	Native/ West Bengal, Orissa, Andhra Pradesh, Chilka Lake, West Godavari.
Climbing perch	*Anabas testudineus* (Bloch, 1792)	20.0 NG	Native/ throughout India.
Family: OSPHRONEMIDAE			
Indian paradisefish	*Ctenops nobilis* McClelland, 1845	10.0 TL male	Native/ North-eastern Bengal, Bihar, Assam and Sikkim.
Sunset gourami, Honey gourami	*Trichogaster chuna* (Hamilton, 1822)	7.0 TL male	Native/ Ganges river systems, Assam.
Giant gourami, Banded gourami	*Trichogaster fasciata* Bloch & Schneider, 1801	10.0 unsexed	Native/ throughout India.
Dwarf gourami	*Trichogaster lalius* (Hamilton, 1822)	8.8 cm TL male	Native/ Assam, West Bengal, Bihar, Uttar Pradesh, Ganges, Yamuna, Chilka lake, Western Ghats, Maharashtra.

Common Name	Scientific Name	Total length (TL) Standard length (SL) in cm	Distribution
Order: BELONIFORMES Family: BELONIDAE	*Xenentodon cancila* (Hamilton, 1822)	40.0 NG unsexed	Native/throughout India.
Family: ZENARCHOPTE-RIDAE (viviparous halfbeaks)	*Zenarchopterus dispar* (Valenciennes, 1847)	19.0 TL unsexed	Native/occurs throughout India and Andaman Islands.
	Zenarchopterus ectuntio (Hamilton, 1822)	18.0 TL male	Native/ Hooghly river in West Bengal.
Order: CYPRINODONTIFORMES Family: APLOCHEILIDAE			
Dwarf panchax, Green panchax	*Aplocheilus blockii* Arnold, 1911	6.0 cm TL male	Native/ Kutch, south Kerala, Tamil Nadu, Coromandel coast.
Day's panchax	*Aplocheilus dayi* Steindachner, 1892	9.0 TL male	Native/throughout India.
Malabar killie stripped panchax	*Aplocheilus lineatus* (Valenciennes, 1846)	10.0 TL unsexed	Nativc/ Western and south eastern region, Western Ghats, Peninsular India.
Panchax minnow, Blue panchax	*Aplocheilus panchax* (Hamilton, 1822)	9.5 unsexed	Native/ northern India, Western Ghats Rivers, Maharashtra, Kerala, Coromandel coast, Laccadive, Andaman Islands, Tripura.
	Aplocheilus pervus (Sundararaj, 1916)	6.26 TL male	Native/ Coromandel coast in Peninsular India.
Order: GOBIIFORMES Family: GOBIIDAE			
Grey goby	*Acentrogobius griseus* (Day, 1876)	8.0 SL male	Endemic/ Andhra Pradesh and Tamil Nadu coast, Ennur backwaters in Tamil Nadu, Chilka Lake.

Common Name	Scientific Name	Total length (TL) Standard length (SL) in cm	Distribution
Tank goby	*Glossogobius giurus* (Hamilton, 1822)	50.0 SL unsexed	Native/throughout India.
	Pseudapocryptes elongatus (Cuvier, 1816)	20.0 NG unsexed	Native/throughout India.
Family: OXUDERCIDAE			
	Stigmatogobius sadanundio (Hamilton, 1822)	9.0 SL male	Native/ Sundarbans, West Bengal, Godavari estuary.
Order: MUGILIFORMES Family: MUGILIDAE			
Gangetic bola	*Johnius gangeticus* Talwar, 1991	12.0 SL male	Endemic/ Ganga river including its estuary.
Corsula mullet	*Rhinomugil corsula* (Hamilton, 1822)	30.0 NG unsexed	Native/ Uttar Pradesh, Bihar, West Bengal, Orissa, Tamil Nadu, Western Ghats.
Yellow -tail mullet	*Sicamugil cascasia* (Hamilton, 1822)	12.0 SL male	Native/ Ganga, Yamuna and Brahmaputra river systems.

2.3 An Explicit Endeavour of some Indian Freshwater Ornamentals

2.3.1 Rosy barb

Figure 2.1

Pethia conchonius (Hamilton, 1822)

Common Name(s):

English—Rosy barb, red barb.

IUCN Red List Status—Least Concern.

Systematic Position:

Kingdom: ANIMALIA

Phylum: CHORDATA

Subphylum: CRANIATA

Grade: TELEOSTOMI

Class: OSTEICHTHYES

Subclass: ACTINOPTERYGII

Order: CYPRINIFORMES

Family: CYPRINIDAE

Subfamily: BARBINAE

Genus: *Pethia*

Species: *P. conchonius*

Scientific Name: *Pethia conchonius* (Hamilton, 1822)

Description:

- Dorsal and ventral profiles of the body equally convex. Mild notch over the nape.
- Barbels absent. Lateral line is incomplete.
- Dorsal fin with moderately strong serrated spine.
- Greenish grey colour on the dorsal side, which become silvery on sides.
- A large black blotch on the middle of caudal peduncle.
- Fins orange coloured, dorsal fin tip tinged with black.
- In male during breeding season, colour intensifies to a deep rosy or purplish red.

Sexual Dimorphism:

- The male has a brighter pinkish body colour and black colour on their fins.
- Adult males are smaller, slimmer and more colourful than females. During breeding season, males develop tubercles on the head and snout.
- The female is less brightly coloured and without any black colour on their fins. But during breeding season becomes more luminous.
- The female is slightly plumper.

Species Details:

- Distribution:

  India (Ganges, Brahmaputra, Mahanadi, Godavari, Krishna and Cauvery river systems and other west -flowing rivers); Pakistan, Nepal, Bangladesh and Myanmar.
- Habitat: Inhabits lakes and fast flowing hill streams.
- Food and feeding: Omnivorous. Food consists of worms, insects, crustaceans, plant matter and organic debris. In aquarium, bloodworm, *Daphnia*, *Artemia*, dried flakes and granules are taken.
- Feeding frequency: Several feedings per day.
- Maximum length: 14.0 cm TL (Total Length) male/unsexed.
- Life span: 5 years.

Aquarium aspects:

- Minimum tank size: 20 gallons (76L).

- Suitable for Nano tank: Yes.
- Tank set-up: Well planted set-up with a dark substrate.
- Aquarium water temperature: 17.8 to 22.8°C; 64-72° F.
- Breeding temperature: 73° to 77° F (22°-25° C).
- Aquarium water conditions: pH 6 to 8; hardness 5-19 dGH.
- Substrate type: Any.
- Lighting needs: Moderate -Normal lighting.
- Water movement: Moderate.
- Water region: All.
- Water changes: Monthly.

Behaviour:

Social behaviour-- It is an active, peaceful species, well suited for community aquarium. Very hardy and is considered as one of the hardiest barbs. This fish is undemanding. Can be aggressive toward other fish. It is a schooling species and should be kept in a group of at least 8-10 specimens.

Reproductive--It is an egg-scattering fish, exhibiting no parental care. When in good condition they will often spawn. They are very prolific breeders but they need a breeding tank with shallow water. Spawning usually occurs in the early morning and lasts several hours resulting in several hundred eggs. Eggs are deposited in groups of plants.

Remarks: It is of commercial importance in the aquarium trade. This is one of the barbs used to develop hybrid versions of "tiger barb".

2.3. 2 Pool barb

Figure 2.2

Puntius sophore **(Hamilton, 1822)**

Common Name(s):

English—Pool barb, spot fin swamp barb, stigma barb.

IUCN Red List Status: Least Concern (LC).

Systematic Position:

Order: CYPRINIFORMES

Family: CYPRINIDAE

Subfamily: BARBINAE

Genus: *Puntius*

Species: *P. sophore*

Scientific Name: *Puntius sophore* (Hamilton, 1822)

Description:

- Dorsal profile more convex than the ventral profile.
- Barbels absent.

- Dorsal fin spines: 3; dorsal fin soft rays: 8-9; anal fin spines: 3; anal fin soft rays: 5.
- Dorsal spine weak but entire.
- Colour of the body silvery, with a diffused spot at the caudal fin base.
- A smaller black spot close to the gill-opening.
- A dark diffused central spot at the base of dorsal fin.
- Fins almost colourless.
- Sexual Dimorphism: During breeding season, scarlet lateral band may be seen in mature males. In mature females this band is less distinct.

Species Details:

- Distribution:

  India: Throughout India. Bangladesh, Nepal, Pakistan, Myanmar, Yunnan China, Bhutan and Afghanistan.
- Habitat: Inhabit rivers, streams, ponds in plains and sub montane regions. Large rivers with sand, mud, gravel, pebble, cobble and boulders, in tanks, beels and paddy-fields.
- Food and feeding: Omnivorous, feeds on plant matter and small invertebrates.
- Maximum length: 20.0 cm NG male/unsexed.

Aquarium aspects:

- Aquarium water pH: 8; water temperature: 22° C; breeding temperature 20°-30° C.

Behaviour:

Social—An active, schooling fish.

Reproductive-- It is an egg-scattering species and breeds naturally during monsoon season, at a temperature range of 20-30° C. They do not show parental care and parents may eat the eggs. Breeding takes place close to the bottom of the water bodies near or within areas of dense plant growth.

Remarks: It is a moderately favourite aquarium fish.

2.3.3 Arulius barb

Figure 2.3

Dawkinsia arulius (Jerdon, 1849)

Common Name(s):

English—Aruli barb, longfin barb.

IUCN Red List Status: Endangered (EN).

Systematic Position:

Order: CYPRINIFORMES

Family: CYPRINIDAE

Subfamily: BARBINAE

Genus: *Dawkinsia*

Species: *D. arulius*

Scientific Name: *Dawkinsia arulius* (Jerdon, 1849)

Description:

- Body slender and elongated.
- Colour of the body is dark, brownish olive on the back, which is lighter on the sides and white on the ventral surface.
- There are three vertical black blotches, one in the middle of the body, second one is above the anal fin and the third blotch is on the base of the caudal fin.
- The fins are thin and transparent.
- The young fishes have less pronounced colours than adults.

Sexual Dimorphism:

- Adult males have more ardent or intense colouration than adult females.
- During spawning time, the males have noticeable tubercles on the head.
- Adult females are less colourful, a little longer and are heavier bodied.

Species Details:

- Distribution:

  Cauvery River basin of south east India, Karnataka, Kerala and Tamil Nadu. Distributed in the upper reaches of the Tungabhadra river system in the upper reaches of Karnataka and the Cauvery river basin, in Wayanad of Kerala, the upper and middle reaches of Karnataka and the Nilgiris Hills of Tamil Nadu.
- Habitat: Large streams, rivers and lakes.
- Food and feeding: It is a foraging omnivore. The food includes a variety of worms, insects, crustaceans, plant materials and other organic debris. In aquarium, it is easily fed with blood worms, *Daphnia* and *Artemia* along with good quality dried flakes and granules.
- Maximum length: 12.0 cm TL male/ unsexed.
- Life span: 4-8 years.

Aquarium aspects:

- Minimum tank size: 40-55 gallons.
- Aquarium substrate: Sand, fine gravel, some small boulders and hardy plants.
- Water temperature: 19°-25° C.
- Aquarium water conditions: pH 6.0-7.5; hardness 36-179 ppm.

Behaviour:

Social behaviour-- It is a peaceful, community species and is of schooling behaviour by nature. This is a hardy and fast- moving fish. It thrives well in a well-maintained tank.

Reproductive behaviour--It is an egg-scattering free spawner which exhibits no parental care. It spawns often. A mature tank, dimly lit, base covered with mesh, water slightly acidic to neutral pH, is required for good spawning.

Remarks: It is in aquarium trade and collections are made from the wild. This could be a threat to the species.

2.3.4 Mascara barb

Figure 2.4

Dawkinsia assimilis (Jerdon, 1849)

Common Name(s):

English—Mascara barb.

IUCN Red List Status: Vulnerable (VU).

Systematic Position:

Order: CYPRINIFORMES

Family: CYPRINIDAE

Subfamily: Barbinae

Genus: *Dawkinsia*

Species: *D. assimilis*

Scientific Name: *Dawkinsia assimilis* (Jerdon, 1849)

Description:

- A black band is present across each caudal fin lobe.
- No distinct markings present on body in advance of anal fin origin.
- A caudal blotch is present.
- Lower lip is continuous and mouth is inferior.
- Adult males from some specific region are with branched dorsal rays prolonged into filament-like extensions.
- Sexual Dimorphism: Adult males develop a more intense colour pattern than females and exhibit tubercles on the head at the time of spawning. Adult females are less colourful, little larger and heavier bodied.

Species Details:

- Distribution:

  India: Nethravati, Chalakudy and Kallada rivers in Kerala and also in Kanjirappally waterfall.
- Habitat: Inhabits slow-flowing areas close to the banks with muddy substrate and clear fast flowing water with rocky substrate.

- Food and feeding: Omnivorous. Food include worms, insects, crustaceans, plant material and other organic debris. They may be fed with bloodworms, *Daphnia, Artemia,* dried flakes and granules in captivity.
- Maximum length: 11.2 cm male/unsexed.

Aquarium Aspects:

- Water conditions: Temperature: 19-25° C; pH: 6.0-7.0; hardness: 36-179 ppm.

Behaviour:

Social—It is a peaceful community and schooling fish. Mature males may show aggressive behaviour.

Reproductive—Egg scattering species showing no parental care. Eggs hatch in 24-28 hours. Aquarium breeding is possible.

Remarks: It is a favourite aquarium fish.

2.3.5 Putitor mahseer

Figure 2.5

Tor putitora (Hamilton, 1822)

Common Name (s):

English—Putitor mahseer, Himalayan mahseer, golden mahseer.

IUCN Red List Status: Endangered (EN).

Systematic Position:

Order: CYPRINIFORMES

Family: CYPRINIDAE

Subfamily: Barbinae

Genus: *Tor*

Species: *T. putitora*

Scientific Name: *Tor putitora* (Hamilton, 1822)

Description:

- Head is comparatively big in length.
- Two pairs of barbels as long as eye.
- Thick lips with a continuous labial groove.
- Dorsal fin spines: 3; dorsal fin soft rays: 9; anal fin spines 2-3; anal fin soft rays: 5.
- Dorsal fin spines bony, strong but smooth.
- In adults, the body above the lateral line is generally golden with light pinkish and silvery below. A broad light greyish blue or purplish lateral line is seen, fins are reddish.

Species Details:

- Distribution:

India: Occurs all along the base of the Himalayas including Kashmir, Jammu, Himachal Pradesh, Uttar Pradesh, Maharashtra, Darjeeling, in Brahmaputra River system and central Western Ghats. Also, in Afghanistan, Pakistan, Nepal, Bangladesh, Bhutan and Myanmar.

- Habitat: Inhabits streams, riverine pools, lakes, rapid streams with rocky bottom.
- Food and feeding: Omnivorous. Feeds on zooplankton, plant matter, dipteran larvae and fish. Juveniles feed on phytoplankton and fingerlings prefer to feed on algae.
- Maximum length: 275.0 cm TL male/unsexed.

Aquarium aspects:

- Water conditions: pH 7.4-8.0; temperature 13°-30° C.

Behaviour:

Reproductive—To spawn, this fish migrates from the lower to the middle reaches of streams and rivers. Spawning takes place during low phase of flood. Fry of *Tor putitora* locate themselves among stones in the marginal areas of the stream.

Remarks: It is a favourite aquarium pet.

2.3.6 Zebra danio

Figure 2.6

Danio rerio (Hamilton, 1822)

Common Name(s):

English—Zebra danio, zebra fish.

Trade name: Zebra danio.

IUCN Red List Category: Least Concern.

Systematic Position:

Order: CYPRINIFORMES

Family: CYPRINIDAE

Subfamily: DANIONINAE

Genus: *Danio* (synonym: *Brachydanio*)

Species: *Danio rerio*

Scientific name: *Danio rerio* (Hamilton, 1822)

Description:

- Slender, compressed body with five uniform, pigmented, horizontal, blue stripes present on the side of the body and extend to the end of the caudal fin.
- Mouth is directed upwards with a barbel at the end of each lip.
- Anal fin is distinctively striped. Sides and fins are marked with broad blue longitudinal stripes, and the dorsal fin is outlined in white.
- Sexual Dimorphism: The male is torpedo-shaped, with gold stripes

between the blue stripes. The female has a larger, whitish belly and silver stripes instead of gold. Mature female exhibits a small genital papilla in front of the anal fin origin. Sexually mature female is usually rounder bellied, slightly less colourful and little larger than male. Male fish is with streamlined and slim body.

Species Details:

- Distribution:

  Native to India (Arunachal Pradesh, Assam, Karnataka, Meghalaya, Orissa, Uttar Pradesh, West Bengal). Also occur in Bangladesh, Pakistan and Nepal.
- Habitat: It is an annual species. Adults inhabit streams, canals, ditches, ponds and beels. Occur in slow-moving to stagnant standing water bodies, particularly rice-fields and lower reaches of streams, rivulets at foot hills.
- Food and feeding: This fish is omnivorous. Foods include zooplankton, phytoplankton, insects, insect larvae, worms, small crustaceans, brine shrimps, *Paramecia* in the wild. In the aquarium, they eat any prepared or live food. Flakes, pellets, tablets, live foods, vegetables and meaty foods are accepted.
- Feeding frequency: Several feedings per day.
- Maximum length: 3.8 cm SL male/ unsexed.
- Life span: 3.5 to 5.5 years.

Aquarium aspects:

- Minimum tank size: 10 gallons.
- Suitable for Nano tank: Yes.
- Substrate type: Any.
- Lighting requirements: Moderate to normal lighting.
- Aquarium water conditions: Temperature: 64.0° to 75.0 °F (17.8° to 23.9°C); pH: 6.0-8.0; hardness: 2-20 dGH.
- Water movement: Weak.
- Water region: All.
- Water changes: Monthly.

Behaviour:

Social behaviour--This fish is hardy, very peaceful, with playful attitude. A good community fish and are best kept in a school of 5 or more of their own kind.

Reproductive behaviour-- It is an egg scattering spawner that exhibits no parental care. It is a prolific breeder and can easily breed in aquariums. In favourable conditions, the female will release her eggs in water and the male will fertilize the eggs. The eggs will sink to the bottom and fall through the marbles, out of their parent's reach.

Remarks: It is one of the most popular aquarium fishes.

2.3.7 Flying barb

Figure 2.7

Esomus danrica (Hamilton, 1822)

Common Name(s):

English—Flying barb.

IUCN Red List: Least Concern.

Systematic Position:

Order: CYPRINIFORMES

Family: CYPRINIDAE

Subfamily: DANIONINAE

Genus: *Esomus*

Species: *E. danrica*

Scientific Name: *Esomus danrica* (Hamilton, 1822)

Description:

- Body is silvery with a black elongated line along the length of the body, from the mouth up to the caudal peduncle. Backside is olive green.
- Fins are golden.
- Two pairs of barbels present, the lower pair is longer and reaches the middle of the body.
- The pectoral fins are much elongated.
- Sexual Dimorphism: Males are smaller and slender than the females.

Species detail:

- Distribution:

  Found throughout India. Found also in Pakistan, Nepal, Bangladesh, Afghanistan, Sri Lanka and Myanmar.
- Habitat: It is a benthopelagic species, usually found in ponds, tanks, ditches and canals. This species has also been found in brackish water.
- Food and feeding: They prefer to eat insects close to the surface. In the aquarium, a diet of chopped earthworms or artificial food may be given.
- Maximum length: 13.0 cm TL male/unsexed.
- Life span: 3-5 years.

Aquarium aspects:

- Aquarium water conditions: Temperature: 20-25°C; pH 7.6; hardness: soft to medium; water: well aerated.
- An aquarium with a tight- fitting lid with no gaps is recommended.

Behaviour:

Social behaviour-- It has an exceptional ability of jumping.

Reproductive behaviour-- A large thickly planted aquarium should be provided for breeding. Water columns may be kept at 15 cm. Eggs are partially sticky and fall to the bottom of the tank among the pebbles.

Remarks: It is a popular aquarium fish.

2.3.8 Blue-dotted hill trout

Figure 2.8

Barilius bakeri Day, 1865

Common Name(s): Malabar baril, blue-dotted hill trout, royal danio.

IUCN Red List Category: Least Concern.

Systematic Position:

Order: CYPRINIFORMES

Family: CYPRINIDAE

Subfamily: DANIONINAE

Genus: *Barilius*

Species: *B. bakeri*

Scientific Name: *Barilius bakeri* Day, 1865

Description:

- Body moderately elongate, compressed, abdomen rounded.
- Head sharply pointed, snout compressed and pointed.
- Mouth anterior, eyes large and superior.
- Lips thin and simple. Upper jaw longer than lower.
- The body of the fish is of olive- green in colour.
- A horizontal row of bluish spots is present across the flanks.
- Sexual Dimorphism: Males are more brightly coloured than the females. They develop breeding tubercles on the head and orange pigmentation in the lower portion of the body. Adult females are less brightly coloured. They are larger in size and thick-bodied than the males.

Species Details:

- Distribution: It is endemic to the Western Ghats in Kerala and is native to India (Karnataka, Kerala and Tamil Nadu).
- Habitat: Inhabits rapid and micro-tributaries. In small mountain streams with bed rock, boulders, cobbles and gravel as major substrate.
- Food and feeding: In nature, they are surface-feeders preying on aquatic and terrestrial insects, small fishes and benthic invertebrates.
- Maximum length: 15.0 cm TL male/unsexed.

Aquarium aspects:

- Aquarium water conditions: Temperature—18-26° C; pH 6.0-7.5; hardness-18-179 ppm.

Behaviour:

Social behaviour--It is a schooling fish. This species is an extremely fast swimmer and vigorous feeder.

Reproductive behaviour—Spawning in captivity, can be stimulated by cool water changes. But there are no reports of raising fry.

Remarks: This fish is collected and exported for the International Aquarium Trade.

2.3.9 Bata

Figure 2.9

Labeo bata (Hamilton, 1822)

Common Name(s):

English—Bata, Bata labeo.

IUCN Red List Status: Least Concern (LC).

Systematic Position:

Order: CYPRINIFORMES

Family: CYPRINIDAE

Subfamily: LABEONINAE (labeos)

Genus: *Labeo*

Species: *L. bata*

Scientific Name: *Labeo bata* (Hamilton, 1822)

Description:

- Elongate body, dorsal profile more convex than ventral.
- A pair of very short maxillary barbels present.
- Lips are thin and continuous.
- A tubercle is present inside the lower jaw.
- Dorsal fin arises nearer to snout than to the caudal base.
- Caudal fin deeply forked.
- Dorsal fin spines: 2-3; dorsal fin soft rays: 9-10; anal fin spines: 2; anal fin soft rays: 5.
- Colour of the body silvery, dorsal side dark, fins orangish, with fine black dots.

Species Details:

- Distribution:

  Throughout India except Kerala.
- Habitat: Ponds, rivers and rivulets.
- Food and feeding: Omnivorous. Protozoa, algae and tiny fishes are taken as food.
- Maximum length: 61.0 cm TL male/unsexed.

Aquarium Aspects:

- Water conditions: Temperature: 28-31° C; pH: 7.4-8.3.

Behaviour:

Social—Bottom dwelling, slow moving, peaceful fish.

Reproductive—Breeding time is between the months of June and October especially during monsoon time in rivers.

Remarks: It is often preferred for home aquaria.

2.3.10 Roho labeo

Figure 2.10

Labeo rohita (Hamilton, 1822)

Common Name (s):

English—Roho labeo, Rohu.

IUCN Red List Status: Least Concern (LC).

Systematic Position:

Order: CYPRINIFORMES

Family: CYPRINIDAE

Subfamily: LABEONINAE

Genus: *Labeo*

Species: *L. rohita*

Scientific Name: *Labeo rohita* (Hamilton, 1822)

Description:

- Dorsal profile more convex than the ventral.
- Snout obtuse, depressed and without labial lobe.
- Thin and very short maxillary pair of barbels.
- Dorsal fin is situated between caudal fin base and snout.
- Caudal deeply forked.
- Dorsal fin spines: 3; dorsal fin soft rays: 12-13; anal spines: 2; anal fin soft rays: 5.
- Sides of the body silvery and belly is dull white. Dorsal side brownish blue or reddish blue. Fins are pinkish.

Species Details:

- Distribution:

  India: Throughout India. Bangladesh, Pakistan and Myanmar.
- Habitat: Inhabit rivers.
- Food and feeding: Herbivorous. Feeds on plants. As juveniles eat zooplankton but as grow eat phytoplankton.

Aquarium Aspects:

- Water conditions: Temperature: 25.3-29° C; pH: 8.23-8.80; hardness: 459.7 mg/l-462.83 mg/l.

Behaviour:

Social—Usually remains solitary. A diurnal species.

Reproductive—Spawning takes place in flooded rivers, in middle reaches of rivers, well above tidal reaches. In reservoirs and in bundh-type tanks also. Fecundity is from 226,000 to 2,794,000.

Remarks: This fish is often preferred for home aquariums.

2.3.11 Reticulate loach

Figure 2.11 A & B

Botia lohachata Chaudhuri, 1912

Common Name(s): Reticulate loach, Y loach, Yo Yo loach, Pakistani loach.

IUCN Red List Status: Not Evaluated.

Taxonomic Position:

Order: CYPRINIFORMES

Family: BOTIIDAE

Genus: *Botia*

Species: *B. lohachata*

Scientific Name: *Botia lohachata* Chaudhuri, 1912

Description:

- Body oblong, short, moderately deep. Abdomen rounded.
- Head and body marked with vertical bands.
- Four pairs of barbels, protruding from the mouth area.
- Silvery, gold colouration in young turns grey to green in adults and with black irregular stripes.
- Three unconnected "Y" markings typically occur in adults.
- Very small scales are present which are embedded in their skin.
- Sexual Dimorphism: Males are slenderer than females. Females are plumper when full of eggs. Males may show a red colour around the barbels and mouth.

Species Details:

- Distribution: It is native to Ganga drainage and River Gandak (India). Found in Indus River system, Pakistan, Nepal and Bangladesh.
- Habitat: Inhabits creeks with rocky and sandy bottoms.
- Food and feeding: It is an omnivorous fish. Younger fish in aquarium will eat most prepared food whereas adults prefer live food such as tubifex, blackworms and small snails. They eat live, frozen, flakes or pellet food.

They are micro-predators and prefer live food once in a week and can consume large amount of food.

- Feeding frequency: Take small amounts of food several times a day.
- Maximum length: 15.4 cm TL male/ unsexed.
- Life span: 8-12 years.

Aquarium aspects:

- Minimum tank size: 30 gallons.
- Aquarium water conditions: Temperature: 72-86° F, 19-27° C; pH 6.0-7.5; hardness 36-215 ppm. Well oxygenated water with some degree of current is required.
- Substrate: Sandy or gravel substrate, smooth river rocks, a few drift wood or bogwood and some plants.
- Water movement: Weak.
- Water region: Bottom.
- Water changes: Weekly.

Behaviour

Social behaviour- It is a good community tank fish, semi-aggressive, active scavenger and enjoys the company of its own species. It prefers caves, holes and other places to hide, especially when it sleeps. This fast swimmer. is mostly active in the evening and at night. Excellent diggers, excellent jumpers also.

Reproductive behaviour- It is an oviparous fish. Distinct pairing forms during breeding. Successful captive breeding has not been recorded yet.

Remarks: *Botia lohachata,* is the only current legal imports to Australia and is the only species of the genus found in store.

2.3.12 Stone loach

Figure 2.12

Nemacheilus corica (Hamilton, 1822)

IUCN Red List Status: Least Concern (LC).

Systematic Position:

Order: CYPRINIFORMES

Family: NEMACHEILIDAE

Genus: *Nemacheilus*

Species: *N. corica*

Scientific Name: *Nemacheilus corica* (Hamilton, 1822)

Description:

- Dorsal originates nearer to the snout and anterior to the pelvic fins.
- Third and fourth pectoral rays produced reaching the base of the ventral fin.

- Eyes placed in the middle of the head.
- Caudal fin is forked.
- Dorsal fin spines: 2; dorsal fin soft rays: 8; anal fin spines: 2; anal fin soft rays: 5.
- Body is bluish with a silvery band along the middle of the body.
- About 13 black blotches along the middle of the side with smaller ones above.

Species Details:

- Distribution:

  India: In the foothills of the Himalayas, in the drainages of the Ganges, from Darjeeling, through Kumaon to Himachal Pradesh, Punjab and Nainital Lake. Bangladesh, Nepal and Pakistan.
- Habitat: Small, clear hill-streams with sandy bottom, foot-hills, ditches, large rivers with high turbid monsoon flow and with diverse substrate like sand, mud, gravel, cobble and boulders.
- Maximum length: 4.2 cm SL male/ unsexed.

Aquarium aspects:

- Water conditions: Temperature 15-24° C; pH 6.0-8.0; hardness 36-268 ppm; dissolved oxygen high.
- Water changes: weekly

Behaviour

Social—Remains best in the presence of conspecifics.

Reproductive—Captive breeding is not reported.

Remarks: This fish has good export value.

2.3.13 Gangetic Mystus

Figure 2.13

Mystus cavasius (Hamilton, 1822)

Common Name(s):

English—Gangetic mystus.

IUCN Red List Status: Least Concern (LC).

Systematic Position:

Order: SILURIFORMES

Family: BAGRIDAE

Genus: *Mystus*

Species: *M. cavasius* (Hamilton, 1822)

Scientific Name: *Mystus cavasius* (Hamilton, 1822)

Description:

- Body elongate and compressed.

- A conical head and narrow occipital process present.
- Barbels 4, maxillary barbels extend posteriorly beyond the caudal fin base, but young fishes do not possess barbels that extend beyond the anal fin.
- Dorsal spine weak, feebly serrated.
- Colour is greyish with a more or less well-defined mid-lateral longitudinal stripe.
- A dark spot is present just anterior to the first dorsal spine.
- Dorsal, adipose and caudal fins have melanophores.
- Adipose fin large and originate from just behind dorsal.
- Dorsal spine: 1; dorsal soft rays: 7; anal spine: 0; anal soft ray: 10-11.

Species Details:

- Distribution:

  Uttar Pradesh, Bihar, West Bengal, Assam, Orissa, Maharashtra, Tamil Nadu and Western Ghats in India. And in Pakistan, Nepal, Sri Lanka and Myanmar.
- Habitat: Inhabits tidal rivers and lakes, beels, canals, ditches, ponds and inundated fields.
- Food and feeding: Feeds on plant matter, insects and crustaceans.
- Maximum length: 40.0 cm SL male/unsexed.

Aquarium Aspects:

- Water conditions: Temperature: 25.6-31.6 °C; pH: 7.5-8.65; dissolved oxygen: > 3.5 mg/l.

Behaviour:

Reproductive—They are oviparous.

Remarks: Population of this species is declining now because of trading, habitat destruction, and catching.

2.3.14 Striped dwarf catfish

Figure 2.14 A & B

Mystus vittatus (Bloch, 1794)

Common Name(s):

English—Striped dwarf catfish.

IUCN Red List Status: Least Concern (LC).

Systematic Position:

Order: SILURIFORMES (catfish)

Family: BAGRIDAE (Bagrid catfish)

Genus: *Mystus*

Species: *M. vittatus* (Bloch, 1794)

Scientific Name: *Mystus vittatus* (Bloch, 1794)

Description:

- Body elongate and slightly compressed.
- Maxillary barbels extending beyond pelvic fins, often to the end of the anal fin.
- Dorsal spine weak, finely serrated on its inner edge.
- Adipose fin small, inserted much behind rayed dorsal fin but anterior to the anal fin.
- Dorsal spines (total): 1; dorsal soft rays (total): 6-7; anal spines: 0; anal soft rays: 12-13.
- Body delicate grey-silvery to shining golden, with several, about 5 , pale blue or dark brown to deep black longitudinal stripes on side.
- The fins are glassy with dark tips.
- A narrow dusky spot is present on the shoulder.

Species Details:

- Distribution:

  Western Ghats and throughout India, Pakistan, Bangladesh, Sri Lanka, Nepal, and Myanmar. Reported from Malaysia, Laos, Bhutan, Viet Nam, and Cambodia.
- Habitat: Adults fishes live in standing and flowing waters. Inhabit marginal vegetation in lakes and swamps with a mud substrate.
- Food and feeding: Omnivorous. Feeds on plants, shrimps, insects, molluscs and fish.
- Maximum length: 21.0 cm SL male/unsexed.

Aquarium Aspects:

- Water conditions: Temperature: 22-28° C; pH: 6.0-7.5; dGH range: 4-25.

Behaviour:

Social—Adults are not gregarious and can be seen in clear waters moving about solitarily. Of mild temperament and can be maintained in small aquariums with other species of smaller fish. A bottom dweller.

Reproductive—This egg-layer fish that spawns in shallow water and rainwater pools. Makes sound during spawning.

Remarks: Due to trading and habitat destruction, population of this species is declining.

2.3.15 Pangas catfish

Figure 2.15 A & B

Pangasius pangasius (Hamilton, 1822)

Common Name(s):

English—Pangas catfish.

IUCN Red List Status: Least Concern (in 2009), now endangered.

Systematic Position:

Order: SILURIFORMES

Family: PANGASIIDAE (shark catfishes)

Genus: *Pangasius*

Species: *P. pangasius*

Scientific Name: *Pangasius pangasius* (Hamilton, 1822)

Description:

- Body streamlined and compressed, without any scale. Mouth is wide and eyes small.
- Dorsal fin situated far forward with one or two spines and 5-7 soft rays.
- A small adipose fin is present.
- Two pairs of barbels present. Maxillary barbel reaches to gill aperture.
- A dark grey colour on its back, a silvery belly is present.
- Bright yellow caudal fin in adults.
- Dorsal spine: 2; dorsal soft rays: 7; anal spine: 0; anal soft rays: 29-32.

Species Details:

- Distribution:

  Large rivers of Indian subcontinent and Myanmar.
- Habitat: Inhabit large rivers and estuaries. Found in high estuary (freshwater tidal zone) as juveniles, moving to brackish water as sub-adults, and finally as adults to river mouths and inshore areas.
- Food and feedings: Omnivorous and bottom feeder. Feeds on snails, other molluscs, insects, fishes, crustaceans and plants.
- Maximum length: 300 cm SL male/unsexed.
- Life span: 10 years-20 years.

Aquarium Aspects:

- Water conditions: pH: 6.0-7.5; dGH-25; temperature: 23 -28° C.

Behaviour:

Social— This fish prefers to live in group.

Reproductive-Eggs stick to the plants or roots of plants. After 24 hours of breeding, the eggs grow into fry. It has been reported that the fish attains maturity in the estuary, then migrates and breeds in the freshwater and the young one drift into the tidal stretch of the river where they grow and attain maturity.

Remarks: It has been reported that population of this fish is on decline due to over exploitation, habitat degradation and destruction of breeding ground.

2.3.16 Asian stone catfish

Figure 2.16

Hara jerdoni Day, 1870

Common Name(s):

English—Asian stone catfish, Sylhet Hara.

IUCN Red List Status: Least Concern (LC).

Systematic Position:

Order: SILURIFORMES

Family: ERETHISTIDAE

Genus: *Hara*

Species: *H. jerdoni*

Scientific Name: *Hara jerdoni* Day, 1870

Description:

- Small, delicate, streamlined, slightly depressed body anteriorly.
- The posterior part of the body from the end of the dorsal fin up to the caudal fin is laterally compressed.
- A very straight whip-like tail.
- Snout depressed, pointed bluntly.
- Rayed dorsal with a strong serrated spine.
- Pectoral spine very long.
- Median longitudinal groove on head long.
- Dorsal fin spine: 1; dorsal fin soft rays: 5; anal fin rays: 10.
- Two sets of antennae like barbels present.
- Brownish body with irregular cross bands.
- The skin is covered with tiny tubercles.
- Sexual Dimorphism: Male fish has longer fins and barbels than the female fish.

Species Details:

- Distribution:

  Native to North eastern India; Bangladesh.
- Habitat: Inhabits slow-moving hill steams with sandy bottom, rivers and beels with soft substrates.
- Food and feeding: Omnivorous and nocturnal feeder. Accepts bloodworms, *Daphnia* and dried pellets in captivity.
- Maximum length: 4.0 cm TL male/unsexed.

Aquarium Aspects:

- Minimum tank size: 10 gallons.
- Water conditions: Temperature: 18-24° C (64-75° F); pH: 5.6-7.6; hardness: 8-15 GH.

- Calm, cool, well-oxygenated, well-filtered water and planted aquarium is needed.

Behaviour:

Social—Very peaceful and remain well in communities of other small peaceful species. Very sociable fish. It is also very inactive.

Reproductive—Egg-layer fish. Fry are extremely delicate. Aquarium breeding is difficult.

Remarks: It is a popular aquarium fish.

2.3.17 Butter catfish

Figure 2.17

Ompok bimaculatus (Bloch, 1794)

Common Name(s): Butter catfish.

IUCN Red List Status: Near Threatened (NT).

Systematic Position:

Order: SILURIFORMES

Family: SILURIDAE

Genus: *Ompok*

Species: *O. bimaculatus*

Scientific Name: *Ompok bimaculatus* (Bloch, 1794)

Description:

- Two pairs of barbels present. Maxillary pair longer than the head.
- A short dorsal without spine much nearer to snout.
- Anal rays long, continuous.
- Caudal forked with upper lobe longer.
- Colour silvery shot with purple, a black blotch on the shoulder and another at the base of the caudal peduncle. One or two faint – looking black lateral bands the upper part of body.
- Anal and caudal fins golden yellow.
- Sexual Dimorphism: Males have a serrated edge to the pectoral spine, which females lack. Mature females are thicker-bodied than males.

Species Details:

- Distribution:

  Throughout India, submontane region, and Myanmar.
- Habitat: Inhabit quiet, shallow, muddy water, sandy streams, rivers, tanks, canals, beels and inundated fields.
- Food and feedings: Vegetable matter and fish. In the aquarium, the fish takes most live, frozen and dried foods. Earthworms, pieces of prawn, mussel and dried pellets.

- Maximum length: 45.0 cm SL male/unsexed.

Aquarium Aspects:

- Minimum tank size: 680 litres.
- Aquarium Region: Open swimming space, along with some hiding places.
- Temperature: 68°-79° F (20°-26° C).
- pH: 6.0-8.0.
- Hardness: 4-28 dGH.

Behaviour:

Social—It is a peaceful fish and shy one but not safe for small fishes. It is compatible with medium to large sized characins, catfish and cichlids. Prefer hiding places. When active, try to remain in open water.

Reproductive—In captivity breeds only through the use of hormone injections.

Remarks: Popular fish in public aquariums.

2.3.18 Wallago

Figure 2.18

Wallago attu (Bloch & Schneider, 1801)

Common Name(s):

English--Wallago, wallago catfish, helicopter catfish.

IUCN Red List Status: Near Threatened (NT).

Systematic Position:

Order: SILURIFORMES

Family: SILURIDAE

Genus: *Wallago*

Species: *W. attu*

Scientific Name: *Wallago attu* (Bloch & Schneider,1801)

Description:

- Elongate and strongly compressed body.
- Head large and broad, snout depressed.
- Mouth very deeply cleft, its corner reaches far behind the eyes.
- Barbels two pairs. Maxillary barbels extending to anterior margin posterior of anal fin. Mandibular barbels to angle of mouth.
- Eyes small.
- Dorsal fin small without spine. Anal fin very long, free from caudal fin.
- Dorsal soft rays: 5; anal soft rays: 77-97.
- Teeth in jaws set in wide bands, vomerine teeth in two small patches.
- Colour uniform silvery, grey becoming dull white below.
- Sexual Dimorphism: Pectoral spine is well developed, broad and strongly

denticulated in males, whereas weakly developed, narrow and feebly serrated in female.

Species Details:

- Distribution:

  Major rivers of India-Ganga, Indus, Narmada, Godavari, Krishna and Mahanadi. Pakistan to Viet Nam and Indonesia.

  Habitat: Inhabits large rivers, lakes and tanks. Found in deep, still or slow-flowing water with a mud or silt substrate. Mostly hides under holes in river banks and canals.
- Food and feeding: Predatory and carnivorous in nature. Adults are also piscivorous. Juveniles feed mainly on insects. Adults feed on smaller fish, crustaceans and molluscs. In captivity, they are seen to forage actively from top to the bottom of the tank.
- Maximum length: 240 cm TL male/ unsexed.

Aquarium Aspects:

- Minimum tank size: 5500 litres.
- Substrate: Gravel.
- Tank type: Very heavy tight- fitting cover to prevent the fish from escaping.
- Temperature: 19°- 29° C, 66-84° F.
- pH: 6.0-7.6.
- Hardness: 1-20 dGH.

Behaviour:

Social—Sluggish in nature, stays on muddy bottom. It should be kept alone as a single specimen, as it may consume any fish that fits into its mouth.

Reproductive—No reports of captive breeding in aquaria.

Remarks: It is often found in freshwater public aquariums.

2.3 19 Indian mottled eel

Figure 2.19

Anguilla bengalensis (Gray, 1831)

Common Name(s):

English—Indian mottled eel, Indian long fin eel, long-finned eel, river eel, African mottled eel.

IUCN Red List Status: Near Threatened (NT).

Systematic Position:

Order: ANGUILLIFORMES

Family: ANGUILLIDAE (Freshwater eels)

Species: *A. bengalensis*

Subspecies*: Anguilla bengalensis bengalensis; Anguilla bengalensis labiate.*

Scientific Name: *Anguilla bengalensis* (Gray, 1831)

Description:

- Body elongate, head flattened dorsally.
- Head is conical in shape.
- Mouth terminal, lower jaw prominent.
- Narrow bands of teeth on jaws. Broad band on vomer.
- Dorsal fin soft rays 250-305, anal fin soft rays 220-250.
- Colour of body brownish above and yellowish on the sides and beneath.
- Sometimes black spots present on the upper surface of the body.

Species Details:

- Distribution:

  Kerala, Tamil Nadu, Andhra Pradesh, Arunachal Pradesh, Manipur, Meghalaya, Orissa, West Bengal, Assam, and Nagaland in India. And in Bangladesh, Nepal, Myanmar, Sri Lanka, Pakistan and the East-Indies.
- Habitat: Found in freshwaters, also occurs in estuaries and in the sea during early life and near maturity. Inhabits freshwater streams, pools and reservoirs, mud substrates of tanks and in deep rock pools of rivers.
- Food and feeding: Carnivorous and nocturnal predator in nature. Adults are piscivorous. Fry consume small bottom-dwelling invertebrates, and juveniles feed on insects and other small aquatic insects. Feeds on small crustaceans, worms, molluscs, annelids, crustaceans, macrophytes and microphytes.
- Maximum length: 200 cm TL male/unsexed.

Behaviour:

Social—Forages actively at night, feeding mainly on a wide range of bottom invertebrates.

Reproductive—Breeds in sea. After the larval stages, the adult fishes migrate to the freshwater and spend their life in paddy fields, lakes and lagoons. Catadromous species; migrates to sea as adults for breeding after 10-15 years in freshwater. Breeds in far offshore areas and then die. The life cycle includes five stages, leptocephalus, glass eel, elver, yellow eel and silver eel stages. The "leptocephali" larvae, reach near offshore and after few weeks they metamorphose to tiny and transparent "glass eel" stage and enters rivers.

Remarks: Common in public aquariums.

2.3.20 Bronze featherback Figure 2.20 A & B

Notopterus notopterus (Pallas, 1769)

Common Name(s):

English—Bronze featherback, Asian Knifefish.

IUCN Red List Category: Least Concern.

Systematic Position:

Order: OSTEOGLOSSIFORMES

Family: NOTOPTERIDAE

Genus: Notopterus

Species: *N. notopterus*

Scientific Name: *Notopterus notopterus* (Pallas, 1769)

Description:

- Body dark bronze – like in young age that becomes lighter with age.
- Elongated and flat body with an arched back or convex, concave dorsal head.
- The dorsal fin is very small.
- It has a continuous fin along the underside formed by a joining of the caudal and anal fin.
- Dorsal soft rays: 7-9; anal soft rays: 97-111.
- Sexual Dimorphism: Not known.

Species Details:

- Distribution:

  Asia: Indus, Ganges, Brahmaputra, Mahanadi, Krishna, Cauvery and other river basins in southern India, Irrawaddy, and Salween; Mekong, Chao Phraya, Mekong, all coastal river basins of Peninsular Thailand and Malaysia; Sumatra and Java.
- Habitat: It inhabits fresh and brackish waters, and appears to thrive well in lentic waters. Adults inhabit standing and sluggish waters of lakes, floodplains, canals and ponds. Common in tanks throughout the greater parts of India.
- Food and feeding: Omnivorous. In the wild, it takes insects, fish, crustaceans and some young roots of aquatic plants. In the aquarium, they can be fed small fish, worms or small shrimps. They take flake foods or pellets occasionally.
- Feeding frequency: Daily.
- Maximum length: 60.0 cm SL male/unsexed.
- Life span: 15 years.

Aquarium Aspects:

- Minimum tank size: 40 gallons.
- Suitable for Nano tank: No.
- Substrate type: Sand.
- Lighting needs: Low -subdued lighting.
- Temperature: 75.0° to 82.0° F (23.9° to 27.8° C).

- Range pH: 6.0-6.5.
- Hardness range: 2-12 dGH.
- Water movement: Weak.
- Water changes: Weekly.

Behaviour:

Social—They are usually peaceful, though semi-aggressive and cannot tolerate fishes of their own kind.

Reproductive—It breeds in stagnant or running water in the rainy season. A ripe female bears relatively fewer eggs, they are laid in small clumps on submerged vegetation. Fingerlings are available in upper reaches of Cauvery in July-August.

Remarks: A favourite fish in public aquariums.

2.3.21 Cuchia

Figure 2.21

Monopterus cuchia (Hamilton, 1822)

Common name(s):

English—Cuchia, Gangetic swamp eel.

IUCN Red List Status: Least Concern (LC).

Systematic Position:

Order: SYNBRANCHIFORMES

Family: SYNBRANCHIDAE

Subfamily: SYNBRANCHINAE

Genus: *Monopterus*

Species: *M. cuchia*

Scientific Name: *Monopterus cuchia* (Hamilton, 1822)

Description:

- An eel-shaped fish with a tapering tail which is flattened in a vertical plane.
- A rudimentary dorsal fin is present which is present slightly before anus. Pelvic, ventral, anal and caudal fins are absent.
- Upper jaw is slightly longer. Lips are fleshy.
- Distinct scales longitudinally arranged.
- Several spots are present all over the tail.

Species Details:

- Distribution:

  Asia: India (West Bengal, Orissa and Punjab), Bangladesh, Pakistan, Nepal, and Myanmar.
- Habitat: Inhabits fresh and brackish waters. Lives in ditches, canals, bheries and beels.

- Food and feeding: Carnivorous. Feeds mainly on worms, aquatic insects, molluscs, small fishes, tadpoles.
- Maximum length: 70.0 cm TL male/unsexed.

Behaviour:

Social—In summer days they live inside mud-holes near paddy in mud fields (Sen, 1992). Adults may undergo hibernation during cold season. During day time, of pre-monsoon and monsoon seasons, it hides under crevices, floating weeds, stones, mud etc., and comes out at night, for scavenging in search of food (Barman et al., 2013).

Reproductive—The male fish builds nest or burrow, and guards the eggs.

Remarks: It is traded as ornamental fish.

2.3.22 Lesser spiny eel

Figure 2.22

Macrognathus aculeatus (Bloch, 1786)

Common Name(s):

English—Lesser spiny eel, peacock eel, peacock spiny eel, mountain spiny eel.

IUCN Red List Status: Not Evaluated.

Systematic Position:

Order: SYNBRANCHIFORMES

Family: MASTACEMBELIDAE

Genus: *Macrognathus*

Species: *M. aculeatus*

Scientific Name: *Macrognathus aculeatus* (Bloch, 1786)

Description:

- An eel-shaped fish with tapering head and tail.
- Narrow cleft of mouth, distinct lateral line.
- Snout is elongated and fleshy.
- Lower jaw is much shortened.
- Dorsal and anal fins with a row of isolated spines followed by a long continuous soft portion, both fins extending to but not confluent with caudal fin.
- Pelvic fins absent.
- Caudal fin rounded.
- Body with series of obliquely oriented bars. A fine yellow line extends from the tip of the snout through the eye and to the rear of the fish.
- The upper part of the dorsal fin has three obvious eye-spots.
- Colour greenish or brownish above, fading to yellowish below. Smaller ones with 4 to 9 large black ocelli along the base of the dorsal.

- Sexual Dimorphism: Mature females may appear plumper than males.

Species Details:

- Distribution:

  Asia: India, Bangladesh, Nepal, Thailand, Malay Peninsula, Borneo and Indonesia and Viet Nam.
- Habitat: Adults are found in medium to large-sized rivers. Occur in lowland wetlands and peats.
- Food and feeding: Carnivorous. Can be given bloodworms and small shrimp. Pellets, flakes can be given with bloodworms, small dried shrimp, earthworms and other live food.
- Feeding frequency: Once or twice a day.
- Maximum length: 38.0 cm TL male/ unsexed.
- Life span: 8-12 years.

Aquarium Aspects:

- Minimum tank size: 20 gallons (75 L).
- Substrate: Soft, sandy substrate with pieces of woods and rocks.
- Specific gravity: 1-1.010.
- pH: 6.5-7.5.
- Temperature: 23°-28° C (73.4°-82.4° F).
- Water hardness: 10-15 dGH.

Behaviour:

Social—They are crepuscular and nocturnal burrowers. They may prey on small fish. They are usually bottom dwellers during night but can also come to the surface or mid- level for food during the day.

Reproductive—About a minimum of 1000 eggs are produced. They have been bred in captivity. About 1000 clear eggs are laid around the base of plants and rocks. They hatch in 3 days, and the fry are free swimming.

Remarks: It is in freshwater aquarium trade.

2.3.23 Barred spiny eel

Figure 2.23

Macrognathus pancalus Hamilton, 1822

Common Name(s):

English—Barred spiny eel.

IUCN Red List Status: Least Concern (LC).

Systematic Position:

Subclass: ACTINOPTERYGII

Order: SYNBRANCHIFORMES

Family: MASTACEMBELIDAE

Genus: *Macrognathus*

Species: *M.pancalus*

Scientific Name: *Macrognathus pancalus* Hamilton, 1822

Description:

- Body cylindrical, elongated, eel-shaped.
- Eyes small, cleft of mouth narrow.
- A long fleshy snout with a tri-lobed extremity.
- First dorsal with 24-26 free spines. Dorsal soft rays are 30-42.
- The soft dorsal and anal are separated from the caudal by a small notch.
- No pelvic fins.
- Colour greenish olive along the back. Yellowish beneath. Fins yellowish with black spots.
- White spots on the body.
- Sexual Dimorphism: Males are smaller, slenderer with a more yellow belly. Females are larger, heavier and have a white belly.

Species Details:

- Distribution:

  India (Peninsular India, North Bengal, Tripura, Maharashtra, Chilka lake and Western Ghats), Bangladesh and Pakistan.
- Habitat: Found in slow and shallow waters of rivers of plains and estuaries. Also inhabits canals, streams, beels, ponds and inundated fields.
- Food and feeding: Carnivorous. Feeds on aquatic insects, crustaceans, annelids, small forage fish.
- Maximum length: 18.0 cm TL male/unsexed.

Aquarium Aspects:

- Minimum tank size: 15 gallons (60 L).
- Water conditions: pH: 6-8; temperature: 20°-27° C (68°-81° F).
- Substrate: Sandy.

Behaviour:

Social--They like to bury, only parts of their head and their typical nostrils are kept out of the substrate. Then they like cavities in the aquarium as hiding places. They are predominantly nocturnal, though often visible during day time. They generally stay in their hideouts carefully watching everything around them.

Reproductive--It stays on the bottom and spawns in the upper water-level. Slender and smaller males pursue the female. They spawn in groups consisting of one female and several males. Females attach the sticky eggs to floating plants close to the water surface. The parents do not care for the fry.

Remarks: This species is suitable for a fresh water or lower-end-brackish water community tank.

2.3.24 Elongate glass perchlet

Figure 2.24

Chanda nama Hamilton, 1822

Common Name(s):

English—Elongate glass perchlet, elongate glassy perchlet.

IUCN Red List Status: Least Concern.

Systematic Position:

Series: PERCOMORPHA

Subseries: OVALENTARIA

Family: AMBASSIDAE (CHANDIDAE)—Asiatic glassfishes

Genus: *Chanda*

Species: *C. nama*

Scientific Name: *Chanda nama* Hamilton, 1822

Description:

- Body is strongly compressed and laterally flat.
- The dorsal and abdominal profiles are equally convex.
- A considerable rise is present from the occiput to the base of the first dorsal fin.
- Lower jaw is much longer than the upper jaw.
- More or less transparent body.
- Colour of the body is yellowish olive. Body is covered with minute black dots.
- On the shoulder dots are collected into an oblong patch. Summit of the head and top of the eyes are black.
- Fins are orange, the upper half of the first dorsal fin is deep black.
- Upper edge of the second dorsal fin is dark, caudal fin is dark with a light margin.
- Anal fin is with a black mark over the bases of the spines.
- Sexual Dimorphism: Unknown.

Species Details:

- Distribution:

  Throughout India, Nepal, Bangladesh, Pakistan, and Myanmar.
- Habitat: Standing and running waters; clear streams, canals, beels, ponds and inundated paddy fields. During rainy season found in huge numbers in both fresh and brackish waters.

- Food and feeding: It is a predominantly carnivorous fish. Insect parts, fish scales, zooplankton, insect larvae, insects. It is a surface as well as column feeder.
- Maximum length: 11.0 cm TL male/unsexed.

Behaviour:

Social—It is very timid in nature, on slightest provocation, it hides among the leaves and fronds of submerged plants.

Reproductive—Breeding time is April to May, June to August, in India. First maturity at 25 mm length. Highly fecund fish.

Remarks: It is a popular freshwater aquarium fish.

2.3.25 Indian glass fish

Figure 2.25

Parambassis ranga (Hamilton, 1822)

English—Indian glassy fish, Indian glassy perch, Indian x-ray fish.

IUCN Red List Status: Least Concern (LC).

Systematic Position:

Series: PERCOMORPHA

Subseries: OVALENTARIA

Family: AMBASSIDAE (CHANDIDAE)

Genus: *Parambassis*

Species: *P. ranga*

Scientific Name: *Parambassis ranga* (Hamilton, 1822)

Description:

- Body deep and laterally compressed.
- Dorsal and abdominal profiles both are very convex, but the profile over the eyes is slightly concave.
- Head is compressed and short.
- Two dorsal fins united at their base.
- This fish has a striking transparent body revealing its bones and internal organs.
- Colour of the body, olive, vertical fins with grey borders.
- Sexual Dimorphism: Males have blue edging on the dorsal and anal fins. Body is deeper yellow than females. During breeding season, these differences are more apparent. In males, the swim bladder(visible) has a pointed back edge. A dark edge is also developed to the dorsal fin of male fish.

Species Details:

- Distribution:

  Asia: India, Bangladesh, Myanmar, Thailand, Pakistan, Malaysia and Nepal.

- Habitat: Inhabit sluggish and standing water, in impoundments, rivers, streams, canals, ditches, beels, ponds and derelict waterbodies. Found in abundance during rainy season.
- Food and feeding: Carnivorous, mainly earthworms, invertebrates, worms and crustaceans. Accepts most live, frozen and dried food in captivity.
- Maximum length: 8.0 cm TL male/unsexed.

Aquarium Aspects:

- Minimum tank size: 25 gallons (94.0 litres) for 6 specimens.
- Substrate: Dark plants along with some floating vegetation. Rocks and driftwood can be used.
- Temperature: 68°-86° F (20°-30° C).
- pH: 6.5-8.0.
- Hardness: 7-19 dGH.
- Water region: Middle to bottom (demersal).
- Specific gravity: 1-1.010 (can be kept in very mild brackish or freshwater).
- Lighting: Low, morning sunlight is preferred.

Behaviour:

Social—It is a peaceful and shy fish. They should not be kept with vigorous or aggressive species.

Reproductive--Oviparous, egg- scatterers. Breeds everywhere during the rains. Builds a nest and guards its young. Produces up to 500 eggs. In captivity breeding is not too difficult but raising the fry is difficult. A heavily planted aquarium is required. Eggs stuck to plant leaves and stems. They will hatch in around 24 hours and will be seen hanging from the plants. They become free swimming in another 3-4 days. They should be fed heavily with brine shrimp nauplii.

Remarks: This species has been kept in aquarium for many years. These fishes are often injected with colour dyes into the muscles and sold to hobbyists as 'painted' in the name of 'disco fish'.

2.3.26 Gangetic leaf fish

Figure 2.26

Nandus nandus (Hamilton, 1822)

Common Name((s)

English—Gangetic leaf fish.

IUCN Red List Category: Least Concern.

Taxonomic Position:

Order: ANABANTIDAE

Family: NANDIDAE

Genus: *Nandus*

Species: *N. nandus*

Scientific Name: *Nandus nandus* (Hamilton, 1822)

Description:

- Body laterally compressed, deeper with a nearly straight belly and arched back.
- Snout pointed with very much protractile mouth, with deep cleft of mouth.
- Caudal rounded.
- Colour of the body greenish brown with bronze or brassy reflections in live fishes.
- Three vertical broad bands or patches on the body.
- Narrow bands or spots across the fins.
- Sexual Dimorphism: Males have larger fins than the females. Body colour of male fish are darker than female fish.

Species Details:

- Distribution:

  Throughout India, Bangladesh, Nepal and Pakistan.
- Habitat: Inhabits freshwaters. Found in rivers and also in agricultural lands. Found in dried -up beds of tanks, beels, bheries etc.
- Food and feeding: It is a predator fish. Prefers live food like insects, bloodworms and small crustaceans. Also take pellets, flakes along with live food in aquarium.
- Feeding time: Night preferably.
- Maximum length: 20.0 cm TL male/ unsexed.
- Life span: 5-8 years.

Aquarium aspects:

- Minimum tank size: 45 gallons.
- Aquarium water conditions: Temperature: 22.8°-27.2° C (73°-81° F); pH: 6.5-7.5. hardness: 8-15 dGH.
- Tank requirements: Needs a spacious tank with hiding places. Rocks and wood should be kept in the tank. An addition of 0.5-1% marine salt in the tank water is best for their health.

Behaviour:

Social-- This fish preys on smaller fishes in the tank, so it is better to keep them with other large robust fishes.

Reproductive-- This fish is gonochoristic. A spawning site is cleaned by the male which courts the female. Female spawns around 200-400 demersal eggs which are then fertilized by the male. In aquarium also, the male show parental care.

Remarks: It is in aquarium trade.

2.3.27 Golden cobra snakehead Figure 2.27

Channa aurantimaculata Musikasinthorn, 2000

Common Name(s):

English—Golden cobra snakehead.

IUCN Red List Status: Data Deficient (DD).

Systematic Position:

Order: ANABANTIFORMES—labyrinth fishes

Family: CHANNIDAE—snakeheads

Genus: *Channa*

Species: *C. aurantimaculata*

Scientific Name: *Channa aurantimaculata* Musikasinthorn, 2000

Description:

- Dorsal fin soft rays: 45-47; anal fin soft rays: 28-30.
- Upper half of body dark brown to black with 7-8 large irregular orange blotches.
- Pectoral fin with a black blotch at the base and 5 broad blackish bands.
- Sexual Dimorphism: After reaching a size of about 15 cm, males begin to develop a more extended and intensely patterned dorsal fin.

Species Details:

- Distribution:

  India: Endemic to the Brahmaputra River basin in Assam and Arunachal Pradesh, north eastern India.
- Habitat: Inhabits tropical monsoon rainforest, having heavy rainfall and high humidity. Found in forest stream, swamps and ponds connected with the Brahmaputra River (Musikasinthorn, 2000).
- Food and feeding: Carnivorous.
- Maximum length: 19.08 SL male.

Aquarium Aspects:

- Water conditions: Temperature: 10-28° C; pH: 6.0-8.0; hardness: 36-357 ppm.
- Dimly lit aquarium with a layer of surface vegetation.

Behaviour:

Social—It may behave aggressively territorial towards conspecifics.

Reproductive—It is a paternal mouthbrooder. It can be bred in captivity.

Remarks: It is a favourite aquarium fish.

2.3.28 Snakehead Figure 2.28

Channa gachua (Hamilton, 1822)

Common Name(s):

English—Snakehead.

IUCN Red List Status: Least Concern (LC).

Systematic Position:

Order: ANABANTIFORMES—labyrinth fishes

Family: CHANNIDAE—snakeheads

Genus: *Channa*

Species: *C. gachua*

Scientific Name: *Channa gachua* (Hamilton, 1822)

Description:

- Margins of dorsal, anal and caudal white.
- Relatively small-sized fish.
- Sexual Dimorphism: Male fish has more intense colouration and develops more-extended dorsal and anal fins than female. Female fish is a bit longer than the male, and is deeper-bodied when gravid.

Species Details:

- Distribution:

  India: Kerala and Maharashtra, Bhimtal and Naukuchiatal lakes. Afghanistan in the west, to Indonesia through South and Central Asia.
- Habitat: Inhabit hill streams, medium to large rivers, brooks, rapid-running mountain streams, stagnant water-bodies and sluggish flowing canals (Taki, 1978).
- Food and feeding: Carnivorous and nocturnal feeder. Feeds on insects, crustaceans, and small fishes (Lim et al., 2008). In aquarium may be fed, Chironomus larvae, small earthworms, chopped prawn and mussels.
- Feeding frequency: Once or twice a day.
- Maximum length: 28.8 cm TL male/unsexed.

Aquarium Aspects:

- Temperature: 18-28° C; pH: 5.0-8.0; hardness: 18-357 ppm.

Behaviour:

Social—A community fish, peaceful but males may show aggressive behaviour during spawning season.

Reproductive—Exhibit parental care. Male fish broods the eggs and fry in his mouth. This fish can be bred in captivity.

Remarks: It is in aquarium trade.

2.3.29 Great snakehead

Figure 2.29

Channa marulius (Hamilton,1822)

Common Name(s):

English—Great snakehead, bullseye snakehead.

IUCN Red List Status: Least Concern (LC).

Systematic Position:

Order: ANABANTIFORMES

Family: CHANNIDAE -snakeheads

Genus: *Channa*

Species: *C. marulius* (Hamilton, 1822)

Scientific Name: *Channa marulius* (Hamilton, 1822)

Description:

- Body sub-cylindrical with depressed head.
- A large head resembling a snake head.
- Deep gape, and toothed mouth.
- Very large scales.
- Caudal fin rounded.
- No barbels.
- Colour greyish brown becoming lighter below.
- Four to five roundish black blotches on the middle of the body below the lateral line (most distinguishing character from all other *Channa* species).
- Young ones with a brilliant orange coloured lateral band.

Species Details:

- Distribution:

  India—Maharashtra, Tamil Nadu, Andhra Pradesh, West Bengal, Tripura, Bhimtal, Naukuchiatal lakes and Western Ghats. China, south to Thailand and Cambodia and Pakistan.
- Habitat: Found in deep pools in rivers, sometimes in lakes, flooded forest, waters with submerged aquatic vegetation. Also occur in sluggish or standing water in canals, lakes and swamps.
- Food and feeding: Predatory carnivorous. Eats fish, frogs, insects, earthworms and tadpoles. Also take water birds, snakes, rodents, young fish, *Chironomus* larvae, small earthworms and chopped prawns.
- Maximum length: 46.0 cm-183 cm TL male/unsexed.

Aquarium Aspects:

- Substrate: Rocky bottom, with lots of open space.
- Aquarium with a tightly-fitting hood as they can easily escape.
- A gap should be left between this and the water surface as they require access to a layer of humid air.
- Temperature: 22-28° C / 72-82° F.
- pH: 6.0-8.0.
- Hardness: 36-357 ppm.

Behaviour:

Social—They remain best in a species-specific aquarium. Though juveniles and young adults are peaceful, but become aggressive when become mature.

Reproductive—Builds floating nest with weeds and leaves. Orange-yellow eggs are laid and deposited in the nest. Eggs hatch in 36 to 48 hours and fry remain in the nest for about 10 days after hatching. Fry are guarded by the parents for a month.

Remarks: Suitable for public aquaria.

2.3.30 Striped snakehead

Figure 2.30

Channa striata (Bloch, 1793)

Common Name(s):

English—Striped snakehead, common snakehead, chevron snakehead, snakehead murrel, mudfish.

IUCN Red List Status: Least Concern (LC).

Systematic Position:

Order: ANABANTIFORMES

Family: CHANNIDAE

Genus: *Channa*

Species: *C.striata*

Scientific Name: *Channa striata* (Bloch, 1793)

Description:

- Body is sub-cylindrical in shape.
- Head is depressed and looks like a snake-head, lower jaw is longer. Gape of mouth deep.
- No barbels present.
- Caudal fin roundish.
- Dorsal fin spines: 0; dorsal fin soft rays: 38-43; anal fin spines: 0; anal fin soft rays: 23-27.
- Very large scales present.
- Dark greyish to black colour above, yellowish-white beneath, conspicuous greyish bands descend from side to abdominal region. Fins are greyish.

Species Details:

- Distribution:

  Asia: India (In Hooghly, Damodar and Bhagirathi river systems in West Bengal), Pakistan to Thailand and south China.
- Habitat: Inhabits ponds, streams, rivers, stagnant and muddy waters of plains, swamps, in lowland rivers, deep still water, freshwater plains, also found in medium to large rivers, brooks, flooded fields and stagnant waters.

- Food and feeding: Carnivorous. Feed on fish, frogs, snakes, insects, earthworms, tadpoles and crustaceans.
- Maximum length: 100.0 cm SL male/unsexed.

Aquarium Aspects:

- Water conditions: Temperature: 23-27° C; pH: 7.0-8.0; hardness: 20 dGH.

Behaviour:

Social—Solitary territorial fish. Some population migrate between permanent waters and flooded areas created by seasonal rains. They can breathe air from the surface of water.

Reproductive—It is a monogamous fish. Reproduces up to 2 times per year. Breeding seasons are June- August and October -December. About 4,000 eggs are produced. Breed in ditches, ponds and flooded paddy fields. Males and females take part in construction of nests with the help of water plants. Both parents guard the eggs. Fry are reddish orange and are guarded by both parents.

Remarks: It is used as an aquarium fish and also used in the control of densities of Nile Tilapia.

2.3.31 Climbing perch

Figure 2.31

Anabas testudineus (Bloch, 1792)

Common Name(s):

English—Climbing perch.

IUCN Red List Status: Data Deficient (DD).

Systematic Position:

Order: ANABANTIFORMES (labyrinth fishes)

Suborder: ANABANTOIDEI

Family: ANABANTIDAE

Genus: *Anabas*

Species: *A. testudineus*

Scientific Name: *Anabas testudineus* (Bloch, 1792)

Description:

- Scales large and regularly arranged.
- Mouth relatively large.
- Jaws, prevomer and parasphenoid bones with fixed conical teeth.
- A single dorsal fin longer than anal with 17 to 18 spines. Caudal fin rounded.
- Lateral line interrupted.
- Colour dark to pale greenish, very pale below. Back dusky to olive.
- Head with longitudinal stripes ventrally.

- Posterior margin of operculum with a dark spot. Iris golden reddish.
- Four wide cross bands on body.
- Possess an accessory air-breathing organ.
- Sexual Dimorphism: Mature males are darker in colour. Develop tubercles on the pectoral fins when breeding. During spawning males develop a reddish hue to the body, particularly on the pectoral and ventral fins and a blackish diamond -shaped marking on the caudal peduncle. Mature females are slightly larger and thicker-bodied than males. Females have faint reddish colour, caudal peduncle spot is oblong and diffused.

Species Details:

- Distribution:

  Throughout India to Wallace line including China.
- Habitat: Inhabit canals, lakes, ponds, swamps and estuaries. Adults found in medium to large rivers, brooks, flooded fields and stagnant water bodies, also in areas with dense vegetation. Can tolerate extreme unfavourable water conditions.
- Food and feeding: Predatory omnivore. Feeds on macrophytes, shrimps, fish fry, invertebrates. In aquarium, larger live and frozen foods such as earthworms or shellfish, pellets and flakes are eaten.
- Maximum length: 25.0 cm TL male/ unsexed.

Aquarium Aspects:

- Substrate: Soft sandy bottom with drift wood.
- Lighting: Dim.
- Temperature: 15-30°C.
- pH: 5.5-8.0.
- Hardness: 36-447 ppm.

Behaviour:

Social—Not suitable for the general community tank due to predatory nature. Can survive for several days or weeks out of water if the air breathing organs are kept moist.

Reproductive—The eggs are deposited in open water at the onset of the rainy season. During a single spawning about 5000-35,000 eggs may be laid. The eggs are not adhesive and rise to the surface where they float for approximately 24 hours before hatching. The fry released are about 2-3 mm long and free swimming within two days of hatching. Guards eggs at the surface of hypoxic waters.

Remarks: It is an interesting and moderately popular species in the aquariums.

2.3.32 Dwarf gourami

Figure 2.32A & B

Trichogaster lalius (Hamilton, 1822)

Common Name(s):

English—Dwarf gourami.

IUCN Red List Category: Least Concern.

Systematic Position:

Order: PERCIFORMES

Family: OSPHRONEMIDAE

Genus: *Trichogaster*

Species: *T. lalius*

Scientific Name: *Trichogaster lalius* (Hamilton, 1822)

Description:

- Small, compressed body with sparkling, almost translucent blue colour with fine stripes of red or deep orange.
- Fins are relatively large with rounded ends.
- Ventral fins are threadlike and carry touch-sensitive cells that are extremely perceptive.
- It is a labyrinth fish and can breathe atmospheric oxygen.
- Sexual Dimorphism: The male fish is more colourful and have a smaller belly than the female. The male fish has diagonal stripes of alternating blue and red colours. The male's dorsal fin is pointed. The female has a silvery coloured body and rounded or curved dorsal fin. They carry touch-sensitive cells on their thread-like pelvic fins.

Species Details:

- Distribution:

  Widely distributed in India (lowland Ganges and Brahmaputra basins, Arunachal Pradesh, Assam, Bihar, Manipur, Uttaranchal, Uttar Pradesh and West Bengal). Also found in Bangladesh and Nepal.
- Habitat: Slow moving streams, rivulets and lakes with plenty of vegetation, rice fields, irrigation channels and other agricultural lands.
- Systems: Freshwater.
- Food and feeding: It is an omnivore fish that prefers algae-based foods and meaty foods. In the wild, they eat small invertebrates, algae and other aufwuchs. In the aquarium they feed live, fresh, flakes, pellets, vegetables and meaty food, freeze-dried bloodworms, tubifex and brine shrimps.
- Feeding frequency: Daily, once or twice a day.
- Maximum length: 8.8 cm TL male/ unsexed.
- Life span: 4-8 years.

Aquarium aspects:

- Minimum tank size: 10 gallons.

- Suitable for Nano tank: Yes.
- Substrate type: Any.
- Aquarium water conditions: Temperature: 72°-82.0° F (22.2° to 27.8° C); Breeding temperature: 80.0°-84° F (26°-29° C); pH 6.0-8.0; hardness: 5-18 dGH.
- Water movement: Weak.
- Water changes: Weekly.

Behaviour:

Social behaviour: It is a good community fish that can be kept with other peaceful fish. Pairs generally swim together.

Reproductive behaviour: The male fish builds a floating bubble nest in which the eggs are laid. Males use plants, twigs and other debris to hold the nest together better. Elaborate nests are built. During spawning the water level in the tank should be reduced to 7-10 cm. Males take care of the eggs and hatchlings.

Remarks: Different colour variants are developed by the breeders. These are flame red, powder blue, turquoise/neon blue, rainbow, red, sunset, red blushing. It is one of the most popular aquarium fish. It is readily available in stores and online and is moderately priced.

2.3.33 Freshwater garfish

Figure 2.33

Xenentodon cancila (Hamilton, 1822)

Common Name(s):

English—Needlefish, freshwater garfish, silver needlefish, Asian freshwater needlefish, needle nose halfbeak, freshwater gar, needle nose garfish.

IUCN Red List Status: Least Concern (LC).

Systematic Position:

Order: BELONIFORMES

Family: BELONIDAE

Genus: *Xenentodon*

Species: *X. cancila*

Scientific Name: *Xenentodon cancila* (Hamilton, 1822)

Description:

- Elongated body with long, beak-like jaws filled with teeth.
- The dorsal and anal fins are positioned far back along the body close to the tail.
- A deep median longitudinal groove on upper surface of head.
- Greenish-grey above and whitish along the abdomen.
- A silvery streak with a dark margin extends along the body.
- Upper part of the body with fine black spots.

- Dorsal and caudal tipped darker.
- Sexual Dimorphism: The male fish have anal and dorsal fins with a black edge.

Species Details:

- Distribution:

  Throughout India, Sri Lanka and Mekong.
- Habitat: Inhabits freshwater, primarily rivers in North Bengal. It occurs in clear, gravelly, perennial streams and ponds of Terai and Duars.
- Food and feeding: Carnivorous. In the wild, feed on crustaceans, fish, frog. Often acts as predator. In the aquarium, they can be fed, meals of live shrimp, fish, crickets, tadpoles. They take live food but no flakes or pellets are taken.
- Feeding frequency: Weekly.
- Maximum length: 40.0 cm TL male/unsexed.
- Life Span: 3-8 years.

Aquarium Aspects:

- Minimum tank size: 55 gallons.
- Suitable for Nano tank: No.
- Substrate type: Small Gravel.
- Lighting needs: Low-subdued lighting.
- Temperature: 71.0° to 82.0° F (21.7° to 27.8°C).
- Breeding temperature: 79.0° F.
- Range pH: 7.0-7.5.
- Hardness range: 5-20 GH.
- Water movement: Weak.
- Water region: Top.

Behaviour:

Social—They are not aggressive but are greedy predators who bite first to know whether it is a food. They are best in schools of three or more.

Reproductive—Oviparous. Small numbers of eggs get deposited among plants, after spawning in the morning. The eggs take 10 days to hatch. The eggs adhere to the plants by sticky threads. They have been bred in captivity.

Remarks: Often found in public aquariums.

2.3.34 Striped panchax Figure 2.34

Aplocheilus lineatus (Valenciennes, 1846)

Common Name(s):

English: Striped panchax.

IUCN Red List Status: Least Concern (LC).

Systematic Position:

Order: CYPRINODONTIFORMES—killifishes

Family: APLOCHEILIDAE—Asian rivulines

Genus: *Aplocheilus*

Species: *A. lineatus*

Scientific Name: *Aplocheilus lineatus* (Valenciennes, 1846)

Description:

- Body elongate, slender with a slightly arched back.
- The head is flattened on top with a pointy snout and upturned mouth.
- Barbels are absent.
- The dorsal fin is set far towards the posterior half of the body.
- Basic body colour is coppery bronze with sparkling yellow-gold scales along the sides and reaching up to the fins.
- Females and young ones have several broad dark vertical bars along the posterior half of the body.
- Sexual Dimorphism: Males are larger, more colourful and their anal fin are more pointed than that of the females.

Species Details:

- India: Widely distributed in Peninsular India.
- Habitat: Inhabit streams, reservoirs at high altitudes and in rivers, wells of the plains, low-lying paddy fields, swamps and brackish waters (Talwar and Jhingran, 1991).
- Food and feeding: Carnivorous. They feed on insect larvae, small aquatic insects, small crustaceans and river worms. In captivity, they can be fed with flakes, small pellets and freeze- dried foods and may also be given some fresh or frozen brine shrimp or blood worms.
- Feeding frequency: Several feedings per day.
- Maximum length: 10.0 cm TL male/unsexed.
- Life span: 4 years.

Aquarium Aspects:

- Minimum tank size: 20 gallons (76 litres).
- Suitable for Nano tank: Yes.
- Suitable substrate: Small gravel.
- Lighting: Moderate to normal lighting.
- Temperature: 22.2°-25.0° C (72.0°-77.0° F); pH: 6.0-7.5; hardness: 5-20 dGH.
- Water movement: Weak.

- Water region: Top.
- Water changes: Bi-weekly.

Behaviour:

Social—Semi-aggressive. They are quarrelsome amongst themselves, so they may be kept alone or in a group of 4 or more.

Reproductive—Breeding in captivity is easy. Spawning continues for several weeks and 50 to 300 eggs may be laid daily. Eggs may be laid in fine leafed plants or in a clean spawning mop.

Remarks: The striped panchax is a popular aquarium fish. It is hardy and favourite with aquarists. One of the most popular varieties, the "Golden Wonder Killifish" is very popular in the aquarium trade.

2.3.35 Blue panchax

Figure 2.35

Aplocheilus panchax (Hamilton,1822)

Common Name(s):

English—Blue panchax.

IUCN Red List Status: Least Concern (LC).

Systematic Position:

Order: CYPRINODONTIFORMES

Suborder: APLOCHEILOIDEI

Family: APLOCHEILIDAE

Genus: *Aplocheilus*

Species: *A. panchax*

Scientific Name: *Aplocheilus panchax* (Hamilton, 1822)

Description:

- Body elongated, with a pointed snout and an up-turned mouth.
- Lower jaw longer, pectorals nearly as long as head.
- Ventral small without any elongated ray.
- Upper surface of the body greyish, becoming dull white on sides and beneath.
- Fins yellowish, lower third of dorsal covered with a large black spot.
- Dorsal, caudal and anal fins with orange margins.
- A white spot found on the head.
- One of the most popular varieties, the orange-tail killifish from India has a bright orange tail.
- Sexual Dimorphism: The male and female fishes look almost alike though body colouration of male fish is sometimes a little darker than the female.

Species Details:

- Distribution:

  India--Northern India, Western Ghats Rivers, Maharashtra, Kerala, Coromandel coast, Laccadive, Andaman Islands and Tripura. Bangladesh, Myanmar, Nepal, Pakistan and Sri Lanka.
- Habitat: It inhabits clear, shallow and brackish waters at low altitudes.
- Food and feeding: Carnivorous. Accepts most standard fish foods. In captivity, they eat almost all the floating prepared foods which are small enough to be engulfed by the fish. Prefer frozen bloodworms and brine shrimp. Accept flake foods and pellets occasionally.
- Feeding frequency: Several feedings a day.
- Maximum length: 9.0 cm TL male/unsexed.
- Life span: 5 years.

Aquarium Aspects:

- Minimum tank size: 20 gallons.
- Suitable for Nano tank: Yes.
- Substrate type: Small gravel.
- Temperature: 20°-25° C (68.0°-70.0F).
- Range pH: 6.5-7.5.
- Hardness: Soft-moderate; 5-12 dGH.

Behaviour:

Social—Generally peaceful. They can be kept with other peaceful fish of similar size or larger.

Reproductive—Egg layer and this fish is relatively easy to bred. Eggs are laid on plants and usually hatch in about 14 days.

Remarks: An aquarium fish, widely used in controlling mosquito population.

2.3.36 Tank goby Figure 2.36

Glossogobius giuris (Hamilton, 1822)

Common Name(s):

English—Tank goby.

IUCN Red List Status: Least Concern (LC).

Systematic Position:

Order: GOBIIFORMES--gobies

Family: GOBIIDAE—gobies

Genus: *Glossogobius*

Species: *G. giuris*

Scientific name: *Glossogobius giuris* (Hamilton, 1922)

Description:

- Elongate body with a vertically compressed, pointed head.
- Lower jaw is pointed, snout also pointed.
- Two dorsal fins are placed closely.
- Pelvic fins united forming a disc.
- Dorsal fin spines:7; dorsal fin soft rays: 8-9; anal fin spines: 1; anal fin soft rays: 8-9.
- Colour of the body, olive to dusky green above, lighter below. Two alternating rows of 4 to 6 blotches on body along the lateral line.
- Dorsal fins with small spots forming longitudinal stripes.
- Pelvic fins, pectoral fins and caudal fin are grey.

Species Details:

- Distribution:

  India, Africa to Oceania.
- Habitat: Inhabit freshwater and estuaries also enter the sea. Found in canals, ditches, ponds, clear to turbid streams with rock, gravel or sand bottoms.
- Food and feeding: Carnivorous. Feed on small insects, crustaceans and small fishes.
- Maximum length: 50.0 cm SL male/ unsexed.

Behaviour:

Reproductive—Spawning occurs in freshwater and eggs and larvae are washed into sea by the river current.

Remarks: It is a popular aquarium fish.

Figure 2.1 Rosy barb

Figure 2.2 Pool barb

Figure 2.3 Arulius barb

Figure 2.4 Mascara barb

Figure 2.5 Putitor mahseer

Figure 2.6 Zebra danio

Figure 2.7 Flying barb

Figure 2.8 Blue-dotted hill trout

Figure 2.9 Bata

Figure 2.10 Rohu

Figure 2.11 A Reticulate loach

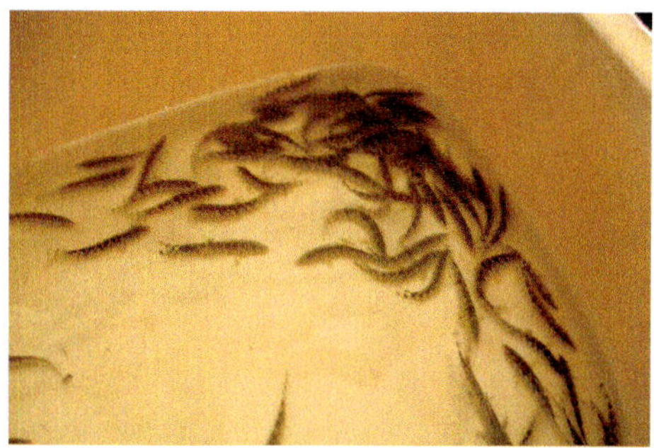

Figure 2.11 B Reticulate loaches in a plastic tank

Figure 2.12 Stone loach

Figure 2.13 Gangetic mystus

Figure 2.14 A Striped dwarf catfish

Figure 2.14 B Striped dwarf catfishes

Figure 2.15 A Pangas catfish

Figure 2.15 B Pangas catfish

Figure 2.16 Asian stone catfish

Figure 2.17 Butter catfish

Figure 2.18 Wallago

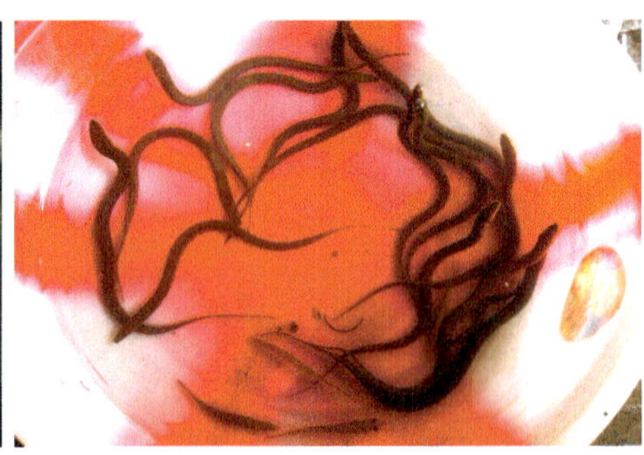

Figure 2.19 Indian mottled eel

Figure 2.20 A Bronze featherbacks

Figure 2.20 B Bronze featherback

Figure 2.21 Cuchia

Figure 2.22 Lesser spiny eel

Figure 2.23 Barred spiny eel

Figure 2.24 Elongate glass perchlet

Figure 2.25 Indian glass fish

Figure 2.26 Gangetic leaf fish

Figure 2.27 Golden Cobra Snakehead

Figure 2.28 Snakehead

Figure 2.29 Great snakehead

Figure 2.30 Striped snakehead

Figure 2.31 Climbing perch

Figure 2.32 A Dwarf gourami

Figure 2.32 B Dwarf gourami for sale

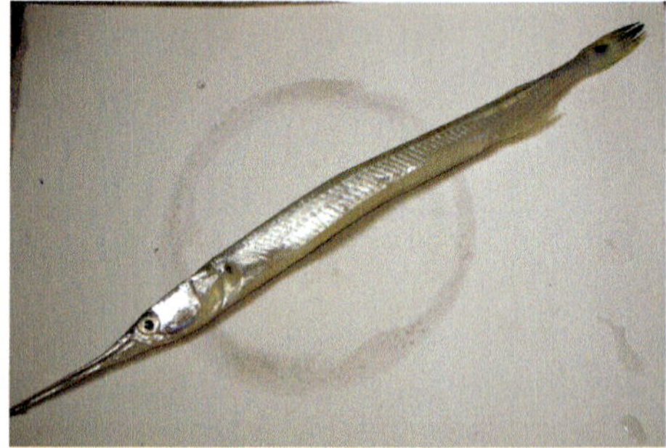

Figure 2.33 Freshwater garfish

Figure 2.34 Striped panchax

Figure 2.35 Blue panchax

Figure 2.36 Tank goby

2.4 Freshwater Ornamental Fishes in the World Trade

A large number of fishes, inhabiting freshwaters of various regions, all over the world, are valued as aquarium fishes. Such trade-friendly ornamental fishes are listed below.

Table 2.2: List of freshwater ornamental fishes of the world

Order	Family	Common Name Scientific Name	Maximum Length (cm)	Distribution
(AROWANAS)				
OSTEOGLOSSI-FORMES	OSTEOGLOSSIDAE	Silver arowana *Osteoglossum bicirrhosum*	90.0	South America.
		Black arowana *Osteoglossum ferreirai*	90.0	South America.
		Asian bonytongue *Scleropages formosus*	90.0	Asia.
	NOTOPTERIDAE	Clown knifefish *Chitala ornata*	100.0	Freshwaters of Cambodia, Laos, Thailand, Vietnam.
(CATFISHES)				
SILURIFORMES (catfishes)	CALLICHTHYIDAE (armoured catfishes)	Brown-point shield skin longirostris *Aspidoras fuscoguttatus*	4.0	Inland waters in South America, upper Parana River basin in Brazil and Peru.
		Aspidoras lakoi	4.0	Inland waters in South America and in the upper Parana River basin in Brazil.
		Banded Corydoras *Scleromystax barbatus*	10.0	Coastal drainages in South America from Rio de Janeiro to Santa Catarina, Brazil.
		Barred-tail Corydoras *Corydoras cochui*	2.5	Inland waters in South America, upper Araguaia River basin Brazil.
		Black stripe corydoras *Corydoras bondi*	4.7	Inland waters in South America.

Order	Family	Common Name Scientific Name	Maximum Length (cm)	Distribution
		Blacktop Corydoras *Corydoras acutus*	4.4	Inland waters in South America.
		Blue Corydoras *Corydoras nattereri*	5.4	Coastal rivers in South America.
		Blue spotted Corydoras *Corydoras melanistius*	5.1	Inland waters in South America.
		Britski's catfish *Brochis britskii*	8.9	South America- upper Paraguay River basin in Brazil.
		Bronze Corydoras *Corydoras aeneus*	6.35	South America.
		Loach catfish *Aspidoras rochai*	4.0	Inland waters in south America, and in coastal rivers in Ceara, Brazil.
		Six ray Corydoras, false Corydoras *Aspidoras pauciradiatus*	2.9	Inland waters of South America, in upper Araguaia, River basin in Brazil.
		Emerald catfish *Brochis splendens*	10.0	Amazon Basin in South America.
		Hognosed brochis *Brochis multiradiatus*	6.6	South America- western Amazon basin in Ecuador and Peru.
		Cascarudo *Callichthys callychthys*	20.0	All major river drainages of South America.
		Adolfo's catfish *Corydoras adolfoi*	5.7	Inland waters in South America.
		Masked Corydoras, bandit Corydoras *Corydoras metae*	4.8	Inland waters of South America, in the Meta River basin in Colombia.
		Caracha *Corydoras atropersonatus*	4.5	In inland waters in South America, in the upper Amazon River basin in Ecuador and Peru.

Order	Family	Common Name Scientific Name	Maximum Length (cm)	Distribution
		Ehrhardt's Corydoras *Corydoras ehrhardi*	4.1	Coastal rivers in the Santa Catarina and Parana states of Brazil from the Iguazu River basin.
		Evelyn's cory *Corydoras evelynae*	4.0	Inland waters in South America, and is found in the Amazon River basin in Brazil.
		Corydoras geoffroy	7.0	In coastal rivers in South America, and is found in Suriname and French Guiana.
		Corydoras latus	5.2	In inland waters in South America, and is found in the Beni River basin in Bolivia.
		Loxozonus cory *Corydoras loxozonus*	4.9	Inland waters in South America, in the Meta River basin in Colombia.
		Corydorus nain *Corydoras nanus*	4.5	Inland waters of South America.
		Long nosed arched cory *Corydoras narcissus*	6.5	Inland waters in South America.
		Corydoras ornatus	4.9	Inland waters in South America.
		Corydoras osteocarus	4.0	South America.
		Corydoras polystictus	3.2	Inland waters in South America.
		Scleromystax prionotos	5.3	Inland waters in South America.
		Corydoras semiaquilus	6.0	Inland waters in South America.
		Corydoras septentrionalis	4.9	Inland waters in South America.
		Corydoras simulatus	4.9	Inland waters in South America.

Order	Family	Common Name Scientific Name	Maximum Length (cm)	Distribution
		Corydoras undulatus	4.4	Inland waters in South America.
		Dwarf Corydoras *Corydoras hastatus*	3.5	Inland waters in South America.
		Elegant Corydoras *Corydoras elegans*	5.1	Inland waters in South America.
		False network catfish *Corydoras sodalis*	4.9	Inland waters in South America.
		False spotted catfish *Corydoras leucomelas*	4.5	Inland waters in South America.
		Green gold catfish *Corydoras melanotaenia*	5.8	Inland waters in South America.
		Guapore Corydoras *Corydoras guapore*	4.1	Inland waters of South America.
		Mosaic Corydoras, reticulated Corydoras *Corydoras haraldschultzi*	5.9	Brazil and Bolivia.
		Panda Corydoras *Corydoras panda*	5.1	In rivers of South America.
		Pastaza Corydoras *Corydoras pastazensis*	7.1	South America-in Pastaza river basin in Ecuador.
		Peppered Corydoras *Corydoras paleatus*	7.6	Lower Parana River basin and coastal rivers in Uruguay and Brazil.
		Pink Corydoras *Corydoras axelrodi*	5.0	Inland waters in South America.
		Pygmy Corydoras *Corydoras pygmaeus*	3.3	Tropical inland waters in South America.
		Sailfin Corydoras *Corydoras macropterus*	8.7	Coastal rivers in South America.
		Salt and pepper catfish *Corydoras habrosus*	3.6	Inland waters in South America.

Order	Family	Common Name Scientific Name	Maximum Length (cm)	Distribution
		Schwartz's catfish *Corydoras schwartzi*	5.1	Inland waters in South America.
		Spotted Corydoras, longnose Corydoras *Corydoras ambiacus*	6.1	Inland waters in South America.
		Sterba's Corydoras *Corydoras sterbai*	6.8	Guapore River region between Bolivia and Brazil.
		Sychr's catfish *Corydoras sychri*	4.3	Inland waters in South America.
		Tailspot Corydoras *Corydoras caudimaculatus*	5.0	Inland waters in South America.
		Three stripe Corydoras, leopard catfish *Corydoras trilineatus*	6.1	Inland waters in South America.
		Xingu Corydoras *Corydoras xinguensis*	5.1	Inland waters in South America-Xingu river, river basin in Brazil.
		Julii Corydoras *Corydoras julii*	6.4	Eastern Brazil.
		Flagtail catfish *Dianema urostriatum*	12.5	Amazon Basin in South America.
	LORICARIIDAE (armoured sucker-mouth catfish, plecos)	Bristlenose pleco, bushynose pleco *Ancistrus* spp.	13.0	Freshwater habitats in South America and Panama.
		Gold nugget pleco *Baryancistrus* spp.	34.0	Clearwater in the basins of the Amazon and Orinoco in Brazil and Venezuela.
		Whiptail catfish *Rineloricaria* spp.	26.0	Endemic to Columbia.
		Twig catfish *Farlowella* spp.	16.0	South America.
		Panaque *Panaque* spp.	30.0	Tropical South America-in fast flowing streams and rivers.

Order	Family	Common Name Scientific Name	Maximum Length (cm)	Distribution
		Pineapple pleco, orange cheek pleco *Pseudorinelepsis* spp.	35.6	Brazil and Peru-in the upper Amazon basin.
		Common pleco, sucker mouth catfish, Amazon sailfin catfish *Hypostomus punctatus*	30.0	South America-coastal drainages of south eastern Brazil and Uruguay.
		Common pleco, suckermouth catfish *Hypostomus plecostomus*	50.0	Tropical North eastern South America.
		Leopard sailfin pleco, clown sailfin pleco *Pterygoplichthys gibbiceps*	50.0	Brazil, Ecuador, Peru and Venezuela-in the Orinoco and Amazon basins.
		Zebra pleco *Hypancistrus zebra*	6.4	Brazil-Xingu river.
		Sailfin catfish *Pterygoplichthys multiradiatus*	50.0	Orinoco river basin in Venezuela.
		Golden dwarf sucker, golden oto *Macrotocinclus affinis*	5.0	Southeast Brazil.
		Zebra dwarf, zebra oto *Otocinclus cocama*	4.4	Peru.
		Goldspot ancistrus albino *Ancistrus* sp.	12.0	South America and Panama.
	CLARIIDAE (air-breathing catfish)	Walking catfish *Clarias batrachus*	50.0	Southeast Asia.
	ASPREDINIDAE (Banjo catfish)	Banjo catfish *Bunocephalus coracoideus*	12.0	Amazon River basin.
	DORADIDAE (talking catfish, thorny catfish)	Raphael catfish *Platydoras costatus*	24.0	Northern South America down to Argentina.
		Striped Raphael *Platydoras armatulus*	20.0	Amazon, Ornico, Suriname and other rivers of South America.

Order	Family	Common Name Scientific Name	Maximum Length (cm)	Distribution
		Spotted raphael *Agamyxis pectinifrons*	15.0	Amazon basin.
	MOCHOKIDAE (squeakers and up-side-down catfish)	Cuckoo squeaker *Synodontis multipunctatus*	27.5	Lake Tanganyika, Africa.
		Decorated squeaker, clown catfish *Synodontis decorus*	31.8	Cameroon, the Democratic Republic of the Congo.
		Even spotted squeaker *Synodontis petricola*	10.0	Burundi, Zambia, the Democratic Republic of Congo, Lake Tanganyika.
		Feather fin squeaker *Synodontis eupterus*	30.0	Basius of the White Nile, Volta, Niger Rivers and the Chad Basin.
		Lake Malawai Syno *Synodontis njassae*	19.2	Lake Malawi.
		Polka dot syno *Synodontis angelicus*	18.0	Congo Basin of Cameroon, the Democratic Republic of the Congo, and the Republic of Congo.
		Common Syno, false upside- down catfish *Synodontis nigrita*	33.5	Northern Africa.
		Upside-down catfish *Synodontis nigriventris*	9.6	Congo Basin of Cameroon the Democratic Republic of Congo and the republic of Congo.
	PANGASIIDAE (shark catfish)	Iridescent shark *Pangasius hypothalamus*	102.0	Rivers of Southeast Asia.

Order	Family	Common Name Scientific Name	Maximum Length (cm)	Distribution
		Chao Phraya giant shark, giant pangasius *Pangasius sanitwongsei*	300.0	Chao Phraya and Mekong basins in Indo China.
	(sea catfish)	Colombian shark catfish *Ariopsis seemanni*	39.5	Pacific-draining rivers and estuaries in Central and South America.
		Berney's shark catfish *Arius berneyi*	3.8	South Pacific coast areas of Northern Australia and New Guinea.
		Australian shark catfish *Neoarius graeffei*	60.0	Freshwater rivers of Australia and Papua New Guinea
	SILURIDAE (sheat fishes)	Glass catfish *Kryptopterus vitreolus*	6.5	Asia: Peninsular and south eastern Thailand.
		Kryptopterus bicirrhis	15.0	Asia: Mekong Chao Phraya basins; Malay Peninsula, Sumatra and Borneo.
		Kryptopterus macrocephalus	9.7	Asia: Indonesia, Malaysia and Brunei Darussalam.
		Butter catfish, two-spot catfish *Ompok bimaculatus*	45.0	Asian countries, such as India Bangladesh, Pakistan, Sri Lanka and Myanmar.
	BAGRIDAE (Bagrid catfishes)	Crystal-eyed catfish *Hemibagrus wyckii*	71.0	Asia from Thailand to Indonesia.
		Asian red tail catfish *Hemibagrus wyckioides*	130.0	Mekong basin, Chao Phraya, Mae Klong and Peninsular Thailand River system.

Order	Family	Common Name Scientific Name	Maximum Length (cm)	Distribution
	PIMELODIDAE (long whiskered catfish)	*Leiarius marmoratus*	100.0	South America.
		Leiarius pictus	60.0	South America.
		Red tail catfish *Phractocephalus hemioliopterus*	120.0	Venezuela.
		Spotted pimelodus, pictus catfish *Pimelodus pictus*	30.0	Amazon and Orinoco River basin.
		Ornate pimelodus *Pimelodus ornatus*	38.5	South America.
		Sturgeon catfish *Platystomatichthys sturio*	40.0	Amazon Basin.
		Barred sorubim *Pseudoplatystoma fasciatum*	104.0	Suriname, Corantijn, Essequibo.
		Tiger sorubim *Pseudoplatysoma tigrinum*	130.0	Amazon Basin in South America.
		Firewood catfish, planiceps, shovelnose catfish. *Sorubimichthys planiceps*	178.0	South America.
		Lima shovelnose catfish *Sorubium lima*	90.0	South America.
	ICTALURIDAE (North American catfish)	Blue catfish *Ictalurus furcatus*	117.0	North America.
		Black bullhead *Ameiurus melas*	66.0	North America: Great Lakes of northern Mexico.
		Stonecut *Noturus flavus*	31.0	North America.
(CHARACINS)				
CHARACI-FORMES (characins)	CHARACIDAE (tetras) Tetras	Black phantom tetra *Hyphessobrycon megalopterus*	3.6	Upper Paraguay basin and upper Madeira basin in Brazil and Bolivia.
		Black neon tetra *Hyphessobrycon herbertaxelrodi*	3.2	Paraguay basin of southern Brazil.

Order	Family	Common Name Scientific Name	Maximum Length (cm)	Distribution
		Black tetra, skirt tetra *Gymnocorymbus ternetzi*	5.0	Paraguay River basin of south -central Brazil, Paraguay and northeast Argentina.
		Bleeding heart tetra *Hyphessobrycon erythrostigma*	6.5	Upper Amazon River basin.
		Bloodfin tetra *Aphyocharax anisitsi*	5.0	Parana River basin in South America.
		Blue tetra *Knodus borki*	5.0	Peru.
		Bucktooth tetra *Exodon paradoxus*	5.0	Amazon Basin and Guyana.
		Buenos Aires tetra *Hyphessobrycon anisitsi*	7.5	South America- in the La Plata region of Argentina, Paraguay and south eastern Brazil.
		Cardinal tetra *Paracheirodon axelrodi*	5.0	Upper Orinoco and Negro Rivers in South America.
		Cave tetra, blind tetra *Astyanax mexicanus*	12.0	Lower Rio Gran de and the Neueces and Pecos Rivers in Texas; central and eastern parts of Mexico.
		Congo tetra *Phenacogrammus interruptus*	7.5	Central Congo River Basin in Africa.
		Diamond tetra *Moenkhausia pittieri*	6.0	Lake Valencia in Venezuela, South America.
		Ember tetra *Hyphessobrycon amandae*	2.0	South America: Araguaia River basin.

Order	Family	Common Name Scientific Name	Maximum Length (cm)	Distribution
		Emperor tetra *Nematobrycon palmeri*	5.0	Atrato and San Juan river basins in western Colombia.
		Garnet tetra, pretty tetra *Hemigrammus pulcher*	6.5	Near Iquitos in the Peruvian Amazon, Brazil and Colombia.
		Glowlight tetra *Hemigrammus erythrozonus*	3.75	Essequibo River, Guyana, South America.
		Golden pristella tetra *Pristella maxillaris*	5.0	Amazon and Orinoco basins; coastal rivers in the Guianas.
		Green neon tetra *Paracheirodon simulans*	2.5	Upper Orinoco and Negro Rivers in South America.
		Head and tail tetra *Hemigrammus ocellifer*	4.0	Rivers of Guyana, Suriname, French Guiana, and the Amazon Basin of Brazil and Peru.
		Lemon tetra *Hyphessobrycon pulchripinnis*	5.0	South America.
		Long-fin tetra *Brycinus longipinnis*	12.5	Western coastal regions of Africa from the Gambia to Democratic Republic of the Congo.
		Neon tetra *Paracheirodon innesi*	3.75	Orinoco and Amazon basin of South America.
		Jelly bean tetra *Ladigesia roloffi*	4.0	Sierra Leone and Liberia in Africa.

Order	Family	Common Name Scientific Name	Maximum Length (cm)	Distribution
		Penguin tetra, blackline penguin fish *Thayeria boehikei*	6.5	Amazon river basin, Araguaia river in Peru and Brazil respectively.
		Red eye tetra *Moenkhausia sanctaefilomenae*	4.0	Sao Francisco, upper Parana, Paraguay and Uruguay river basins in eastern and central South America.
		Red Phantom tetra *Hyplessobrycon sweglesi*	3.2	South America: Orinoco River basin.
		Rummy-nose tetra *Hemigrammus rhodostomus*	5.0	South America.
		Disk tetra *Myleus schomburgkii*	40.0	Middle and lower Amazon River basin, Naay River, upper Orinoco River basin in Brazil, Peru, Venezuela and Suriname.
		Splashing tetra *Copella arnoldi*	3.4	South America: lower Amazon, coastal Guianas to mouth of Orinoco River.
		Serpae tetra *Hyphessobrycon serape*	5.0	Amazon River, drainage in Brazil, Peru, Paraguay, Bolivia and northern Argentina.
		Silvertip tetra *Hasemania nana*	3.0	Blackwater rivers in Brazil.
		X-ray tetra *Pristella maxillaris*	4.5	Amazon and Orinoco basins, coastal rivers in the Guianas.
		Yellow phantom tetra *Hyphessobrycon roseus*	4.5	Maroni and Oyapock Rivers, French Guinea, South America.

Order	Family	Common Name Scientific Name	Maximum Length (cm)	Distribution
	GASTEROPELECIDAE (hatchetfishes)	Common hatchetfish *Gasteropelecus sternicla*	6.5	South America in the Peruvian and middle Amazon, the Guianas and Venezuela.
		Silver hatchetfish *Gasteropelecus levis*	3.5	South America: lower Amazon River basin in Brazil.
		Marbled hatchetfish *Carnegiella strigata*	5.0	South America.
		Black-winged hatchetfish *Carnegiella marthae*	4.0	South America.
	LEBIASINIDAE (pencilfishes)	Barred pencilfish *Nannostomus espi*	3.0	Mazaruni River in Guyana.
		Becked pencilfish, golden pencilfish *Nannostomus beckfordi*	6.0	Brazil, Guyana, French Guiana and Suriname.
		Three-lined pencilfish *Nannostomus trifasciatus*	6.0	Amazon basin, Colombia, Peru, northern Bolivia, the Guianas, in the Rio Negro.
		Hockeystick pencilfish *Nannostomus eques*	3.5	South America-Brazil, Peru, Colombia, and Guyana.
		Dwarf pencilfish *Nannostomus marginatus*	2.5	South American Continent-Brazil, Guyana, Colombia, Suriname, and Peru.
		Coral-red pencilfish *Nannostomus mortenthaleri*	3.0	Small tributary of the Nanay River, and Tiger River, Peru.
		Nannostomus nigrotaeniatus	3.3	South America: Venezuela.
		One-lined pencilfish *Nannostomus unifasciatus*	4.0	Amazon basin, in Brazil, the Guiana shield, Colombia, Venezuela, and northern Bolivia.

Order	Family	Common Name Scientific Name	Maximum Length (cm)	Distribution
	SERRASALMIDAE (pacus, piranhas and silver dollars)	Silver dollar *Metynnis argenteus*	15.0	Tapajos River Basin in Brazil.
		Red-bellied pacu *Colossoma bidens*	120.0	Amazon and Orinoco basins in tropical South America.
		Red piranha *Pygocentrus nattereri*	33.0	South America, in the Amazon, Paraguay, Parana and Essequibo basins, coastal rivers of north eastern Brazil.
	CANODONTIDAE	Payara *Hydrolycus scomberoides*	117	Amazon basin in tropical South America.
		Disk tetra *Myleus sckomburgkii*	40.0	Middle and lower Amazon River basin, Nanay River, Upper Orinoco River basin in Brazil, Peru, Venezuela.
		Pink-tail chalceus *Chalceus macrolepidotus*	25.0	South America: Negro and Orinoco River basins and coastal rivers in Guyana, Suriname, and French Guiana.
	HEPSETIDAE	African pike characin *Hepsetus odoe*	28.0	Western sub-Saharan Africa.
	ALESTIDAE	Goliath tigerfish *Hydrocynus goliath*	150.0	Congo River Basin and Lake Tanganyika.
	CTENOLUCIIDAE	Hujeta gar *Ctenolucinua hujeta*	30.0	South America mainly Colombia.
	PROCHILODONTI-DAE	Kissing prochilodus *Semaprochilodus insignis*	27.5	Central and western parts of the Amazon basin.

Order	Family	Common Name Scientific Name	Maximum Length (cm)	Distribution
	ANOSTOMIDAE	Marbled headstander *Abramites hypselonotus*	14.0	Streams and rivers of the Orinoco and Amazon river systems.
		Banded leporinus *Leporinus fasciatus*	27.0	Amazon basin in South America.
		Striped headstander *Anostomus anostomus*	16.0	South America, Orinoco River, in the Guayana Region of the upper Amazon.
	DISTICHODONTI-DAE	Six-banded distichodus *Distichodus sexfasciatus*	75.0	Africa.
	ERYTHRINIDAE	Red wolf fish *Erythrinus erythrinus*	20.0	South America-the Amazon and Orinoco basins in the Guianas.
		Red tail barracuda *Acestrorhynchus falcatus*	30.0	Rivers of Guyana, Suriname, and French Guiana, and throughout much of the Amazon and Orinoco River basins.
	CHILODONTIDAE	Spotted headstander *Chilodus punctatus*	8.0	North and South America.
(CICHLIDS)				
CICHLIFORMES	CICHLIDAE Lake malawi cichlids	Sunshine peacock cichlid *Aulonocara baenschi*	15.0	Lake Malawi East Africa.
		Red fin hap *Copadichromis borleyi*	16.0	Lake Malawi in East Africa.
		Eureka red peacock *Aulonocara jacobfreibergi*	15.0	Lake Malawi.
		Spilo *Champsochromis spilorhynchus*	40.0	Freshwater lakes in Malawi, Mozambique and Tanzania.

Order	Family	Common Name Scientific Name	Maximum Length (cm)	Distribution
		Blue dolphin cichlid *Cyrtocara moori*	25.0	Lake Malawi.
		Afra cichlid, dogtooth cichlid *Cynotilapia afra*	10.0	Lake Malawi.
		Rusty cichlid, lavender cichlid *Iodotropheus sprengerae*	10.0	Boadzulu and Chinyankiwazi islands and Chinyamwezi island region of Lake Malawi.
		Fuellebom's cichlid, blue mbuna *Labeotropheus fuellebomi*	18.0	Lake Malawi.
		Electric yellow cichlid *Labidochromis caeruleus*	20.0	Central western coastal region of Lake Malawi.
		Electric blue cichlid *Sciaenochromis fryeri*	11.5	Rocky coastal areas of Lake Malawi.
		Malawi eyebiter *Dimidiochromis compressiceps*	23.0	Lake Malawi.
		Hongi, Red-top kimpumpa *Labidochromis* sp.	13.0	Eastern Africa.
		Auratus cichlid, Malawi golden cichlid *Melanochromis auratus*	10.0	Southern region of Lake Malawi.
		Chipokee cichlid *Melanochromis chipokae*	12.0	Lake Malawi.
		Blue johanni cichlid, Maigano *Melanochromis cyaneorhabdos*	10.0	Lake Malawi.
		Pearl of Likoma *Melanochromis panjohnsonae*	10.0	Lake Malawi.

Order	Family	Common Name Scientific Name	Maximum Length (cm)	Distribution
		Aurora *Melanochromis aurora*	10.0	Lake Malawi.
		Red zebra cichlid *Metriaclima estherae*	10.0	Rock dweller or mbuna in Lake Malawi.
		Cobalt blue cichlid *Maylandia callainos*	10.0	Nikhata Bay, Lake Malawi.
		Kenyi cichlid *Maylandia lombardoi*	13.0	Rocky shores of Mbenji Island, Lake Malawi in East Africa.
		Fusco *Nimbochromis fuscotaeniatus*	25.0	Lake Malawi and Lake Malombe.
		Livingston's cichlid *Nimbochromis livingstonii*	25.0	Lake Malawi, upper shire River and Lake Melombe.
		Kaligono *Nimbochromis polystigma*	30.0	Lake Malawi, in the Great Rift Valley system of Africa.
		Venustus cichlid, giraffe cichlid *Nimbochromis venustus*	25.0	Lake Malawi.
		Red empress cichlid *Protomelas taeniolatus*	15.0	Lake Malawai.
		Bumblebee cichlid, hornet cichlid *Pseudotropheus crabro*	15.0	Lake Malawi.
		Yellow-tail acei *Pseudotropheus acei*	18.0	Lake Malawi.
		Johanni cichlid *Pseudotropheus johannii*	10.0	East Mozambique coastline of Lake Malawi.
		Dwarf Mbuna *Pseudotropheus demasoni*	10.0	Lake Malawi, Tanzanian waters.
		Saulosi *Pseudotropheus saulosi*	7.5	Lake Malawi.

Order	Family	Common Name Scientific Name	Maximum Length (cm)	Distribution
		Malawi barracuda *Rhamphochromis macrophthalmus*	23.0	Lake Malawi.
		Tyrannochromis macrostoma	38.0	Lake Malawi in rocky shallows.
	Lake Tanganyika cichlids	Giant cichlid, emperor cichlid *Boulengerochromis microlepis*	90.0	Lake Tanganyika in Africa.
		Frontosa cichlid *Cyphotilapia frontosa*	40.0	Lake Tanganyika in East Africa.
		Featherfin cichlid *Cyathopharynx furcifer*	21.0	Lake Tanganyika off rocky shores.
		Variabilichromis moori	10.3	Lake Tanganyika.
		Frontosa cichlid *Cyphotilapia gibbernosa*	40.0	Lake Tanganyika in East Africa.
		Frontosa cichlid *Cyphotilapia* sp. North	40.0	Lake Tanganyika.
		Julie cichlid *Julidochromis dickfeldi*	11.0	Lake Tanganyika.
		Masked Julie *Julidochromis marlieri*	15.0	Lake Tanganyika.
		Julie cichlid *Julidochromis ornatus*	8.5	Extreme northern and southern shorelines of Lake.
		Julie cichlid *Julidochromis regain*	15.0	Lake Tanganyika.
		Julie cichlid *Julidochromis transcriptus*	10.0	North western shore of Lake Tanganyika.
		Herring cichlid, sardine cichlid *Cyprichromis* spp.	11.0	Open waters of Lake Tanganyika.
		Ectodus descampsii	10.4	Lake Tanganyika.

Order	Family	Common Name Scientific Name	Maximum Length (cm)	Distribution
		Chalinochromis brichardi	12.0	Africa: endemic to Lake Tanganyika.
		Neolamprologus multifasciatus	3.8	Lake Tanganyika.
		Lamprologus ocellatus	5.0	Lake Tanganyika.
		Neolamprologus similis	3.8	Lake Tanganyika.
		Lobochilotes labiatus	40.0	Lake Tanganyika.
		Lyretail cichlid, fairy cichlid *Neolamprologus brichardi*	10.0	Lake Tanganyika.
		Lemon cichlid *Neolamprologus leleupi*	11.5	Lake Tanganyika.
		Featherfin *Ophthalmotilapia ventralis*	15.0	Africa: South end of Lake Tanganyika.
		Threadfin cichlid *Petrochromis trewavasae*	20.0	Lake Tanganyika with rocky substrate.
		Zaire texas *Petrochromis* sp.	18.0	Endemic to Lake Tanganyika.
		Simochromis pleurospilus	10.0	Southwestern portion of Lake Tanganyika.
		Tropheus duboisi	13.0	Lake Tanganyika.
		Blunt head cichlid *Tropheus moorii*	14.0	Lake Tanganyika.
		Tropheus polli	14.0	Central eastern portion of the coast of Lake Tanganyika.
	Lake Victoria cichlids	*Astatotilapia aenocolor*	13.0	Freshwaters of Northern and Eastern Africa.
		Haplochromis *Astatotilapia elegans*	10.0	Uganda—Lake George, Lake Edward and Kazinga channel.

Order	Family	Common Name Scientific Name	Maximum Length (cm)	Distribution
		Zebra obliquidens *Astatotilapia latifasciata*	11.5	Lake Kyoga in Uganda.
		Blue Victoria mouth brooder *Astatotilapia nubila*	11.5	Several lakes and stream in eastern and central Africa.
		Astatotilapia piceatus	10.0	Lake Victoria.
		Astatotilapia schubotziellus	10.0	Lake Victoraia.
		Astatotilapia sp. "Red Tail"	11.5	Lake Victoria.
		Astatotilapia sp." Spot Bar"	13.0	Lake Victoria.
		Hippo Point Salmon *Ptyochromis* sp.	14.0	Lake Victoria.
	Miscellaneous African cichlids (non-Rift Lake)	African butterfly cichlid *Anomalochromis thomasi*	8.0	Sierra Leone, Liberia and Guinea-in-smaller streams.
		Jewel cichlid, two-spotted jewel cichlid *Hemichromis bimaculatus*	4.5	West Africa-rivers, streams, canals.
		Steatocranus spp.	10.7	Congo River Basin.
		Steatocranus irvine	10.7	Volta River in Ghana.
		Lifalili jewel cichlid, blood-red jewel cichlid *Hemichromis lifalili*	10.0	The Democratic Republic of the Congo, the lower and central Congo River basin.
		Kribensis krib *Pelvicachromis pulcher*	12.5	Nigeria and Cameroon.
		Zebra tilapia *Tilapia buttikoferi*	30.8	West African species.
		Guenther's Mouth brooder *Chromidotilapia guentheri*	18.0	Africa.

Order	Family	Common Name Scientific Name	Maximum Length (cm)	Distribution
		Alluaud's haplo *Astatoreochromis alluaudi*	19.0	Lakes of East Africa.
		Pale usisya aulonocara *Aulonocara steveni*	8.9	Malawi.
		Benthochromis tricoti	16.5	Lake Tanganyika in Africa.
	Dwarf cichlids (apisto-grammus, rams and others)	Agassiz's dwarf cichlid *Apistogramma agassizi*	4.2	Southern tributaries of Amazon river.
		Zebra acara *Nannacara adoketa*	10.0	Two side rivers of Rio Negro in Brazil.
		Yellow dwarf cichlid *Apistogramma borellii*	3.9	Paraguay River and Parana River, South America.
		Cockatoo dwarf cichlid *Apistogramma cacatuoides*	8.0	South America-Amazon River basin.
		Panda dwarf cichlid *Apistogramma nijsseni*	8.0	Blackwaters of a small stream in the Ucayali River drainage,
		Three-stripe dwarf cichlid *Apistogramma trifasciata*	3.8	South America.
		Apistogramma eremnopyge	0.35	Tributary of Rio Itaya near Iquitos, Peru.
		Two-stripe dwarf cichlid *Apistogramma bitaeniata*	7.5	South America-Blackwater rivers of the Amazon Basin in Brazil and Peru.
		Checkerboard cichlid *Dicrossus filamentosus*	3.8	Rio Negro and Orinoco basins of Colombia, Venezuela and Brazil.

Order	Family	Common Name Scientific Name	Maximum Length (cm)	Distribution
		Blue ram, German ram *Mikrogeophagus ramirezi*	10.0	Orinoco River basin, in the Savannah's of Venezuela and Colombia in South America.
		Bolivian ram *Mikrogeophagus altispinosa*	8.0	Amazon River basin in Brazil and Bolivia.
		Dwarf flag cichlid *Laetacara curviceps*	12.0	Rivers, streams of the Amazon basin and the lower Negro River basin.
	American cichlids (cichlasomas, acaras, angelfish, discus and others)	Midas cichlid, red devil cichlid *Amphilophus citrinellus*	24.4	San Juan River and watersheds in Costa Rica and Nicaragua.
		Poor man's tropheus *Hypsophrys namatopus*	12.0	Central America, Atlantic coasts of Costa Rica and Nicaragua.
		Red Devil cichlid *Amphilophus labiatus*	24.0	Central America-Lake Managua and Lake Nicaragua.
		Firemouth cichlid *Cichlasoma meeki*	15.0	Central America.
		Jack Dempsey cichlid *Cichlasoma octofasciatus*	25.0	North and Central America (From Mexico to Honduras).
		Jaguar cichlid, managuense cichlid *Parachromis managuensis*	35.0	Freshwater in Central America, from Honduras to Costa Rica.
		Mayan cichlid *Cichlasoma urophthalmus*	39.4	Middle America Eastern Mexico to Nicaragua.
		Convict cichlid *Archcentrus nigrofasciatus*	12.0	Central America.

Order	Family	Common Name Scientific Name	Maximum Length (cm)	Distribution
		T-bar cichlid *Cryptoheros sajica*	12.0	Central America-Pacific slope of Costa Rica.
		Severum *Heros severus*	20.0	Upper Ornico and upper Rio Negro basins in South America.
		Green terror *Andinoacara rivulatus*	30.0	The Pacific side of South America.
		Blue acara *Andinoacara pulcher*	16.0	Various freshwater habitats in Venezuela and Trinidad.
		Thread-finned acara *Acarichthys heckelii*	13.4	Rivers in the Amazon and Essequibo basins in tropical South America.
		Eartheater cichlid *Geophagus altifrons*	22.5	Amazon River Basin, Brazil.
		Greenstreaked Eartheater, cupid cichlid *Biotodoma cupido*	11.7	Rivers in the Amazon and Essequibo basins.
		Keyhole cichlid *Cleithracara maronii*	7.1	Tropical South America, Orinoco basin and river basins in Guianas.
		Angelfish *Pterophyllum scalare*	15.0	Amazon basin in Peru, Colombia and Brazil.
		Altum angelfish *Pterophyllum altum*	18.0	Venezuela, Colombia and Northern Brazil.
		Spotted angelfish *Pterophyllum leopoldi*	5.0	Amazon River, Essequibo River, Rupununi River.

Order	Family	Common Name Scientific Name	Maximum Length (cm)	Distribution
		Common discus, red discus *Symphysodon discus*	20.0	Amazon basin.
		Blue discus, green discus *Symphysodon aequifasciatus*	20.0	Rivers of the eastern and central Amazon basin down river from the Purus arch.
		Chocolate cichlid *Hypselecara temporalis*	15.0	South America.
		Oscar *Astronotus ocellatus*	45.0	Tropical South America.
		Uaru, waroo *Uaru amphiacanthoides*	25.0	Amazon basin, South America.
		Peacock bass, butterfly peacock bass *Cicla orinocensis*	62.0	Rio Negro and Orinoco River basins in northern South America.
		Texas cichlid, Rio Grande cichlid *Herichthys cyanoguttatus*	33.0	United States.
	Other cichlids	Orange chromide *Etroplus maculatus*	8.0	Southern India and Sri Lanka.
		Blood parrot cichlid Hybrid	20.0	
		Flowerhorn cichlid Hybrid	25.0	
		Flowerhorn King kamfa Hybrid	20.0	
		Wolf cichlid *Parachromis dovii*	72.0	Central America.
(CYPRINIDS)				
CYPRINIFORMES (carps, loaches, minnows and relatives)	CYPRINIDAE (minnows, carps and loaches)	Arulius barb *Dawkinsia arulius*	12.0	Kaveri River basin of southeast India.
		Bigspot barb,Duncker's barb *Barbodes dunckeri*	13.0	Malaysia inhabiting acidic swamps.

Order	Family	Common Name Scientific Name	Maximum Length (cm)	Distribution
		Black ruby barb *Pethia nigrofasciatus*	6.5	Sri Lanka.
		Checker barb *Oliotus oligolepis*	5.0	Endemic to western central Sumatra, Indonesia.
		Cherry barb *Puntius titteye*	4.8	Sri Lanka.
		Clipper barb *Enteromius callipterus*	7.5	Chad basin in Nigeria and Cameroon.
		Clown barb *Barbodes everetti*	15.0	Borneo and Sumatra.
		Denison's barb, red line torpedo barb *Sahyadria denisonii*	15.0	Western Ghats in India.
		Fiveband barb, pentazona barb *Desmopuntius pentazona*	5.0	Southeast Asia, including Cambodia, Indonesia, Malaysia, Singapore and Viet Nam.
		Gold barb *Barbodes semifasciolatus*	8.0	Red River basin in southeast Asia.
		Golden barb *Pethia gelius*	5.0	Inland waters in Asia, in Pakistan, India, Bangladesh.
		Greenstripe barb *Puntius vittatus*	5.0	Inland waters in Asia, in Pakistan, India and Sri Lanka.
		Indian Glass Barb *Laubuca laubuca*	6.7	India, Bangladesh, Pakistan, Sri Lanka, Myanmar, Nepal and Indonesia.
		Melon barb *Puntius fasciatus*	15.0	Rivers in Goa, Karnataka, Kerala, Tamil Nadu in the Western Ghats of South India.

Order	Family	Common Name Scientific Name	Maximum Length (cm)	Distribution
		Odessa barb *Puntius padamya*	8.0	Central Myanmar.
		One spot barb *Puntius terio*	15.0	Inland waters in Asia, in Pakistan, India, Bangladesh, Myanmar.
		Panda barb *Puntius fasciatus*	12.5	Rivers in Goa, Karnataka, Kerala and Tamil Nadu in the Western Ghats of South India.
		Pool barb *Puntius sophore*	18.0	Inland waters in Asia, in India, Nepal, Bangladesh, Myanmar, Bhutan, Afghanistan, Pakistan and Yunnan, China.
		Partipentazona barb *Puntigrus partipentazona*	3.8	Southeast Asia-Mekong, Mae klong, and Chao Phraya basins of Thailand, the Malay Peninsula and coastal streams of southeast Thailand and Cambodia.
		Rosy barb *Pethia conchonius*	13.0	Southern Asia from Afghanistan to Bangladesh.
		Shortfin barb *Barbus brevipinnis*	4.6	Southern Africa-Sabie-Komati and the Steelpoort-Limpopo river systems.
		Snakeskin barb *Desmopuntius rhoboocellatus*	5.08	Borneo-in western and central Kalimantan.

Order	Family	Common Name Scientific Name	Maximum Length (cm)	Distribution
		Spanner barb, T-barb *Barbodes lateristriga*	20.0	Malay Peninsula and the island of Borneo.
		Spotted barb, common barb *Barbodes binotatus*	19.5	Southeast Asia.
		Spotted sail barb, dwarf barb *Pethia phutunio*	7.8	In inland waters in Asia-in Pakistan, India, Bangladesh and Myanmar.
		Swamp barb *Puntius chola*	15.0	In inland waters in Asia-India, Bangladesh, Pakistan, Nepal, Sri Lanka, Bhutan and Myanmar.
		Ticto barb *Barbus ticto*	10.0	India, Bangladesh, Nepal,Pakistan, Myanmar, Thailand and Sri Lanka.
		Tic-tac-toe barb *Puntius Stoliczkanus*	5.0	India, Bangladesh, Pakistan, Nepal, Myanmar, Laos, Thailand, China and Sri Lanka.
		Tiger barb, Sumatra barb *Puntius tetrazona*	7.5	Throughout the Malay Peninsula, Sumatra and Borneo in Indonesia, Cambodia.
		Tinfoil barb *Barbonymus schwanenfeldii*	36.0	Southeast Asia.
	Cold-water cyprinids Subfamily: Cyprininae	Goldfish and its varieties *Carassius auratus*	15.0	East Asia.
		Koi, common carp *Cyprinus carpio*	30.0	Lakes and large rivers in Europe and Asia.

Order	Family	Common Name Scientific Name	Maximum Length (cm)	Distribution
	Subfamily: Acheilognathinae	Bitterling *Rhodeus amarus*	10.0	Europe.
		White cloud mountain minnow *Tanichthys albonubes*	3.8	China.
	Subfamily: Leuciscinae	Red shiner *Cyprinella lutrensis*	9.0	Mississippi River basin.
		Common dace *Leuciscus leuciscus*	40.0	Rivers and streams of Europe north of the Alps as well as in Northern Asia.
		Rosy red minnow, fathead minnow *Pimephales promelas*	7.5	North America from central America south along the rockies to Texas and east to Virginia and the North eastern United States.
		European minnow *Phoxinus phoxinus*	14.0	Eurasia from Britain and Spain to eastern Siberia.
		Southern red belly dace *Phoxinus erythrogaster*	10.0	Western New York to Minnesota and south to Oklahoma, Arkansas, and Alabama.
		Tench *Tinca tinca*	70.0	Throughout Eurasia from Western Europe including the British Isles east into Asia as far as Ob and Yenisei Rivers.
		Golden orfe *Leuciscus idus*	100.0	Larger rivers, ponds, and lakes across northern Europe and Asia.

Order	Family	Common Name Scientific Name	Maximum Length (cm)	Distribution
	CATOSTOMIDAE	Chinese sucker *Myxocyprinus asiaticus*	60.0	Asia: China (Yangtze River basin).
		Chinese hill stream loach *Beaufortia kweichowensis*	7.5	Rivers in China.
	Danios and other danionins Subfamily: Danioninae	Bengal danio, Sind danio *Devario devario*	10.0	India, Pakistan, Nepal, Bangladesh, and Afghanistan.
		Giant danio *Devario aequipinnatus*	10.0	West coast of India, Sri Lanka, Nepal.
		Malabar danio *Devario malabaricus*	14.0	West coast of India, Sri Lanka.
		Pearl danio *Danio albolineatus*	6.5	Sumatra, Myanmar and Thailand.
		Queen danio *Devario regina*	7.8	India, Myanmar, Thailand, north western Malaya, and the Mekong River basin.
		Spotted danio *Danio nigrofasciatus*	3.7	Northern Myanmar.
		Turquoise danio *Danio kerri*	5.0	Islands of Langkawi and Ko Yao Yai in Malaysia.
		Zebra danio *Danio rerio*	5.0	Himalayan region-steams of the south eastern Himalayan region.
	Rasboras Subfamily: Danioninae	Harlequin rasbora *Trigonostigma heteromorpha*	5.0	Malaysia, Singapore, Sumatra and southern Thailand.
		Lambchop rasbora *Trigonostigma espei*	5.0	Asia: Thailand and Cambodia.

Order	Family	Common Name Scientific Name	Maximum Length (cm)	Distribution
		Blackline rasbora, red-tailed rasbora *Rasbora borapetensis*	6.0	Asia-in the Mekong, Chao Phraya, Mae Klong basins and the northern Malay Peninsula.
		Red-striped rasbora *Rasbora pauciperforata*	7.5	Freshwaters of southeast Asia.
		Volcano rasbora *Rasbora vulcanus*	4.0	Sumatra.
		Dwarf rasbora *Boraras maculatus*	2.5	Asia: Malay Peninsula and Indonesia.
		Clown rasbora *Rasbora kalochroma*	10.0	Asia: Malay Peninsula and Indonesia (Borneo and Sumatra).
		Scissortail rasbora *Rasbora trilineata*	13.0	Asia: Mekong and Chao Phraya basins; Malay Peninsula, Sumatra and Borneo.
	Subfamily: Labeoninae	Eyespot rasbora *Brevibora dorsiocellata*	2.3	Asia: Malay Peninsula and Indonesia.
		Brilliant rasbora *Rasbora einthovenii*	9.0	Asia: Malay Peninsula and Indonesia.
	Other cyprinids Subfamily: Barbinae	Bala shark *Balantiocheilus melanopterus*	40.0	Malay Peninsula, Sumatra and Borneo.
	Subfamily: Labeoninae	Black shark *Labeo chrysophekadion*	90.0	Mekong and Chao Phraya river basins, Malay Peninsula, Sumatra, Java and Borneo.

Order	Family	Common Name Scientific Name	Maximum Length (cm)	Distribution
		Flying fox fish *Epalzeorhynchos kalopterus*	11.94	Fast-flowing foothill rivers and streams of the Thai-Malay Peninsula, Borneo, Java and Sumatra in southeast Asia.
		Red-tailed black shark *Epalzeorhynchos bicolor*	15.0	Thailand.
		Rainbow shark *Epalzeorhynchos frenatum*	15.0	Basins of Mekong, Chao Phraya, Xe Bangfai and Maeklong in Indochina.
		Cambodian log sucker *Garra cambodgiensis*	15.0	Southeast Asia.
		Siamese algae eater *Crossocheilus siamensis*	16.0	Mainland southeast Asia, including the Chao Phraya and Mekong basins, Malaya Peninsula.
	GYRINOCHEILIDAE	Chinese algae eater *Gyrinocheilus aymonieri*	28.0	Large parts of southeast Asia.
(LOACHES AND RELATED CYPRINIFORMS)				
CYPRINIFORMES	BOTIIDAE	Bengal loach, Queen loach *Botia dario*	15.0	Brahmaputra and Ganges basins in Bangladesh, Bhutan and north India.
		Blue botia *Yasuhikotakia modesta*	25.0	Large rivers in Thailand, Laos, Cambodia and Vietnam.
		Burmese border loach *Botia kubotai*	13.0	Ataran River (tributary of Salween river) in Thailand and Salween River system in Myanmar.

Order	Family	Common Name Scientific Name	Maximum Length (cm)	Distribution
		Clown loach *Chromobotia macracanthus*	20.0	In inland waters in Indonesia on the islands of Sumatra and Borneo.
		Dwarf loach, pygmy loach *Yasuhikotakia sidthimunki*	6.0	Mae Klong basin in Thailand, and the Ataran River on the Thai-Myanmar border.
		Dojo loach, weather loach *Misgumus anguillicaudatus*	30.5	East Asia.
		Horseface loach *Acantopsis choirorhynchus*	30.0	Streams, rivers of southeast Asia, from India to Indonesia.
		Kuhli loach, coolie loach *Pangio kuhlii*	10.0	Indonesia and the Malay Peninsula.
		Longnose loach *Acantopsis octoactinotos*	11.0	Indonesia.
		Skunk loach *Yasuhikotakia morleti*	10.0	Mekong River Basin in Indochina, and Chao Phraya and Mae Klong basins in Thailand.
		Yoyo loach *Botia almorhae*	16.0	Slow-running and still waters of the Ganges basin in northern India and possibly Nepal.
		Zebra loach *Botia striata*	9.0	Rivers and streams in the Western Ghats of India.
(KILLIFISHES)				
CYPRINODONTI-FORMES	APLOCHEILIDAE	Golden wonder, striped panchax *Aplocheilus lineatus*	10.0	Fresh and brackish waters of India and Sri Lanka.

Order	Family	Common Name Scientific Name	Maximum Length (cm)	Distribution
		Blue Lyretail *Fundulopanchax gardneri*	5.0	Benue and Cross River basins of Nigeria and Cameroon.
		Bluefin Notho, Rachow's Notho, Rainbow Notho *Nothobranchius rachovii*	6.0	Mozambique and South Africa.
		Clown killifish *Epiplatys annulatus*	3.5	Guinea, Liberia, Sierra Leone in West Africa.
	CYPRINODONTI-DAE	Arabian Killifish *Aphanius dispar*	7.0	Shores of the Red Sea south to Ethiopia, the Gulf of Aden, the Arabian Sea, along the Persian Gulf east to Pakistan and India.
		American flagfish *Jordinella floridae*	6.0	North America: Florida, USA.
	FUNDULIDAE	Golden Topminnow *Fundulus chrysotus*	8.5	United States-throughout the southeast, from Kentucky and Ohio south into Florida.
		Red-seam killifish *Aphyosemio calliurum*	5.0	Africa.
(KNIFEFISHES)				
GYMNOTI-FORMES	APTERONOTIDAE	Black ghost knifefish *Apteronotus albifrons*	50.0	South America-from Venezuela to the Paraguay-Parana River, Amazon Basin.
		Brown ghost knifefish *Apteronotus leptorhynchus*	26.9	South America.
		Elephant nose *Gnathonemus petersii*	25.0	Rivers of West and Central Africa.

Order	Family	Common Name Scientific Name	Maximum Length (cm)	Distribution
		Electric eel *Electrophorus electricus*	200.0	Amazon and Orinoco River basins in South America.
		Glass knifefish *Eigenmannia virescens*	140.0	South America.
(LIVE-BEARERS)				
CYPRINODONTI-FORMES	POECILIIDAE	Guppy *Poecilia reticulata*	5.0	Northeast South America.
		Endler's livebearer *Poecilia wingei*	3.8	Paria Peninsula in Venezuela.
		Black molly *Poecilia sphenops*	5.0	Freshwater streams and coastal, brackish and marine waters of Mexico.
		Sailfin molly *Poecilia latipinna*	5.0	Fresh, brackish, salt and coastal waters from North Carolina to Texas and the Yucatan Peninsula of Mexico.
		Dalmatian molly Hybrid	5.0	
		Lyretail molly Variety	5.0	
(PLATIES AND SWORDTAILS)		Southern platy *Xiphophorus maculatus*	6.0	Area of North and Central America stretching from Veracruz, Mexico to northern Belize.
		Variable platy/ variegated platy *Xiphophorus variatus*	7.0	North eastern Mexico.
		Green swordtail *Xiphophorus helleri*	16.0	North and Central America.

Order	Family	Common Name Scientific Name	Maximum Length (cm)	Distribution
		Western mosquitofish *Gambusia affinis*	7.0	Southern parts of Illinois and Indiana, throughout the Mississippi River, Gulf coast, north eastern parts of Mexico.
		Eastern mosquito fish *Gambusia holbrooki*	6.4	Eastern and southern United States.
		Celebes half beak *Nomoramphus liemi*	8.0	Freshwaters of the southern half of the island of Sulawesi.
		Cuban lima *Limia vittata*	8.0	Central America: endemic to Cuba.
		Tanganyika pearl killifish *Lamprichthys tanganicanus*	15.0	Africa: endemic to the shore regions of Lake Tanganyika.
		Knife livebearer *Alfaro cultratus*	8.0	Central America.
		Least killifish *Heterandria formosa*	3.0	United States.
(LABYRINTH FISHES)				
ANABANTI-FORMES	OSPHRONEMIDAE	Chocolate gourami *Sphaerichthys osphronemoides*	7.0	Malay Peninsula, Sumatra and Borneo.
		Croaking gourami *Trichopsis vittata*	6.5	Still waters in Southeast Asia.
		Dwarf gourami *Colisa lalia*	5.0	South Asia.
		Giant gourami *Osphronemus goramy*	60.0	Southeast Asia.
		Paradise fish *Macropodus opercularis*	7.5	East Asia-from Korean Peninsula to northern Vietnam.

Order	Family	Common Name Scientific Name	Maximum Length (cm)	Distribution
		Pearl gourami *Trichopodus leerii*	11.5	Southeast Asia.
		Pigmy gourami, sparling gourami *Trichopsis pumila*	3.8	Southeast Asia.
		Kissing gourami *Helostoma temminckii*	15.0	Thailand to Indonesia.
		Moonlight gourami *Trichopodus microlepis*	15.0	Indochina.
		Snakeskin gourami *Trichopodus pectoralis*	20.0	Southeast Asia.
		Three spot gourami *Trichopodus trichopterus*	10.0	South eastern Asia.
		Siamese fighting fish *Betta splendens*	7.5	Mekong basin of Laos, Cambodia, Vietnam, and Thailand.
		Giant betta *Betta anabatides*	8.2	Asia: Indonesian Borneo.
(DARTERS)				
PERCIFORMES	PERCIDAE Subfamily: Etheostomatinae	African striped darter *Nannocharax fasciatus*	6.6	Africa: from Senegal to Cameroon, including the Lake Chad basin.
		Banded darter *Etheostoma zonale*	7.8	Eastern United States.
		Gilt darter *Percina evides*	9.0	In a number of states in the Mississippi River drainage of the United States.
		Orange throat darter *Etheostoma spectabile*	0.74	Central and eastern United States-parts of the Mississippi River basin and Lake Erie Basin.

Order	Family	Common Name Scientific Name	Maximum Length (cm)	Distribution
		Rainbow darter *Etheostoma caeruleum*	0.76	Freshwater streams in North America.
		Splendid darter *Etheostoma barrenense*	6.4	Barren River system in South -central Kentucky and north-central Tennessee.
(GOBIES)				
GOBIIFORMES	OXUDERCIDAE	Black toraja goby *Mugilogobius sarasinorum*	8.0	Lake Poso in Sulawesi, Indonesia.
		Bumblebee goby *Brachygobius doriae*	4.2	Fresh and brackish waters of Indonesia, Brunei and Singapore, Malaysia, Cambodia, Thailand and Vietnam.
	GOBIIDAE	Dragon goby, violet goby *Gobioides broussonetii*	81.0	Marine, fresh and brackish waters near the Atlantic coast of North and South America from South Carolina in the United States of America to northern Brazil.
	OXUDERCIDAE	Emerald river goby *Ctenogobius smaragdus*	12.0	Western Atlantic: South Carolina, USA to Pernambuco, Brazil, and Cuba.
	ELEOTRIDAE	Empire gudgeon *Hypseleotris compressa*	12.0	Australia and south-central New Guinea.
		Empire goby *Hypseleotris maculatus*	12.0	Brackish water in Australia, New Guinea.

Order	Family	Common Name Scientific Name	Maximum Length (cm)	Distribution
	OXUDERCIDAE	Knight goby *Stigmatogobius sadanundio*	9.0	South Asia from India to Indonesia including Sri Lanka and Andaman island.
	BUTIDAE	Marbled sleeper *Oxyeleotris marmorata*	65.0	Fresh and brackish waters of the Mekong and Chao Phraya basins, Cambodia, Thailand, Malaysia, Singapore, Indo China, Philippines and Indonesia.
	GOBIIDAE	*Rhinogobius duospilus*	4.5	China and Vietnam-in fresh and brackish waters.
	ELEOTRIDAE	Peacock goby *Tateurndina ocellicauda*	7.5	Eastern part of Papua New Guinea.
		Purple sleeper goby *Mogurnda mogurnda*	20.0	Australia and New Guinea.
		Striped sleeper goby *Dormitator maculatus*	70.0	Fresh and brackish waters of Bahamas and North Carolina to Brazil.
(GARS)				
LEPISOSTEI-FORMES	LEPISOSTEIDAE			
		Spotted Gar *Lepisosteus oculatus*	70.0	North America.
		Florida Gar *Lepisosteus platyrhincus*	70.0	USA-Georgia and throughout peninsular Florida.

Order	Family	Common Name Scientific Name	Maximum Length (cm)	Distribution
		Longnose Gar *Lepisosteus osseus*	120.0	East coast of North and Central America. Kansas, Texas and southern New Mexico.
		Shortnose Gar *Lepisosteus platostomus*	62.5	United States.
		Alligator Gar *Atractosteus spatula*	300.0	Freshwaters of North America.
		Cuban Gar *Atractosteus tristoechus*	200.0	Central America.
		Tropical Gar *Atractosteus tropicus*	125.0	Central America.
(BICHIRS AND REEDFISHES)				
POLYPTERI-FORMES (bichirs; reedfishes)	POLYPTERIDAE	Gray bichir, Senegal bichir *Polypterus senegalus*	70.0	Africa.
		Ornate bichir *Polypterus ornatipinnis*	60.0	Lake Tanganyika and the Congo River basin in Central and East Africa.
		Retropinnis bichir *Polypterus retropinnis*	34.0	Central Congo River basin and Ogoone River in Africa.
		Reedfish *Erpectoichthys calabaricus*	37.0	West and Central Africa.
(PUFFERFISH)				
TETRADONTI-FORMES	TETRADONTIDAE Subfamily: Tetradon-tinae	Amazon puffer *Colomesus asellus*	12.8	Amazon, Essequito and Orinoco basins in tropical South America.
		Fahaka puffer *Tetraodon lineatus*	43.0	West, Northeast and East Africa.
		Figure 8 pufferfish *Tetraodon biocellatus*	6.0	Southeast Asia.

Order	Family	Common Name Scientific Name	Maximum Length (cm)	Distribution
		Green spotted puffer *Tetraodon nigroviridis*	15.0	South and Southeast Asia.
		Giant freshwater pufferfish *Tetraodon mbu*	75.0	Congo River in Africa, east coast of Lake Tanganyika.
		Dwarf pufferfish *Carinotetraodon travancoricus*	2.9	Kerala and southern Karnataka in Southwest India.
		Congo pufferfish *Tetraodon miurus*	15.0	Congo River in Africa, including rapids.
(RAINBOWFISHES)				
ATHERINI-FORMES	MELANOTAENIIDAE Subfamily: Melanotae-niinae	Salmon-red rainbowfish *Glossolepis incises*	15.0	Lake Sentani in West Papua, Indonesia.
		Tami River rainbowfish *Glossolepis pseudoincisus*	7.5	West Papua in Indonesia and Papua New Guinea.
		Lake Wanam rainbowfish *Glossolepis wanamensis*	15.0	Lake wanan near Lae in Papua New Guinea.
		Threadfin rainbowfish *Iriatherina werneri*	5.0	Freshwater swamps or vegetated flowing waters in tropical northern Australia and New Guinea.
		New Guinea rainbowfish *Melanotaenia affinis*	13.0	North Papua New Guinea.
		Boeseman's rainbowfish *Melanotaenia boesemani*	13.0	Bird's Head Peninsula, West Papua, Indonesia.
		Duboulayi's rainbowfish *Melanotaenia duboulayi*	15.0	Eastern Australia.

Order	Family	Common Name Scientific Name	Maximum Length (cm)	Distribution
		Australian rainbowfish *Melanotaenia fluviatilis*	15.0	Australia.
		Lake kurmoi rainbowfish *Melanotaenia parva*	8.0	West Papua in Indonesia.
		Madagascar rainbowfish *Bedotia madagascariensis*	8.0	Madagascar.
		Neon rainbowfish *Melanotaenia praecox*	7.5	Mamberamo River basin in West Papua in Indonesia.
		Eastern rainbowfish *Melanotaenia splendida splendida*	15.0	Australia.
(SPINY EELS)				
SYNBRANCHI-FORMES	MASTACEMBELIDAE	Fire eel *Mastacembelus erythrotaenia*	100.0	Southeast Asia.
		Spotfinned spiny eel, peacock eel *Macrognathus siamensis*	30.0	Rivers of southeast Asia.
		Tire track eel *Mastacembelus armatus*	90.0	Riverine fauna of India, Pakistan, Sri Lanka, Thailand, Viet Nam, Indonesia and other parts of South East Asia.
		Lesser spiny eel *Macrognathus aculeatus*	35.0	Freshwaters of Southeast Asia.
(SUNFISHES AND REALITIVES)				
PERCIFORMES	CENTRARCHIDAE	Orange spotted sunfish *Lepomis humilis*	15.0	Freshwaters of North America. Middle and eastern United States.

Order	Family	Common Name Scientific Name	Maximum Length (cm)	Distribution
		Pumpkin seed *Lepomis gibbosus*	28.0	Freshwaters of North America.
		Dollar sunfish *Lepomis marginatus*	10.0	Southern Atlantic coastal drainages from North Carolina to Florida, to Texas.
		Red breast sunfish *Lepomis auritus*	30.0	River systems of eastern Canada and the United States.
		Red ear sunfish *Lepomis microlophus*	43.2	South eastern United States.
		Long ear sunfish *Lepomis megalotis*	24.0	Eastern North America.
		Blackbanded sunfish *Enneacanthus chaetodon*	10.0	United States from New Jersey to central Florida.
		Blue-spotted sunfish *Enneacanthus gloriosus*	9.5	South eastern and eastern United States.
		Bluegill *Lepomis macrochirus*	30.0	Streams, rivers, lakes and ponds in North America.
		Green sunfish *Lepomis cyanellus*	30.0	Wide area of North America and northern Mexico.
		Warmouth *Lepomis gulosus*	25.0	Eastern United States.
		Pygmy sunfish *Elassoma* spp.	4.0	South America.
(OTHER FISHES)				
MYLIOBATI-FORMES	POTAMOTRYGONI-DAE	Ocellate river stingray *Potamotrygon motoro*	50.0	Rio de la Plata, Amazon, Mearins and Orinoco basins in tropical and subtropical South America.

Order	Family	Common Name Scientific Name	Maximum Length (cm)	Distribution
		Black devil stingray *Potamotrygon leopoldi*	75.0	Xingu River basin in Brazil with rocky bottoms.
SYNGNATHI-FORMES	SYNGNATHIDAE	Freshwater pipefish *Doryichthys martensii*	15.0	Indonesia, Malaysia, Brunei Darussalam, and Thailand.
ACIPENSERI-FORMES	ACIPENSERIDAE	Starlet *Acipenser ruthenus*	125.0	Eurasia.
		Shovelnose sturgeon *Scaphirhynchus platorynchus*	100.0	United States of America.
BLENNIIFORMES	BLENNIDAE	Freshwater blenny *Salaria fluviatilis*	15.0	In some rivers of Africa, Europe, and Asia (Turkey and Israel).
BELONIFORMES	BELONIDAE	Freshwater needlefish *Xenentodon cancila*	40.0	Freshwaters and brackish waters in South and Southeast Asia.
ESOCIFORMES	ESOCIDAE	Pickerel *Esox americanus*	40.0	North America.
PLEURONECTI-FORMES	SOLEIDAE	Freshwater sole *Brachirus panoides*	20.0	Asia: Thailand to Indonesia.
SCORPAENI-FORMES	GASTEROSTEIDAE	Brook stickleback *Culaea inconstans*	5.0	The US and Canada.
BELONIFORMES	ADRIANICHTHYI-DAE	Medaka *Oryzias latipes*	3.6	East Asia.
PERCIFORMES	PERCIDAE	European perch *Perca fluviatilis*	60.0	Europe and northern Asia.
(LUNGFISHES)				
CERATODON-TIFORMES	NEOCERATODON-TIDAE	Australian lungfish *Neoceratodus forsteri*	200.0	Freshwater: Australia.
	LEPIDOSIRENI-DAE	South American lungfish *Lepidosiren paradoxa*	125.0	Freshwater: Brazil and Paraguay.
	PROTOPTERIDAE	African lungfish *Protopterus* spp.	125.0	Freshwater: Africa.

2.5 Detail Endeavour of Popular Ornamental Freshwater Fishes of the World Trade

Arowana

Common Name:

English—Arowana, bonytongue.

Systematic Position:

Kingdom: ANIMALIA

Phylum: CHORDATA

Class: OSTEICHTHYES

Subclass: ACTINOPTERYGII

Order: OSTEOGLOSSIFORMES--**bonytongues**

Suborder: OSTEOGLOSSOIDEI

Family: OSTEOGLOSSIDAE--bonytongues

Subfamily: OSTEOGLOSSINAE

- Genus: *Osteoglossum*
- Genus: *Scleropages*

Scientific Names:

Osteoglossum bicirrhosum (Cuvier,1829)

Osteoglossum ferreirai Kanazawa, 1966

Scleropages formosus (Muller & Schlegel,1840)

Scleropages leichardti Gunther,1864

Scleropages jardinii (Saville-Kent,1892)

Scleropages inscriptus Roberts,2012

The family includes several extant species from South America, one from Africa, several from Asia and two from Australia.

2.5.1 Silver arowana

Figure 2.37

Osteoglossum bicirrhosum (Cuvier, 1829)

Common Name:

English—Silver arowana, arawana, bony-tongued fish.

IUCN Red List Status: Not Evaluated.

Systematic Position:

Subfamily: OSTEOGLOSSINAE

Genus: *Osteoglossum*

Species: *O. bicirrhosum*

Scientific Name: *Osteoglossum bicirrhosum* (Cuvier, 1829)

Description:

- Body covered with very big scales that reflect blue, red and green highlights as they grow older.
- Dorsal and anal fins are almost fused with the caudal fin.
- At the extremity of the lower jaw, 2 mandibular barbels are present.
- Body colour of the adults is silvery, juveniles with blue glints and a yellow-orange bar.
- Mouth open in three pieces.
- Many of their oral bones are covered with teeth which include the jaw, palate, tongue and pharynx.
- Sexual Dimorphism: Male fish is slender and with a long anal fin whereas the female fish is plump and with a short anal fin.

Species Details:

- Distribution:

  South America: Amazon River basin, Rupununi and Oyapock Rivers.
- Habitat: Freshwater, benthopelagic. It can remain in water with low oxygen levels. Found in white and black water floodplains. They prefer the very calm waters in swamps and shallow flooded areas.
- Food and feeding: It is an omnivore and generally prefer to eat fish at the surface. It also jumps out of the water to feed on large insects. The superior position of the mouth helps it to capture prey while swimming from below. They eat fish, insects, insect larvae, feeder fish, shrimps and all kinds of live foods. Large flakes, pellets are also eaten in captivity.
- Feeding frequency: Daily. Depends on what food they are provided. Arowana pellets can be given daily.
- Maximum length: 90.0 cm TL male/unsexed.
- Life span: 20 years.

Aquarium Aspects:

- Minimum tank size: 55 gallons for juveniles, 300 gallons for an adult.
- Suitable for Nano tank: No.
- Substrate type: Small gravel-dark gravel is suitable.
- Lighting needs: Moderate-normal light. No sudden lighting changes should be followed.
- Temperature: 75.0° to 86.0° F (23.9° to 30.0° C).
- Range pH: 6.5-7.0; hardness range: 8-12 dGH.
- Water movement: Weak.
- Water region: Top.
- Water changes: Weekly.

- Aquarium set-up: The aquarium should have a tight fitting and secured strong lid.

Behaviour:

Social—Not a good community fish. They are semi-aggressive in temperament.

Reproductive—Large amount of live food and proper water conditions are necessary to induce spawning. The male fish brood the eggs in its mouth. The eggs develop after 50-60 days and produce fry which have egg sac. They feed entirely on their own.

Remarks: It is one of the most amazing and interesting fish which can get very large.

2.5.2 Black arowana

Figure 2.38

Osteoglossum ferreirai Kanazawa, 1966

Common Name:

English—Black arawana, black arowana.

IUCN Red List Status: Least Concern.

Systematic Position:

Subfamily: OSTEOGLOSSINAE

Genus: *Osteoglossum*

Species: *O. ferreirai*

Scientific Name: *Osteoglossum ferreirai* Kanazawa 1966

Description:

- Body is elongated with a tapered tail.
- The juveniles are black with yellow markings down the length of the body, head and the tail.
- At the length of 15 cm, the fish develops a dark iridescent steel grey to blue colouration. The dorsal, caudal and anal fins have yellow and red outlining.
- Adults are very similar to the silver arowana.

Species Details:

- Distribution:

  South America: Negro River basin.
- Habitat: Freshwater, benthopelagic. Inhabits blackwater. It is a sedentary species and during the dry season it lives in backwaters, marginal lagoons and small tributaries. During the high water, it is found in flooded forests.
- Food and feeding: Shrimps, insects, smaller fishes, animals that float on the water surface and bigger animals.
- Maximum length: 90.0 cm TL male/unsexed.
- Life span: 20 years.

Aquarium Aspects:

- Water conditions: Temperature: 24.0- 30.0° C (75.2-86.0° F); pH: 5-7; dGH: 0-214.29 ppm.

Behaviour:

Social—Peaceful fish. As they can jump out of the water to catch prey, they are often called water monkey or monkey fish.

Reproductive—The females spawn at the time of high water. Eggs are quite large about 0.5 inches (1.27 cm) in dimeter. Female fish deposits them on the substrate ready for the male to fertilize. The male mouth brood about 210 eggs and young ones about 7cm length are fully released.

Remarks: A prized aquarium fish with an extreme demand.

2.5.3 Asian bonytongue

Figure 2.39

Scleropages formosus (Muller & Schlegel, 1840)

Common Name:

English—Asian bonytongue, dragonfish.

IUCN Red List Status: Endangered (EN).

Systematic Position:

Subfamily: OSTEOGLOSSINAE

Genus: *Scleropages*

Species: *S. formosus*

Scientific Name: *Scleropages formosus* (Muller & Schlegel, 1840)

Description:

- Body long with large elongated pectoral fins. Dorsal and anal fins located far back on the body.
- Scales large, cycloid.
- With 1 pair of barbels.
- Caudal fin is quite large.
- The mouth is oblique with a very wide gape.
- Lower jaw is prominent and has two barbels at its tip.
- Teeth are present on the jaws, vomer, palatines, pterygoids, and parasphenoid bones and tongue.
- Sexual Dimorphism: Difficult to differentiate but mature females may be plumper than males.

Species Details:

- Distribution:

  Asia: Southern Myanmar to Malay Peninsula and Indonesia, eastern Thailand to Cardamom Range.

- Habitat: Inhabits tannin stained blackwater streams and in forest covered streams including peat adjacent areas.
- Food and feeding: Carnivorous. Young individuals feed on insects at the water surface, adults feed on shrimps, crickets, earthworms, fishes and small frogs. In captivity, high quality meaty food is to be provided. Prepared food includes shrimps, lean pork, frozen fish, koi fry and pelleted food.
- Feeding frequency: Once a day.
- Maximum length: 90.0 cm TL male/unsexed.
- Life span: 8-18 years.

  Some naturally occurring colour varieties of Asian bonytongues are recognized as haplotypes.
- Green—Colour is dark green on the back, silvery or golden-green on its sides. Occurs in Indonesia, Vietnam, Myanmar, Thailand, Cambodia and Malaysia.
- Silver Asian—Silvery or whitish on the ventral side with dark greenish or bluish patches seen through the lateral scales. Occurs in different parts of the Island of Borneo in Indonesia.
- Red-tailed golden—Brilliant metallic gold lateral scales, gill covers, bellies, pectoral and pelvic fin membranes. Back is dark. Occurs in Northern Sumatra, Indonesia.
- Gold cross back, blue Malayan or Bukit Merah Blue—Metallic gold crossing the back completely. Occurs in State of Pahang and the Bukit Merah area in Perak, Peninsular Malaysia.
- Red, super red, blood red or chili red—The back is dark brown; gill covers lateral scales and fin membranes are metallic red with hue varying from gold-tinged to deep red. Occurs in the upper part of the Kapuas River and nearby lakes in Western Borneo, Indonesia.

Aquarium Aspects:

- Minimum tank size: 250 gallons.
- Water conditions: pH: 6.5-7.5; temperature: 24°-30° C (75.2°-86°G F); hardness: 8-18 dGH.

Behaviour:

Social—They are territorial. A tight fitting cover is needed for the tank to prevent escape of this fish. They feed at night.

Reproductive—This arowana reaches sexual maturity after 3-4 years. The females produce large 30-100 eggs. The eggs are fertilized and the arowanas show great parental care with paternal mouthbrooding. The fertilized eggs as well as the larvae are brooded within the male's mouth.

Remarks: These popular aquarium fish are seen as the symbols of good luck and prosperity particularly by Asian people. The resemblance of this fish with the

Chinese dragon is behind the belief. These arowanas are favourite due to positive Feng sui associations with water and the colours red and gold.

International trade in these fishes is controlled under the "Convention on the International Trade in Endangered species of Wild Flora and Fauna (CITES), under which it was placed on Appendix 1, the most restrictive category, in 1975.

High value of this fish as aquarium fish has impact on its conservation. In 1989, CITES started allowing Asian Arowanas to be traded, provided certain criteria were fulfilled such as they were bred in captivity on a fish farm for at least two generations. There are two ways, that the captive-bred arowanas, legal for trade under CITES, are documented. First fish farms provide each buyer with a certificate of authenticity and a birth certificate. Second, each specimen receives an implemented microchip, called a passive integrated transponder, which identifies individual animals (Cheng, 2009; https://en.wikipedia.org/wiki/Asian_arowana,2019). There are more than 150 CITES-registered Asian arowana farms in Singapore, Malaysia, Indonesia and other Southeast Asian countries producing Asian arowanas for commercial purpose.

2.5.4 Suckermouth catfish

Figure 2.40

Hypostomus plecostomus (Linnaeus, 1758)

Common Name(s):

English—Suckermouth catfish, common pleco, algae sucker/eater, pleco, janitor fish.

IUCN Red List Status: Not Evaluated.

Systematic Position:

Order: SILURIFORMES--catfishes

Family: LORICARIIDAE—suckermouth armoured catfishes

Subfamily: HYPOSTOMINAE

Genus: *Hypostomus*

Species: *H. plecostomus*

Scientific Name: *Hypostomus plecostomus* (Linnaeus, 1758)

Description:

- Body short and of robust built.
- Head big, the eyes are small and high set.
- Upper parts of head and body encased in longitudinal 4 rows of scutes.
- Lower surface of head and abdomen naked.
- Caudal peduncle not depressed.
- Dorsal fin spines: 1; dorsal soft rays total: 7; anal spines: 1; anal soft rays: 3-5.
- There is a membrane on the eyes of this fish. This covers the eye during the daytime and it opens during the night.

- Body colour is grey-brown with green tint. It has a pattern made of dark spots and stripes.

Species Details:

- Distribution:

  South America: Guyana coastal drainages.
- Habitat: Tropical waters. It is found in oxygen-rich lotic waters in brackish waters of river estuaries.
- Food and feeding: Omnivorous. Feeds on algae and small crustaceans. In aquarium, spinach, lettuce, cucumber and live foods like earthworms, bloodworms, insect larvae and small crustaceans may be given.
- Feeding frequency: Once in the evening per day.
- Maximum length: 50.0 cm SL male/unsexed.
- Life span: 12-15 years.

Aquarium Aspects:

- Minimum tank size: 80-150 gallons.
- Tank set-up: Driftwood and caves.
- Temperature: 24.0°-26.0° C (75.0°-78.0° F).
- pH: 6.5-8.0.

Behaviour:

Social—Peaceful as young but becomes territorial as adult. It is a bottom dweller, stays in caves during the day. At night, it becomes active. It generally mows on its fins between the tank plants and shelters.

Reproductive—Eggs are deposited on smooth rocks or in depressions. One or both parents guard the eggs. Captive breeding is not yet possible.

Remarks: In an aquarium trade, it is the most popular freshwater catfish and is often purchased for its ability to clean algae from the fish tanks.

2.5.5 Glass catfish

Figure 2.41

Kryptopterus bicirrhis (Valenciennes, 1840)

Common Name(s):

English—Glass catfish.

IUCN Red List Status: Least Concern (LC).

Systematic Position:

Order: SILURIFORMES

Family: SILURIDAE (sheatfishes)

Genus: *Kryptopterus*

Species: *K. bicirrhis*

Scientific Name: *Kryptopterus bicirrhis* (Valenciennes, 1840)

Description:

- Most of its body is not transparent, it is mostly opaque and light greyish. Head part is transparent.
- Body elongate, with dorsal profile arched and there is a nuchal concavity.
- Dorsal fin is rudimentary and pectoral fin length greater than head length.
- The maxillary barbels reaches up to the anal fin.
- Anal soft rays are 55-58.
- Sexual Dimorphism: Adult males appear to be slimmer than females.

Species Details:

- Distribution:

  Asia: Mekong and Chao Phraya basins; Malay Peninsula, Sumatra and Borneo.
- Habitat: Found in large rivers with turbid waters. Inhabits fast flowing water and occurs along the shores, flooded fields, lowland streams to peat adjacent in large school.
- Food and feeding: Omnivore and micro-predator. This diurnal predator feeds on pelagic hemipterans and some small fishes, worms, crustaceans, insects, blood worms, tubifex, white worms and brine shrimp. Accepts small sized granules, all kinds of frozen and prepared food in aquarium.
- Maximum length: 15.0 cm SL male/unsexed.

Aquarium Aspects:

- Minimum tank size: 25 gallons.
- Temperature: 75.0°-81.0° F (20.0°-27.0° C); pH: 6.5-7.5; hardness: 36-268 ppm.
- Light: Poor lighting.
- Water region: Middle.

Behaviour:

Social—It is timid and peaceful fish. This fish should be kept in a group of 10-12 fish. Some peaceful fishes of same size may be kept with this species in the tank.

Reproductive—In the wild, the breeding process is a seasonal one, particularly rainy season. The glass catfish female scatter about several hundreds of eggs on the plants. Captive breeding is a rare event. Simulating the rainy season, this fish may be provoked to breed in the tank.

Remarks: It is a common ornamental fish in freshwater aquariums.

2.5.6 Striped glass catfish

Figure 2.42

Kryptopterus macrocephalus (Bleeker, 1858)

Common Name(s):

English—Striped glass catfish.

IUCN Red List Status: Least Concern (LC).

Systematic Position:

Order: SILURIFORMES

Family: SILURIDAE

Genus: *Kryptopterus*

Species: *K. macrocephalus*

Scientific Name: *Kryptopterus macrocephalus* (Bleeker, 1858)

Description:

- Body is opaque, transparent with either striped or mottled pattern.
- Head is large and is longer than other species of the genus *Kryptopterus*.
- Sexual Dimorphism: Adult males are much slimmer than adult females. The posterior edge of the pectoral fin spine is serrated in males and smooth in females.

Species Details:

- Distribution:

  Asia: Indonesia, Malaysia and Brunei Darussalam.
- Habitat: Inhabits tannin-stained, very acidic peat swamps, shady spots in running waters.
- Food and feeding: Omnivorous, larger fishes may prey smaller ones. Feeds on crustaceans, invertebrates and smaller fishes in nature. In captivity, takes sinking-dried food, live and frozen bloodworm, tubifex and small earthworms.
- Maximum length: 9.7 cm SL male/unsexed.
- Life span: 6-8 years.

Aquarium Aspects:

- Minimum tank size: 90 litres.
- Water conditions: Temperature: 22.0-27.0° C (71.6-80.6° F); pH: 4.0-7.0; hardness: 18-143 ppm, dGH: 10-20.
- Substrate: Should be underlain with at least ½ "of peat, with plenty of driftwood.

Behaviour:

Social—Timid and peaceful in nature, though may predate on small fishes. It is gregarious in nature and form schools.

Reproductive—No report of captive breeding.

Remarks: It is a common freshwater aquarium fish.

2.5.7 Black bullhead

Figure 2.43 A & B

Ameiurus melas (Rafinesque, 1820)

Common Name (s):

English—Black bullhead, black bull head catfish.

IUCN Red List Status: Least Concern (LC).

Systematic Position:

Order: SILURIFORMES

Family: ICTALURIDAE (North American freshwater catfishes)

Genus: *Ameiurus*

Species: *A. melas*

Scientific Name: *Ameiurus melas* (Rafinesque, 1820)

Description:

- Dorsoventrally flattened body with a slightly humped back.
- Very broad, robust head with pigmented barbels.
- Barbels are long and are completely pigmented, nostril whiskers resemble horns.
- The caudal fin is rounded and occasionally has a pale vertical stripe at its base.
- Anal fin rays: 19-25; caudal fin rays: 15-18.
- Pectoral spine is nearly smooth.
- It is black or dark brown on the dorsal side of its body and yellow or white on the ventral side. Fins are black pigmented. The colour varies greatly depending on location and water characters.
- The anal fin has a grey base and tail has a pale bar.

Species Details:

- Distribution:

  North America: Great Lakes to northern Mexico.
- Habitat: Inhabits pools, backwaters, in slow moving water currents over soft substrate in creeks and small to large rivers. Found in impoundments, oxbows and ponds.
- Food and feeding: Carnivorous. This nocturnal feeder as young consume immature insects, leeches and crustaceans whereas adults feed on clams, snails, plant material and fishes.
- Maximum length: 66.0 cm TL male/unsexed.
- Life span: 5- 10 years.

Aquarium Aspects:

- Water conditions: Temperature: 8.0°-30.0°C; pH range: 6.5-8.0; dGH range: 4-25.

Behaviour:

Social—Adults are very inactive during daylight hours. Feeding usually takes place after dark only.

Reproductive—Black bull heads breed once yearly. They breed from May to July. They are monogamous in breeding habit. The nest is prepared by the female by clearing debris, gravel and silt found on the bottom. The male fish wraps the caudal fin around the head of the female and then the female releases her eggs. In an hour, the spawning can take place up to five times. The female fans the eggs in between spawning. The eggs are guarded by the parents. Eggs hatch in 4 to 10 days. Free swimming fry remain close to the adult male for around two weeks.

Remarks: Common in freshwater public aquariums.

2.5.8 Neon tetra

Figure 2.44

Paracheirodon innesi (Myers,1936)

Common Name(s):

English—Neon tetra.

IUCN Red List Status: Not Evaluated.

Systematic Position:

Order: CHARACIFORMES (Characins)

Family: CHARACIDAE (Characins)

Genus: *Paracheirodon*

Species: *P. innesi*

Scientific Name: *Paracheirodon innesi* (Myers, 1936)

Description:

- Small and slim fish.
- Body with a light-blue back and a silver-white belly.
- There is an iridescent blue horizontal stripe along each side of the body from the anterior most part to the base of the adipose fin.
- An iridescent red stripe is present at the middle of the body that extends posteriorly to the base of the caudal fin.
- Rest of the body is transparent.
- The blue and red colour is visible during the day, whereas at night they become grey or black.
- Sexual Dimorphism: The female fish has a bent iridescent stripe and the male has a straight stripe. The female has a slightly larger belly than the male.

Species Details:

- Distribution:

  South America: Blackwater or clearwater stream tributaries of the Solimoes River.

- Habitat: Found in acidic blackwater streams, and also in transparent clearwater streams. Inhabits slow-moving tributaries of main rivers. Found in blackwaters beneath dense forest canopies. Inhabit middle water layers.
- Food and feeding: Omnivorous. They eat all types of small or finely ground foods, live, fresh and flakes. Good quality flake and pellets should be provided. Fishes, shrimps, worms and meaty foods are taken.
- Feeding frequency: Several feedings per day.
- Maximum length: 2.2 cm SL male/unsexed.
- Life span: 5-10 years.

Aquarium Aspects:

- Minimum tank size: 10 gallons (38 litres).
- Suitable for Nano tank: Yes.
- Substrate type: Any.
- Lighting needs: Moderate to normal lighting.
- Temperature: 68.0-77.0° F (20.0 to 25.0° C).
- Breeding temperature: 75.0° F.
- Range pH: 5.0-8.0.
- Hardness range: 3-25 dGH.
- Water movement: Moderate.
- Water region: All.
- Water changes: Bi-weekly.

Behaviour:

Social-Peaceful. It is fairly active and social fish. Can be kept in a small school of at least 6.

Reproductive—Neon tetras have been bred in captivity. These egg layers breed at the age of 9 months. It is better to condition the fish prior to breeding, with small live food. They can breed as pairs or in a school. They spawn early in the morning. About 130 clear eggs are laid. The eggs get scattered and should be removed to a separate tank immediately. The eggs will hatch in about 24 hours with the fry becoming free-swimming after 3-4 days.

Remarks: This fish has been in the hobby for a long time and is considered as one of the most popular fishes of the ornamental fish trade.

2.5.9 Silver dollar

Figure 2.45

Metynnis argenteus Ahl, 1923

Common Name(s):

English—Silver dollar.

IUCN Red List Status: Not Evaluated.

Systematic Position:

Order: CHARACIFORMES

Family: SERRASALMIDAE

Genus: *Metynnis*

Species: *M. argenteus*

Scientific Name: *Metynnis argenteus* Ahl, 1923

Description:

- It is a deep-bodied fish with a round, almost disc like shape, and is very compressed laterally.
- The silver colour of the body has slight green and blue tints with some hints of red.
- Sometimes red colouration is seen on the fins and around the throat.
- Sexual Dimorphism: The anal fin of the male is edged in red, sometimes, they have small dots on their sides. Red colouration of the fins deepens and contrasting black colours appear. Mature females are longer and fuller in the belly than males. During breeding, males develop two large very prominent black spots, one above the other, just behind the base of the pectoral fins.

Species Details:

- Distribution:

  South America, Guyana, the Amazon east of Rio Negro to Paraguay. It is endemic to the Tapajos River Basin in Brazil.
- Habitat: Tropical climate in the sides of weedy rivers, spends most of its time in the mid to upper level of water.
- Food and feeding habit: They are primarily herbivorous. They eat vegetable matter, including a variety of plants, lettuces, cress, chickweed and large vegetable flakes, spirulina. Sometimes bloodworms and brine shrimps are also eaten. In aquaria, they take flakes, pellets, lettuce, spinach, peas, carrots, cucumbers, fruits and boiled potatoes.
- Feeding frequency: Several feedings per day.
- Maximum length: 14.0 cm SL male/unsexed.
- Life span: 10 years or longer.

Aquarium Aspects:

- Minimum tank size: 75 gallons.
- Substrate type: Any.
- Lighting needs: Low-subdued lighting.
- Temperature: 75.0°-82.0° F (23.9° C to 27.8° C).
- Breeding temperature: 79°-82° F (26°-28° C).

- Range pH: 5.5-7.5; breeding pH 6.0-7.0.
- Hardness range: 4-18 dGH.
- Water movement: Moderate.
- Water region: Middle.

Behaviour:

Social—It is a very peaceful and shoaling species. These fishes are peaceful enough and can be kept with much smaller fish than themselves and can also cope in the company of much larger fish.

Reproductive—They are prolific, group-spawning egg- scatterers. Eggs are scattered among fine-leaved plants as well as directly on the substrates. These fishes easily breed in captivity. The eggs are not adhesive and fall to the bottom of the tank. About 2000 eggs are produced. The parents neither look after the eggs nor eat them. The eggs hatch in about 3 days.

Remarks: Silver dollar is a popular aquarium fish in the market.

2.5.10 Black pacu

Figure 2.46

Colossoma macropomum (Cuvier, 1816)

Common Name(s)

English—Pacu, Black Pacu, Tambaqui, Black-finned Pacu, Cachama.

IUCN Red List Status: Not Evaluated.

Systematic Position:

Order: CHARACIFORMES

Family: SERRASALMIDAE

Genus: *Colossoma*

Species: *C. macropomum*

Scientific Name: Colossoma macropomum (Cuvier, 1816)

Description:

- Very large and rounded in shape with small eyes and mouth.
- The body of the adult fish is very deep and laterally compressed with a slight arch on the back.
- Body is covered with iridescent scales which are dark silver grey in colour with hues of red and blues, fins are black. Lower half of body are blackish.
- This fish has squarer, straighter teeth.
- Sexual Dimorphism: Mature males are more brightly coloured than females.

  The dorsal fin of the male is more pointed than that of the females and the anal fin of the male has a toothed edge.

Species Details:

- Distribution:

  South America: Amazon and Orinoco basins as wild form; pisciculture form largely distributed in South America.
- Habitat: Adults stay in flooded forests during the first 5 months of flooding. Young and juveniles live in black waters of flood plains until their sexual maturity.
- Food and feeding: This fish is an omnivorous. Adults feed on fruits, grains, insects, detritus and plant matter. In captivity, they take a wide variety of food, pellets, live foods, vegetable foods and meaty foods. Juveniles consume zooplankton, insects, snails and decaying plants.
- Feeding frequency: Several feedings per day.
- Maximum length: 108 cm TL male/ unsexed.
- Life span: 20-25 years.

Aquarium Aspects:

- Minimum tank size: 634 gallons (2400 litres).
- Suitable for Nano tank: No.
- Lighting arrangements: Moderate -normal.
- Temperature: 74.0° to 82.0° F (23.3° to 27.8° C).
- Range pH: 5.0-7.8.
- Hardness range: 2-15 dGH.
- Brackish: No.
- Water movement: Moderate.
- Water region: Middle.
- Water changes: Bi-weekly.

Behaviour:

Social—They are peaceful and remain solitary and not aggressive. Its tankmates should be neither too aggressive nor too timid.

Breeding—Breeding is difficult. Captive breeding is not successful in the home aquarium. In fisheries, artificial breeding is possible by the use of hormone.

Remarks: It is a favourite pet of public aquariums.

2.5.11 Red-bellied piranha

Figure 2.47

Pygocentrus nattereri Kner, 1858

Common Name(s):

English—Red-bellied piranha, Red piranha.

IUCN Red List Status: Not Evaluated.

Systematic Position:

Order: CHARACIFORMES

Family: SERRASALMIDAE

Genus: *Pygocentrus*

Species: *P. nattereri*

Scientific Name: *Pygocentrus nattereri* Kner, 1858

Description:

- The body is grey with silver-flecked scales.
- Have a reddish tinge to the belly when fully grown, the juveniles are silver coloured with darker spots.
- Blackish spots are sometimes seen behind the gills and the anal fin is usually black at the base.
- The pectoral and pelvic fins may vary from red to orange.
- Sexual Dimorphism: Females have slightly deeper red colour on their bellies than the males.

Species Details:

- Distribution:

  South America: Amazon River basin, Paraguay- Parana River basin, north eastern Brazilian coastal rivers and Essequibo River basin.
- Habitat: In creeks and interconnected ponds in Mato Grosso, Brazil, in Rio Machado and Rio Negro.
- Food and feeding: They are carnivorous, occasionally omnivorous. In the wild, they feed on various food items like fish, whole and in pieces, fish fins and scales, insects, snails and plants. In captivity, they take whole dead fish, variety of other meaty foods such as prawn, mussels and fish flesh. They also feed on live foods such as feeder fish, earthworms, and silver shrimps. They do not take flake foods or pellet foods. Though little vegetable food may be taken, they take meaty food mostly.
- Feeding frequency: Daily.
- Maximum length: 50.0 cm SL male/unsexed.
- Life span: 10-20 years.

Aquarium Aspects:

- Minimum tank size: 40 gallons (151 litres) for single fish; for a group –100 gallons or more.
- Suitable for Nano tank: No.
- Substrate type: Small gravel.
- Lighting needs: Low-subdued lighting.
- Temperature: 74.0° to 82.0° F (23.3° to 27.8° C).
- Range pH: 6.5-7.5.
- Hardness Range: 4-18 dGH.

- Brackish: No.
- Water movement: Moderate.
- Water Region: All.
- Water Changes: Bi-weekly. 30-50 % water changes every other week.

Behaviour:

Social—It is aggressive. Sturdy tankmates may survive with this species, still it is best to keep it in a single species tank. This ferocious predator is not a candidate for a community aquarium.

Breeding—Though breeding in captivity is difficult, it has been bred in home aquariums. In the wild, spawning usually takes place in the wet season. A bowl-shaped spawning pit is dug by the pair. After laying eggs, the female swims away, the male guards the nest. The eggs remain in clusters and attach to the bottom vegetation. The eggs should be raised in a separate tank. Free-swimming fry are fed with newly hatched brine shrimp. In the wild they take macroscopic plants. Fry are cannibalistic.

Remarks: These fishes are the most common piranha for sale, but they are no longer legal to keep in all states.

CICHLIDS

2.5.12 Kenyi cichlid

Figure 2.48

Maylandia lombardoi (Burgess, 1977)

Common Name(s):

English—Kenyi cichlid, lombardoi mbuna, Kenyi mbuna, Kennyi mbuna.

IUCN Red List Status: Least Concern (LC).

Systematic Position:

Order: PERCIFORMES

Family: CICHLIDAE (cichlids)

Subfamily: PSEUDOCRENILABRINAE

Genus: *Maylandia*

Species: *M. lombardoi*

Scientific Name: *Maylandia lombardoi* (Burgess, 1977)

Description:

- Elongated and muscular body.
- A well-developed pharyngeal set of teeth that are in the throat, along with their regular teeth.
- Spiny rays are present in the back parts of the anal, dorsal, pectoral and pelvic fins.
- One nostril is present on each side.
- There are faded vertical bars and at the dorsal fins that have hints of colour.

- Juveniles have same colouration as females, they are smaller only.
- Sexual Dimorphism: The male fish has a golden yellow body with faint vertical bars and has egg spots on the anal fin along with a lighter coloured stomach area. The females are with several blue- black vertical bars/ bands extending into the dorsal fin.

Species Details:

- Distribution:

  East Africa: Mbenji Island, Lake Malawi in East Africa.
- Habitat: Rocky shores.
- Food and feeding: Omnivorous. Feeds on algae, plankton, aufwuchs, tiny benthic organisms such as crustaceans, invertebrates and some zooplanktons. In captivity, they are given frozen or live brine shrimp, *Mysis* larvae, high quality flakes, pellets, spirulina and bloodworms, live feeder guppies and daphnia. These are supplemented with vegetable food such as spinach, zucchini and peas.
- Feeding frequency: Daily. Several small feedings several times a day instead of one large feeding.
- Maximum length: 8.7 cm SL male/unsexed.
- Life span: 10 years.

Aquarium Aspects:

- Minimum tank size: 50 gallons.
- Suitable for Nano tank: No.
- Substrate type: Any.
- Lighting needs: Moderate-normal lighting.
- Temperature: 76.0° to 82.0° F (24.4° to 27.8° C).
- Range pH: 7.8-8.6.
- Hardness range: 10-15 dGH.
- Water movement: Moderate.
- Water region: Middle.
- Water changes: Weekly.

Behaviour:

Social-It is not a community fish. Generally kept with one male and several females in a large tank. They are to be kept in a species- specific tank. If kept with other large Mbunas, a very large aquarium should be used. They are also considered as aggressive.

Reproductive—The Kenyi cichlid is a ovophile mouth brooder. The female after laying eggs, pick them up in her mouth to be fertilized. The female can carry about 50 eggs for 3 weeks. Enough hiding places should be provided for the female for successful breeding. They breed in captivity.

Remarks: These fish are sold for the aquarium hobby.

2.5.13 Frontosa cichlid

Figure 2.49

Cyphotilapia frontosa (Boulenger, 1906)

Common Name(s):

English—Frontosa, Humphead cichlid.

IUCN Red List Status: Least concern.

Systematic Position:

Family: CICHLIDAE

Subfamily: PSEUDOCRENILABRINAE

Genus: *Cyphotilapia*

Species: *C. frontosa*

Scientific Name: *Cyphotilapia frontosa* (Boulenger, 1906)

Description:

- This fish has a heavy body with large rounded head.
- A wide underslung mouth is present.
- Both the sexes have the cranial hump on the forehead.
- The nuchal hump is absent on juveniles.
- The body has six or seven broad black stripes on a white background.
- The fins are white or light blue in colour.
- Sexual Dimorphism: The mature male is larger and there may be a more pronounced cranial hump than the female.
- Some geographic variants of frontosa cichlid are:
- Burundi six-stripe frontosa–Body is very high and a large nuchal hump is present. Good blue colouration and nice banding are present.
- Zaire Blue Frontosa—Intense blue colouration of the body with the band on the head running between the eyes and down the gill plate.
- Tanzanian six-stripe Frontosa—It looks very similar to the Burundi six-stripe with a large nuchal hump. In cross-breeding of the two, the offspring resulted are with slanted bands.
- Tanzanian seven-stripe Frontosa or Frontosa "Kigoma"—A seventh band is present that is like a patch on the face than an actual band. A large nuchal hump is present which is blue and the dorsal fin has gold accents. This fish is in great demand in the trade.
- Kipili Frontosa—It is somewhat blue but not as much as other varieties. It is very rarely exported.
- Zmbian blue Frontosa or Blue Face Frontosa—There is banding with some blue on the face. The band on the head fades just above the eye.

- Samazi Frontosa—It has a high body, and colour difference is present among sexes. The male has nice blues whereas the female has none or very little.
- Kavalla Frontosa—It has six nice bands and blue accents but often gets dark with the mood of the fish. The dorsal fin is yellow with a sheen of multiple colours. This is an extremely rare variety.

Species Details:

- Distribution:

  Africa: endemic to Lake Tanganyika, distributed in the northern half of the lake.
- Habitat: Rock-dwellers. They live off shore on the sloping lake bottom. They live in groups at depths of 35 to 170 feet or more, dwell among rocks.
- Food and feeding: Carnivorous. They eat all types of live foods. They feed on fish, shellfish such as, snails, molluscs. In captivity, they take pellets, variety of meaty foods such as feeder fish, worms, crustaceans (Mysis, shrimp and krill), frozen foods (brine shrimp), vitamins and supplements are to be added. Flake foods are not consumed.
- Feeding frequency: Several feedings per day. Instead of a single large feeding, several small feedings per day are given.
- Maximum length: 33.0 cm TL male/unsexed.
- Life span: 25 years.

Aquarium Aspects:

- Minimum tank size: 70 gallons for a single specimen. For a group 125-200 gallons.
- Suitable for Nano tank: No.
- Substrate type: Sand/gravel.
- Lighting needs: Moderate-normal lighting.
- Temperature: 74.0° to 79.0° F (23.3° to 26.1° C).
- Breeding temperature: 77.0° F.
- Range pH: 8.0-8.5.
- Hardness Range: 12-15 dGH.
- Water movement: Moderate.
- Water region: All.
- Water changes: Weekly.

Behaviour:

Social—These fishes are gregarious and prefer to remain in groups. Either 8-12 individuals or a group of one male with three or more females are to be kept. They are fairly non-aggressive community fish and can be kept with other durable fish.

Reproductive—Egg-layer fish, that form matriarchal families. Monogamous, one male spawns with several females. Mouth brooder female spawns in cave in the wild. In captivity, they spawn on the substrate. Captive breeding needs some criteria such as 100 gallons or more -sized aquaria, alkaline, medium hard water with a pH about 8, about 10 GH, and a temperature between 77°-82.4° F (25°-28° C). About 50 eggs are deposited by the females, in a sand pit between the rocks. After fertilization, the eggs hatch in about 3 days. The male defends the territory and the female takes the eggs into her mouth. Female takes care of the fry between 4 to 6 weeks.

Remarks: Very popular aquarium fish. This handsome fish is highly regarded by cichlid keepers.

2.5.14 Jaguar cichlid

Figure 2.50

Parachromis managuensis (Gunther, 1867)

Common Name(s):

English—Jaguar cichlid, managuense cichlid, managua cichlid, Aztec cichlid, spotted guapote, jaguar guapote.

IUCN Red List Status: Not Evaluated.

Systematic Position:

Order: CICHLIFORMES

Family: CICHLIDAE

Subfamily: CICHLASOMATINAE

Genus: *Parachromis*

Species: *P. managuensis*

Scientific Name: *Parachromis managuensis* (Gunther, 1867)

Description:

- Elongated, oval shaped body.
- The base colour of the body of this fish is silvery mixed with a light blue-green to light purple hue. Head is with light tan-yellow to burgundy-red colour.
- Mature males have jaguar-patterning over the body and fins. Females may or may not have bars while also forming a line of several large black dots across the body.
- Young fish both the male and female have several dark bars extending from the top of the back terminating around the middle.
- Two dark bars are present just behind their eye.
- Sexual Dimorphism: The male fish is larger in size and have more vertical broken bars when young. Bars disappear with age. The female fish have more prominent dark broken horizontal bars, and have thinner vertical bars.

Species Details:

- Distribution:

  On the Atlantic slope of Central America from the Ulna River in Honduras to the Matina River in Costa Rica.
- Habitat: Inhabit turbid waters with muddy bottoms and are found in warm, low oxygenated water in springs, ponds and lakes.
- Food and feeding: Carnivorous and in the wild feeds on other smaller fish and large invertebrates. In aquarium, they accept small fishes, insects, earthworms and tadpoles, freeze dried and dry foods. Flake food, pellets are accepted.
- Feeding frequency: Daily.
- Maximum length: 55.0 cm TL male/unsexed.
- Life span: 15 years.

Aquarium Aspects:

- Minimum tank size: 125 gallons.
- Substrate type: Small gravel.
- Lighting needs: Moderate-normal lighting.
- Temperature: 75.0°-82.0° F (23.9° to 27.8° C).
- Breeding temperature: 82.0° F.
- Range pH: 7.0-8.7.
- Hardness range: 10-15 dGH.
- Water movements: Moderate.
- Water region: All.
- Water changes: Bi-weekly.

Behaviour:

Social—Jaguar cichlid is a predatory, territorial and aggressive fish and cannot be considered as a community fish. It can be kept with other large Central and South American cichlids.

Reproductive—These fish can be bred in captivity. Before spawning male and female fish dig an area behind a rock. After hatching, eggs are deposited in that depression. Both the parents guard the eggs and the resulting fry.

Remarks: This large and beautiful cichlid is a popular pet with the hobbyist.

2.5.15 Convict cichlid

Figure 2.51

Amatitlania nigrofasciata (Gunther, 1867)

Common Name(s):

English—Convict cichlid, zebra cichlid.

IUCN Red List Status: Not Evaluated.

Systematic Position:

Order: CICHLIFORMES

Family: CICHLIDAE

Subfamily: CICHLASOMATINAE

Genus: *Amatitlania*

Species: *A. nigrofasciata*

Scientific Name: *Amatitlania nigrofasciata* (Gunther, 1867)

Description:

- Body blue-grey with about 8-9 vertical dark stripes on it. There is a dark blotch on the operculum.
- Eyes dark coloured.
- Sexual Dimorphism: Males are larger (6.3 to 6.6 cm) than females. Body colour grey with light black stripes. They have more pointed ventral, dorsal and anal fins which often extend into filaments. Females are smaller (4.2 to 5.5 cm) than males. They have more intense black bands across the body and pink to orange colouration in the ventral region and on the dorsal fin.
- Some strains have been developed through selective breeding. The leucistic strain developed lacks the dark barring of the wild type. White convicts, pink convicts, gold convicts and *A. nigrofasciata* "Kongo" have been developed. This leucistic colouration is caused by a mutation in an autosomal gene and is recessively inherited.

Species Details:

- Distribution:

  Central America: Pacific slope, from Riosucio, El Salvador to Rio Suchiate, Guatemala; Atlantic Slope, from Rio Patuca, Honduras to Rio Jutiapa, Guatemala.
- Habitat: Found in flowing water from small creeks and streams to the shallows of large and fast flowing rivers. Inhabits rocky habitats and in various creeks and crevices, among roots and debris. Also found in warm pools of springs and their effluents.
- Food and feeding: Omnivorous. Foods include worms, crustaceans, insects, fish and plant matter. It readily takes a wide variety of food such as vegetable- based food with spirulina, blanched lettuce or other veggies and beef heart.
- Feeding frequency: Several feedings per day.
- Maximum length: 10.0 cm SL male/unsexed.
- Life span: 8-10 years.

Aquarium Aspects:

- Minimum tank size: 20 gallons to 50 gallons.

- Substrate type: Sand.
- Lighting needs: moderate-normal lighting.
- Temperature: 74.0°-82.0° F (23.3° to 27.8° C).
- Range pH: 6.0-8.0.
- Hardness range: 6-8 GH.
- Water movement: Moderate.
- Water region: All.
- Water changes: Bi-weekly.

Behaviour:

Social—It is an aggressive fish which can be kept with either fishes of the same size or larger fishes of the same temperament. They become very aggressive at the time of breeding.

Reproductive—They breed readily in captivity. Both the male and female clean a place and dig a depression in the substrate near a rock or cave. About 20-40 eggs are laid by the female inside the depression and the male will fertilize them. The process will continue till the fertilization of 100-300 eggs. The female will fan the eggs, whereas the male guards them. The young hatch in about 48-72 hours depending on the temperature and pH of the tank water.

Remarks: These small and beautiful fish are very easy to keep.

2.5.16 Severum

Golden severum Figure 2.52

Heros severus Heckel, 1840

Common Name (s): Golden severum, gold severum, banded cichlid, severum cichlid.

IUCN Red List Status: Not Evaluated.

Systematic Position:

Family: CICHLIDAE

Subfamily: CICHLASOMATINAE

Genus: *Heros*

Species: *H. severus*

Scientific Name: *Heros severus* Heckel, 1840

Description:

- It has high, oval shaped and laterally compressed body, with pointed anal and dorsal fins.
- The captive bred colour morph of the severum cichlid, the golden severum has pale yellow or gold-coloured body.
- Anal, pelvic and pectoral fins are all yellow.
- The tail and dorsal fin is generally white with yellow specks.

- Eyes are yellow.
- The original severum cichlid has a greenish body with yellowish gold body.
- Juveniles have eight dark and prominent black vertical bands along the sides, which get faded in the adults.
- Well-developed pharyngeal teeth on the throat, along with their regular teeth.
- Spiny rays on the back parts of the dorsal, anal, pectoral and pelvic fins.
- One nostril on each side.
- In the aquarium hobby, there are more colourful severums. These are all captive bred and through selective breeding there are variants like Golden Severum, Red Severum, Green Severum, Yellow Severum.
- Sexual Dimorphism: The males are larger in size and have more pointed anal and dorsal fins. Mature males if well fed may develop a nuchal hump. Some males may develop markings on gill area. The female fish has a dark spot on the dorsal and lacks patterning on the head.

Species Details:

- Distribution:

  South America: Orinoco River basin, in the upper Orinoco River drainage in Colombia and Venezuela. Amazon River basin, in the upper Negro River basin.
- Habitat: Rivers, lakes, flooded areas with vegetation under water.
- Food and feeding: Omnivorous fish. They eat earthworms, bloodworms, mealworms, marine crustaceans. In captivity, they eat pellets, flakes, live foods, vegetable and meaty foods. Vitamins and supplements are to be added.
- Feeding frequency: Several feedings per day.
- Maximum length: 20.0 cm SL male/unsexed.
- Life span: 10 years.

Aquarium Aspects:

- Minimum tank size: 45 gallons for a single fish, 100 gallons for a pair.
- Substrate: Sand.
- Lighting needs: Low-subdued lighting.
- Temperature: 74.0° to 84.0° F (23.3° to 28.9° C).
- Breeding temperature: 78.8°-80.6° F (26°-27° C).
- Range pH: 6.0-6.5.
- Hardness range: 4-6 dGH.
- Water movement: Moderate.
- Water region: Middle.

- Water changes: Bi-weekly.

Behaviour:

Social—They are more or less peaceful and less aggressive though require enough space if kept with other fish of the same size and temperament.

Reproductive—They are open breeders. Once the eggs hatch, they are in the pre-swimming larval stage. The parents take them in their mouth. As they become free swimming, they are allowed to hunt for foods like freshly hatched brine shrimp, ground up flake or pellet food, micro worms and *Daphnia*. They are cared for up to 6 weeks.

Remarks—It is a beautiful cichlid derived from the common and popular severum cichlid.

2.5.17 Green terror

Figure 2.53 A & B

Andinoacara rivulatus (Gunther, 1860)

Common Name(s):

English—Green terror, gold saum.

IUCN Red List Status: Not Evaluated.

Systematic Position:

Order: CICHLIFORMES

Family: CICHLIDAE

Subclass: CICHLASOMATINAE

Genus: *Andinoacara*

Species: *A. rivulatus*

Scientific Name: *Andinoacara rivulatus* (Gunther, 1860)

Description:

- It is deep-bodied fish with a prominent forehead. Pointed dorsal and anal fins are present.
- This fish is marked with various striping on its body and face.
- Body with an overall green and blue metallic sheen. In addition, they have other colours like purple, pink, red and / or electric blue.
- Adults have white or gold-orange edging to the tail and dorsal fins.
- Juveniles lack the bright iridescent blue, green and orange colouration and long flowing fins of adult fish.
- Sexual Dimorphism: Male fish of green terror grow larger than females and may reach lengths of up to 30 cm (12 inches). Females are sexually mature at around 12 cm (4.7 inches). The adult males are the most colourful of this species. Have a brilliant red edging on their tail fin. Males have blue anal fins whereas females have green anal fins. The male has a green and blue metallic sheen, a blue anal fin, and a red band at the edge of the anal

fin. Females do not have a red band along the edge of its tail fin. Mature males will develop a rounded hump on their heads. Females become more aggressive than the males during spawning.

Species Details:

- Distribution:

  South America: Pacific slope from the Esmeraldas River in Ecuador to the Tumbes River in Peru.
- Habitat: In small and larger coastal streams.
- Food and feeding: It is an omnivorous feeder and like to eat all types of food. It eats both the live and frozen blood worm, tubifex, white worms, brine shrimps, *Gammarus*, crickets, worms, fish fillet, prawns, mussels, meat, feeder fish like goldfish and other rich feed, flakes and pellet foods.
- Feeding frequency: Several feedings per day.
- Maximum length: 40.0 cm SL; in aquarium: males 25-31 cm; females 18-23 cm.
- Life span: 10 years.

Aquarium Aspects:

- Minimum tank size: 40 gallons (151 litres)- 40 gallons for a single fish and 70 gallons for a pair.
- Substrate type: Large gravel.
- Lighting needs: Moderate -normal lighting.
- Temperature: 70.0°-79.0° F (25°-26° C).
- Breeding temperature: 77.0°-79.0° F (25°-26° C).
- Range pH: 6.5-8.0.
- Hardness range: 5-13 dGH.
- Water movement: Moderate.
- Water region: All.
- Water changes: Bi-weekly.

Behaviour:

Social—It is highly to moderately aggressive cichlid that can be kept with other large South American cichlids. The female green terror is more aggressive than the males and should be kept singly unless breeding. They are peaceful depending on their personality and the environment in which they grow up.

Breeding—This fish is an egg layer and open breeder, the female deposits the eggs and the male fertilizes them in the open water. The parents clean a flat rock for spawning and lay up to 400 eggs. The fry hatch in 3-4 days and are free swimming by 11 days. During this time the parents are attentive and protective of their young.

Remarks: A favourite fish for the public aquaria.

2.5.18 Blood parrot cichlid

Figure 2.54

These were the earliest defined type of cichlid hybrid. This hybrid fish was first created in Taiwan, in the mid to late 1980's. It is produced from pairing of Midas Cichlid (*Amphilophus citrinellus*) or Red Devil (*Amphilophus labiatus*) with the Redhead Cichlid (*Vieja synspilum*).

Common Name (s):

English—Blood parrot cichlid, parrot cichlid, Red blood parrot, Red parrot cichlid, Red parrot fish.

Systematic Position:

Order: CICHLIFORMES

Family: CICHLIDAE

Description:

- A beak-shaped mouth, an abnormal spine and an occasionally missing caudal fin.
- Have a round body, a beak shaped head with an up-side down triangle mouth.
- They are found in bright orange, red, yellow, brown and tan colours.
- Other colours such as green, blue and purple are produced by dying the fish. This unfortunately shortens life expectancy.
- Sexual Dimorphism not observed.

Details:

- Food and feeding: They consume pelleted foods with blood worm and shrimps. Pellets should be of smaller size. They may be given "Hickari Cichlid", "New life Spectrum cichlid"/ "Cichlasoma Formula 1-3 mm" depending on the size of the fish. Pellets may be soaked in water.
- Maximum length: 15-24 cm.
- Life span: 5-15 years.

Aquarium Aspects:

- Minimum tank size: 30 gallons for one, 10 more gallons for each additional Blood Parrot.
- Water condition: Temperature: 76.0°-86.0° F (24.0°-30.0° C); pH 6.0-8.5.

Behaviour:

Social—When young, they are very shy and prefer to hide under hiding places. They are territorial as adults and not typically suitable for small community fish.

Reproductive—At the size of 5.0-7.5 cm, they start to pair off and breed. All most all males are sterile. Females can cross with any male Central American Cichlid or any Hybrid Central American Cichlid and the eggs are viable.

Remarks: These fishes are the victims of cruel cosmetic surgery where the fish is dyed or mutilated to enhance its beauty. This causes great shortening of their life span.

2.5.19 King Kong parrot Figure 2.55

King Kong Parrot represents an early stage in the transition to Flowerhorn breeding. Man- made hybrid fish created with the genes of South American Cichlids.

Description:

- It is longer, with more oval shaped body than Blood Parrots and has less-protruding head, with a reddish orange colour.
- Dorsal fin is shorter than the anal fin.
- Have normal working mouths.
- Generally, shows deeper colours than Blood Parrots and usually grow larger than Blood Parrots.
- Red Mammon are the highest grade of King Kong Parrot available.
- They have longer bodies.
- Deeper orange/red colour.
- Sunken eyes and large nuchal hump or "kok".

Details:

- Food and feeding: Omnivorous. Eat all most all prepared and frozen foods. "Kikari Cichlid" or "New Life Spectrum Cichlid Formula 1-3 mm" depending on the size of the fish.
- Maximum length: 15-30 cm

Aquarium Aspects:

- Minimum tank size: Kong Kong parrot-- 55 gallons; Red Mammon Parrot—75 gallon.
- Water conditions: Temperature—76.0°-86.0° F (24.0°-30.0° C); pH—6.0-8.5.

Behaviour:

Social—They cannot be kept with other small community fishes.

Reproductive—All male King Kong Parrots are sterile. Females can cross with any male Central American Cichlid or any Hybrid Central American Cichlid and the eggs are viable just like with female Blood Parrots.

Remarks: They are popular public aquarium fish.

2.5.20 Flowerhorn cichlid Figure 2.56A

It is a man-made cichlid developed from the deliberate cross breeding of many types of cichlids by Malaysian fish keepers. The flowerhorn has been carefully bred and blood lines meticulously not crossed in order to produce a healthy fish.

Description:

- It has a big bulky body.

- Unpaired fins and bright coloured head protuberance or kok, is termed as nuchal hump.
- Its colouration and patterning are fluid and will change and evolve until it becomes a fully mature adult.
- Sexual Dimorphism: The male fish has a pronounced hump, in the female it is small and unpronounced.
- Flowerhorn breeding dates back to 1993. In 1995, the blood parrots were crossbred with the Human Face Red God of Fortune, producing a new breed called the Five-Colours God of Fortune. Seven-colours Blue Fiery Mouth was crossbred with the Jin gang Blood Parrots. This crossbreeding led to the first generation of Hua luo han flowerhorn hybrids. Subsequent flowerhorn introductions followed then.
- Varieties: The luo hans are the original flowerhorn hybrid stock. The four main derived varieties are Zhen Zhu, Golden Monkey, Kamfa, and the golden base group, which includes Faders and the Golden Trimac.
- Golden Monkey: It is the type of flowerhorn which is an original luo han-based fish. There are hints of red on the body, while their fins tend to be red.
- Kamfa: Origin directly from the luo han, characteristics are fantail, a water-coloured head bump, smaller lips, white or yellow eyes, sunken eyes. The heads are blue, grey, or purple. They tend to have a large squarish body, smaller lips than related strains. They have head flowers.
- The King Kamfa: It has an intense black "Flower row" and have stronger white pearling.
- The Kamfamalau is a hybrid of Kamfa and malau fish.
- Zhen Zhu: It has a rounded tail, large mouth, prominent head flower.
- The **Super Red Dragon Flowerhorn** (Figure 2.56 B): This fish is bright red. It has a large overwhelmingly red hump on the head. It has a black lateral stripe. The fish may have white speckles on the body. The fins are gold or yellow tipped with black. They often have red eyes.
- The **Thai Silk Flowerhorn** (Figure 2.56 C): It is one of the newest strains of flowerhorn cichlids. It has metallic blue or white colour metallic grey body with white patterning. The fins may be a darker shade of the same colour.

Details:

- Food and feeding: Carnivorous. They eat protein- based food, flakes, pellets, frozen foods, live foods like diced shrimp, diced earthworms, tubifex, bloodworms. Live foods and meaty foods mostly, vegetables also.
- Feeding frequency: Several times a day.
- Maximum length: 40.64 cm.

- Life span: 10-12 years.

Aquarium Aspects:

- Minimum tank size: 55 gallons.
- Substrate type: Sand/ Gravel Mix.
- Lighting needs: Moderate- normal lighting.
- Temperature: 80.0°-89.0° F (26.7° to 31.7° C).
- Breeding temperature: 82.0° F with a pH of 7.0.
- Range pH: 6.5-7.8.
- Hardness range: 9-20 dGH.
- Water movement: Moderate.
- Water region: All.
- Water changes: Bi-weekly.

Behaviour:

Social—They are not community fish as are very territorial and aggressive. It is best to keep them alone, as a pair kept in a large tank and with other large and aggressive tankmates in a very large tank.

Reproductive—Flowerhorn males are fertile. The females are to be provided with plenty of places to hide so that the male cannot see her. Directing the waters to flow from the male's side to the female's side will help the male fertilize the eggs. Parents guard their eggs. Fry are large and easy to care for.

Remarks: Unique and large cichlid flower. They are exclusively domestic and only found in aquariums.

Angelfish

Angelfish as it is commonly known to the aquarists and also to the general public, is a freshwater cichlid fish belonging to the genus *Pterophyllum*. There are three species of *Pterophyllum*, all with greatly laterally compressed and round bodies, and elongated triangular dorsal and anal fins.

Systematic Position:

Class: OSTEICHTHYES

Subclass: ACTINOPTERYGII

Order: PERCIFORMES

Family: CICHLIDAE

Subfamily: CICHLASOMATINAE

Genus: *Pterophyllum*

Species: 1. *Pterophyllum altum*

2. *Pterophyllum leopoldi*

3. *Pterophyllum scalare*

Aquarium Varieties:

Most strains of angelfish available in the trade are the results of many years of selective breeding (https://en. Wikipedia.org/wiki/Pterophyllum).

2.5.21 Altum angelfish

Figure 2.57

Pterophyllum altum Pellegrin, 1903

Common Name(s):

English—Orinoco angelfish, Altum angel, deep angelfish.

IUCN Red List Status: Not Evaluated.

Systematic Position:

Subfamily: CICHLASOMATINAE

Genus: *Pterophyllum*

Species: *P. altum*

Scientific Name: *Pterophyllum altum* Pellegrin, 1903

Description:

- It is the largest of the three described angelfish species, both in length and height.
- This species has a notch on the upper part of its snout (rather than a flat or slightly rounded forehead), just above the nares, and a steeply rising forehead.
- Extended dorsal and anal fins along with the tail fin are present. In mature fish the tail fin can develop streamers on the outside corners.
- The pelvic fins are also very long and delicate.
- Body of this fish is silver coloured with brownish or red coloured bars.
- The fins may have red striations, some red spots and a blue-green cast may be present to the dorsal fin. A black operculum spot is often present.
- Sexual Dimorphism: During breeding season only, the papilla on the male is pointed and on the female it is blunt.

Species Details:

- Distribution:

  South America: Amazon River basin, in the upper Negro River drainage; Orinoco River basin, in tributaries of the upper Orinoco River (Inirida and Atabapo rivers) to Puerto Ayacucho.
- Habitat: Inhabits rivers.
- Food and feeding: These fishes are omnivorous. They eat all kinds of live, fresh and flake foods. They eat flakes, pellets, brine shrimps, mosquito larvae, lettuce and spinach.
- Feeding frequency: Several feedings per day.
- Maximum length: 18.0 cm TL male/unsexed in captivity. About 40 cm in the wild.

- Life span: 15 years.

Aquarium Aspects:

- Minimum tank size: 55 gallons (208 litres).
- Substrate type: Any.
- Lighting requirements: Low-subdued lighting.
- Temperature: 78.0° to 84.0° F (25.6° to 28.9° C).
- Breeding temperature: 86.0°-87.8° F (30.0°-31.0° C).
- Range pH: 4.5-5.8.
- Hardness range: 2-10 dGH.
- Water movement: Moderate.
- Water region: All.
- Water changes: Weekly.

Behaviour:

Social—This species of *Pterophyllum* is more peaceful fish than other two species. They are semi-aggressive, community fish. But as they mature, they form pairs, form a nuclear family, and defend the territory of the breeding area.

Reproductive—Breeding is difficult in the aquarium. Breeding in captivity may be achieved in water with a pH of 5.5-6.0, a water hardness, 2-4 dGH and a warm water temperature of 84-88° F (29-31° C). These egg layer fish spawn on submerged roots and tree branches in the wild. If the eggs are not eaten by parents, the larvae and fry are carefully guarded. The eggs hatch in a few days and fry become free swimming in a week.

Remarks: This species is a challenging but rewarding angelfish to keep. It is sensitive to changes in water chemistry, pollutants and are subject to infections.

2.5.22 Leopold's angelfish

Figure 2.58

Pterophyllum leopoldi (Gosse, 1963)

Common Name(s):

English-- Teardrop angelfish, Leopold's angelfish, dwarf angelfish, Roman-nosed angelfish.

IUCN Red List Status: Not Evaluated.

Systematic Position:

Subfamily: CICHLASOMATINAE

Genus: *Pterophyllum*

Species: *P. leopoldi*

Scientific Name: *Pterophyllum leopoldi* (Gosse, 1963)

Description:

- The pre-dorsal notch is absent.

- Fins are normally shorter than body length.
- A black blotch is present at the dorsal insertion of the 4th vertical bar.
- It is gold with silver stripes.
- Has paler stripes than the other two *Pterophyllum* species.
- It is the smallest of all species of angelfish.

Species Details:

- Distribution:

  South America: Amazon River basin, along the Solimoes Amazon River between about Manacapuru and Santarem; Rupununi River in the Essequibo River drainage in Guyana.
- Habitat: Inhabits rivers.
- Food and feeding: Omnivorous. Consume *Artemia, Daphnia,* insect larvae, blackworms, frozen bloodworms and shrimp. Also take, good quality flakes and frozen bloodworm and mosquito larvae.
- Feeding frequency: Twice a day.
- Maximum length: 5.0 cm SL male/ unsexed.

Aquarium Aspects:

- Minimum tank size: 55 gallons or larger.
- Water conditions: Temperature: 27.0-31.0° C (81.0-88.0° F); pH 4.8-6.2; hardness: 1-5 dGH; nitrates: low.
- Substrate: Sandy.
- Lighting: Dim light.
- Water changes: Regular partial water changes.

Behaviour:

Social—It is the most aggressive of all three species. Becomes aggressive at the time of breeding. It has strong personality and can interact with people.

Reproductive—During spawning, defend their territory vigorously.

Remarks: This species is one of the most colourful and aesthetically attractive freshwater fish species. It is often misidentified as *P. dumerilii* when imported in the aquarium trade.

2.5.23 Freshwater angelfish

Figure 2.59

Pterophyllum scalare (Schultze, 1823)

Common Name(s):

English—Silver angelfish, freshwater angelfish, common angelfish.

IUCN Red List Status: Not Evaluated.

Systematic Position:

Subfamily: CICHLASOMATINAE

Genus: *Pterophyllum*

Species: *P. scalare*

Scientific Name: *Pterophyllum scalare* (Schultze, 1823)

Description:

- Body is laterally compressed and of distinctive diamond-shaped and with pointed snout.
- Dorsal and anal fin spiny rays increase in length from the anterior to posterior part of the fin.
- Dorsal, anal and tail fin are long and flowing.
- First branched rays are also very long.
- In mature fish the tail fin may develop streamers on the outside corners.
- The pectoral fins are very long and delicate.
- Body colour is silvery with dark vertical bars.
- Its colour is deeper in clear water.

Species Details:

- Distribution:

  South America: Amazon River basin, in Peru, Columbia, and Brazil, along the Ucayali, Solimoes and Amazon rivers; rivers of Amapa (Brazil), Rio Oyapock in French Guyana; Essequibo River in Guyana.
- Habitat: It inhabits swamps or flooded grounds with dense aquatic and riverine vegetation. The water may be either clear or silty.
- Food and feeding: They are omnivorous. In captivity they feed on flake food, small pellets, live foods like shrimps, fishes, worms, vegetable food, meaty foods.
- Feeding frequency: Several feedings per day.
- Maximum length: 7.5 cm SL male/ unsexed to 15.24 cm.
- Life span: 10-15 years.

Aquarium Aspects:

- Minimum tank size: 30 gallons (114 litres). For a pair 55 gallons tank is required and much larger tank for a community.
- Substrate type: Sand/ Gravel Mix.
- Lighting requirement: Low-subdued lighting.
- Temperature: 75.0-82.0° F (23.9 to 27.8° C).
- Breeding temperature: 80.0-85.0° F.
- Range pH: 6.0-7.5.
- Hardness range: 2-10 dGH.
- Water movement: Moderate.
- Water region: All.

- Water changes: Weekly.

Behaviour:

Social—They are semi-aggressive because though they are a community fish, but may become aggressive towards small fish. When mature, they pair off and form a strong nuclear family. And defend a territory of the breeding area.

Breeding—They are egg layers and open breeders. Spawning takes place on the submerged leaves in the wild. The pair needs very clean water and need to be conditioned to spawn. The parents either eat the eggs, or guard the larvae and fry carefully. The eggs hatch in a few days and fry become free swimming in a week.

Remarks: This is the hardiest species and also the easiest one to breed in captivity, among the three species of angelfish. This very attractive fish is very popular with both the beginner and long-time aquarists.

2.5.24 Discus

Figure 2.60

Discus are the commonly known fish belonging to the *Symphysodon* genus of cichlids, native to the Amazon river basin in South America.

Systematic Position:

Subclass: ACTINOPTERYGII

Order: CICHLIFORMES

Family: CICHLIDAE

Tribe: HEROINI

Genus: *Symphysodon* Heckel, 1840

Species:

- *Symphysodon aequifasciatus* Pellegrin, 1904
- *Symphysodon discus* Heckel, 1840
- *Symphysodon tarzoo* Lyons, 1959

Description:

- Relatively large disc-shaped body without the extended finnage.
- They are the most laterally compressed of all the cichlids.
- The sides of the fish are often patterned in shades of green, red, brown and blue.
- Through selective breeding there are many different varieties of body shapes, colours and patterns, captive bred varieties are also variants of the wild form.
- Sexual Dimorphism: During breeding season, the male fish have a more pointed papillae while the female has rounded. Males may be larger and some may have a more pointed dorsal fin and thicker lips.

Varieties of discus:

Golden, pearl pigeon, milky way, red rainbow, golden ruby spotted snake, green mamba, giant snake skin, checker board turquoise.

Species:

1. ***Symphysodon aequifasciatus*** Pellegrin, 1904

   (Blue discus or brown discus)

   Distribution:

   South America: eastern Amazon River, lower reach of Amazon River and affluent of east of the confluence of the Negro and Solimoes Rivers.

   Habitat: In deep, rocky areas in crevices and among roots.

   Maximum length: 13.7 cm SL male/unsexed.

2. ***Symphysodon discus*** Heckel, 1840

   (Red discus or Heckel discus)

   Distribution:

   South America: Amazon River basin in Brazil, near the mouth of the Negro River, in the lower Abacaxis River, and in the lower Trombetas River.

   Habitat: Freshwater, benthopelagic

   Maximum length: 12.3 cm SL male/unsexed

3. ***Symphysodon tarzoo*** Lyons, 1959

   (Green discus)

   Distribution:

   South America: western Amazon river drainages west of the Purus Arch.

   Habitat: Freshwater benthopelagic.

   Maximum length: 13.2 cm SL male/ unsexed.

Species Details:

- Food and feeding: They are carnivorous and eat all types of live and frozen foods. Both frozen and live brine shrimp, bloodworms, chopped beef heart and white worms are preferred. They eat high quality flake food and pellets. They also take vegetables and meaty foods.
- Feeding frequency: Several feedings per day.
- Maximum length: 25.40 SL.
- Life span: 10 years

Aquarium Aspects:

- Minimum tank size: 50 gallons (189 litres). About 50 gallons for one fish.
- Substrate type: Sand/ Gravel Mix.
- Lighting needs: Moderate-normal lighting.
- Temperature: 82.0 to 88.0° F (27.8 to 31.1° C).
- Breeding temperature: 82.0-88.0° F (27.7-31.0° C).
- Range pH: 6.0-6.5.
- Hardness range: 10-15 dGH.

- Water movement: Moderate.
- Water region: All.
- Water changes: Weekly.

Behaviour:

Social—These fishes are peaceful and highly social. They are schooling fish and do not like to be kept alone. They are slow feeders and are not predatory. They do not burrow in the substrate. But they can get aggressive while spawning.

Reproductive—These fishes are egg layers and attach their eggs to plants, driftwood, rocks and other objects in the aquarium. Females lay about 200-400 eggs which hatch in about 60 hours. Discus form nuclear families.

Remarks: This fish is known as the king of the aquarium. Perhaps the most beautiful of all tropical fish in the trade. Discus is also one of the more difficult tropical fish to keep in aquarium and hence not recommended for beginners.

2.5.25 Oscar

Figure 2.61 A, B & C

Astronotus ocellatus (Agassiz, 1831)

Common Name(s)

English—Tiger Oscar, velvet cichlid, Red Oscar, marble cichlid.

IUCN Red List Status: Not Evaluated.

Systematic Position:

Family: CICHLIDAE

Subfamily: ASTRONOTINAE

Genus: *Astronotus*

Species: *A. ocellatus* (Agassiz, 1831)

Scientific Name: *Astronotus ocellatus* (Agassiz, 1831)

Description:

- Body oval shaped with large mouth and thick lips.
- Dorsal spines: 12-14; dorsal soft rays: 19-21; anal spines: 3; anal soft rays: 15-17.
- Body colour dark with bright orange opercular margin and ventral parts of the lateral sides of the body.
- Often a black rounded blotch with orange margin is present at caudal fin base and on the dorsal fin.
- An adult wild Oscar (Figure 2.61 A) is dark coloured with orange around the gills, on the sides towards the back and with an orange-ringed black spot at the base of the caudal fin.
- Other colour varieties are red Oscar with solid red body and the speckled red Oscar with a speckled solid red body and black fins.
- The tiger Oscar is more similar to the wild form but with deep red colour. There are Tiger Oscar (Figure 2.61 B), Red Tiger Oscar, albino tiger Oscar,

albino red Oscar, other colour morphs and Leucistic Oscar (Figure 2.61 C). Some Oscars are artificially coloured or painted. All the various lutino and long finned varieties are recently developed.

- Sexual Dimorphism: It is difficult to distinguish male and female Oscars externally.

Species Details:

- Distribution:

  South America: Orinoco and Amazon in Peru, Colombia and Brazil, French Guiana, reported from Argentina.
- Habitat: They inhabit quiet shallow waters. Slow-moving waters of rivers, canals and ponds with muddy or sandy bottoms.
- Food and feeding: They are omnivorous. Feeds on small fish, crayfish, worms and insect larvae. They also eat live, meaty, pellets, dead or frozen foods.
- Feeding frequency: Several feedings per day.
- Maximum length: 45.7 cm TL male/unsexed.
- Life span: 10 years.

Aquarium Aspects:

- Minimum tank size: 100 gallons.
- Substrate type: Sand.
- Lighting requirements: Moderate- normal lighting.
- Temperature: 72.0-77.0° F (22.2 to 25.0° C).
- Breeding temperature: 78.8-86° F (26.0-30.0° C).
- Range pH: 6.5-7.2.
- Water movement: Weak.
- Water region: All.
- Water changes: Weekly.

Behaviour:

Social—It is aggressive as well as predatory.

Breeding—This fish is an egg layer and open spawner. They are extensively bred in captivity. Oscars become sexually mature at about 4 ½ inches. In captivity, both the male and female fish clean a suitable spawning site, often with a flat rock or branches in a circular nest excavated in shallow water. Eggs are deposited and are guarded by both parents. Eggs hatch in 3 or 4 days and parent move the fry to a shallow pit in the sand where they remain for 6 or 7 days.

Remarks: The Oscar is one of the most popular cichlids in the aquarium hobby.

2.5.26 Tinfoil barb

Figure 2.62

Barbonymus schwanenfeldii (Bleeker, 1854)

Common Name(s):

English—Tinfoil barb, River barb.

IUCN Red List Category: Least Concern (LC).

Systematic Position:

Subclass: ACTINOPTERYGII

Order: CYPRINIFORMES

Family: CYPRINIDAE

Subfamily: CYPRININAE

Genus: *Barbonymus*

Species: *B. schwanenfeldii*

Scientific Name: *Barbonymus schwanenfeldii* (Bleeker, 1854)

Description:

- Body is torpedo-shaped with a high back and a forked tail.
- Silver or golden yellow body is covered with reflective scales that give the tinfoil effect.
- The red dorsal fin is having a black blotch at the tip.
- Pectoral, pelvic and anal fins are red. Caudal fin is red with white margin and a black sub-marginal stripe along each lobe.
- The natural form of Tinfoil barb has a distinct blackline near the edge of each lobe of the tail fin, which is absent on the Red-tail Tinfoil barb. Both tinfoil barbs have a black marking on their dorsal fin, but only the tinfoil barb has red on this fin as well.
- Juveniles are drab-coloured with scales of tin-plated look.
- Sexual Dimorphism: No visible external differences between the sexes, though females are significantly rounder during spawning season.

Species Details:

- Distribution:

  Mekong and Chao Phraya basins, Malay Peninsula, Sumatra and Borneo.
- Habitat: It is found in rivers, streams, canals and ditches. It occurs in medium-sized rivers, especially tributaries, and enters flooded fields.
- Food and feeding: It is omnivorous but prefers to be herbivorous. It generally consumes aquatic macrophytes and submerged land plants, as well as filamentous algae and sometimes insects. It also consumes small fishes, worms and crustaceans. Consumes flakes, pellet foods also.
- Feeding frequency: Several feedings per day.
- Maximum length: 35.0 cm SL male/ unsexed.

- Life span: 8-10 years.

Aquarium Aspects:

- Minimum tank size: 100 gallons (379 litres).
- Suitable for Nano tank: No.
- Substrate type: Any.
- Lighting needs: Moderate-normal lighting.
- Temperature: 72.0 to 77.0° F (22.2 to 25.0° C).
- Range pH: 6.5-7.5.
- Hardness range: 2-10 dGH.
- Water movement: Strong.
- Water region: All.
- Water changes: Monthly.

Behaviour:

Social—It is not an aggressive species, but they will eat small fish if they have the opportunity. In nature, they are found swimming in large schools. In the aquarium, they should be kept with some of their own kind. Groups of 5 or more will help to keep them happy.

Reproductive—They are egg-layers that scatter their eggs. The female releases thousands of eggs but none of parent care for the young. They do not breed in aquaria.

Remarks: It is commercially important in the aquarium hobby trade. Golden Tinfoil barb is one of the most notable colour morphs of the Tinfoil barb. This is an attractive and impressive fish for large public aquarium.

2.5.27 Goldfish

Figure 2.63 A, B & C

Carassius auratus (Linnaeus, 1758)

Common Name(s):

English-Goldfish.

IUCN Red List Category: Least Concern (LC).

Systematic Position:

Order: CYPRINIFORMES

Family: CYPRINIDAE

Subfamily: CYPRININAE

Genus: *Carassius*

Species: *C. auratus*

Scientific Name: *Carassius auratus* (Linnaeus, 1758)

Description:

- Body is stout and thick-set. Head broadly triangular in shape.

- Caudal peduncle thick and short.
- The caudal fin is either rounded or slightly forked, other fins generally stand fully erect and the edge of the dorsal fin is slightly concave. The dorsal fin is with long base.
- Its body tapers smoothly from its back and belly to the base of its caudal fin.
- Dorsal fin spines: 3-4; dorsal fin rays: 14-20; anal fin spines: 2-3; anal soft rays: 4-7.
- Barbels absent on upper jaw.
- Body is metallic orange-red without any silver patch.
- Goldfish have four kinds of cone cells and are able to distinguish between four different primary colours.
- Goldfish have strong associative learning abilities as well as social learning skills.
- Sexual Dimorphism: The male fish is smaller and slenderer than the female. They have breeding tubercles on gill covers.

Species Details:

- Distribution:

  Asia: central Asia and China and Japan. Native to East Asia from the Amur River to the Pearl River basin including the Korean Peninsula and Taiwan. It had been domesticated in China more than 1,000 years ago.
- Habitat: Rivers, lakes, reservoirs, ponds and ditches with stagnant or slow-moving water. It occurs in eutrophic fresh and brackish waters, well vegetated ponds and canals.
- Food and feeding: Omnivorous fish. Eat all kinds of fresh, frozen, and flake foods. Eat live fishes, shrimps, worms and vegetable food.
- Feeding frequency: Several feedings per day.
- Maximum length: 48.0 cm TL male/unsexed.
- Life span: 10-15 years. Maximum reported age: 41 years.

Aquarium Aspects:

- Minimum tank size: 15 gallons.
- Suitable for Nano Tank: Sometimes.
- Substrate Type: Any.
- Lighting Needs: Moderate to Normal lighting.
- Temperature: 65.0-72.0° F (18.3-22.2° C). Can tolerate colder temperatures.
- Breeding temperature: 68.0° F (20.0° C).
- Range pH: 6.0-8.0.
- Hardness Range: 5-19 dGH.

- Water movement: Moderate.
- Water region: All.
- Water changes: Weekly.

Behaviour:

Social- Being very social fish thrive well in a community. They are scavengers also. Goldfishes are very active and friendly.

Reproductive- Common Goldfishes are egg layers and egg-scatterers. They spawn easily under right conditions and suitable temperature. Cold water temperatures during the winter months are necessary for proper ova development. Spawning takes place in shallow water among weeds, sometimes among willow roots. Often in meadows inundated by spring flood. Spawning activity begin just before dawn to mid-afternoon. Females spawn 3-10 lots of eggs at intervals of 8-10 days. Juveniles need high temperature to grow. Eggs are sticky and get attached to water plants or submerged objects. Spawning takes place several times during spawning period.

Goldfish Varieties

Though the original goldfish was golden in colour, selective breeding over centuries has produced several colour variations. Different body shapes and eye configurations are also observed. Some of which are much less hardy and can thrive only in aquariums, except Shubunkin which is hardier. Large varieties of goldfish have been originated from China. There are about 200 breeds of goldfish recognised in China.

1. Single-Tailed Varieties

These goldfish varieties have a single caudal fin and anal fin. They are fast swimmers and have long streamlined bodies. They have originated from the common goldfish.

1.1 Long-bodied

- **Common goldfish**: It is the most common type of goldfish. It is the direct descendant of the Prussian carps. All varieties of goldfish are developed from this variety. It is also called feeder fish or feeder goldfish. Common goldfish are available in a variety of colours, such as red, orange, yellow white, black and calico goldfish. The most common variation is a shiny-orange, while second most common variation being a mix of white and red. The black colour is uncommon whereas rare colours are yellow and pure white.
- **Comet goldfish** (Figure 2.63 A): This goldfish variety is most common in the United States. It is similar to the common goldfish except for having long, deeply-forked tail. It is developed from the common goldfish stock in the United States.
- **Shubunkin**: These are hardy, single-tailed goldfish, with nacreous scales with a calico pattern.

- London Shubunkin—This goldfish is with a stout body and also short, rounded finnage, similar to the common goldfish.
- American Shubunkin—It is with a slimmer body shape with deeply forked, pointed tailfins, and longer finnage all round.
- Bristol Shubunkin—It is a slim-bodied goldfish with well-developed finnage, having a tail which is large, moderately forked and rounded at the end. It is a large, heart-shaped tail.

1.2 Egg-shaped bodied

- **Nymph goldfish**: They have single caudal fin and anal fin. This rare type of goldfish is mutated from the fantails or can be mutated in the ryukins and other fancy goldfish. All single-tailed mutated fancy goldfish are considered as nymph goldfish. It is closely related to the tamasaba. It is developed from the fantail.

2. Double-Tailed Varieties

They are the most popular and the most expensive types of goldfish. Fancy in goldfish means they have double caudal fins and anal fins. There are two types of fancy goldfish.

2.1 Streamlined bodied:

- **Wakin goldfish**—It resembles the common goldfish but has double caudal fins and anal fins. All fancy goldfish are developed from this variety.
- **Jikin goldfish**—It resembles the wakin goldfish but their double caudal fins and anal fins splay outwards. It is developed from the wakin goldfish.

2.2 Egg-shaped bodied:

This is the most popular type of goldfish. There are of two types.

2.2.1 Dorsal-finned varieties

- **Fantail goldfish**—This goldfish variety has an egg-shaped body, a high dorsal fin, double caudal and anal fins, and no shoulder hump. It is the western form of the nyukin.
- **Ryukin goldfish**—This variety has a short, deep body with a noticeable shoulder hump. It resembles the fantail goldfish except with a shoulder hump. It has been bred in Ryukin islands. It needs a taller aquarium because of its larger hump.
- **Oranda goldfish** (Figure 2.63 B)—It has a prominent raspberry like head (also referred to as wen or head growth) that encases the whole head except for the eyes and mouth. It resembles the fantail except it has a head growth. Orandas are not kept in the aquarium with rough objects than can harm their head growths.
- Pearlscale goldfish (Figure 2.63 C)—This goldfish variety has a spherical body with finnage which resembles the fantail goldfish finnage. They possess whitish pearl-like raised scales which look attractive. When these

scales fall off, they will grow only into regular scales. Due to their sphere-shaped body, these fishes are susceptible to constipation. Hence their food should be soaked in water before feeding. The aquarium used for them should be free from rough objects so their scales are not damaged.

- **Telescope goldfish**—This goldfish variety resembles fantails except that it has protruding eyes. This fish has a zoomed and limited vision and therefore are not kept in an aquarium with rough objects that can harm or even can cause blindness to their protruding eyes.
- **Curled-gill goldfish**—This may be any goldfish type with curled-gills or their gills are turned outwards. This is an uncommon variety of fancy goldfish.
- **Veiltail goldfish**—It has all elongated fins that resemble veil. These are slow-moving creatures and are prone to fin nipping because of their elaborate fins.
- **Tosakin goldfish**—It has undivided double tail fins that are curled at the ends. This distinctive breed of fancy goldfish, has a large tail fin that spreads out horizontally (like a fan) behind the fish. The two halves of this divided tail are attached at the centre /middle forming a single fin.
- **Butterfly tail**—It is a variant form of the telescope goldfish. It has protruding eyes and distinguished by the butterfly-shaped caudal fins when viewed from above.

3. Dorsalfin-Less Varieties

- **Eggfish goldfish**—It resembles the wakin goldfish. The only difference is that it has no dorsal fin. It has been developed from the fantail goldfish.
- **Pompom goldfish**—It resembles the eggfish, the only difference is that it has larger nasal septa. The name of this goldfish comes from the cheerleader's pompom balls. It has been developed from the eggfish goldfish. Pompom goldfishes with dorsal fin are also available in the trade.
- **Lionhead goldfish**—It is similar to the eggfish goldfish only difference is that it has a soft, spongy head growth. It has been developed from the eggfish goldfish. This fish should not be kept in an aquarium with rough objects that can harm its head growth.
- **Ranchu goldfish**—This fancy goldfish resembles the lionheads. Though they have more-arched backs and have much shorter tails that are tucked in at a 45-degree angle.
- **Celestial eye goldfish**—This fancy goldfish variety has a breed-defining pair of upturned, telescope eyes with pupils gazing skyward. It has double tails. As their eyeballs are turned permanently upwards, they are also called "stargazers" or "sky-gazers". It has been developed from the dorsal-less telescope eye goldfish. These fishes should not be kept in the aquarium with rough objects as these may harm or hurt their eyes.
- **Bubble eye goldfish**—This fancy goldfish has its eye accompanied by two large fluid-filled sacs. These sacs are fragile and can get easily ruptured.

These fishes are developed from the eggfish goldfish. If the fluid-filled sacs get ruptured, they will grow back again though not immediately. And the newly grown one may become asymmetrical. The aquarium for these fishes should not have rough objects as they may hurt and rupture the sacs, which may cause infections.

Pond Goldfish

Most goldfish varieties can thrive in outdoor ponds. Only telescope, celestial eye and bubble eye are not capable of living in ponds, due to their fragile eyes.

Goldfish Tail Types

Single tail types:

- Wild—This is the standard single tail of any goldfish. It resembles fantail though not doubled.
- Comet-tailed—It resembles a ribbon-tail though not doubled. It is much longer and more deeply forked than common-tailed.
- Heart-shaped tailed—This is a single heart-shaped tail.

Double tail types:

- Fan tailed—This is a tail of common goldfish that is doubled, and is the standard tail type of fancy goldfish.
- Veil tailed—It is more than two times that of fantail.
- Ribbon tailed—It is the tail of comet goldfish that is doubled. This is a longer tail and longer fork than fantail.

Other tail types

- Broad tailed—This is the broad finnage type of fancy goldfish.
- Butterfly tailed—This is butterfly finnage or with tails that are flattened horizontally.
- Angled tail—With an angled tail peduncle.

Colour Pigments of Goldfish

There are three main colour pigments such as:

- Erythrophores—contain red pigments.
- Xanthophores—yellow pigments
- Melanophores—black pigments.

Colour combinations and appearance of the three main colour pigments are--

- Red—Many erythrophores present, xanthophores and melanophores are absent.
- Orange—A balanced number of erythrophores and xanthophores are present. Melanophores absent.
- Yellow—Many xanthophores, erythrophores and melanophores are absent.

- Black—Many melanophores are present.
- Gray—The topside of the goldfish has some melanophores, but the number of melanophores decreases downside.
- White—No pigments present.
- Brown—All three pigments present but in balance.
- Blue—Melanophores are situated deep within the skin whereas erythrophores and xanthophores are absent.

2.5.28 Koi, common carp

Figure 2.64 A & B

Cyprinus rubrofuscus Lacepede, 1803

Common Name(s): Koi, Nishikigoi.

IUCN Red List: Vulnerable.

Systematic Position:

Family: CYPRINIDAE

Genus: *Cyprinus*

Species: *C. rubrofuskus*

Scientific Name: *Cyprinus rubrofuscus* Lacepede, 1803

Synonym: *Cyprinus carpio haematopterus* Linnaeus, 1758

Koi are coloured form of Amur carp (*Cyprinus rubrofuscus*) that are kept for decorative purposes in outdoor koi ponds or water gardens. They were developed from common carp in Japan in the 1820s.

Description:

- Large body with a greater range of colour and pattern combinations.
- Scales are large and thick.
- Koi have clearly visible 2 pairs of barbels near their mouths.
- Scales large and thick.
- In Koi, the caudal fin is joined.
- The dorsal fin of Koi is attached along the whole length of back.
- Sexual Dimorphism: Males have spawning tubercles which are absent in females. Females are broader and rounded in shape than males.

Species Details:

- Distribution:

  East Asia: Amur to Red River drainages in Mongolia, Russia (Amur), eastern China, Lao PDR and Vietnam. It is introduced elsewhere.
- Habitat: Found in freshwater, but tolerant to slightly brackish water.
- Food and feeding: Omnivorous fish. They eat a wide variety of food, lettuce, peas, water melons etc.

- Maximum length: 3 ft SL male/ unsexed.
- Life Span: 50 years or more.

Aquarium Aspects:

- Minimum tank size: 100-350 gallons.
- Water conditions: Temperature: 15-25° C (59-77° F).

Behaviour:

Social: It is a very hardy and community fish.

Reproductive: The spawning season of this fish is in the spring and summer. After the female koi releases her eggs, these eggs sink to the bottom of the pond. The sticky outer shell of the egg, helps it to stick to a place. One or more males fertilize them. After 4-7 days, the fry will be hatched from the egg.

Remarks: This fish is considered as a lucky symbol in Japanese culture. It represents luck, prosperity and good fortune in Japan.

Koi Varieties:

Koi varieties are available in different colours, pattern and scalation. Some of the major colours are white, black, red, yellow, blue and cream. The most notable category is Gosanke, which is made up of the Kohaku, Taisho Sanshoku and Showa Sanshoku varieties. The major varieties are as follows:

- Kohaku—The first ornamental variety developed in Japan (late 19th Century) is a white-coloured fish with large markings on top. Gin Rin Kohaku (Figure 2.64 A & B) is white splashed with brilliant reddish orange markings.
- Taisho Sanshoku—Similar to Kohaku, except the presence of small black markings.
- Showa Sanshoku—Black with red and white markings.
- Tancho—With a solitary red patch on its head.
- Chagoi—From pale-olive green, brown to copper or bronze or orange.
- Asagi—Light blue above and red below.
- Utsurimono—Black with white, red or yellow markings in a zebra pattern.
- Bekko—White, red or yellow with black markings.
- Goshiki—Dark fish with red pattern.
- Shusui—Pale, sky-blue/grey above and red and orange below.
- Kinginrin—Metallic scales present.
- Kawarimono—With several categories.
- Ogon (Figure 2.64 B)—With one metallic colour only. The most common colours are gold, platinum and orange.
- Platinum Ogon—Koi with a solid, metallic white colour. A solid, white, non-metallic Koi is referred as a "Shiro Muji".

- Kumonryu (Figure 2.64 B)—Black with curling white markings. There is a solid white base and areas of black pattern overlying the white.
- Ochiba—Light blue/grey with copper, bronze or yellow pattern.
- Koromo—White fish with blue or black-edged scales.
- Ghost koi—Hybrid of Ogon and wild carp.
- Butterfly koi—Hybrid of Koi and Asian carp with long flowing fins.
- Doitsu-goi—Originated from cross breeding of various varieties with scale-less German carp.
- Gosanke—Made up of the Kohaku, Taisho, Sanshoku and Showa Sanshoku varieties. It is the most popular category of Koi.

2.5.29 Guppy

Figure 2.65

Poecilia reticulata Peters, 1859

Common Name(s):

English—Guppy, million fish, rainbow fish.

IUCN Red List Status: Not Evaluated.

Scientific Name:

Order: CYPRINODONTIFORMES-- Killifishes

Family: POECILIDAE--livebearers

Subfamily: POECILIINAE (livebearers)

Genus: *Poecilia*

Species: *P. reticulata*

Scientific Name: *Poecilia reticulata* Peters, 1859

Description:

- It is a small fish with an elongated body.
- Being a member of live-bearing tooth carps, it has teeth in both upper and lower jaws.
- Dorsal spines: 0; dorsal soft rays: 7-8; anal spines: 0; anal soft rays: 8-10.
- Sexual Dimorphism: The female guppy is both longer (3-6 cm) and stouter than the male. Have less rich colouration. Generally, have a spawning patch, a small dark blotch at the base of the caudal fin. The male guppy is smaller (1.5-3.5 cm) and slender. Generally, have a longer, more colourful tail fin. The male has a gonopodium, the elongated anterior anal-fin, primarily formed from the third, fourth, and fifth rays.

Different types of guppies:

- Different types of guppies have been developed by the breeders through the years. Three major characteristics have been considered, for differentiating one type of guppy from another. These are—colour of the body and fins; body pattern and tail pattern and fin types.

- Selectively bred guppies come in all colours and combinations of colours. Here, ground colour indicates upper body colour, and cover indicates the lower body. There are 4 basic ground colours (wild, albino, blond and blue) and colours developed by the combinations of four. The cover can be any imaginable colour of combination of colours. Different guppy types based on the colours result in—albino, AOC, AOC bicolour, black, blue, blue green bicolour, bronze, green, half black AOC, half black green, half black purple, half black red, half black yellow, multi, purple, red, red bicolour, round tail, snakeskin, swordtail, yellow.
- Patterns: Body patterns include-
    - Tuxedo—The front and back halves of the body are actually two different colours.
    - Cobra—Some vertical barring as well as rosettes.
    - Snakeskin—Some rosettes, but it looks more like chain-link pattern displayed across the base colour.
    - Tail fins can be solid-coloured but they can also come in different patterns. Tail patterns are--
    - Leopard—The fin is in series of spots.
    - Mosaic—Spots that connect to one another to create a somewhat irregular pattern.
    - Grass—Tiny dots resemble grass seeds.
    - Lace—Fine, web-like tail pattern. It occurs most often on guppies with snakeskin pattern on their bodies.
- Tail fin types: Selective breeding has resulted in the development of a number of distinct tail types. These are—
    - Fan tail, Delta/Triangle tail, Veil tail, Flag tail, Double sword tail, Top sword tail, Bottom sword tail, Lyre tail, Pin or Needle tail, Spade or Coffer tail, Spear tail, Round tail.
- Red Eye guppies:
    - Real Red Eye (RRE)—Normal coloured guppies with red eye.
    - Real Red Eye Albinos (RREA)—Albino gyppies with slightly paler red eyes.

Species Details:

- Distribution:

  South America: Venezuela, Barbados, Trinidad, northern Brazil and the Guyana.
- Habitat: They occur in a wide variety of aquatic bodies, with preference to quiet and densely vegetated waters. They inhabit warm springs and their effluents, weedy ditches, canals, highly turbid water in ponds, canals and ditches at low elevations to clear mountain streams at high elevations.

- Food and feeding: Omnivorous. They eat detritus with algae, plant matter, insects and insect larvae in the wild. In the aquarium, eat all types of fresh, frozen and flake foods.
- Feeding frequency: Several feedings per day.
- Maximum length: 5.0 cm SL male/unsexed; 6.0 cm SL female.
- Life span: 2 years.

Aquarium Aspects:

- Minimum tank size: 5 gallons.
- Suitable for Nano tank: Yes.
- Substrate type: Small gravel.
- Lighting: Moderate to normal lighting.
- Temperature: 68.0-82.0° F (20.0-27.8° C).
- Range pH: 6.8-8.5.
- Hardness range: 10-28 dGH.
- Water movement: Weak.
- Water region: All.
- Water changes: Bi-weekly.

Behaviour:

Social—They are extremely active and are also very friendly, with peaceful temperament.

Reproductive—Guppies are highly prolific live bearers and breed readily in the home aquarium without any special attention. But they should be well fed and given proper care. These fishes are polyandrous, where females mate with multiple males. Female guppies are attracted to brightly coloured males, especially ones with orange spots on the flanks. Females give birth to 20-40 live young. No parental care is shown.

Remarks: It is one of the world's most widely distributed tropical fish and at the same time, also one of the most popular freshwater aquarium fish species. In the studies and research on ecology, evolution and behaviour, guppies are used as model organism.

Molly

Poecilia is a genus of fishes collectively known as "mollies". While there may be many species under this genus, three species under this genus are favoured by aquarists. These are *Poecilia sphenops, Poecilia latipininna* and *Poecilia velifera.*

2.5.30 Common molly

Poecilia sphenops Valenciennes, 1846

Common Name(s):

English—Common molly, short-finned molly.

IUCN Red List Status: Not Evaluated.

Systematic Position:

Order: CYPRINODONTIFORMES--killifishes

Family: POECILIDAE--livebearers

Subfamily: POECILINAE (livebearers)

Genus: *Poecilia*

Species: *P. sphenops*

Scientific Name: *Poecilia sphenops* Valenciennes, 1846

Description:

- These fishes are elongated and laterally compressed.
- Teeth are present in both upper and lower jaws.
- Fins are small. The dorsal fin is situated above and behind the anal fin.
- The wild fish is dull silvery in colour.
- Sexual Dimorphism: The female fish is larger (12 cm) and rounder when fully grown. The male fish is smaller (8 cm), slender and has a gonopodium.
- Varieties of molly: Several colour varieties and different body shapes have been developed through selective breeding.
    - Short-finned or common molly: The wild form is of plain silvery colouration with brown and green hues. Small fish with rounded and compact fins.
    - Black molly: This fish is pure jet black all over. It is one of the favourite aquarium fish.
    - White molly: A white coloured variety.
    - Golden molly: It is also known as "24 carat".
    - Balloon molly: It is extremely short and round. Its appearance is due to a deformed spine of genetic defect.
    - Lyretail: This variety has an altered caudal fin structure.
    - Dalmatian molly: Silver coloured body with black speckles.

Species Details:

- Distribution:

  Central and South America: Mexico to Colombia.
- Habitat: It inhabits freshwater streams and coastal brackish and marine waters of Mexico.
- Food and feeding: Omnivorous. Food include lots of algae and vegetation. In the wild, they feed on algae, rotifers, small crustaceans and aquatic insects. In aquarium, they take flakes, pellet food, live foods, vegetables, green algae, dried food.
- Feeding frequency: Several feedings per day.
- Maximum length: 7.5 cm SL male/unsexed.

- Life span: 3 years.

Aquarium Aspects:

- Minimum tank size: 20 gallons.
- Suitable for Nano tank: Yes.
- Substrate: Any.
- Lighting: Moderate-normal.
- Temperature: 70.0-82.0° F (21.1-27.8° C).
- Breeding temperature: 79.0° C.
- pH range: 7.0-8.5.
- Hardness range: 20-30 dGH.
- Water movement: Moderate.
- Water region: All.
- Water changes: Monthly.

Behaviour:

Social—It is a peaceful fish and an active swimmer.

Reproductive—It breeds readily in the aquarium. The fish reaches sexual maturity at a very young age. It produces 20 to 150 young after 28 days of gestation.

Remarks: It is one of the most popular aquarium fish due its hardiness and adaptable nature. I t is also a popular feeder fish due to its high growth rate and brood number.

2.5.31 Sailfin molly

Poecilia latipinna (Lesueur, 1821)

Common Name(s):

English—Sailfin molly

IUCN Red List Status: Least Concerned.

Systematic Position:

Family: POECILIDAE--livebearers

Subfamily: POECILINAE (livebearers)

Genus: *Poecilia*

Species: *P. latipinna*

Scientific Name: *Poecilia latipinna* (Lesueur, 1821)

Description:

- The body is oblong, laterally compressed, with small, dorsally flattened head and upturned mouth.
- The caudal peduncle is broad and a large, rounded caudal fin is sometimes tipped with black.

- The pelvic fin originates at a point anterior to the dorsal fin.
- The body is generally light grey. Several rows of spots occur along the sides, back and dorsal fin. Often, these spots blend together, forming stripes.
- The dorsal fin has 15 fin rays.
- Sexual Dimorphism: Females are larger in size and plainly coloured. In mature males, the dorsal fin is greatly enlarged and the caudal fin is colourful. Male fish with a gonopodium.
- Colour morphs of *P. latipinna*: Selective breeding has resulted in different colour morphs. These are—yellow, orange, albino, and melanic/speckled. Midnight molly is a selectively bred black variety molly.

Species Details:

- Distribution:

  North America: Cape Fear drainage in North Carolina, USA to Veracruz, Mexico.
- Habitat: Inhabits still or gently flowing warm waters of small creeks and drains. Found in ponds, lakes, sloughs, coastal waters, tidal ditches, brackish canals, in vegetated backwaters and pools.
- Food and feeding: Omnivorous. Feeds on a variety of zoobenthos and detritus in the wild. Diet includes algae, other plant materials, Blanched spinach, cucumber, aquatic invertebrates, larvae of mosquitoes, rotifers, small crustaceans (copepods and ostracods), and aquatic insects.
- Maximum length: 15.0 cm TL male/unsexed; 17.5 cm TL female.
- Life span: Males less than 2 years; Females more than 2 years.

Aquarium Aspects:

- Minimum tank size: 50 gallons.
- Suitable for Nano tank: Yes.
- Substrate type: Any.
- Lighting needs: Moderate to normal lighting.
- Temperature: 21.1 to 27.8° C (70.0-82.0° F).
- Breeding temperature: 79.0° F.
- Range pH:7.0-8.5.
- Hardness range: 20-30 dGH.
- Water movement: Moderate.
- Water region: All.

Behaviour:

Social—It is a peaceful fish and a good community one and should be kept with their own kind or with members of other livebearers.

Reproductive—This fish breeds readily in the home aquarium. One male and at least 2-3 females are kept. Gestation period is 28 days. Female produces 10 to 100 young ones.

Remarks: A popular aquarium pet.

2.5.32 Yucatan molly

Poecilia velifera (Regan, 1914)

Common Name(s):

English—Yucatan molly, Mexican sailfin molly, Giant sailfin molly.

IUCN Red List Status: Vulnerable.

Systematic Position:

Order: CYPRINODONTIFORMES

Family: POECILIIDAE

Subfamily: POECILIINAE

Genus: *Poecilia*

Species: *P. velifera*

Scientific Name: *Poecilia velifera* (Regan, 1914)

Description:

- It looks like sailfin molly, *P. latipinna*, but larger in size and with a higher and longer dorsal fin in males.
- Dorsal fin has 20 fin rays.
- The spots on the dorsal fin are lighter in colour and round.
- Sexual Dimorphism: The male fish has a larger dorsal fin. When spread for display, these have distinct fan of trapezoid shape. The upper edge of the dorsal fin is distinctly higher than the lower. The male possesses a gonopodium.

Species Details:

- Distribution:

  Central America: Yucatan-Quintana Roo and lower Usumacinta-Laguna de Terminos systems in Mexico.
- Habitat: Inhabits coastal waters.
- Food and feeding: Omnivorous. Diet includes zoobenthos and detritus in the wild. In aquarium, accepts live, frozen or dried, vegetable matter, worms, crustaceans, insects and insect larvae.
- Maximum length: 15.0 cm TL male/unsexed.
- Life span: 3 years.

Aquarium Aspects:

- Minimum tank size: 35 gallons.
- Temperature: 72.0-82.0° F (22.0-28.0° C).
- pH: 7.0-8.5.

- Hardness: 15-35 dGH.
- Light: Strong light.
- Water changes: Regular.

Behaviour:

Social—It is a peaceful community fish.

Reproductive—Breeding slightly difficult as compared to other livebearers. One male with several females are kept.

Remarks: This brightly coloured fish is one of the most popular fish available.

Platy

2.5.33 Southern platy

Xiphophorus maculatus (Gunther, 1866)

Common Name(s):

English—Southern platy, common platy or moonfish.

IUCN Red List Status: Not Evaluated.

Systematic Position:

Order: CYPRINODONTIFORMES

Family: POECILIDAE

Subfamily: POECILIINAE

Genus: *Xiphophorus*

Species: *X. maculatus*

Scientific Name: *Xiphophorus maculatus* (Gunther, 1866)

Description:

- It is a laterally compressed stocky fish, with relatively compact fins and a fan-shaped tail.
- Have teeth in both upper and lower jaws.
- Ventral rays of caudal fin of males not elongated into a sword. Distinct line of black pigment mid-ventrally on caudal peduncle absent.
- The wild colourations for pure-bred platy species are many and varied but depend upon the river systems from which they originate.
- A crescent-shaped dark spot is generally present at the base of the tail.
- Sexual Dimorphism: Females are more elongate than the males. Females are rather plain looking and mature males are more colourful. Mature males develop gonopodium.
- Colouration: Platies are found in a wide range of colouration. Many beautiful colour variations and hybrids have been produced, and now it is difficult to find pure-bred original strains. Common platy has three ground colours of the body, wild-colour, albino and blond. Cover colours are red, blue, marigold and black.

- Some of the true -bred types, which are popular aquarium fish, are as follows:

Type	Varieties	Colouration
Solid coloured fish	Red platy, coral platy, golden platy	Deep red, brick red, bright yellow colour
Moonfish	Red moon platy, blue moon platy, blue platy, gold moon platy, gold crescent platy	Particular colour and distinctive black crescent shaped dot on the tailfin
Wagtail	Red wagtail platy, gold wagtail platy, black wagtail platy, mixed wagtail platy	With a specific colour and deep black tails.
Tuxedo	Black tuxedo platy, green tuxedo platy	With specific colour and black colouration below the lateral line from the gill to the tail.
Salt and Pepper	Salt and pepper platy	With a bit of all colours and breeds these colours true. If the hybrid variety has sword, it is called hybrid swordtail.
Mixed	Mixed platy	With various colours, without any particular pattern, but it does not breed true.

- Popular platy varieties developed from the variable platy are as follows:
  - Redtail platy: With bright red tail.
  - Yellowtail platy: With yellow tailfin.
  - Sunset platy: With a tail containing many colours, with usually yellow being prominent and have a yellow dorsal fin and a red tail fin.
  - Rainbow platy: With a tail of multiple colours in which no one colour is prominent.
  - Hawaii platy: With an entirely black body with a yellow dorsal fin and red tail fin.
  - Marigold platy: With yellow on the dorsal fin and top of the body with the lower half and the tail fin being orange.

Species Details:

- Distribution:

  North and Central America: Ciudad Veracruz, Mexico to northern Belize.
- Habitat: Adults occur in warm springs, canals and ditches with typically slow-moving water with silt bottoms, dense vegetation and weedy banks.
- Food and feeding: Omnivorous. In the wild, these fish feed on worms, crustaceans, insects and some plant matter. In the aquarium, they eat flakes, pellet food, live foods (fishes, shrimps, worms), vegetable matter.
- Feeding frequency: Several feedings per day.

- Maximum length: 4.0 cm TL male/unsexed, 6.0 cm TL female.
- Life span: 3-5 years.

Aquarium Aspects:

- Minimum tank size: 10 gallons.
- Suitable for Nano tank: Yes.
- Substrate type: Any.
- Lighting needs: Moderate-normal lighting.
- Temperature: 70.0-77.0° F (21.1 to 25.0° C).
- Range pH: 6.8-8.0.
- Hardness range: 10-28 dGH.
- Water region: All.
- Water changes: Monthly.

Behaviour:

Social—It is very peaceful and an excellent community fish.

Reproductive—The young southern platy reach sexual maturity at four months of age, and breeds readily in the home aquarium. After 24-30 days of gestation, 20-80 young are born.

Remarks: Several colour varieties of this species are popular aquarium fish. They are frequently used for genetic research.

2.5.34 Variable platyfish

Xiphophorus variatus (Meek, 1904)

Common Name(s):

English—Variable platyfish, variegated platy, Variatus platy.

IUCN Red List Status: Least Concern (LC).

Systematic Position:

Family: POECILIDAE

Subfamily: POECILIINAE

Genus: *Xiphophorus*

Species: *X. varietus*

Scientific Name: *Xiphophorus variatus* (Meek, 1904)

Description:

- Short, stocky fish, though slimmer and more elongated than southern platy.
- Wild colour or ground colour is grey or olive green, with black marbling or spots on the sides of the caudal peduncle.
- The tip of the fifth anal fin ray is without claw.
- The fourth pectoral ray is serrated.

- Sexual Dimorphism: Males have more pointed or needle like anal fin. Females have a more triangular anal fin. Males show blackish blotches on the dorsal fin.

Species Details:

- Distribution:

  North America: Endemic to Mexico from southern Tamaulipas to northern Veracruz.
- Habitat: Inhabits warm springs, weedy canals and ditches.
- Food and feeding: Omnivorous. They have a high need of lots of algae and other vegetation. Feeds on worms, crustaceans, insects and plant matter. Brine shrimp, tubifex or blood worms may be given as treat in captivity.
- Maximum length: 7.0 cm TL male/unsexed.

Aquarium Aspects:

- Minimum tank size: 10 gallons.
- Suitability for Nano tank: Yes.
- Substrate: Any.
- Lighting needs: Moderate-normal lighting.
- Temperature: 70.0 to 77.0° F (21.0-25.0° C).
- Range pH: 6.8-8.0.
- Hardness range: 10-28 dGH.
- Water movement: Weak.
- Water region: All.
- Water changes: Monthly.

Behaviour:

Social—Variable platy is very peaceful, active and a very good community fish.

Reproductive—If well fed and provided good care, this fish breeds readily in aquarium. After 24 days gestation, up to 100 young are born.

Remarks: It is a prolific breeder and very popular aquarium fish. Often used in genetic research.

2.5.35 Green swordtail

Xiphophorus hellerii Heckel, 1848

Common Name(s):

English—Green swordtail.

IUCN Red List Status: Not Evaluated.

Systematic Position:

Order: CYPRINODONTIFORMES

Family: POECILIIDAE

Subfamily: POECILIINAE

Genus: *Xiphophorus*

Species: *X. hellerii*

Scientific Name: *Xiphophorus hellerii* Heckel, 1848

Description:

- Moderately sized, sturdy and elongated body.
- Lower lobe of the male's caudal fin is elongated.
- Teeth is present in both their upper and lower jaws.
- Dorsal spines: 0; dorsal soft rays: 11-14; anal spines: 0; anal soft rays: 8-10.
- The wild form is olive green in colour, with a red or brown lateral stripe and speckles on the dorsal and sometimes caudal fins.
- The male's sword is yellow in colour and with lower black edge.
- Sexual Dimorphism: Female is larger than the male fish but lacks the sword. The female has a fan shaped anal fin, body is rounder and have a spawning patch at breeding time. The male has a sword shaped anal fin, called gonopodium. Male is also slimmer.
- There are many captive-bred colour varieties, these are black, red, green, albino and many patterns.
- Popular cross-bred varieties with black markings and a black tail are called, wagtail. Varieties those have black markings on about ¾ of the body, but not on the tail, are called, Tuxedo. Types are—Green sword, Green wagtail swordtail, Green tuxedo wagtail swordtail, Red swordtail, Red wagtail swordtail, Red tuxedo wagtail swordtail, Variegated swordtail, Hybrid swordtail, Marigold swordtail, Black Nubian swordtail, Pineapple swordtail, Neon swordtail, and Koi swordtail.

Species Details:

- Distribution:

  North and Central America: Rio Nantla, Veracruz in Mexico to north-western Honduras.
- Habitat: Adults inhabit rapidly flowing streams and rivers, usually heavily vegetated ones. They are also found in warm springs and their effluents weedy canals and ponds.
- Food and feeding: Omnivorous. Feed on worms, crustaceans, insects and plant matter. They eat all kinds of fresh, frozen and flake foods. In aquarium, commercially prepared fish foods, supplementary live and vegetable -based food, brine shrimp, tubifex, blood worms are preferred.
- Feeding frequency: Several feedings per day.
- Maximum length: 14.0 cm TL male unsexed; 16.0 cm TL female; common length: 2.8 cm TL male/unsexed.
- Life span: 5 years.

Aquarium Aspects:

- Minimum tank size: 20 gallons.
- Suitable for Nano tank: Yes.
- Substrate type: Any.
- Lighting needs: Moderate to normal lighting.
- Temperature: 64.0-82.0° F (17.8 to 27.8° C).
- Range pH: 7.0-8.3.
- Hardness range: 12-30 dGH.
- Water movement: Moderate.
- Water region: Middle.
- Water changes: Monthly.

Behaviour:

Social—These fishes are peaceful in nature. They swim in loosely-grouped schools. A good community fish species. One male to several females are the best ratios. Males get aggressive with each other.

Reproductive—If well fed and cared for, this fish breeds readily in captivity. They start breeding at the edge of 3 months. A female fish has a tendency to undergo sex reversal under suitable environmental conditions. Gestation period lasts for 24 to 30 days. Fry 20 to 200 are born.

Remarks: It is one of the most popular tropical aquarium fish. Various hybrid forms are available in the trade because of its hardiness and suitability for community tanks. Used for genetic research.

Mosquitofish

2.5.36 Western mosquitofish

Gambusia affinis (Baird & Girard, 1853)

Common Name(s):

English—Western mosquitofish, gambezi.

IUCN Red List Status: Least Concern.

Systematic Position:

Order: CYPRINODONTIFORMES--killifishes

Family: POECILIIDAE--livebearers

Subfamily: POECILIINAE (livebearers)

Genus: *Gambusia*

Species: *G. affinis*

Scientific Name: *Gambusia affinis* (Baird & Girard, 1853)

Gambusia is a large genus of fish in the family Poeciliidae having 45 recognised species.

Description:

- Body is short, with flattened head. Its mouth is pointed upward.
- A large abdomen and rounded dorsal and anal fins present.
- Eight horizontal scale rows between back and abdomen.
- Dorsal spines: 0; dorsal fin rays: 7-9; anal spines: 0; anal fin rays: 9-10.
- Sexual Dimorphism: Mature males are smaller (4 cm) than mature females (7 cm). The anal fin of adult male (gonopodium) is pointed. The anal fin of adult female resembles the dorsal fin. Adult female has a gravid spot on the posterior of her abdomen.

Species Details:

- Distribution:

  North and Central America: Mississippi River basin from central Indiana and Illinois in USA south to Gulf of Mexico and slope drainages west to Mexico.
- Habitat: In lower reaches of streams. Adults are found standing in slow-flowing water. Most commonly found in vegetated ponds and lakes, backwaters, quiet pools of streams and in brackish water.
- Food and feeding: Larvivore fish. Pelagic and surface predator also. This fish feed on zooplankton, small insects, detritus, larvae of insects. Their primary source of food is mosquito larvae. Adult females can consume hundreds of mosquito larvae in one day.
- Maximum length: 5.1 cm TL male/unsexed; 7.0 cm TL female.
- Life span: 3 years.

Aquarium Aspects:

- Minimum tank size: 12.5 litres.
- Temperature: 18.0-27.0° C (65.0-80.0° F).
- Range pH: 6.0-8.0.
- Range dGH: 5-30.

Behaviour:

Social: They are both competitive and aggressive and reflect both the physiological and morphological characteristics of a generalist predatory fish species. Mosquitofish are strongly attracted to their conspecifics with more yellow pigments. Large amounts of yellow pigmentations in male mosquitofish are generally related to higher social ranks in mosquitofish populations.

Reproductive: The species is viviparous. Internal fertilization is possible because the anal fin of the male is modified into a copulatory tube- like structure. The female fish carries about 30 alevins and gestation period is of 24 days to a month.

Remarks: A common freshwater aquarium fish in the market.

2.5.37 Giant gourami

Figure 2.66 A & B

Osphronemus goramy Lacepede, 1801

Common Name(s):

English—Giant gourami, common gourami.

IUCN Red List Status: Least Concern (LC).

Systematic Position:

Order: ANABANTIFORMES—labyrinth fishes

Family: OSPHRONEMIDAE—gouramies and fighting fishes

Subfamily: OSPHRONEMINAE (giant gouramies)

Genus: *Osphronemus*

Species: *O. goramy*

Scientific Name: *Osphronemus goramy* Lacepede, 1801

Description:

- Oval shaped, laterally compressed, deep body with a blunt head.
- In young fish, head is flat and pointed.
- In adults they develop a "nuchal hump" or a swollen knob or lump on the forehead, along with thick lips and a thick chin.
- Dorsal spines: 12-14; dorsal soft rays: 10-13; anal spines: 9-13; anal soft rays: 18-21.
- Soft rayed portion of anal fin is greatly enlarged, its distal margin is parallel to distal margin of caudal fin.
- Caudal fin rounded or obtusely rounded.
- Pelvic fins with first soft ray prolonged into a thread-like tentacle reaching posteriorly to or beyond hind margin of caudal fin.
- Ventral fins are threadlike.
- They like other gouramies, have special labyrinth organ.
- Juveniles are with 8-10 dark bars and yellow fins. The bars are a silvery blue/grey to black on a golden/ yellow background.
- They become rather plain when become adult.
- Colour of the body is brownish black or even an overall pinkish to white.
- Sexual Dimorphism: The male fish has longer and more pointed dorsal and anal fins. A larger nuchal hump or a swollen knob/lump develops on the forehead of the male. The dorsal and anal fin of the females are rounder and they have thicker lips than males.

Species Details:

- Distribution:

  Asia: Sumatra, Borneo, Java, the Malay Peninsula, Thailand and Indo-China (Mekong basin).

- Habitat: Found in swamps, lakes, and rivers, among vegetation. Found also in flooded forest medium to large rivers and stagnant water bodies including sluggish canals.
- Food and feeding: Omnivorous. Feeds on both plants and animals such as some aquatic algae, weeds, fish, frogs, earthworms and dead animals also. In captivity, they consume all kinds of foods. High quality flake, pelleted food, cooked meat, bread, boiled potatoes and other vegetables, occasionally brine shrimps (live or frozen), blood worms may be given. It can be a voracious predator.
- Feeding frequency: Several feedings per day.
- Maximum length: 70.0 cm SL male/unsexed.
- Life span: 20-25 years.

Aquarium Aspects:

- Minimum tank size: 200 gallons.
- Suitable for Nano tank: No.
- Substrate type: Any.
- Lighting needs: Moderate to normal lighting.
- Temperature: 68.0 to 86.0° F (20.0 to 30.0° C).
- Range pH: 6.5-8.0.
- Hardness range: 5-25 dGH.
- Water movement: Moderate.
- Water region: All.
- Water changes: Weekly.

Behaviour:

Social—It is a peaceful fish. With age, males tend to become aggressive to one another and may fight.

Reproductive—The male builds a ball-shaped nest with bits of plants just below the surface of water. These are generally 40 cm wide and 30 cm deep, with an entrance, circular in shape and 10 cm diameter, always points toward where the water is deepest. The nests are built during the months of April and May, though spawning takes place throughout the year. The nest is anchored to reed stems at 15-25 cm below the water surface. About 8-10 days are required to build the nest by the male. About 1,500 to 3000 eggs are released. The eggs float to the top. The male collects the eggs in his mouth and place them in the nest. About 40 hours are taken by the eggs to hatch. The male guards the offspring for about 14 days after the spawn. For captive breeding a huge tank is required.

Remarks: It is a highly prized fish and are used in public aquaria. The size of the fish and ease of care makes it sought after.

2.5.38 Three-spot gourami Figure 2.67

Trichopodus trichopterus (Pallas, 1770)

Common Name(s):

English—Three- spot gourami, blue gourami, opaline gourami, gold gourami.

IUCN Red List Status: Least Concern.

Systematic Position:

Subclass: ACTINOPTERYGII

Order: ANABANTIFORMES—labyrinth fishes

Family: OSPHRONEMIDAE

Subfamily: LUCIOCEPHALINAE

Genus: *Trichopodus*

Species: *T. trichopterus*

Scientific Name: *Trichopodus trichopterus* (Pallas, 1770)

Description:

- Body elongate and somewhat laterally compressed.
- Fins are relatively large and rounded.
- Ventral fins are threadlike and carry touch-sensitive cells which are extremely perceptive.
- The body is of hazy, whitish-blue in colour, occasionally with an irregular striped patterning.
- The wild three spot gourami is a brownish fish with yellow gill covers and marked with irregular darker bars on the shoulder.
- There are two dark spots, one at the middle of the body and the other at the base of the tail. The eye is considered as the third spot.
- White spots extend on to the fins.
- Sexual Dimorphism: The male Blue gourami has a longer and more pointed dorsal fin while the female has a shorter and rounded dorsal fin.

Varieties:

- Gold and blue gourami
- Opaline gourami with nipped fins
- Golden gourami

Species Details:

- Distribution:

  Asia: Mekong basin in Laos, Yunnan, Thailand, Cambodia and Viet Nam; Southeast Asia.
- Habitat: Inhabits marshlands, swamps to peatlands and also in flowing waters. Well adapted to impounded water bodies. Can tolerate slightly polluted waters.

- Food and feeding: Omnivorous. In the wild, they take crustaceans, insect larvae and zooplankton. In the aquarium, they eat all types of live, fresh and flake foods. White worms, blood worms, brine shrimps, fresh vegetables etc.
- Feeding frequency: Daily.
- Maximum length: 15.0 cm SL male/unsexed.
- Lifespan: 4-6 years.

Aquarium Aspects:

- Minimum tank size: 35 gallons.
- Suitable for Nano tank: Sometimes.
- Substrate type: Any.
- Lighting needs: Moderate-normal lighting.
- Temperature: 73.0 to 82.0° F (22.8 to 27.8° C).
- Breeding temperature: 80.0° F (26.0° C).
- Range pH: 6.0-8.8.
- Hardness Range: 5-35 dGH.
- Water movement: Weak.
- Water region: All.
- Water changes: Weekly.

Behaviour:

Social—It is a good community fish when small, but not very peaceful like other gouramies.

Reproductive—The male builds a bubble nest, when ready to breed. It also begins to entice the female by swimming back and forth and also flaring fins and raising tail. The female lays about 800 eggs. The male protects the eggs and fry. After fry becomes free swimming, the male is removed.

Remarks: This fish is a beautiful and very hardy and popular in aquarium trade. They use their pelvic fins to feel the environment and their tank mates. They are well known as hydra feeder.

2.5.39 Siamese fighting fish

Figure 2.68

Betta splendens Regan, 1910

Common Name (s): Siamese fighting fish, Betta, split-tailed betta, libby betta.

IUCN Red List Status: Vulnerable (VU).

Systematic Position:

Order: ANABANTIFORMES

Family: OSPHRONEMIDAE—gouramies and fighting fishes

Subfamily: MACROPODUSINAE

Genus: *Betta*

Species: *B. splendens*

Scientific Name: *Betta splendens* Regan, 1910

Description:

- This fish has stout and elongated body and short rounded fins.
- Reddish bars on operculum.
- Dorsal fin spines: 1.
- In the wild form, the body colouration is a dull green and brown.
- They have a special labyrinth organ.
- Bettas nowadays are available in many brilliant colours and colour patterns with incredible fins.
- Both the sexes have been developed through selective breeding.
- Two primary mutation forms, a xanthphorous form (excess of yellow pigmentation) and a black form, and from these multiple varieties have been developed.
- Common colours are: super red, super blue, super yellow, opaque, super black, super white, orange, marble, candy, nemo, galaxy nemo, koi, alien, copper, cellophane, gold, galaxy koi.
- Rare colours are: super orange, metallic, turquoise, lavender, mustard gas, grizzle, green.
- Colour patterns are: solid, bi-colour, Cambodian, butterfly, marble, piebald, full mask, dragon, multicolour, pastel, koi, nemo.
- Some Bettas seen in aquarium trade are: Giant Betta, slim betta, brown red dwarf fighter, wine-red betta, chameleon betta, crescent betta, spotfin betta, spotted betta, Penang betta, blue betta, Borneo betta, Howong betta.
- Bettas have been selectively bred for longer and differently shaped finnage. Some of these are:
    - Plakat betta, short-finned betta.
    - Round tail betta, single tail betta.
    - Veil tail betta.
    - Crown tail betta, fringe tail betta.
    - Comb tail betta.
    - Half-sun betta.
    - Half-moon betta.
    - Delta tail.
    - Super delta tail.
    - Half-moon plakat.
    - Over-half-moon.
    - Rose tail.
    - Double-tail.

- Elephant ear.
- Spade tail.

- Sexual Dimorphism: The male is brightly coloured and possess long colourful fin. Female is rather drab in colour and has shorter fins. A female develops a pale colour to the belly when filled with eggs, and may display horizontal striping when ready to spawn.

Species Details:

- Distribution:

  Asia: Mekong basin.
- Habitat: Inhabits standing water of floodplains, canals, rice fields, medium to large rivers.
- Food and feeding: Omnivorous. They eat zooplankton, mosquito and other insect larvae, crustaceans and aquatic insects. In captivity, they eat all kinds of live, fresh and dry protein foods. High quality protein pellets or flake foods, brine shrimp (live or frozen), blood worms.
- Feeding frequency: Several feedings per day. And feed once or twice a day.
- Maximum length: 6.5 cm TL male/unsexed.
- Life span: 2-3 years.

Aquarium Aspects:

- Minimum tank size: One male 3 gallons, a pair 5-10 gallons.
- Suitable for Nano tank: Yes.
- Substrate type: Any.
- Lighting needs: Moderate to normal lighting.
- Temperature: 75.0 to 86.0° F (23.9 to 30.0° C).
- Breeding temperature: 79.0-82.0° F (26.0-28.0° C).
- Range pH: 6.0-8.0.
- Hardness range: 5-35 dGH.
- Water movement: Weak.
- Water region: All.
- Water changes: Weekly.

Behaviour:

Social—It is a good community fish. Male fishes are completely intolerant of each other and fight to death. Females are less aggressive towards each other but are very hierarchal. They can be kept with each other. The betta is peaceful with other peaceful fish of similar size but similar looking colourful fish should not be kept with.

Reproductive—They are bubble nest builders. Males build bubble nests usually beneath a leaf. The number of eggs can be 200. The eggs sink and the male picks them up in his mouth and place them in his nest. The female must be removed after the eggs are produced, otherwise she will eat them. In about 24-36 hours, the eggs hatch. Fry remain in the nest for the next 2 to 3 days, then they become free swimming.

Remarks: Often called as "the jewel of Orient", this fish is greatly prized for the gorgeous looks and interesting behaviours. They are one of the oldest known freshwater fish to be kept in aquaria.

2.5.40 Spotted gar

Figure 2.69

Lepisosteus oculatus Winchell, 1864

Common Name(s):

English—Spotted gar.

IUCN Red List Status: Not Evaluated.

Systematic Position:

Order: LEPISOSTEIFORMES--gars

Family: LEPISOSTEIDAE--gars

Genus: *Lepisosteus*

Species: *L. oculatus*

Scientific Name: *Lepisosteus oculatus* Winchell, 1864

Description:

- Body cylindrical, dart-like with dark colouration.
- Large number of dark spots on the body, head and fins.
- Elongated mouth with many needle-like teeth.
- In the spotted gar, the distance from the eye to the gill cover is greater than two-thirds the snout length, whereas this distance in the Florida gar is less than two-thirds.
- Diamond-shaped, thick, enamel (ganoid) scales are present.
- Sexual Dimorphism: Females are generally larger than males.

Species Details:

- Distribution:

  North America: Lake Erie and south Lake Michigan drainages south through Mississippi River basin to Gulf Slope drainages from lower Apalachicola River in Florida to Nueces River in Texas, USA.
- Habitat: Inhabits shallow, open waters of depth 3-5 m, quiet, clear pools and backwaters of lowland creeks, small to large rivers, oxbow lakes, swamps and sloughs, occasionally found in brackish waters.
- Food and feeding: Voracious predator. Feeds on different fishes and crustaceans.

- Maximum length: 150 cm TL male/unsexed.
- Life span: 8-10 years.

Aquarium Aspects:

- Water condition: Temperature: 12.0-20.0° C; pH 6.0-8.0; hardness 90-357 ppm.

Behaviour:

Social—Primarily nocturnal. Best kept alone, or kept with fishes of same-size and non-aggressive ones that cannot be fit into its mouth.

Reproductive—More than one male gather in shallow water near vegetation to compete for the larger females. Spawning occurs in between the months of May and July. Eggs are scattered among aquatic vegetation. Females allow more than one male to fertilize their eggs. Females splash and make quick movements at the moment the eggs are deposited. About 20,000 eggs may be deposited by a single female. Parental care is not shown. There is no record of reproduction of these fishes in captivity.

Remarks: A popular fish in public aquariums.

2.5.41 Florida gar

Figure 2.70

Lepisosteus platyrhincus Dekay, 1842

Common Name(s):

English—Florida gar.

IUCN Red List Status: Least Concern.

Systematic Position:

Order: LEPISOSTEIFORMES--gars

Family: LEPISOSTEIDAE--gars

Genus: *Lepisosteus*

Species: *L. platyrhincus*

Scientific Name: *Lepisosteus platyrhincus* Dekay, 1842

Description:

- Body colour dark with several dark spots on the anterior part of its body and head.
- Snout wider and lower jaw wider.
- Plates on ventral surface of the isthmus lacking.
- Small fins are set far back on their elongated torpedo-shaped body.
- The snout is elongated, with the nostrils located at the tip.
- Irregular round black spots on the top of the head and over the body and fins.
- The body colour is olive-brown along the back and upper sides, belly may be white to yellow.

- The young fish may have dark stripes along back and sides.
- It has a broad snout with a single row of irregularly spaced canines on the upper and lower jaws.
- Sexual Dimorphism: Females reach larger sizes than males.

Species Details:

- Distribution:

  North America: USA from Savannah River drainage, Georgia to Ocklockonee River drainage, Florida and Georgia, throughout the Peninsular Florida.
- Habitat: Inhabits sluggish mud or sand-bottomed pools of quiet lowland streams and lakes. They often congregate in spring-fed rivers of Florida.
- Food and feeding: Carnivorous. Young fish feed on zooplankton, insect larvae and small fish. Adults feed primarily on fish, some crustaceans and insects. This fish floats silently near the surface of water, when comes upon a fish, it propels slowly forward secures the prey with its sharp teeth.
- Feeding frequency: Two to three small meals 6 days a week.
- Maximum length: 132 cm TL male /unsexed.

Aquarium Aspects:

- Minimum tank size: 180 gallons.
- Substrate: Sand or mixed sand and gravel substrate with plenty of vegetation.
- Water currents: Medium.
- Lighting: Filtered light is preferred.
- Water conditions: Temperature: 22.0-26.0° C (72.0-80.0° F); pH: 7.0-7.5; dGH: 5-12.

Behaviour:

Social—Temperament is semi-aggressive. Their tank mates should be other large fish species with a peaceful to semi-aggressive temperaments.

Reproductive—Groups of males and females come in shallow, weedy water. Spawning takes place mostly during April and May, which may continue into October. The female spawns by distributing her adhesive eggs into shallow pools, weedy backwaters. Multiple males fertilize the eggs. The hatched young ones have adhesive organs at the end of their snouts and they get attached to vegetation till about 2 cm in length.

Remarks: It is often seen in public aquaria. Not suitable for home aquariums due to their large size and large tank requirements.

Figure 2.37 Silver arowana

Figure 2.38 Black arowana

Figure 2.39 Asian bony tongue

Figure 2.40 Sucker mouth catfish

Figure 2.41 Glass catfish

Figure 2.42 Striped glass catfish

Figure 2.43 A Black bullhead

Figure 2.43 B Black bullhead

Figure 2.44 Neon tetra

Figure 2.45 Silver dollar

Figure 2.46 Black pacu

Figure 2.47 Red-bellied piranha

Figure 2.48 Kenyi cichlid

Figure 2.49 Frontosa cichlid

Figure 2.50 Jaguar cichlid

Figure 2.51 Convict cichlid

Figure 2.52 Severum

Figure 2.53 A Green terror

Figure 2.53 B Green terror

Figure 2.54 Blood parrot cichlid

Figure 2.55 King Kong parrot

Figure 2.56 A Flower horn cichlid

Figure 2.56 B Super red dragon flower horn

Figure 2.56 C Thai Silk Flower

Figure 2.57 Altum angelfish

Figure 2.58 Leopold angelfish

Figure 2.59 Freshwater angelfish

Figure 2.60 Discus

Figure 2.61 A Wild Oscar

Figure 2.61 B Tiger Oscars

Figure 2.61 C Tiger Oscar and Leucistic Oscar

Figure 2.62 Tinfoil barb

Figure 2.63 A Goldfish-Comet, Oranda

Figure 2.63 B Goldfish-Oranda

Figure 2.63 C Pearlscale goldfish

Figure 2.64 A Koi- Gin Rin Kohaku

Figure 2.64 B Koi- Gin Rin Kohaku, Ogon, Kumonryu

Figure 2.65 Guppy

Figure 2.66 A Giant gourami

Figure 2.66 B Giant gourami

Figure 2.67 Three-spot gourami

Figure 2.68 Siamese fighting fish

Figure 2.69 Spotted gar

Figure 2.70 Florida gar

DIVISION 2: BRACKISH WATER ORNAMENTAL FISHES

2.6 Brackish Water Ornamental Fishes of India

Brackish water is the admixture of freshwater and salt water, with salinity 0.5 to 2.0 ppt and specific gravity between 1.005 and 1.010. It includes the estuaries at the confluence of the river system with the sea; large lagoons, brackish water lakes and backwaters; intertidal mud, sand or salt flats; intertidal salt marshes and salt meadows; salt pans and salines as well as brackish water impoundments and the vast areas of mangrove swamps (Nandi, 2001). **Estuary** is defined as a semi-enclosed coastal body of water which has a free connection with the open sea, and within it, sea water is mixed with freshwater (Odum, 1971). The salinity of the estuarine waters varies between 0.5 and 35%. Estuarine water is extremely variable in its salinity (Rao, 2001; De, 2011).

There are about 1.44 million ha of total estuarine water resources of India. State-wise, it is estimated (in lakh ha): West Bengal, 2.10; Odisha,4.17; Andhra Pradesh,0.79; Tamil Nadu, 0.56; Puducherry, 0.01; Kerala, 2.43; Karnataka, 0.08; Goa, 0.12; Maharashtra, 0.01; Gujarat, 3.76; and Andaman and Nicobar, 0.37). The major estuaries, viz., Hooghly-Matlah, Mahanadi, Godavari, Krishna, Cauvery, Vellar and brackish water lakes of Chilka and Pupicat on east coast and the estuaries of the Narmada, Tapi, Mahi, Mandovi-Zuari and backwaters of Kerala and Vembanad lake on west coast are important fishery resources (De, 2011). In view of the brackish water fishery resources in the country, a list of potential brackish water ornamental fishes has been presented (Table.2.3).

Table 2.3: List of potential brackish water ornamental fishes of India

Order	Family	Common Name Scientific Name	Maximum Length (Cm)	Distribution
Myliobatiformes	Dasyatidae	Feathertail stingray *Pastinachus sephen* (Forsskal, 1775)	183 WD unsexed	Native/ Palk Bay, Gulf of Mannar, coastal waters, of Chilka Lake.
		Whiptail stingray *Pateobatis bleekeri* (Blyth, 1860)	105WD Male	Native/East coast and its estuaries, WB, Orissa, Mumbai coast, Palk Bay and Gulf of Mannar.
		Maculabatis gerrardi (Gray, 1851)	100WD unsexed	Native/Indian Ocean, river mouths and coastal waters.
		Leopard stingray *Himantura uarnak* (Gmelin, 1789)	200 WD, unsexed	Native/ along the coast, Pak Bay, and Gulf of Mannar.
		Scaly stingray *Brevitrygon imbricate* (Bloch & Schneider, 1801)	22.0 WD female	Native, both sides of the Indian coast.
Carcharhiniformes	Carcharhinidae	Ganges shark *Glyphis gangeticus* (Muller &Henie, 1839)	204 TL Male	Native/ West Bengal coast, Chilka lake and Pondicherry coast. Ganges-Hooghly river system.
		Spadenose shark *Scoliodon laticaudas* Muller & Henle, 1838	100 TL unsexed	Native/ West and southeast coast, Hooghly estuary, Chilka Lake.
Pristiformes	Pristidae	Small tooth sawfish *Pristis microdon* Latham, 1794	460 SL unsexed	Native/ Orissa, Tamil Nadu, Travancore and Mumbai coast.
		Smooth-tooth sawfish *Pristis pectinate* Latham, 1794	600 TL unsexed	Native/ Bay of Bengal, Chilka Lake and Hooghly estuary.

Order	Family	Common Name Scientific Name	Maximum Length (Cm)	Distribution
Cyprinodontiformes	Aplocheilidae	*Aplocheilus parvus* (Sundara Raj, 1916)	6.25 TL male	Asia: India and Sri Lanka.
Pleuronectiformes	Soleidae	Oriental Sole *Brachirus orientalis* (Bloch & Schneider, 1801)	38.0 TL Male/unsexed	Indo-West Pacific.
Aulopiformes	Synodontidae	Bombay duck *Harpodon nehereus* (Hamilton, 1822)	30 NG unsexed	Native/northwest coast, estuaries of West Bengal, Maharashtra, Gujarat, Tamil Nadu, Orissa, Andhra Pradesh.
Acanthuriformes	Sillaginidae	Gangetic Whiting *Sillaginopsis panijus* (Hamilton, 1822)	3.5 NG, unsexed	Native/ Hooghly estuaries, Tamil Nadu coast and Chilka Lake.
Angulliformes	Muraenidae	Freshwater moray *Gymnothorax tile* (Hamilton, 1822)	60.0 TL unsexed	Native/ throughout Indian coast and enters brackish waters and freshwaters of Hooghly.
	Ophichthidae	Rice-paddy eel *Pisodonophis boro* (Hamilton, 1822)	88.0 SL, mixed	Native/ coast of India and Hooghly estuary, Chilka Lake.
		Long-fin snake eel *Pisodonophis cancrivorus* (Richardson, 1848)	108 TL unsexed	Native/Chilka Lake, east coast.
	Moringuidae	Purple spaghetti eel *Moringua raitaborua* (Hamilton, 1822)	44.0 SL Male	Native/Gangetic estuaries and freshwaters in West Bengal.
Siluriformes	Bagridae	Long-whiskered catfish *Mystus gulio* (Hamilton, 1822)	20.0 NG unsexed	Native/throughout India, estuaries of West Bengal, Chilka Lake, backwaters of Chennai and Cochin.

Order	Family	Common Name Scientific Name	Maximum Length (Cm)	Distribution
	Clariidae	Valenciennes clariid *Clarias dussumieri* Valenciennes, 1840	50.0 TL unsexed	Endemic/ Goa, Karnataka, Kerala, Pondicherry and Tamil Nadu.
	Ariidae	*Ketengus typus* Bleeker, 1846	25.0 TL (male)	Native/seas and estuaries of West Bengal, Orissa and Andaman.
		Soldier catfish *Osteogeneiosus militaris* (Linnaeus, 1758)	39.9TL unsexed	Native/ along continental shelf, Chilka Lake.
	Plotosidae	Canine catfish eel *Plotosus canius* Hamilton, 1822	150NG unsexed	Native/ Chilka Lake, Godavari estuary, Kerala and Maharashtra.
Clupeiformes	Engraulidae	Gangetic hairfin anchovy *Setipinna phasa* (Hamilton, 1822)	40.0 TL, male	Native/ Ganga river system, Orissa river and estuaries.
		Coila reynaldi Valenciennes, 1848	15.0 TL, male	Native/Andhra coast and estuaries in West Bengal.
	Clupeidae	Kelee shad *Hilsa kelee* (Cuvier, 1829)	40.0 TL unsexed	Native/ southeast and southwest coast of India.
		Day's round herring *Dayella malabarica* (Day, 1873)	6.0 SL Male	Endemic/south western India, rivers, estuaries and coastal waters.
		Toli shad *Tenualosa toli* (Valenciennes, 1847)	60.0 TL male	Native/Rivers and along both eastern and western coasts.
Cypriniformes	Cyprinidae	*Horadandia brittani* Rema Devi & Menon, 1992	2.04 SL, male	Endemic/ coastal flood plain of the southern states of Kerala, Tamil Nadu and Karnataka.

Order	Family	Common Name Scientific Name	Maximum Length (Cm)	Distribution
	Balitoridae	Travancore loach *Travancoria jonesi* Hora, 1941	8.4 SL male	Endemic/ Kerala and Western Ghats.
Beloniformes	Hemiramphidae	Congaturi halfbeak *Hyporhamphus limbatus* (Valenciennes, 1847)	20.5 SL unsexed	Native/coastal waters of India.
	Zenarchopteridae	Hooghly half beak *Zenarchopterus striga* (Blyth, 1858)	?	Native/ Hooghly estuary, West Bengal, Zuari river in Goa.
Mugiliformes	Mugiliidae	Long arm mullet *Osteomugil cunnesius* (Valenciennes, 1836)	35.0 TL unsexed	Native/east coast, Chilka Lake and Godavari estuary.
		Large scale mullet *Planiliza macrolepis* (Smith, 1846)	60.0 SL unsexed	Native/coastal waters of Kerala, Orissa and West Bengal, Chilka Lake.
		Flathead mullet *Mugil cephalus* Linnaeus, 1758	100 TL unsexed	Native/ coastal waters in ascends tidal rivers, Chilka Lake, Karnataka.
Elopiformes	Megalopidae	Oxeye tarpon *Megalops cyprinoides* (Broussonet, 1782)	150 TL unsexed	Native/ F.W. and estuaries all along the coast of India.
Blenniformes	Blennidae	Gossamer blenny *Omobranchus ferox* (Here, 1927)	6.7 TL unsexed	Native/ Mangrove swamps and river estuaries.
Spariformes	Lobotidae	Four-barred tigerfish *Datnioides polata* (Hamilton, 1822)	30.0 SL male	Native/estuary of the Ganges.

Order	Family	Common Name Scientific Name	Maximum Length (Cm)	Distribution
Kurtiformes	Kurtidae	Indian humphead *Kurtus indicus* Bloch, 1786	12.6 TL male	Native/East coast.
Syngnathiformes	Syngnathidae	Chilka seahorse *Hippocampus fuscus* Ruppell, 1838	14.4 OT male	Chilka lake, Orissa.
Perciformes	Leiognathidae	Dussumier's ponyfish *Karalla dussumieri* (Valenciennes, 1835)	14.0 TL	Native/ Gulf of Mannar off Mandapam and Tuticorin in Tamil Nadu, Chilka Lake.
	Kuhliidae	Barred flagtail *Kuhlia mugil* (Forster, 1801)	40.0 SL unsexed	Native/coast and also enters freshwaters and estuaries.
		Rock flagtail *Kuhlia rupestris* (Lacepede, 1802)	56.0 TL unsexed	Native/ estuaries.
	Scatophagidae	Spotted scat *Scatophagus argus* (Linnaeus, 1766)	38.0 TL unsexed	Native/ South India, Chilka Lake, coastal waters, Kerala and Karnataka.
	Latidae	*Lates calcarifer* (Bloch, 1790)	200 TL unsexed	Native/ Chilka Lake and Godavari estuary. In estuarine and marine waters.
	Polynemidae	White salmon *Eleutheronema tetradactylum* (Shaw, 1804)	200 TL, unsexed	Native/ Hooghly estuary, Chilka Lake in Orissa, Godavari estuary.
	Teraponidae	Jarbua terapon *Terapon jarbua*	36.0 TL	Native/ Along the coast, Chilka Lake.
	Toxotidae	Spotted archerfish *Toxotes chatareus* (Hamilton, 1822)	40.0 unsexed	Native/ Rivers, estuaries and coastal waters. Fairly common in the Hooghly estuary.

Order	Family	Common Name Scientific Name	Maximum Length (Cm)	Distribution
		Banded archerfish *Toxotes jaculatrix* (Pallas, 1767)	30.0 TL unsexed	Native/ Common in the Hooghly estuary, West Bengal and Mangrove estuaries.
Cichliformes	Cichlidae	*Etroplus maculatus* (Bloch, 1795)	8.0 unsexed	Native/ Tamil Nadu, Kerala, south Karnataka.
		Etroplus suratensis (Bloch, 1790)	50.0 TL unsexed	Native/ coastal regions of Kerala, Tamil Nadu, Andhra Pradesh, Orissa, Chilka Lake and Godana Tamiraparani River basin.
Gobiiformes	Gobiidae	*Glossogobius giurus* (Hamilton, 1822)	50.0 SL unsexed	Native/ occurs throughout India, estuaries in Tamil Nadu, Kerala, Western Ghats, rivers, Maharashtra, coastal waters, rivers and estuaries.
		Callogobius seshaiyal Jacob & Rangarajan, 1960	3.5 SL male	Endemic/ Vellar estuary in Tamil Nadu and Godavari in Andhra Pradesh.
		Pseudogobius javanicus (Bleeker, 1856)	6.0 SL male	Native/ Mumbai, Chennai, Chilka Lake and Port Blair, Ennore estuary, Madras.
		Pseudogobius poisilosoma (Bleeker, 1849)	6.5 TL male	Native/ Sundarbans in West Bengal and Yenna river, Maharashtra.
		Tropical sand goby *Favonigobius reichei* (Bleeker, 1854)	8.3 TL unsexed	Native/Ennore estuary, Madras, Visakhapatnam and Port Blair.

Order	Family	Common Name Scientific Name	Maximum Length (Cm)	Distribution
		Gobiopterus chuno (Hamilton, 1822)	3.0 SL male	Native/ Hooghly estuary in West Bengal, Chilka lake in Orissa and Ennur backwaters in Tamil Nadu.
		Gobius bontii Bleeker, 1849	7.2 TL male	Native/ coastal waters.
		Sumatra eel goby *Caragobius urolepis* (Bleeker, 1852)	8.5 SL male	Native/ Ennore estuary, Madras.
		Bathygobius ostreicola (Chaudhuri, 1916)	5.0 SL male	Endemic/ Chilka Lake, Godavary estuay and east coast of India.
	Oxudercidae	*Stigmatogobius sadanundio* (Hamilton, 1822)	9.0 SL male/unsexed	Native to south Asia from India to Indonesia including Sri Lanka and the Andaman Islands.
		Periophthalmus barbarous (Linnaeus, 1766)	18.0 TL unsexed	Chilka Lake, mouth of river Hooghly, Sunderbans, West Bengal and Port Blair.
		Boleophthalmus bodderti (Pallas, 1770)	16.0 TL mixed	Native/ Chennai, Mumbai, Visakhapatnam, Sundarbans estuary in West Bengal.
		Bolephthalmus dussumieri Valenciennes, 1837	18.7 TL Male	Native/ coastal waters, Western India, Gangetic delta, Mumbai, Sundarbans in West Bengal.
Subseries: Ovalentaria	Ambassidae			

Order	Family	Common Name Scientific Name	Maximum Length (Cm)	Distribution
		Commerson's glassy perchlet *Ambassis ambassis* (Lacepede, 1802)	15.0 TL Male	Chilka Lake, Western Ghats, and coastal waters.
		Bald glassy perchlet *Ambassis gymnocephalus* (Lacepede, 1802)	5.3 SL unsexed	Indo-West Pacific.
		Myops glassy perchlet *Ambassis miops* Gunther, 1872	8.0 SL unsexed	Indo-West Pacific.
		Nalua chanda *Ambassis nalua* (Hamilton, 1822)	12.5 TL male	Asia and Oceania.

2.7 Brackish Water Fishes in the Global Ornamental Fish Trade

Popular brackish water ornamental fishes available in the world market, and which are kept in the home and public aquaria, are as follows (Table 2.4).

Table 2.4: List of brackish water fishes in the global ornamental fish trade

Order	Family	Common Name Scientific Name	Maximum Length (Cm)	Distribution
Myliobatiformes	Dasyatidae	Sharpnose sting ray *Maculabatis gerrardi*	200	Indo-West Pacific.
Cyprinidontiformes	Anablepidae	Four-eyed fish *Anableps* spp.	32.0	Lowlands in southern Mexico to Honduras and northern South America.
	Cyprnodontidae	American flagfish *Jordanella floridae*	6.0	North America.
Pleuronectiformes	Soleidae	Oriental sole *Brachiurus orientalis*	38.0	Indo-West Pacific.
Siluriformes	Ariidae	Silver-tipped shark *Ariopsis seemanni*	35.05	Pacific draining rivers and estuaries in Central and South America.
		Australian shark catfish *Neoarius graeffei*	60.0	Western Pacific.
		Berney's shark catfish *Arius berneyi*	38.0	South Pacific coast areas of Northern Australia and New Guinea.
Anguilliformes	Muraenidae	*Gymnothorax tile*	60.0	Indo-West Pacific.
Tetradontiformes	Tetradontidae	Common Toadfish *Tetractenos hamiltoni*	14.0	Along Australia's eastern coast, New Zealand waters.
		Figure 8 pufferfish *Tetraodon biocellatus*	8.0	Southeast Asia, Mekong (Cambodia), the Peninsular Malaysia, Borneo.

Order	Family	Common Name Scientific Name	Maximum Length (Cm)	Distribution
		Green spotted puffer *Tetraodon nigroviridis*	15.0	South and Southeast Asia in coastal freshwater and brackish water.
Gobiiformes	Oxudercidae	Bumblebee goby *Brachygobius xanthozonus*	3.8	Thailand and Indonesia.
		Barred mudskipper *Periophthalmus argentilineatus*	15.0	Peninsular Malaysia, North Sulawesi, Australia.
		Atlantic mudskipper *Periophthalmus barbarous*	15.0	West African coastline.
		Knight goby *Stigmatogobius sadanundio*	9.0	India, Bangladesh, Pakistan, Sri Lanka, Thailand, Cambodia, Malaysia, Singapore, Indonesia, Fiji.
	Eleotridae	Water cow *Eleotris picta*	44.0	North to South America.
		Violet goby, dragon goby *Gobioides broussonetii*	7.6-61	Near the Atlantic coast of North and South America.
Cichliformes	Cichlidae	Blackchin tilapia *Saratherodon melanotheron*	26.0	Senegal to Zaire, Tropical Africa.
		Orange chromide *Etroplus maculatus*	9.5	Peninsular India and Sri Lanka.
		Green chromide *Etroplus suratensis*	40.0	Western Indian Ocean: India and Sri Lanka

Order	Family	Common Name Scientific Name	Maximum Length (Cm)	Distribution
Acanthuriformes	Lobotidae	Siamese tigerfish *Datnioides microlepis*	45.0	Southeast mainland of Asia.
		New Guinea tigerfish *Datnioides campbelli*	32.0	Asia and Oceania.
Perciformes	Monodactylidae	African moony *Monodactylus sebae*	25.0	Eastern Atlantic.
	Teraponidae	Jarbua terapon *Terapon jarbua*	36.0	Indo-Pacific.
	Toxotidae	Banded archerfish *Toxotes jaculatrix*	30.0	India and most of tropical Asia, Papua New Guinea, northern Australia.
	Scatophagidae	Green scat, ruby scat *Scatophagus argus*	38.0	Around the Indo-Pacific region, to Japan, New Guinea and southeastern Australia.
	Latidae	Barramundi *Lates calcarifer*	200	Indo-West Pacific.
		Spotted archerfish *Toxotes chatareus*	40.0	Asia and Oceania.
Moroniformes	Moronidae	European bass *Dicentrarchus labrax*	103	Eastern Atlantic, Mediterranean and Black Sea.
Spariformes	Sparidae	Gilthead seabream *Sparus aurata*	70.0	Eastern Atlantic, Mediterranean and Black Sea.

2.8 Explicit Endeavour of some Brackish Water Ornamental Fishes

2.8.1 Sharpnose stingray

Figure 2.71

Maculabatis gerrardi (Gray, 1851)

Common Name(s):

English—Sharpnose stingray, Banded whiptail ray, Bluntnose whiptail ray.

IUCN Red List Status: Vulnerable (VU).

Systematic Position:

Phylum: CHORDATA

Subphylum: CRANIATA

Grade: CHONDRICHTHYOMORPHI

Class: CHONDRICHTHYES-cartilaginous fishes

Division: BATOMORPHI-rays

Order: MYLIOBATIFORMES -stingrays

Family: DASYATIDAE-whiptail stingrays

Subfamily: UROGYMNINAE

Genus: *Maculabatis*

Species: *M. gerrardi*

Scientific Name: *Maculabatis gerrardi* (Gray, 1851)

Description:

- Large sized with a rhomboidal disc and narrowly rounded apex.
- Anterior margin of disc almost straight.
- Snout broadly triangular with small tip.
- There is narrow band of denticle band, not extending to tail base (in juveniles, restricted to central disc). Some enlarged denticles present before and after pearl-like tubercular thorn widely spaced.
- A whip-like slender tail present beyond sting. No cutaneous fold is present on tail.
- Tail is banded beyond sting.
- Surface of disc usually paler greenish grey/ greenish brown with several white spots posteriorly.

Species Details:

- Distribution:

  Indo-West Pacific: Oman to Indonesia; north to Taiwan.
- Habitat: Inhabits the inner continental shelf, over sandy and mud bottoms. Also found in coastal waters and river mouths.
- Food and feeding: Carnivorous. Feeds on bottom crustaceans including shrimp, crabs and small lobsters.

- Maximum length: 200 cm TL male/unsexed.

Behaviour:

Reproductive—Ovoviviparous (aplacental viviparity), with embryo feeding initially on yolk, then receiving additional nourishment from the mother by indirect absorption of uterine fluid enriched with mucous, fat or protein through specialised structures. Bears 2 young in summer. Gives birth to litters of 1-4 pups, born at 18-21 cm WD.

Remarks: A public aquarium pet.

2.8.2 Oriental sole

Figure 2.72

Brachiurus orientalis (Bloch & Schneider, 1801)

Common Name(s):

English—Oriental sole.

Systematic Position:

Grade: TELEOSTOMI

Class: OSTEICHTHYES-bony fishes

Subclass: ACTINOPTERYGII--ray-finned fishes

Order: PLEURONECTIFORMES-flatfishes

Family: SOLEIDAE-soles

Genus: *Brachiurus*

Species: *B. orientalis*

Scientific Name: *Brachiurus orientalis* (Bloch & Schneider, 1801)

Description:

- Body oblong, oval and flat, with a tiny pectoral fin on the eyed-side.
- Mouth small, eyes on the right side with a small curved mouth.
- Dorsal and anal fins joined with the tail fin and uniformly dark.
- The eyed-side is greyish brown with dark patches.
- Scales are rough.

Species Details:

- Distribution:
- Indo-West Pacific: Red Sea and Persian Gulf, then off the west coast of India and Sri Lanka, eastward through the Malay Peninsula and Archipelago to China and northern Australia. Mouth of Vivany River in Madagascar.
- Habitat: Inhabits shallow sand and mud bottoms in coastal waters. Enters brackish waters. Enters freshwaters also.
- Food and Feeding: mainly on bottom-dwelling invertebrates, especially small crustaceans.
- Maximum length: 38.0 cm TL male/unsexed.

Aquarium Aspects:

- Minimum tank size: 55 gal (208 L).
- Substrate type: Sand.
- Lighting: Low-subdued lighting.
- Temperature: 74.0-82.0° F (23.3- 27.8° C).
- Water movement: Weak.
- Water region: Bottom.

Remarks: It is a popular aquarium fish.

2.8.3 Indian mud moray

Figure 2.73

Gymnothorax tile (Hamilton, 1822)

Common Name(s):

English—Indian mud moray, Indian mud moray eel, freshwater moray, freshwater snowflake moray.

IUCN Red List Status: Not Evaluated.

Systematic Position:

Order: ANGUILLIFORMES--Eels and morays

Family: MURAENIDAE--moray eels

Subfamily: MURAENINAE

Genus: *Gymnothorax*

Species: *G. tile*

Scientific Name: *Gymnothorax tile* (Hamilton, 1822)

Description:

- Body moderately elongate and is cylindrical in shape in the front side, and compressed along tail.
- Eyes small, teeth very sharp, more or less in two rows, inner series larger.
- Deep coloured body has white speckles all over.
- Sexual Dimorphism: Not reported.

Species Details:

- Distribution:

  Indo- West Pacific: Andaman Islands to Philippines and Indonesia. Known from India and Hawaii.
- Habitat: Occurs in estuaries and river mouths. Found usually in muddy estuaries, also enters lower portion of rivers in 1-10 m.
- Food and feeding: Mostly feeds as a scavenger. Eats dead fish, shrimps and other such foods. Also acts as an obligate parasite, feeding on smaller fishes, crustaceans and others in nature.

Aquarium Aspects:

- Water conditions: Temperature: 20.0-28.0°C; pH: 7.5-9.0; hardness: 179-536 ppm.

Behaviour:

Social—Highly piscivorous. Even attacks larger fishes. It can be maintained in a group, if the tank is very large and arranged in a way, that each individual has its own refuge.

Remarks: It is among the more commonly seen muraenids in the aquarium trade.

2.8.4 Orange chromide

Figure 2.74

Etroplus maculatus (Bloch, 1795)

Common Name(s):

English—Orange chromide.

IUCN Red List Status: Least concern.

Systematic Position:

Order: CICHLIFORMES

Family: CICHLIDAE

Genus: *Etroplus*

Species: *E.maculatus*

Scientific Name: *Etroplus maculatus* (Bloch, 1795)

Description:

- Slightly elongated, oval-shaped fish.
- Body is yellow to bright orange in colour, greenish back and there are several horizontal lines of orange golden spots.
- Three, short transverse bars are located near the mid-section of the body.
- The lower part of the body may be marked with a large black area.
- The belly is light orange.
- The fins are transparent with orange tint.
- The eyes are dark and large. Pelvic and anal fins are sometimes black.
- Sexual Dimorphism: Male fishes have brighter colours than females. They have a red edge to their anal, dorsal and caudal fins. The females have white stripes on the dorsal and anal fins.

Species Details:

- Distribution:

  Asia: India and Sri Lanka.
- Habitat: This species inhabits brackish water estuaries, coastal lagoons and the lower reaches of rivers. Found among marginal roots and weeds in lagoons.

- Food and feeding: A omnivorous fish. Feeds on zooplankton, and algae. It takes tubifex, tablets, pellets, flakes and colour enhancers in captivity. It acts as a cleaner fish, removing parasites from larger green chromide.
- Feeding frequency: Three times daily.
- Maximum length: 9.5 cm TL male/unsexed.
- Life span: 8 years.

Aquarium Aspects:

- Minimum tank size: 20 gallons.
- Substrate: Any.
- Water region: All.
- Water conditions: Temperature: 20.0-28.0° C; pH: 7.0-8.5; hardness: 179-357 ppm.

Behaviour:

Social—Relatively peaceful fish except in the breeding season. They are somewhat gregarious and tends to form groups.

Reproductive—In nature, it spawns in shallow water. Parent fish excavate a soft depression. About 200 eggs are laid and hatch after 5 days. During this time, the parent take care of them. One of the parents remains with the eggs and other one forages. Fry are guarded by parents. Young orange chromides feed on the mucous coating of their parents. This is essential for the survival of the fry. Captive breeding is possible by making the tank environment favourable.

Remarks: This species is popular with the hobbyists. And is kept frequently in the aquaria. It is collected for the international aquarium trade.

2.8.5 Spotted archerfish

Figure 2.75

Toxotes chatareus (Hamilton, 1822)

Common Name(s):

English: Spotted archerfish.

IUCN Red List Status: Not Evaluated.

Systematic Position:

Order: PERCIFORMES (perch-likes)

Family: TOXOTIDAE (archerfishes)

Genus: *Toxotes*

Species: *T. chatareus*

Scientific Name: *Toxotes chatareus* (Hamilton, 1822)

Description:

- The dorsal profile of the body from the tip of the snout to the dorsal fin is almost straight.
- The dorsal fin is located well back on the body.

- It has a variable colour pattern. The body is typically white or silvery with six or seven black spots of varying size.
- Dorsal spines: 5-6; dorsal soft rays: 12-13; anal spines: 3; anal soft rays: 15-17.

Species Details:

- Distribution:

  Asia and Oceania: Sri Lanka and India to New Guinea and northern Australia.
- Habitat: Inhabits brackish mangrove estuaries but are also found in freshwater rivers and small streams. Found in coastal waters, along the shores of flowing and standing waters, .shaded areas with overhanging vegetation, medium to large-sized rivers, in lakes and billabongs.
- Food and feeding: At the daytime, feed at the surface on floating debris including insects and vegetable matters. Also feed on crustaceans and small fishes. It has the habit of spitting to dislodge the insect prey from tree branches above the water. Maximum shooting range is about 150 cm. Feed primarily on swimming and floating insects.
- Maximum length: 40.0 cm SL male/unsexed.

Behaviour:

Reproductive—Breeding takes place in wet season. Highly fecund females release buoyant eggs. About 20,000 to 150,000 eggs are laid, measuring about 0.4 mm in diameter. Breeding takes place both in fresh and brackish waters.

Remarks: A popular fish in trade.

2.8.6 Spotted scat

Figure 2.76 A & B

Scatophagus argus (Linnaeus, 1766)

Common Name(s):

English—Spotted butterfish, Argus fish, green scat, Red scat.

IUCN Red List Status: Least Concern (LC).

Systematic Position:

Order: PERCIFORMES

Family: SCATOPHAGIDAE

Genus: *Scatophagus*

Species: *S. argus*

Scientific Name: *Scatophagus argus* (Linnaeus, 1766)

Description:

- Body is deep and compressed, squarish in shape, with bluntly pointed and triangular head.
- Rostro-dorsal profile strongly ascending and followed by a deep concavity above the eye.

- Mouth is small, teeth are small and movable.
- Notched dorsal fin with 10- 11 strong spines and 17 soft rays.
- Anal fin with 4 spines and 13-14 soft rays.
- Body dull greenish above with numerous large round brown spots. Belly silvery.
- Two colour versions of Spotted Scat, the Red Scat and the Green Scat are available. The body of Green Scat is a shiny silver with a light greenish cast. The Red Scat is with a reddish cast specially on the back and the fins.
- Sexual Dimorphism: not known.

Species Details:

- Distribution:

  Indo-Pacific: Kuwait to Fiji, north to southern Japan, south to New Caledonia. Reported from Samoa, Tonga and Society Islands.
- Habitat: Found in harbours, brackish estuaries, natural embayments, lower reaches of freshwater streams, and among mangroves.
- Food and feeding: Omnivorous. They feed on a variety of plant matter along with worms, crustaceans and insects. In captivity, they eat any fish food provided. They take flake food, pellets, vegetables (dried seaweed, lettuce), meaty food (brine shrimp, krill, *Mysis*, shrimp and blood worms).
- Feeding frequency: Daily.
- Maximum length: 38.0 cm TL male/unsexed.
- Life span: 20 years.

Aquarium Aspects:

- Minimum tank size: 60 gallons.
- Suitable for Nano tank: No.
- Substrate type: Sand/gravel mix.
- Lighting needs: Moderate-normal lighting.
- Temperature: 68.0-82.0° F (20.0 to 27.8° C).
- Range pH: 7.5-8.5.
- Hardness Range: 12-18 dGH.
- Water movement: Moderate.
- Water region: All.
- Water changes: Weekly, salts to be added.

Behaviour:

Social—Peaceful species that do best in groups. Very lively and outgoing type, with semi-aggressive temperament.

Reproductive—In the wild, they spawn in saltwater reefs. Fry come to the freshwater to grow. They return to a more brackish water later on. Aquarium breeding is not yet reported.

Remarks: It is a popular aquarium fish.

2.8.7 Barramundi

Figure 2.77

Lates calcarifer (Bloch, 1790)

Common Name (s):

English—Barramundi, Asian Sea bass.

IUCN Red List Status: Least concern.

Systematic Position:

Order: PERCIFORMES

Family: LATIDAE

Genus: *Lates*

Species: *L. calcarifer*

Scientific Name: *Lates calcarifer* (Bloch, 1790)

Description:

- Body elongated, compressed, with a large, slightly oblique mouth. The dorsal head profile is clearly concave.
- Upper jaw extends beyond the eye.
- Lower jaw longer, no canines present.
- The lower edge of the preopercular margin is serrated with a strong spine at its angle. Operculum is with serrated flat and a small spine.
- Two dorsal fins united at their base, the first one with 7-9 spines, dorsal soft rays 10-11; anal spines 3; anal soft rays: 7-8.
- Caudal fin rounded.
- Colour grey with green on back, abdomen silvery, golden tinge is present sometimes on the back.

Species Details:

- Distribution:

  Indo-West Pacific: eastern edge of the Persian Gulf to China, Taiwan and southern Japan, southward to southern Papua New Guinea and northern Australia.
- Habitat: Found in coastal waters, estuaries and lagoons, in clear to turbid water. Diadromous in nature, inhabits rivers before returning to the estuaries to spawn.
- Food and feeding: Carnivorous. Feeds on crustaceans, molluscs and smaller fish. Juveniles feed on zooplankton.
- Maximum length: 200 cm TL male/unsexed.

Aquarium Aspects:

- Water temperature: 26.0-32.0° C; pH: 7.5-8.5; salinity: 10-30 ppt.

Behaviour:

Social—They are not aggressive. In aquaria, they are quite tame. Their feeding reflex is sudden and always violent. They can swallow small tank mates.

Reproductive—They are hermaphrodite in a sequential way. In one season, most individuals mature as males and after one spawning season become females. They spawn on the full moon and new moon, at the beginning of an incoming tide. This carries the eggs into the estuary. At the beginning of the monsoon, male fish migrate down river for breeding. Females lay large number of eggs. Neither eggs nor fry get parental care. They require brackish water to develop.

Remarks: Young fishes are popular in the aquarium trade. They are considered as very entertaining during feeding time.

2.8.8 European bass

Figure 2.78

Dicentrarchus labrax (Linnaeus, 1758)

Common Name(s):

English—European bass, European seabass, bass.

IUCN Red List Status: Least Concern.

Systematic Position:

Order: MORONIFORMES—temperate basses

Family: MORONIDAE—temperate basses

Genus: *Dicentrarchus*

Species: *D. labrax*

Scientific Name: *Dicentrarchus labrax* (Linnaeus, 1758)

Description:

- Body rather elongate, mouth terminal and moderately protractile.
- Cycloid scales present on the upper part of head.
- Posterior edge of the operculum finely serrated, lower margin with strong forward-directed denticles.
- Caudal fin moderately forked.
- Dorsal fin spines: 8-10; dorsal fin soft rays: 12-13; anal fin spines:3; anal fin soft rays: 10-12.
- Body silvery grey in colour.
- Young ones with some dark spots on upper part of body. No spots in adults only a diffuse spot on the edge of operculum.

Species Details:

- Distribution:

  Eastern Atlantic: Norway to Morocco, the Canary Islands and Senegal. And in the Mediterranean and in the Black Sea.
- Habitat: Adults are found in coastal waters up to 100 m depth, but

generally found in shallow waters. Inhabit littoral zones of estuaries, lagoons sometimes rivers, coastal waters and river mouths in summer. In colder time migrate offshore and in winter, found in deep waters in the northern zone.

- Food and feeding: Adults are predators and mainly piscivorous. They feed on small fish, crabs, shrimps and molluscs. Juveniles feed on invertebrates.
- Maximum length: 103 cm TL male/unsexed.

Aquarium Aspects:

- Water conditions: Temperature: 5.0-28.0° C; salinity: 3% to full strength sea water.

Behaviour:

Social—Young European bass form school but adult fishes are less gregarious. It is a seasonally migratory species, moving further inshore and north in summer.

Reproductive—Spawn in groups. Near the British Isles, spawning takes place in the spring, and in its southern zone, it takes place earlier. Eggs are pelagic. Eggs are with 1-2 fat drops. Embryo development takes about 3 days at 13-14° C and larval development takes 40 days at 19° C.

Remarks: A popular fish in public aquariums.

2.8.9 Gilthead bream

Figure 2.79

Sparus aurata Linnaeus, 1758

Common Name(s): Gilthead seabream, orate.

IUCN Red List Status: Least Concern (LC).

Systematic Position:

Order: SPARIFORMES—breams and porgies.

Family: SPARIDAE—porgies.

Genus: *Sparus*.

Species: *S. aurata*.

Scientific Name: *Sparus aurata* Linnaeus, 1758

Description:

- Body is tall with snout which is more than twice as long as the eye diameter.
- There is a large black spot on the gill cover.
- Dorsal fin spines: 11; dorsal fin soft rays: 13-14; anal fin spines: 3; anal fin soft rays: 11-12.

Species Details:

- Distribution:

  Eastern Atlantic: British Isles, Strait of Gibraltar to Cape Verde and around the Canary Islands. And the Mediterranean and the Black Sea.
- Habitat: Inhabit seagrass beds and sandy bottoms and the surf zone up to the depths of about 30 m. Adults may be found up to 150 m depth.

- Food and feeding: Carnivorous fish sometimes herbivorous. Eat shellfish such as mussels and oysters.
- Maximum length: 70.0 cm TL male/unsexed.

Behaviour:

Social—Sedentary in habit. Found either as solitary or in small aggregations.

Reproductive—Protandric hermaphrodite fish. This fish matures as male and then in the 2nd or 3rd year, matures as female. The months of October to November is the spawning time and then takes place throughout the period. Incubation takes place in 2 days. Larval stages last for about 50 days.

Remarks: A popular public aquarium fish.

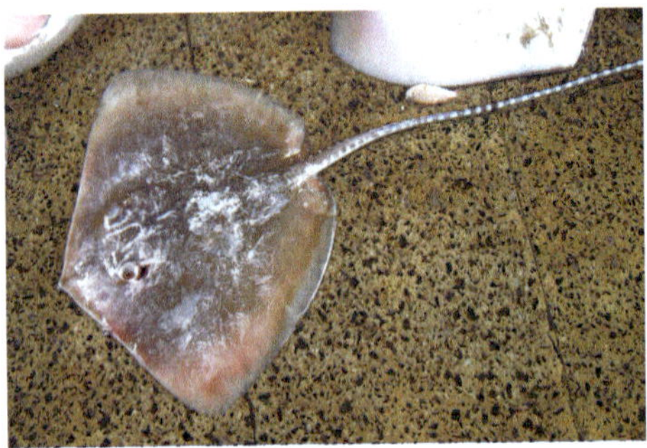

Figure 2.71 Sharpnose stingray

Figure 2.72 Oriental sole

Figure 2.73 Indian mud moray

Figure 2.74 Orange chromide

Figure 2.75 Spotted archerfish

Figure 2.76 A Spotted scat

Figure 2.76 B Spotted scat

Figure 2.77 Barramundi

Figure 2.78 European bass

Figure 2.79 Gilthead bream

DIVISION 3: MARINE ORNAMENTAL FISHES

2.9 Marine Ornamental Fishes of India

About 2500 fish species inhabit Indian oceans and India is the treasury of more than 200 varieties of marine ornamental fishes (Swain et al. 2011). More than 4,000 fish species inhabit coral and rocky areas and also live among marine plants. Out of these reef fishes, about 800 species as traded as ornamental fishes. The most commonly traded family of marine ornamental fishes from India are Pomacentridae, Pomacanthidae, Acanthuridae, Labridae, Gobidae, Chaetodontidae, Callionymidae, Microdesmidae, Serranidae and Blennidae. They are abundant in the Gulf of Mannar, Palk Bay and Gulf of Kutch along the mainland coast and in the Lakshadweep and Andaman Islands. Among all these regions, the lagoons and the reef flats in the Lakshadweep group of islands are the richest both in regard to the abundance of fishes and the number of species (Murty, 1996; Murty et al.,1998; Ramakrishna and Venkataraman, 2001).

2.9.1 Ornamental fishes in the Lakshadweep islands:

There are 36 islands in this region and the lagoons and reef flats harbour rich ornamental fish fauna. Over 300 species belonging to about 35 families are known for their attractive colours and shapes (Venkataraman and Alfred, 2001).

Table 2.5: Ornamental fishes in the Lakshadweep

Common name	Scientific names of dominant species	Family
Surgeon fish	*Acanthurus triostegus* *Ctenochactus strigosus* *Acanthurus matoides* *A.leucosternon,A.lineatus* *A.nigricauda, A.elongatus,* *Zebrasoma veliferum, Naso lituratus, N.unicornis, N.brevirostris, N.tuberosus*	Acanthuridae
Trigger fish	*Rhineacanthus aculeatus, R. rectangulus, Balistapus undulates, Balistoides viridescens*	Balistidae
Parrot fish	*Callyodon taeniurus, C. bataviensis, Cryptotomus spinidens,Callyodon sordidus,Leptoscarus vaigiensis,Callyodon scaber*	Callyodontidae
Butterfly fish	*Chaetodon auriga, C. trifasciatus, C. falcula, C. citrinellus,C.lunula, Megaprotodon strigangulus, Heniochus acuminatus*	Chaetodontidae
Angel fish	*Centropyge multispinis* *Pomacenthodes imperator*	Pomacanthidae
Squirrel fish	*Neoniphon samara* *Myripristis murdjan, Sargocentron punctatissimum,S. candimaculata, S. diadema,* *S. violaceus, S. spinifer,* *Myripristis adustus*	Holocentridae
Wrasses	*Halichoeres scapularis,* *H. centriquadrus, H.marginatus, Thalassoma hardwickii, Gomphosus caeruleus, G. varius, Stethojulius trilineata, S. albovittata, S. strigiventer, Cheilinus trilobatus*	Labridae
Damsel fish	*Dascyllus reticulatus, D. aruanus, Chromis caeruleus, C. chrysurus, Pomacentrus nigricans, P. albicaudatus, P. melanopterus, Abudefduf sexfasciatus, A. lacrymatus, A. dickii, A. biocellatus, A. xanthozona, A. glaucus*	Pomacentridae
Goat fish	*Mulloidichthys samoensis,* *M. auriflamma, Parupeneus barberinus, P. macronemus, P. bifasciatus*	Mullidae
Groupers	*Epinephelus merra, E. hexagonatus, Cephalopholis argus*	Serranidae
Puffer fish	*Canthigaster margaritatus*	Canthigasteridae
Moorish idol	*Zanclus cornutus*	Zanclidae

2.9.2 Marine ornamental fishes from the Andaman Islands:

These islands are situated in the Bay of Bengal and offer habitats like rocky coasts with tidal pools, bays, mudflats and extensive backwaters. About 150 species of ornamental fishes are known to be available here.

Table 2.6: Common marine ornamental fishes from Andamans

Common name	Scientific name of the species	Family
Sea horse/pipe fish	*Hippocampus kuda, Microphis brachiurus*	Syngnathidae
Squirrel fish	*Holocentrus rubrum,*	Holocentridae
Groupers	*Cephalopholis argus, C. leopardus, C.miniate, Epinephelus merra*	Serranidae
Bat fish	*Platax orbicularis*	Ephippidae
Angel fish	*Pomacanthus imperator, P. annularis*	Pomacanthidae
Butterfly fish	*Chaetodon auriga, C. falcula, C. trifasciatus, C. vagabunda, C. lincolatus, Heniochus acuminatus*	Chaetodontidae
Damsel fish	*Abudefduf saxatilis, Amphiprion percula, Chromis dimidiatus, C.caeruleus, Dascyllus aruanas*	Pomacentridae
Wrasses	*Anampses caeruleopunctatus, Gomphosus caeruleus, G. varius, Thalassoma hardwickii*	Labridae
Parrot fish	*Callyodon sordidus*	Callyodontidae
Rabbit fish	*Siganus spinus, S. stellatus, S. virgatus*	Siganidae
Moorish idol	*Zanclus cornutus*	Zanclidae
Surgeon fish	*Acanthurus leucosternon, A.lineatus, A.triostegus, Ctenochaetus strigasus, Naso unicornis, Zebrasoma veliferum.*	Acanthuridae
Scorpion fish	*Pterois volitans*	Scorpaenidae
Trigger fish	*Balistapus undulates, B. viridis*	Balistidae

2.9.3 Ornamental fishes from the Gulf of Mannar and the Palk Bay:

This region has several islands with extensive coral reefs around them. About 100 species of ornamental fishes belonging to about 30 families are known from this region (Varghese et al. 2011).

Table 2.7: Important marine ornamental fishes from Gulf of Mannar and Palk Bay

Common name	Scientific name of the species	Family
Sea horse/ pipe fish	*Micrognathus brevirostris, Syngnathoides biaculeatus*	Syngnathidae
Squirrel fish	*Sargocentron diadema, Holocentrum rubrum*	Holocentridae
Groupers	*Cephalopholis boenack, C. pachycentron, Epinephelus undulosus, E. merra, E. miniatus*	Serranidae

Common name	Scientific name of the species	Family
Goat fish	*Upeneus tragula, U. sulphurous, U. vittatus, U. displurus,U. bensasi,U. oligopilus, U. luzonius, U. sundaicus, Parupeneus indicus,P. pleurospilus, Mulloidichthys flavolineatus*	Mullidae
Bat fish	*Platax tiera, P. orbicularis*	Platacidae
Butterfly fish	*Heniochus acuminatus, Chaetodon auriga, C. trifasciatus, C. plebeius, C. octofasciatus, C. melanotus, C. xanthocephalus, C. collare*	Chaetodontidae
Damsel fish	*Amphiprion sebae, Abudefduf saxatilis, A. septemfasciatus, A. biocellatus*	Pomacentridae
Angel fish	*Pomacanthodes annularis*	Pomacanthidae
Wrasses	*Stethojulis axillaris, S. phaekadopleura, S. interrupta, Cheilinus chlourus, C. trilobatus, C. undulates, Epibulus insidiator, Thalassoma lunare, Cheilio inermis, Halichoeres, nigriscens, H. leucurus*	Labridae
Parrot fish	*Callyodon ghobban, C. dussumieri, Leptoscarus vaigiensis*	Callyodontidae
Rabbit fish	*Siganus oramin, S. javus*	Siganidae
Moorish idol	*Zanclus cornutus*	Zanclidae
Surgeon fish	*Acanthurus matoides, A. gahm, A. strigosus, Naso brevirostris, Zebrasoma veliferum*	Acanthidae
Scorpion fish	*Pterois volitans, P. miles, Gymnapistes dracaena*	
Trigger fish	*Odonus niger, Balistes flavimarginatus*	Balistidae
File fish	*Osbeckia scripta, Paramonocanthus choiocephalis,*	Monocanthidae
Box fish	*Rhyncostacion nasus, Lactoria cornuta*	Ostraciidae
Puffer fish	*Tetradon immaculatus, T. hispidus, T. reticularis*	Tetradontidae
Puffer fish	*Canthigaster margaritatus*	Canthigasteridae

Lakshadweep region is the most important one in regard to abundance of ornamental fishes and in the diversity of species. This region being very shallow, is vulnerable for environmental degradation by human intervention. Therefore, the exploitation on a commercial scale is likely to result not only in quick overexploitation of the fish species but also in the destruction of coral environment.

2.10 Popular Marine Ornamental Fishes in the Trade

A large number of exquisite marine fishes are popular pets in the home and largely in the public aquariums.

Table 2.8: List of marine popular ornamental fishes

CARTILAGINOUS FISHES.

- GRADE CHONDRICHTHYOMORPHI
- Class CHONDRICHTHYES---cartilaginous fishes

Order	Family	Common Name Scientific Name	Maximum Length (Cm)	Distribution
CARCHARHINIFORMES (ground sharks)	CARCHARHINIDAE (requiem sharks)	Blacktip reef shark *Carcharhinus melanopterus*	200	Tropical coral reefs of the Indian and Pacific Oceans.
		Whitelip reef shark *Triaenodon obesus*	213	Entire Indo-Pacific region.
	SCYLIORHINIDAE (cat sharks)	Coral cat shark *Atelomycterus marmoratus*	70.0	Shallow coral reefs across the Indo-West Pacific, from Pakistan to New Guinea.
		Marbled cat shark *Atelomycterus macleaya*	60.0	Eastern Indian Ocean, endemic to Western Australia.
	TRIAKIDAE (hound sharks)	Banded hound shark *Triakis scyllium*	150	North western Pacific Ocean from the southern Russian Far East to Taiwan.
		Brown smooth-hound *Mustelus henlei*	100	Continental shelves of the subtropical eastern Pacific, from northern California to the gulf of California as well as Ecuador and Peru.
		Leopard shark *Triakis semifasciata*	198	Pacific coast of North America, from the U.S. state of Oregon to Mozatlan in Mexico.

Order	Family	Common Name Scientific Name	Maximum Length (Cm)	Distribution
		Gray smooth hound *Mustelus californicus*	116	Continental shelves of the subtropical eastern Pacific, from northern California to the Gulf of California.
	SPHYRNIDAE (hammerhead shark)	Bonnethead shark *Sphyrna tiburo*	213	American coast.
LAMINIFORMES (mackerel sharks)	ODONTASPIDIDAE (sand tiger sharks)	Sand Tiger shark *Carcharias taurus*	330	Indo-West Pacific.
ORECTOLOBIFORMES (carpet sharks)	HEMISCYLLIIDAE (bamboo shark)	Brown banded bamboo shark *Chiloscyllium punctatum*	132	Indo-West Pacific from Japan to northern Australia.
		Whitespotted bamboo shark *Chiloscyllium plagiosum*	83.0	Coral reefs of Pacific Ocean.
		Epaulette shark *Hemiscyllum ocellatum*	107	Shallow tropical waters off Australia and New Guinea.
	GINGLYMOSTOMATIDAE (nurse sharks)	Nurse shark *Ginglymostoma cirratum*	430	Tropical and subtropical coastal waters of Eastern Atlantic, Western Atlantic and Eastern Pacific.
	STEGOSTOMATIDAE (Zebra sharks)	Zebra shark *Stegostoma fasciatum*	354	Throughout the tropical Indo-Pacific.
	ORECTOLOBIDAE (wobbegongs)	Tasselled wobbegong *Eucrossorhinus dasypogon*	125	Shallow coral reefs off northern Australia, New Guinea, and adjacent islands.

Order	Family	Common Name Scientific Name	Maximum Length (Cm)	Distribution
		Japanese wobbegong *Orectolobus japonicus*	118	Tropical western Pacific Ocean from Japan and Korea to Viet Nam and the Philippines.
		Ornate wobbegong *Orectolobus ornatus*	290	Australia and possibly other countries in the Western Pacific Ocean.
		Spotted wobbegong *Orectolobus maculatus*	320	Endemic to Australia.
		Northern wobbegong *Orectolobus wardi*	63.0	Western Pacific Ocean around Australia.
HETERODONTIFORMES (bullhead sharks)	HETERODONTIDAE (bullhead sharks)	Horn sharks *Heterodontus francissi*	122	Endemic to the coastal waters off the western coast of North America, from California to the Gulf of California.
		Port Jackson shark *Heterodontus portusjacksoni*	165	Coastal region of southern Australia, including the waters off Port Jackson.
RAJIFORMES (skates)	RAJIDAE (skates)	Undulate ray *Raja undulata*	100	Eastern Atlantic: southern Ireland and England to Senegal.
PRISTIFORMES guitarfishes and sawfishes	RHINOBATIDAE (guitar fishes)	Blackchin guitar fish *Glaucostegus cemiculus*	180	Warm temperate and tropical waters of the Mediterranean Sea and the eastern Atlantic Ocean.
	RHINIDAE (bowmouth guitarfishes)	Bowmouth guitarfish *Rhina ancylostoma*	300	Indo-Pacific.

Order	Family	Common Name Scientific Name	Maximum Length (Cm)	Distribution
		Eastern fiddler ray *Trygonorrhina fasciata*	126	Endemic to Australia
		Shovelnose guitarfish *Rhinobatos productus*	119	Central California south to the Gulf of California.
		Eastern shovelnose ray *Aptychotrema rostrate*	100	Endemic to subtropical and temperate waters of Australia's east coast.
TORPEDINIFORMES	TORPEDINIDAE	Leopard torpedo ray *Torpedo panthera*	100	Djibouti, Egypt, Eritrea, India, Iran, Oman, Pakistan, Saudi Arabia, Somalia, Sudan and Yemen.
	NARCINIDAE	Lesser electric ray *Narcine bancroftii*	100	Western coastal fringes of the Atlantic Ocean and Caribbean Sea.
MYLIOBATIFORMES (sting rays)	MYLIOBATIDAE (eagle rays)	Javanese cownose ray *Rhinoptera javanica*	150	Indo-West Pacific.
		Bat ray *Myliobatis californica*	180	Eastern Pacific Ocean between the Oregon coast and the Gulf of California, around Galapagos islands.
		Cownose ray *Rhinoptera bonasus*	213	Western Atlantic and Caribbean from New England, United States to southern Brazil.
	DASYATIDAE (whiptail stingrays)	Southern stingray *Hypanus americanus*	67.0 male 150 female	Tropical and subtropical waters of the Western Atlantic Ocean from New Jersey to Southern Brazil.

Order	Family	Common Name Scientific Name	Maximum Length (Cm)	Distribution
		Bluespotted ribbontail ray *Taeniura lymma*	35.0	Tropical Indian and western Pacific Oceans.
		Bluespotted stingray *Neotrygon kuhlii*	70.0	Northern Australia, Kenya, Madagascar. The island of Mauritius, Somalia, the east coast of South Africa, India. The entire continental waters of Asia.
		Atlantic stingray *Hypanus sabinus*	61.0	Atlantic coast of North America from Chesapeake Bay to Mexico.
		Bluntnose stingray *Hypanus say*	100	Coastal waters of the western Atlantic Ocean from the U.S. state of Massachusetts to Venezuela.
	GYMNURIDAE	California butterfly ray *Gymnura marmorata*	100	Colombia, Costa Rica, Ecuador, EI Salvador, Guatemala, Hondurus, Mexico, Nicargua, Panama, Peru and the United States.
	UROLOPHIDAE	Common stingaree *Trygonoptera testacea*	47.0	Eastern Australia
		Striped stingaree *Trygonoptera ovalis*	61.0	Endemic to southwestern Australia.
	PLATYRHINIDAE	Thornback ray *Platyrhinoidis triseriata*	91.0	Tomales Bay to the Gulf of California.
	UROTRYGONIDAE	Round stingray *Urobatis helleri*	58.0	Beaches of southern California.

Order	Family	Common Name Scientific Name	Maximum Length (Cm)	Distribution
		Cortez round stingray *Urobatis maculatus*	42.0	Endemic to Mexico.
		Bulls eyes round stingray *Urobatis concentricus*	47.5	Endemic to Mexico.
		Leopard round stingray *Urobatis pardalis*	46.2	Endemic to the Tropical Eastern Pacific, from Costa Rica to Columbia.
		Yellow stingray *Urobatis jamaicensis*	76.0	Tropical western Atlantic Ocean from North Carolina to Trinidad.
		Chilean round ray *Urotrygon chilensis*	41.9	Chile, Colombia, Costa Rica, Ecuador, El Salvador, Guatemala, Honduras, Mexico, Nicaragua, Panama and Peru.

Teleost Fishes.

- GRADE: TELEOSTOMI
- Class: OSTEICHTHYES—bony fishes
- Subclass: ACTINOPTERYGII—ray-finned fishes
- Division: TELEOSTEOMORPHA
- Subdivision: Teleostei

Order	Family	Common Name Scientific Name	Maximum Length cm)	Distribution
(ANEMONEFISHES)				
Subseries: Ovalentaria	POMACENTRIDAE (Anemonefishes and damselfishes) Subfamily: Amphiprioninae	Cinnamon anemonefish *Amphiprion melanopus*	12.0	Western and southern parts of the Pacific Ocean.
		Clarkii anemonefish *Amphiprion clarkii*	15.0	Tropical waters from the Indian Ocean to the Western Pacific.
		Common clownfish *Amphiprion ocellaris*	11.0	Eastern Indian Ocean and in the western Pacific Ocean.
		Black common clownfish *Amphiprion ocellaris*	11.0	Darwin area of Northern Australia.
		Maroon clownfish *Premnas biaculeatus*	17.0	Indo-Pacific from western Indonesia to Taiwan and the Great Barrier Reef.
		Pink sunk anemonefish *Amphiprion perideraion*	10.0	Northern Australia through the Malay Archipelago and Melanesia.
		Tomato clownfish *Amphiprion frenatus*	14.0	Western Pacific, from the Japan to Indonesia.
		True Percula *Amphiprion percula*	11.0	Warmer waters in the Pacific and Indian Oceans.
		Saddleback anemonefish *Amphiprion polymnus*	13.0	In the centre of the Indo-Pacific area, known as the Coral Triangle. Also found in northern Australia, Melanesia and to the Ryukyu Islands of southern Japan.

Order	Family	Common Name Scientific Name	Maximum Length cm)	Distribution
		Sebae anemonefish *Amphiprion sebae*	16.0	Northern Indian Ocean, from Java to the Arabian Peninsula.
(ANGELFISHES)				
PERCIFORMES	POMACANTHIDAE	Arabian angelfish, Asfur angelfish *Pomacanthus asfur*	40.0	Eastern coasts of Africa.
		Blue ring angelfish, annularis angelfish *Pomacanthus annularis*	30.0	Indo-West Pacific oceans.
		Blueface angelfish *Pomacanthus xanthometopon*	40.0	Coral reefs in the eastern part of the tropical Indo-Pacific.
		Cortez angelfish *Pomacanthus zonipectus*	46.0	Eastern Pacific.
		Emperor angelfish *Pomacanthus imperator*	40.0	Reef in Indian and Pacific Oceans.
		French angelfish *Pomacanthus paru*	41.0	Western Atlantic.
		Gray angelfish *Pomacanthus arcuatus*	60.0	Western Atlantic.
		Half-moon angelfish *Pomacanthus maculosus*	50.0	Throughout the Persian Gulf, the north western Indian Ocean and the Red Sea.
		Koran angelfish *Pomacanthus semicirculatus*	40.0	Indo-West Pacific Ocean.

Order	Family	Common Name Scientific Name	Maximum Length cm)	Distribution
		Majestic angelfish *Pomacanthus navarchus*	30.0	Indo-Pacific region.
		Bellus angelfish *Genicanthus bellus*	18.0	Coral reefs, in the western Pacific and eastern Indian Ocean.
		Blue angelfish *Holacanthus bermudensis*	45.0	Western Atlantic.
		Bluespotted angelfish *Chaetodontoplus caeruleopunctatus*	21.0	Western Pacific Ocean.
		Griffis angelfish *Apolemichthys griffisi*	25.0	Pacific Ocean.
		Passer angelfish, King angelfish *Holacanthus passer*	36.0	Reefs in the eastern Pacific Ocean.
		Personifer angelfish *Chaetodontoplus meridithii*	37.0	Eastern Australia.
		Queen angelfish *Holacanthus ciliaris*	45.0	Western Atlantic Ocean.
		Rock beauty *Holacanthus tricolor*	25.0	Reefs in the tropical western Atlantic Ocean to the northern Gulf of Mexico.
		Royal angelfish *Pygoplites diacanthus*	25.0	Tropical Indo-Pacific oceans.
		Scribbled angelfish *Chaetodontoplus duboulayi*	25.0	Northern Australia and Southern New Guinea.
		Japanese swallow angelfish *Genicanthus semifasciatus*	21.0	Western Pacific.

Order	Family	Common Name Scientific Name	Maximum Length cm)	Distribution
		Yellowtail angelfish *Apolemichthys xanthurus*	15.0	Corals in Western Indian Ocean such as Maldives and the east coast of India.
		Barred angelfish *Centropyge multifasciata*	12.0	Indo-Pacific.
		Bicolor angelfish *Centropyge bicolor*	15.0	Indo-Pacific.
		Blue velvet angelfish *Centropyge deborae*	7.0	One location in Fiji.
		Brazilian flameback angelfish *Centropyge aurantonotus*	8.0	Southern Caribbean Sea and the coastal waters of Brazil.
		Coral beauty angelfish *Centropyge bispinosa*	10.0	Coral reefs and lagoons in Indo-Pacific.
		Cherubfish, pygmy angelfish *Centropyge argi*	8.0	Caribbean and Gulf of Mexico, North to North Carolina.
		Eibli angelfish *Centropyge eibli*	15.0	Reefs in the Indo-Pacific.
		Flame angelfish *Centropyge loricula*	15.0	Tropical waters of the Pacific Ocean.
		Half-black angelfish *Centropyge vroliki*	12.0	Tropical waters of the Indo-West Pacific area.
		Herald's angelfish *Centropyge heraldi*	10.0	Pacific Ocean.
		Keyhole angelfish *Centropyge tibicen*	19.0	Tropical waters of the Indo-Pacific.

Order	Family	Common Name Scientific Name	Maximum Length cm)	Distribution
		Lemonpeel angelfish *Centropyge flavissima*	14.0	Lagoons and reefs in the Indo-Pacific region.
		Multicolor angelfish *Centropyge multicolour*	9.0	Pacific Ocean.
		Orange-back angelfish *Centropyge acanthops*	8.0	East coast of Africa.
		Pacific pygmy angelfish *Centropyge flavicauda*	8.0	Indo-Pacific Ocean.
		Potter angelfish *Centropyge potteri*	10.0	Central Pacific Ocean.
		Rusty angelfish *Centropyge ferrugata*	10.0	Western Pacific Ocean.
		Venustus angelfish *Centropyge venustus*	12.0	Western Pacific.
(ANTHIAS):				
PERCIFORMES	SERRANIDAE (sea basses)	Bicolor anthias *Pseudanthias bicolor*	13.0	Indo-Pacific Ocean.
		Bartlett's anthias *Pseudanthias bartlettorum*	9.0	Pacific Ocean.
		Cooper's anthias *Pseudanthias cooperi*	14.0	Indo-Pacific Ocean.
		Diadem anthias *Pseudanthias parvirostris*	7.0	Indo-West Pacific Ocean.
		Lyretail anthias, Sea goldie *Pseudanthias squamipinnis*	15.0	Western Indian Ocean and in the Pacific Ocean.

Order	Family	Common Name Scientific Name	Maximum Length cm)	Distribution
		Orangehead anthias *Pseudanthias heemstrai*	13.0	Western Indian Ocean.
		Redbar anthias *Pseudanthias rubrizonatus*	12.0	Western Pacific Ocean.
		Squareback anthias *Pseudanthias pleurotaenia*	20.0	Reef in the Pacific Ocean.
		Stocky anthias *Pseudanthias hypseosoma*	19.0	Indo-Pacific Ocean.
		Threadfin anthias *Pseudanthias huchtii*	12.0	Western Central Pacific Ocean.
(BASS AND GROUPERS:)				
PERCIFORMES	SERRANIDAE	African grouper *Cephalopholis taeniops*	69.0	Eastern Atlantic.
		Blacktip grouper *Epinephelus fasciatus*	40.0	Throughout the tropical waters of the Indo-Pacific area.
		Blue and yellow grouper *Epinephelus flavocaeruleus*	90.0	Throughout the tropical waters of the Indian Ocean.
		Blue dot grouper *Cephalopholis argus*	50.0	Indo-Pacific.
		Blue line grouper *Cephalopholis formosa*	34.0	Indo-West Pacific.
		Chalk bass *Serranus tortugarum*	8.0	Western Atlantic.
		Coney grouper *Cephalopholis fulva*	41.0	Western Atlantic.

Order	Family	Common Name Scientific Name	Maximum Length cm)	Distribution
		Golden stripe soapfish *Grammistes sexlineatus*	30.0	Throughout the tropical and subtropical waters of the Indo-Pacific area.
		Harlequin bass *Serranus tigrinus*	29.0	Western Atlantic.
		Leaflip grouper *Pogonoperca punctate*	35.0	Indo-Pacific.
		Marine beta *Calloplesiops altivelis*	20.0	Reef-associated, Indo- Pacific Ocean.
		Miniatus grouper *Cephalopholis miniatus*	45.0	Throughout the tropical waters of the Indo-West Pacific area.
		Orange spot grouper *Cephalopholis panamensis*	30.0	Eastern Pacific Ocean.
		Panther grouper *Cromileptes altivelis*	75.0	Tropical waters of the central Indo-West Pacific region.
		Painted comber *Serranus scriba*	36.0	Eastern Atlantic Ocean, the Mediterranean Sea, and the Black Sea.
		Polleni grouper *Cephalopholis polleni*	43.0	Indo-Pacific Ocean.
		Red flag grouper *Cephalopholis urodeta*	28.0	Tropical waters of the Indo-Pacific area.
		Lyretail grouper *Variola louti*	80.0	Red Sea to South Africa and the Pitcairn Islands, Southern Japan and Australia.
		Saddle grouper *Plectropomus laevis*	125	Indo-Pacific.

Order	Family	Common Name Scientific Name	Maximum Length cm)	Distribution
		Spotted grouper *Epinephelus summana*	52.0	Eastern Indian Ocean.
		Strawberry grouper *Cephalopholis spiloparaea*	30.0	Tropical waters of the Indo-Pacific area, Red Sea excluded.
		V tail grouper *Cephalopholis urodelus*	28.0	Indo-Pacific Ocean.
(BASSLETS AND ASSESSORS)				
Ovalentaria	GRAMMATIDAE (basslets)	Black cap gramma *Gramma melacara*	6.0	Reefs in the tropical western Atlantic Ocean.
		Brazilian gramma *Gramma brasiliensis*	6.0	Reef in the Atlantic Coast of Brazil.
		Royal gramma *Gramma loreto*	5.0	Reef of tropical western Atlantic Ocean.
	PLESIOPIDAE (roundheads)	Blue assessor *Assessor macneilli*	7.0	Southwest Pacific Ocean.
		Yellow assessor *Assessor flavissimus*	7.0	Western Central Pacific.
(BLENNIES AND ENGINEER GOBIES)				
BLENNIIFORMES (Blennies)	CHAENOPSIDAE (tube blennies)	Spinyhead blenny *Acanthemblemaria spinosa*	2.0	Tropical western Atlantic Ocean and Caribbean Sea.
		Sailfin blenny *Emblemaria pandionis*	5.0	Western Atlantic Ocean and the Caribbean Sea.
	BLENNIIDAE (combtooth blennies)	Bicolour blenny *Ecsenius bicolor*	11.0	Indo-Pacific.
		Black combtooth blenny *Ecsenius namieyei*	10.0	Tropical waters of the western Pacific Ocean from the Philippines to Solomon Islands.

Order	Family	Common Name Scientific Name	Maximum Length cm)	Distribution
		Linear blenny *Ecsenius lineatus*	9.0	Throughout the tropical waters of the Indo-West Pacific area.
		Midas blenny *Ecsenius midas*	13.0	Tropical waters of Indo-Pacific area.
		Red Sea mimic blenny *Ecsenius gravieri*	8.0	Western Indian Ocean.
		Tail spot blenny *Ecsenius stigmatura*	6.0	Western Pacific.
		Two-spot blenny *Ecsenius bimaculatus*	4.5	Tropical waters of the western Pacific Ocean.
		Blackline fang blenny *Meiacanthus nigrolineatus*	9.0	Western Indian Ocean.
		Bundoon blenny *Meiacanthus bundoon*	8.0	Pacific Ocean.
		Canary fang blenny *Meiacanthus oualanensis*	5.0	Western Central Pacific, known from Fiji.
		Striped blenny *Meiacanthus grammistes*	12.0	Western Pacific Ocean.
		Black sailfin blenny *Atrosalarias fuscus*	10.0	Tropical waters of the Indian Ocean, Red Sea included.
		Blue & gold blenny *Enchelyurus flavipes*	5.0	Western Pacific Ocean.
		Ember blenny *Cirripectes stigmaticus*	12.0	Indo-Pacific.
		Lawnmower blenny *Salarias fasciatus*	14.0	East Africa and the Red Sea to Samoa and the Islands of Micronesia.

Order	Family	Common Name Scientific Name	Maximum Length cm)	Distribution
		Segmented sailfin blenny *Salarias segmentatus*	10.0	Western Central Pacific.
		Starry blenny *Salarias ramosus*	14.0	Western Central Pacific.
		Molly Miller blenny *Scartella cristata*	12.0	Atlantic, Mediterranean and northwest Pacific Oceans.
		One spot blenny *Crossosalarias macrospilus*	10.0	Western Pacific.
		Red lip blenny *Ophioblennius atlanticus*	19.0	Tropical central Atlantic Ocean.
	LABRISOMIDAE	Diamond blenny *Malacoctenus boehlkei*	6.5	Among corals of the Central western Atlantic Ocean and the Caribbean Sea.
CICHLIFORMES	PHOLIDICHTHYIDAE (convict blenny)	Engineer goby *Pholidichthys leucotaenia*	34.0	West-central Pacific Ocean.
(BOXFISH AND BLOWFISH)				
TETRADONTIFORMES (Plectognaths)	TETRADONTIDAE (puffers)	Dogface pufferfish *Arothron nigropunctatus*	33.0	Tropical waters of the Indian Ocean to the central islands of the Pacific Ocean.
		Golden puffer *Arothron meleagris*	48.0	Indo-Pacific and the Eastern Pacific.
		Immaculate puffer *Arothron immaculatus*	28.0	Indo-West Pacific.
		Map puffer *Arothron mappa*	65.0	Tropical and subtropical waters from the Indian Ocean to the western Pacific Ocean.

Order	Family	Common Name Scientific Name	Maximum Length cm)	Distribution
		Star puffer *Arothron stellatus*	120	Shallow water in the Indo-Pacific region.
		Stars and stripes puffer *Arothron hispidus*	50.0	Indo-Pacific area, Red Sea to the eastern Pacific Ocean.
		Striped dogface puffer *Arothron manilensis*	31.0	Tropical waters of the central Indo-Pacific.
		Hawaiian blue puffer *Canthigaster papua*	10.0	Throughout the tropical waters of the Indian Ocean and of the western Pacific Ocean.
		Hawaiian spotted puffer *Canthigaster jactator*	9.0	Hawaiian Islands.
		Sharpnose pufferfish *Canthigaster rostrate*	12.0	Western Central Atlantic.
		Valentini pufferfish *Canthigaster valentine*	11.0	Tropical and subtropical waters of the Indian Ocean, and Red Sea and until the oceanic islands of the Pacific Ocean.
		Whitebelly puffer *Canthigaster bennetti*	10.0	Indo-Pacific.
	OSTRACIIDAE (boxfishes)	Helmet cowfish *Tetrasomus gibbosus*	30.0	Indo-West Pacific, and Mediterranean Sea.
		Longhorn cowfish *Lactoria comuta*	46.0	Indo-Pacific region.
		Scribbled boxfish *Ostracion solorensis*	12.0	Western Pacific Ocean.
		Yellow boxfish *Ostracion cubicus*	45.0	Reefs throughout the Pacific Ocean and Indian Ocean as well as the south eastern Atlantic Ocean.

Order	Family	Common Name Scientific Name	Maximum Length cm)	Distribution
	DIODONTIDAE (porcupinefish)	Porcupine pufferfish *Diodon holocanthus*	50.0	Atlantic and western Indian Ocean, Pacific Ocean.
		Spotfin porcupinefish *Diodon hystrix*	91.0	Tropical and subtropical waters of the world, including the Mediterranean Sea.
(BUTTERFLYFISHES)				
PERCIFORMES (perches)	CHAETODONTIDAE (butterflyfishes)	Auriga butterflyfish, threadfin butterflyfish *Chaetodon auriga*	23.0	Indo-Pacific.
		Banded butterflyfish *Chaetodon striatus*	16.0	Tropical western Atlantic Ocean.
		Bluecheek butterflyfish, Addis butterflyfish *Chaetodon semilarvatus*	23.0	Western Indian Ocean: Red Sea and Gulf of Aden.
		Copperband butterflyfish *Chelmon rostratus*	20.0	Reefs in Pacific and Indian Oceans.
		Four-eyed butterflyfish *Chaetodon capistratus*	15.0	Western Atlantic Ocean.
		Latticed butterflyfish *Chaetodon rafflesii*	18.0	Indo-Pacific.
		Longnose butterflyfish *Forcipiger flavissimus*	22.0	Tropical waters of the Indo-Pacific area.
		Merten's butterflyfish *Chaetodon mertensii*	12.5	Pacific Ocean.
		Pacific double saddle butterflyfish *Chaetodon ulietensis*	15.0	Central Indo-Pacific region.

Order	Family	Common Name Scientific Name	Maximum Length cm)	Distribution
		Pennant coralfish *Heniochus acuminatus*	25.0	Tropical and subtropical waters of the Indo-Pacific.
		Racoon butterflyfish *Chaetodon lunula*	20.0	Tropical Indo-Pacific region.
		Redback butterflyfish *Chaetodon paucifasciatus*	14.0	Red Sea and the Gulf of Aden, East Africa.
		Reef butterflyfish *Chaetodon sedentarism*	15.0	Western Atlantic Ocean.
		Saddleback butterflyfish *Chaetodon ephippium*	30.0	Indian and Pacific Oceans.
		Schooling bannerfish *Heniochus diphreutes*	21.0	Indo-Pacific.
		Sickle butterflyfish *Chaetodon falcula*	20.0	Indian Ocean from eastern Africa and to east to Indonesia.
		Teardrop butterflyfish *Chaetodon unimaculatus*	20.0	Indian Ocean and Pacific Ocean.
		Tinker's butterflyfish *Chaetodon tinkeri*	15.0	Tropical Pacific Ocean.
(CARDINALFISHES)				
Subseries:Gobiida				
KURTIFORMES (cardinalfishes and nurseryfishes)	APOGONIDAE (cardinalfishes)	Banggai cardinal *Pterapogon kaudemi*	8.0	Banggai Islands of Indonesia.
		Blackstripe cardinalfish *Apogon nigrofasciatus*	10.0	Indo-Pacific.

Order	Family	Common Name Scientific Name	Maximum Length cm)	Distribution
		Bluebarred cardinalfish *Apogon flores*	5.0	Flores, Indonesia.
		Bluestreak cardinalfish *Apogon leptacanthus*	6.5	Red Sea and off Mozambique Island to Samoa and Tonga north to Ryuku Islands, and south to New Caledonia and Micronesia.
		Fragile cardinalfish *Apogon fragilis*	5.0	Indo-West Pacific.
		Flamefish *Apogon maculatus*	11.0	Western Atlantic.
		Frostfin cardinalfish *Apogon hoeveni*	5.0	Indo-West Pacific north to Japan and south to Australia.
		Gilbert's cardinalfish *Apogon gilberti*	5.0	Western Central Pacific.
		Girdled cardinalfish *Archamia zosterophora*	8.0	Western Pacific Ocean.
		Orange-striped cardinalfish *Apogon cyanosoma*	8.0	Indo-West Pacific.
		Ochre-striped cardinalfish *Apogon compressus*	12.0	Indo-West Pacific.
		Ringtailed cardinalfish *Apogon aureus*	14.5	Red Sea and off East Africa to Papua New Guinea, north to Japan and south to Australia.
		Orbic cardinalfish *Sphaeramia orbicularis*	10.0	Indo-Pacific.

Order	Family	Common Name Scientific Name	Maximum Length cm)	Distribution
		Pajama cardinalfish *Sphaeramia nematoptera*	8.5	Western Pacific Ocean.
(CHROMIS):				
Ovalentaria	POMACENTRIDAE (damselfishes) Subfamily: Chrominae	Ambon chromis *Chromis amboinensis*	9.0	Western Pacific.
		Barrier reef chromis *Chromis nitida*	10.0	East coast of Australia.
		Black and gold chromis *Neoglyphidodon nigrosis*	13.0	Indo-West Pacific.
		Black and white chromis *Chromis margaritifer*	9.0	Tropical waters of the central Indo-Pacific region until the oceanic islands of the central Pacific Ocean.
		Black-axil chromis *Chromis atripectoralis*	12.0	Tropical Indo-Pacific.
		Black bar chromis *Chromis retrofasciata*	5.0	Western Pacific Ocean.
		Blue chromis *Chromis cyanea*	13.0	Bermuda, southern Florida and the Caribbean Sea.
		Damselfish *Chromis chromis*	25.0	Eastern Atlantic and Mediterranean.
		Green chromis *Chromis viridis*	10.0	Indo-Pacific including the Red Sea.
		Half and half chromis *Chromis iomelas*	9.0	Pacific Ocean, Great Barrier reef, reef around northern New Guinea, Samoa and the Society Islands.

Order	Family	Common Name Scientific Name	Maximum Length cm)	Distribution
		Limbanghi chromis *Chromis limbanghi*	10.0	Eastern Central Pacific.
		Lined chromis *Chromis lineata*	7.0	Indo-Pacific.
		Paletail chromis *Chromis xanthura*	15.0	Pacific Ocean.
		Spiny chromis *Acanthochromis polyacanthus*	14.0	Western Pacific.
		Sunshine chromis *Chromis insolatus*	16.0	Western Atlantic.
		Yellowspotted chromis *Chromis flavomaculata*	15.0	Western Pacific.
(CLINGFISHES)				
GOBIESOCIFORMES (clingfishes)	GOBIESOCIDAE (clingfishes)	Yellow stripe clingfish *Diademichthys lineatus*	5.0	Tropical waters of Indo-West Pacific region.
(DAMSELFISHES)				
Ovalentaria	POMACENTRIDAE (damselfishes) Subfamily: Pomacentrinae	Ambon damsel *Pomacentrus amboinensis*	10.0	Western Pacific.
		Azure damsel *Chrysiptera hemicyanea*	10.0	Eastern Indian Ocean and western Pacific.
		Black and gold damsel *Neoglyphidodon nigrosis*	13.0	Indo-West Pacific.
		Blackmargined damsel *Pomacentrus nigromarginatus*	9.0	Western Pacific.

Order	Family	Common Name Scientific Name	Maximum Length cm)	Distribution
		Blue damsel, orange tail damsel *Chrsiptera cyanea*	8.5	Indian and western Pacific Oceans.
		Blue and gold damsel *Pomacentrus coelestis*	9.0	Indo-Pacific.
		Blue velvet damsel *Paraglyphidodon oxyodon*	15.0	Pacific Ocean.
		Blueback damsel *Pomacentrus simsiang*	9.0	Indo-West Pacific.
		Bluefin damsel *Neoglyphidodon melas*	18.0	Indo-West Pacific.
		Blueline demoiselle, yellowfin demoiselle *Chrysiptera caeruleolineata*	6.0	Indian and western Pacific Oceans.
		Caerulean damsel *Pomacentrus caeruleus*	8.0	Western Indian Ocean.
		Canary deep- water damsel *Chrysiptera galba*	9.0	Western Pacific Ocean.
		Cloudy damsel *Dascyllus carneus*	7.0	Tropical waters of the Indian Ocean from the eastern coast of Africa to Java Sea.
		Cross' damsel *Neoglyphidodon crossi*	10.0	Western and Central Pacific.
		Domino damsel *Dascyllus trimaculatus*	14.0	Indo-Pacific.

Order	Family	Common Name Scientific Name	Maximum Length cm)	Distribution
		Fiji blue devil damsel *Chrysiptera taupou*	8.0	Western Pacific Ocean from the Coral sea to Samoa.
		Fourstripe damsel *Dascyllus melanurus*	10.0	Western Pacific Ocean.
		Garibaldi damsel *Hypsypops rubicunda*	30.0	Subtropical North eastern part of the Pacific Ocean.
		Honey head damsel *Dischistodus prosopotaenia*	17.0	Eastern Indian Ocean and western Pacific.
		Hawaiian Dascyllus *Dascyllus albisella*	12.5	Eastern Central Pacific.
		Jewel damsel *Microspathodon chrysurus*	20.0	Tropical areas.
		King demoiselle *Chrysiptera rex*	7.0	Eastern Indian Ocean and western Pacific.
		Lemon damsel *Pomacentrus moluccensis*	9.0	Western Pacific Ocean.
		Longfin Gregory *Stegastes diencaeus*	12.5	Western Atlantic.
		Marginated damsel *Dascyllus marginatus*	6.0	Western Indian Ocean.
		Neon damsel *Pomacentrus alleni*	6.0	Eastern Indian Ocean.
		Ocellate damsel *Pomacentrus vaiuli*	9.0	Pacific Ocean.
		Pavo damsel *Pomacentrus pavo*	11.0	Indo-Pacific.

Order	Family	Common Name Scientific Name	Maximum Length cm)	Distribution
		Pink Smith damsel *Pomacentrus smithi*	7.0	Western Central Pacific.
		Rolland's demoiselle *Chrysiptera rollandi*	6.0	Tropical waters of the central Indo Pacific region.
		Sergent major damsel *Abudefduf saxatilis*	22.9	Atlantic Ocean.
		Speckled damsel *Pomacentrus bankanensis*	9.0	Western Pacific.
		Springer's damsel *Chrysiptera springeri*	5.5	Western Pacific Ocean.
		Stark's damsel *Chrysiptera starcki*	7.0	Western Pacific Ocean.
		Talbots damsel *Chrysiptera talboti*	8.0	Eastern Indian Ocean and western Pacific.
		Three stripe damsel *Dascyllus aruanus*	10.0	Tropical waters of the Indo-Pacific region, Red Sea included.
		Three spot damsel *Stegastes planifrons*	13.0	Western Atlantic.
		Tuxedo damsel *Chrysiptera tricinta*	6.0	Western Pacific Ocean.
		Two stripe damsel *Dascyllus reticulatus*	10.0	Tropical waters of the central Indo-Pacific region.
		Yellow damsel *Amblyglyphidodon aureus*	13.0	Central Indo- Pacific.

Order	Family	Common Name Scientific Name	Maximum Length cm)	Distribution
		Yellow threespot Dascyllus *Dascyllus auripinnis*	14.5	Eastern Central Pacific.
		Yellow belly damsel *Pomacentrus auriventris*	5.5	Western Central Pacific.
		Yellow tail Dascyllus *Dascyllus flavicaudus*	12.0	Eastern Central Pacific.
		Yellow tail damsel *Chrysiptera parasema*	8	Indo-Pacific.
		Yellow tail demoiselle *Neopomacentrus azysron*	7.5	Indo-West Pacific.
	PSEUDOCHROMIDAE (dottybacks) Subfamily: Congrogadinae (eelblennies)	Carpet eelblenny *Congrogadus subducens*	45.0	Coastal parts of the Indo-West Pacific.
(DARTFISHES)				
GOBIIFORMES	GOBIIDAE	Blue gudgeon dartfish *Plereleotris hanae*	13.0	Western Pacific Ocean.
		Fire fish *Nemateleotris magnifica*	9.0	Indian and Pacific Oceans.
		Purple fire fish *Nemateleotris decora*	9.0	Tropical waters in the Indo-West Pacific.
		Scissortail dartfish *Ptereleotris evides*	14.0	Indian Ocean and the western Pacific Ocean.
		Zebra barred dartfish *Ptereleotris zebra*	10.0	Indian Ocean and the western Pacific Ocean.

Order	Family	Common Name Scientific Name	Maximum Length cm)	Distribution
(DOTTYBACKS)				
Subseries: Ovalentaria	PSEUDOCHROMIDAE (dottybacks)	Allen's dottybacks *Manonichthys alleni*	12.0	Indonesia and Malaysia.
		Australian multicolour pseudochromis *Ogilbyina novaehollandiae*	10.0	Western Central Pacific, Great Barrier Reef.
		Bicolor pseudochromis *Pseudochromis paccaguellae*	6.0	Western Pacific.
		Blue flavivertex pseudochromis *Pseudochromis flavivertex*	8.0	Western Indian Ocean.
		Bluelined dottyback *Pseudochromis cyanotaenia*	6.1	Western Pacific.
		Brown dottyback, yellow pseudochromis *Pseudochromis aureus*	10.0	South -Western Pacific Ocean and the eastern Indian Ocean.
		Cherry dottyback *Pholidochromis cerasina*	7.9	Western Pacific.
		Dilectus dottyback *Pseudochromis dilectus*	7.62	Western Indian Ocean.
		Dusky dottyback *Pseudochromis fuscus*	10.0	South Western Pacific Ocean and the eastern Indian Ocean.
		Elongate dottyback *Pseudochromis elongatus*	6.4	Western Pacific.

Order	Family	Common Name Scientific Name	Maximum Length cm)	Distribution
		Firetail dottyback *Pseudochromis flammicauda*	5.6	Australia.
		Fridmani Pseudochromis, orchid dottyback *Pseudochromis fridmani*	8.0	Red Sea.
		Longfin dottyback *Manonichthys polynemus*	12.0	Indonesia and nearby islands, in the southern Philippines and Palau.
		Lyretail dottyback *Pseudochromis steenei*	12.0	Indonesia.
		Neon dottyback *Pseudochromis aldabraensis*	10.0	Western Indian Ocean.
		Oblique-lined dottyback *Cypho purpurascens*	7.4	Western Pacific.
		Orangetail dottyback *Pseudochromis coccinicauda*	5.8	Western Indian Ocean.
		Purple stripe pseudochromis *Pseudochromis diadema*	6.0	Coral reefs in the Western Central Pacific Ocean, eastern Malay Peninsula, and western Philippine waters.
		Red dottyback *Labracinus cyclophthalmus*	22.0	Western Pacific.
		Sailfin pseudochromis *Pseudochromis veliferus*	12.0	Western Central Pacific.

Order	Family	Common Name Scientific Name	Maximum Length cm)	Distribution
		Splendid pseudochromis *Pseudochromis splendens*	13.0	Indo-West Pacific.
		Springeri pseudochromis *Pseudochromis springeri*	5.0	Red Sea and eastern Indian Ocean.
		Striped dottyback *Pseudochromis sankeyi*	8.0	Western Indian Ocean.
		Strawberry pseudochromis *Pseudochromis porphyreus*	6.0	Western Pacific.
		Twolined dottyback *Pseudochromis bitaeniatus*	6.9	Western Pacific.
(DRAGONETS) CALLIONYMIFORMES	CALLIONYMIDAE	Mandarinfish *Synchiropus splendidus*	6.0	Pacific ranging from the Ryukyu Islands south to Australia.
		Starry dragonet *Synchiropus stellatus*	12.0	Indian Ocean.
		Ocellated dragonet *Synchiropus ocellatus*	8.0	Southwest Pacific Ocean.
		Spotted mandarin *Synchiropus picturatus*	10.0	Indo-West Pacific.
(EELS) ANGUILLIFORMES	MURAENIDAE	Banded eel *Echidna poyzona*	69.0	Indo-Pacific.
		Black edge moray eel *Gymnothorax saxicola*	60.0	Western Atlantic.
		Blue ribbon eel, black ribbon eel *Rhinomuraena quaesita*	130	Indo- Pacific.

Order	Family	Common Name Scientific Name	Maximum Length cm)	Distribution
		Chain link moray eel *Echidna catenata*	165	Western Atlantic Ocean and from islands elsewhere in the Atlantic.
		Dragon moray eel *Enchelycore pardalis*	92.0	Throughout the Indo-Pacific oceans.
		Golden dwarf eel *Gymnothorax melatremus*	70.0	Western Atlantic Ocean, including the Caribbean Sea and the Gulf of Mexico.
		Golden moray eel *Gymnothorax miliaris*	70.0	Western Atlantic Ocean, including the Caribbean Sea and the Gulf of Mexico.
		Green moray eel *Gymnothorax funebris*	250	Western Atlantic from New Jersey, Bermuda, and the northern Gulf of Mexico to Brazil.
		Jeweled moray eel *Muraena lentiginosa*	61.0	Eastern Pacific.
		Kidako moray eel *Gymnothorax kidako*	91.0	Throughout the western to central Ocean Pacific area.
		Peppered moray *Gymnothorax picta*	135	Indo-Pacific and from islands of the tropical Eastern Pacific.
		Snow flake eel *Echidna nebulosa*	100	Throughout the Indo-Pacific area.
		Tessalata eel or laced moray *Gymnothorax favagineus*	300	Throughout the Indo-West Pacific area.
		Yellow head moray eel *Gymnothorax fimbriatus*	80.0	The tropical waters of the Indo-Pacific area.

Order	Family	Common Name Scientific Name	Maximum Length cm)	Distribution
		Yellow mouth moray eel *Gymnothorax nudivomer*	178	Throughout the Indo-Pacific area.
		Whitemouth moray eel *Gymnothorax meleagris*	120	Throughout the Indo-Pacific area.
		Zebra moray *Gymnomuraena zebra*	150	Throughout the Indo- Pacific area.
	CONGRIDAE	Spotted garden eel *Heteroconger hassi*	40.0	Indo-Pacific.
	OPHICHTHIDAE (snake eels and worm eels)	Banded snake eel *Myrichthys colubrinus*	97.0	Indo-Pacific.
		Spotted snake eel *Myrichthys maculosus*	100	Indo-Pacific.
(FILEFISHES)				
TETRADONTIFORMES (plectognaths)	MONACANTHIDAE (filefishes)	Clown filefish *Cantherhines dumerili*	38.0	Coral reefs of the Indian and Pacific Oceans.
		Colored filefish *Pervagor melanocephalus*	15.0	Indo-West Pacific.
		Fantail orange filefish *Pervagor spilosoma*	18.0	Reefs in the Hawaiian Islands.
		Horseshoe filefish *Meuschenia hippocrepis*	51.0	Eastern Indian Ocean in the temperate waters off southern and western Australia.
		Japanese filefish *Paramonacanthus japonicuss*	13.0	Indo-West Pacific.
		Mimic filefish *Paraluteres prionurus*	11.0	Indian and Pacific Oceans

Order	Family	Common Name Scientific Name	Maximum Length cm)	Distribution
		Orange spotted filefish *Oxymonacanthus longirostris*	13.0	Coral reefs in the Indo-Pacific Oceans.
		Tassle filefish *Chaetodermis penicilligerus*	30.0	Throughout the tropical waters of the Indian Ocean, to the central islands of the Pacific Ocean.
(FLATFISHES)				
PLEURONECTIFORMES (flatfishes)	SOLIEDAE	Banded sole *Soleichthys heterorhinos*	18.0	Eastern Indian Ocean and Western Pacific.
	BOTHIDAE (left eye flounders)	Flowery (Indo-Pacific peacock) flounder *Bothus mancus*	45.0	Shallow waters in the Indo-Pacific, warmer parts of the east Pacific.
		Peacock flounder *Bothus lunatus*	46.0	Warmer parts of the Atlantic including the Caribbean.
(FROGFISHES)				
LOPHIIFORMES (anglerfishes)	ANTENNARIIDAE (frogfishes)	Longlure frogfish *Antennarius multiocellatus*	20.0	Among sponge population in Western Atlantic.
		Giant anglerfish *Antennarius commerson*	38.0	Tropical and subtropical waters from the Indian Ocean to the eastern coasts of the Pacific Ocean.
		Sargassum frogfish *Histrio histrio*	20.0	Among Sargassum seaweed in parts of Atlantic Ocean and the Indo-Pacific Ocean.
		Striated frogfish *Antennarius striatus*	15.0	Indian Ocean to the centre of the Pacific Ocean, and in the Atlantic Ocean.
		Wartskin frogfish *Antennarius maculatus*	10.0	Tropical waters of the Indo-Pacific area.

Order	Family	Common Name Scientific Name	Maximum Length cm)	Distribution
(GOATFISHES)				
PERCIFORMES	MULLIDAE (goatfishes)	Bicolor goatfish *Parupeneus barberinoides*	25.0	Coral reef in Western Pacific Ocean.
		Goldsaddle goatfish *Parupeneus cyclostomus*	51.0	Coral reefs in the Indo-Pacific.
		Manybar goatfish *Parupeneus multifasciatus*	30.0	Coral reefs in Eastern Indian Ocean, Pacific Ocean.
		Yellowback goatfish *Parupeneus barberinus*	41.0	Indian and Pacific Oceans.
(GOBIES AND CLINGFISHES)				
GOBIIFORMES (gobies)	GOBIIDAE (gobies)	Black barred convict goby *Priolepis nocturna*	4.0	Indian Ocean and the western Pacific Ocean.
		Black clown goby *Gobiodon acicularis*	3.9	Tropical coral reefs and eelgrass of the Western Central Pacific Ocean.
		Blue spotted watchman goby *Cryptocentrus pavoninoides*	12.0	Western Central Pacific Ocean.
		Catalina goby *Lythrypnus dalli*	5.0	Eastern Pacific.
		Cave transparent goby *Coryphopterus glaucofraenum*	8.0	Reef in Western Atlantic Ocean and the Caribbean Sea from North Carolina to Brazil.
		Citron clown goby *Gobiodon citrinus*	8.0	Reef in the Indian Ocean and Western Pacific Ocean to Japan, Samoa and the Great Barrier Reef.

Order	Family	Common Name Scientific Name	Maximum Length cm)	Distribution
		Court jester goby *Amblygobius rainfordi*	6.0	Tropical reefs of the Western Pacific Ocean.
		Diagonal bar prawn goby *Amblyeleotris diagonalis*	8.0	Reefs in Indian Ocean and the Western Pacific Ocean.
		Diamond watchman goby *Valenciennea puellaris*	20.0	Indian Ocean and the Western Pacific Ocean.
		Dracula goby *Stonogobiops dracula*	8.0	Around the Seychelles and the Maldives.
		Gold neon eviota goby *Eviota pellucida*	3.0	Throughout the tropical waters of the Pacific Ocean.
		Green banded goby *Tigrigobius multifasciatus*	3.5	Western Atlantic Ocean.
		Green clown goby *Gobiodon atrangulatus*	4.0	Western Pacific around Japan and Fiji.
		Hector's goby *Amblygobius hectori*	8.0	Indian Ocean to the islands of Micronesia.
		Hi fin red banded goby *Stonogobiops nematodes*	5.0	Indian Ocean and Western Pacific Ocean.
		Neon goby *Elacantius oceanops*	5.0	Atlantic and Gulf coast of North America from Florida to Belize.
		Orange marked goby *Amblygobius decussatus*	8.0	Tropical reefs in the Western Pacific Ocean, may be widespread in the central Indo-Pacific area.
		Orange spotted goby *Amblyeleotris guttata*	9.0	Reefs of the Western Pacific Ocean.

Order	Family	Common Name Scientific Name	Maximum Length cm)	Distribution
		Orange stripe prawn goby *Amblyeleotris randalli*	9.0	Tropical reefs of the central Indo-Pacific.
		Pinkspotted shrimp goby *Gobius melanopus*	15.0	Western Pacific Ocean.
		Pinkbar goby *Cryptocentrus aurora*	11.0	Reefs of the western Indian Ocean.
		Red head goby *Elacatinus puncticulatus*	5.0	Reefs from the Gulf of California to Ecuador.
		Red striped goby *Trimma cana*	3.0	Western Pacific Ocean.
		Sleeper banded goby *Amblygobius phalaena*	15.0	Tropical reefs of the Western Pacific Ocean and throughout the central Indo-Pacific area.
		Sleeper blue dot goby *Valenciennea sexguttata*	13.0	Indian Ocean and the Western Pacific Ocean.
		Sleeper gold head goby *Valenciennea strigata*	13.0	Indian Ocean and the Western Pacific Ocean.
		Sleeper railway glider goby *Valenciennea helsdingenii*	25.0	Indo-Pacific.
		Sleeper striped goby *Valenciennea longipinnis*	15.0	Indian Ocean and the Western Pacific Ocean.
		Steinitz goby *Amblyeleotris steinitzi*	8.0	Red Sea through the Indian Ocean to the Western Pacific Ocean.
		Tangaroa goby *Ctenogobiops tangaroai*	5.0	Reefs of the Pacific Ocean.

Order	Family	Common Name Scientific Name	Maximum Length cm)	Distribution
		Tiger watchman goby *Valenciennea wardii*	12.0	Indian Ocean and the Western Pacific Ocean.
		Twospot goby *Signigobius biocellatus*	8.0	Western Pacific Ocean.
		Violet goby *Gobioides broussonnetii*	21.0	Atlantic coast of North and South America.
		Wheeler's watchman goby *Amblyeleotris wheeleri*	8.0	Tropical reefs of the Indian Ocean to the Western Pacific Ocean.
		Yellow watchman goby *Cryptocentrus cinctus*	7.0	Western Pacific.
		Yasha goby *Stonogobiops yasha*	6.0	Western Pacific Ocean.
		Yellow clown goby *Gobiodon okinawae*	3.5	Western Pacific Ocean.
		Yellow priolepis goby *Priolepis aureoviridis*	6.0	Central Pacific Ocean.
(GRUNTERS AND GRUNTS)				
PERCIFORMES	TERAPONIDAE (grunters or tigerperches)	Crescent banded grunt *Terapon jarbua*	36.0	Indo-Pacific.
	HAEMULIDAE (grunts)	Dogfish prientalis *Plectorhinchus lineatus*	72.0	Coral reefs of the Western Pacific Ocean.
		Oriental sweetlips *Plectorhinchus orientalis*	84.0	Indian Ocean and the Western Pacific Ocean.
		Painted sweetlips *Plectorhinchus picus*	84.0	Coral reefs of the Indian Ocean and the Western Pacific Ocean.

Order	Family	Common Name Scientific Name	Maximum Length cm)	Distribution
		Porkfish *Anisotremus virginicus*	41.0	Western Atlantic Ocean.
		Spotted sweetlips *Plectorhinchus chaetodonoides*	74.0	Coral reefs of the Indian Ocean and the Western Pacific Ocean.
		Striped sweetlips *Plectorhinus diagrammus*	51.0	Coral reefs in Western Pacific Ocean.
		Twostriped sweetlips *Plectorhinchus albovittatus*	99.0	Indian Ocean and the Western Pacific Ocean.
(HAMLETS)				
PERCIFORMES (perches)	SERRANIDAE (sea basses) Subfamily: Serraninae	Barred hamlet *Hypoplectrus puella*	15.0	Western Central Atlantic.
		Black hamlet *Hypoplectrus nigricans*	15.0	Central Western Atlantic Ocean and Caribbean Sea.
		Blue hamlet *Hypoplectrus gemma*	13.0	Western Central Atlantic.
		Butter hamlet *Hypoplectrus unicolor*	13.0	Western Central Atlantic.
		Golden hamlet *Hypoplectrus gummigutta*	13.0	Western Central Atlantic.
		Indo hamlet *Hypoplectrus indigo*	14.0	Western Central Atlantic.
		Shy hamlet *Hypoplectrus guttavarius*	13.0	Western Central Atlantic.
(HAWKFISHES)				
PERCIFORMES	CIRRHITIDAE (hawkfishes)	Arc eye hawkfish *Paracirrhites arcatus*	20.0	In reefs in tropical Indo-Pacific.

Order	Family	Common Name Scientific Name	Maximum Length cm)	Distribution
		Blood red hawkfish *Cirrhitichthys fasciatus*	9.5	Western Atlantic.
		Coral hawkfish *Cirrhitichthys oxycephalus*	10.0	Tropical reefs of the Indo-Pacific and the eastern Pacific Ocean.
		Falco's hawkfish *Cirrhitichthys falco*	7.0	Tropical reefs in the Indo-Pacific region.
		Flame hawkfish *Neocirrhitus armatus*	9.0	Tropical reefs of the Pacific Ocean.
		Freckled hawkfish *Paracirrhites forsteri*	22.5	Indo-Pacific.
		Golden hawkfish *Paracirrhites xanthus*	12.0	Eastern Central Pacific.
		Longnose hawkfish *Oxycirrhites typus*	13.0	Tropical reefs of the Indian Ocean and the Pacific Ocean.
		Lyretail hawkfish *Cirrhitichthys polyactis*	14.0	Indo-West Pacific.
		Redspotted hawkfish *Amblycirrhitus pinos*	9.5	Tropical reefs in the Western Atlantic.
		Spotted hawkfish *Cirrhitichthys aprinus*	12.5	Throughout the tropical waters of the Indo-West Pacific region.
		Whitespot hawkfish *Paracirrhites hemistitus*	29.0	Eastern Indian Ocean and in the West and Central Pacific Ocean.
		Yellow hawkfish *Cirrhitichthys aureus*	14.0	Tropical reefs of the Indian Ocean and Western Pacific Ocean.
(HOGFISHES)				
LABRIFORMES (wrasses and relatives)	LABRIDAE (wrasses)	Coral hogfish *Bodianus mesothorax*	20.0	Western Pacific Ocean and the eastern Indian Ocean.

Order	Family	Common Name Scientific Name	Maximum Length cm)	Distribution
		Cuban hogfish *Bodianus pulchellus*	28.5	Atlantic Ocean.
		Hawaiian hogfish *Bodianus bilunulatus*	55.0	Indian Ocean and Western Pacific Ocean.
		Red diana hogfish *Bodianus diana*	60.0	Indian Ocean.
		Spanish hogfish *Bodianus rufus*	40.0	Western Atlantic Ocean.
		Twin spot hogfish *Bodianus bimaculatus*	10.0	Indo-Pacific.
(JACKS)				
CARANGIFORMES (jacks)	CARANGIDAE (jacks and pompanos)	Pilot fish *Naucrates ductor*	70.0	Warm or tropical open seas.
		Golden trevally *Gnathanodon speciosus*	120	Tropical and subtropical waters of the Indian and Pacific Oceans.
		Indian threadfin *Alectis indicus*	165	Tropical Indo-West Pacific Ocean.
		Snubnose Pompano *Trachinotus blochii*	110 FL (Fork Length)	Indo-Pacific.
		Threadfin lookdown *Selene vomer*	48.3	Western Atlantic.
(JAWFISHES)				
Subseries: Ovalentaria	OPISTOGNATHIDAE	Black cap jawfish *Opistognathus lonchurus*	10.0	Western Atlantic Ocean.

Order	Family	Common Name Scientific Name	Maximum Length cm)	Distribution
		Blue dot jawfish *Opistognathus rosenblatti*	9.0	Gulf of California.
		Dusky jawfish *Opistognathus whitehurstii*	14.0	Western Atlantic Ocean.
		Yellowhead jawfish *Opistognathus aurifrons*	10.0	Coral reefs in the Caribbean Sea.
(LIONFISHES)				
SCORPAENIFORMES (mail-cheeked fishes)	SCORPAENIDAE (scorpionfishes) Subfamily: Scorpaeninae	Antenneta lionfish *Pterois antennata*	20.0	Tropical Indian and Western Pacific Oceans.
		Blackfoot lionfish *Paraterois heterura*	23.0	South eastern coast of Africa as well as off Japan and in Indonesia.
		Devil lionfish *Pterois mombasae*	20.0	Tropical Indian Ocean and the Western Pacific.
		Fu Man Chu lionfish *Dendrochirus biocellatus*	13.0	Tropical waters of the Indo-West Pacific region.
		Fuzzy dwarf lionfish *Dendrochirus brachypterus*	18.0	Tropical waters of the Indo-West Pacific, Red Sea included.
		Green lionfish *Dendrochirus barberi*	16.5	Eastern Central Pacific.
		Radiata lionfish *Pterois radiata*	24.0	Indian and Western Pacific Oceans.
		Russell's lionfish *Pterois russellii*	30.0	Indo-Pacific Ocean.
		Volitan lionfish *Pterois volitans*	43.0	Indo-Pacific region.

Order	Family	Common Name Scientific Name	Maximum Length cm)	Distribution
		Zebra lionfish *Dendrochirus zebra*	25.0	Tropical waters of the Indo-West Pacific including the Red Sea.
(MOONFISHES)				
PERCIFORMES	MONODACTYLIDAE (moonfishes)	Silvery mooney *Monodactylus argenteus*	27.0	Indo-West Pacific.
(MOORISH IDOLS)				
ACANTHURIFORMES	ZANCLIDAE (Moorish Idols)	Moorish idol *Zanclus cornutus*	23.0	Tropical and subtropical reefs and lagoons throughout the Indo-Pacific.
(PARROT FISHES)				
LABRIFORMES (wrasses and relatives)	SCARIDAE (parrotfishes)	Bicolor parrotfish *Cetoscarus bicolor*	76.0	Red Sea.
		Princess parrotfish *Scarus taeniopterus*	25.0	Caribbean, South Florida, the Bahamas and Bermuda.
(PIPEFISHES)				
SYNGNATHIFORMES (pipefishes and seahorses)	SYNGNATHIDAE (pipefishes and seahorses)	Banded pipefish *Doryrhamphus dactyliophorus*	20.0	Throughout the tropical waters of the Indo-Pacific region, Red Sea included.
		Dragonface pipefish, Messmate pipefish *Corythoichthys haematopterus*	18.0	Throughout the tropical waters of the Indo-Pacific region.
		Janss' pipefish *Doryrhamphus janssi*	20.0	Tropical waters of the Central Indo-Pacific.
		Yellow multibanded pipefish *Doryrhamphus pessuliferus*	18.0	Western Central Pacific from the Philippines to Australia.

Order	Family	Common Name Scientific Name	Maximum Length cm)	Distribution
(RABBITFISHES)				
PERCIFORMES	SIGANIDAE (rabbitfishes)	Bicolor Foxface *Siganus uspi*	24.0	Coral reefs in Fiji in the Pacific Ocean.
		Foxface rabbitfish *Siganus vulpinus*	23.0	Reefs and lagoons in the tropical Western Pacific.
		Blue-lined Rabbitfish *Siganus doliatus*	25.0	Western Pacific Ocean.
		Magnificent Foxface *Siganus magnificus*	24.0	Eastern Indian Ocean.
		One spot Foxface *Siganus unimaculatus*	18.0	Reefs and lagoons in the central Indo-Pacific.
		Yellow Blotch Rabbitfish *Siganus guttatus*	42.0	Eastern Indian Ocean.
(SCORPIONFISHES)				
SCORPAENIFORMES (mail-cheeked fishes)	SCORPAENIDAE Subfamily: Scorpaeninae	Ambon scorpionfish *Pteroidichthys amboinensis*	12.0	Indian and Pacific Oceans.
		Decoy scorpionfish *Iracundas signifer*	13.0	Reefs in Indian and Pacific Oceans.
		Eschmeyer's scorpionfish *Rhinopias eschneyeri*	19.0	Indo-West Pacific.
		Flasher scorpionfish *Scorpaenopsis macrochir*	13.0	Pacific.
		Lacey scorpionfish *Rhinopias aphanes*	24.0	Western Pacific.

Order	Family	Common Name Scientific Name	Maximum Length cm)	Distribution
		Leaf scorpionfish *Taenianotus triacanthus*	10.0	East African coast and the Red Sea to the tropical Indo-Pacific.
		Mozambique scorpionfish, Papuan scorpionfish *Scorpaenopsis papuensis*	25.0	Coral reefs in Pacific.
		Poss's scorpionfish *Scorpaenopsis possi*	19.3	Indo-Pacific.
		Rogue scorpion *Amblyapistus taenionotus*	15.0	Western Indian Ocean.
		Sea goblin *Inimicus didactylus*	25.0	Eastern Indian and Western Pacific Oceans.
		Stonefish *Synanceja verrucose*	40.0	Pacific and Indian Oceans from the Red Sea to the Great Barrier Reef.
		Weedy scorpionfish *Rhinopias frondosa*	23.0	Indian Ocean and Western Pacific Ocean.
		Yellowspotted scorpionfish *Sebastapistes cyanostigma*	8.0	Indo-Pacific.
(SEAHORSES)				
SYNGNATHIFORMES (pipefishes and seahorses)	SYNGNATHIDAE (pipefishes and seahorses) Subfamily: Hippocampinae (seahorses)	Brazilian seahorse, slender seahorse *Hippocampus reidi*	17.0	Western Atlantic.
		Dwarf seahorse *Hippocampus zosterae*	5.0	Bahamas and parts of the United States.

Order	Family	Common Name Scientific Name	Maximum Length cm)	Distribution
		Long-snouted seahorse *Hippocampus guttulatus*	21.5	Eastern Atlantic.
		Great seahorse *Hippocampus kelloggi*	28.0	Indo-Pacific region.
		Lined seahorse *Hippocampus erectus*	19.0	From the northern point of Nova Scotia, Canada to the southern area of Venezuela in South America.
		Pot-bellied seahorse *Hippocampus abdominalis*	35.0	Southeast Australia and New Zealand.
		Pygmy seahorse *Hippocampus bargibanti*	2.4	Central Indo-Pacific area.
		Short-snouted seahorse *Hippocampus breviceps*	15.0	Coastal waters in south eastern and south eastern Australia.
		Spotted seahorse *Hippocampus kuda*	30.0	Indo-Pacific.
		Thorny seahorse *Hippocampus histrix*	17.0	Indo-Pacific area.
		Tiger tail seahorse *Hippocampus comes*	18.0	Western Central Pacific.
		White's seahorse *Hippocampus whitei*	13.0	Southwest Pacific.
(SHRIPMFISHES)				
SYNGNATHIFORMES	CENTRISCIDAE (shrimpfishes)	Striped razorfish *Aeoliscus strigatus*	15.0	Indo-West Pacific.
(SNIPEFISHES)				

Order	Family	Common Name Scientific Name	Maximum Length cm)	Distribution
SYNGNATHIFORMES	MACRORAMPHOSIDAE (snipefishes)	Longspine snipefish *Macroramphosus scolopex*	20.0	Western Atlantic, east Atlantic, the Mediterranean and the Indo-West Pacific.
(SQUIRRELFISH)				
HOLOCENTRIFORMES (squirrelfishes)	HOLOCENTRIDAE (squirrelfishes)	Big eye soldierfish *Myripristis vittata*	25.0	Indo-Pacific.
		Blackbar soldierfish *Myripristis Jacobus*	25.0	Western Atlantic.
		Glass eye squirrelfish *Heteropriacanthus cruentatus*	30.0	Western and Eastern Atlantic Indo-Pacific, Eastern Pacific.
		Popeye catalufa soldierfish *Pristigenys serrula*	34.0	Eastern Pacific.
		Scarlet squirrelfish *Sargocentron tiere*	33.0	Indo-Pacific.
		Striped squirrelfish *Sargocentron xantherythrum*	18.0	Hawaiian Islands.
(SNAPPERS)				
PERCIFORMES	LUTJANIDAE (snappers)	Black snapper, black and white snapper *Macolor niger*	76.0	Indian Ocean and the Western Pacific Ocean.
		Emperor snapper *Lutjanus sebae*	114	Indian Ocean and the Western Pacific Ocean.
		Threadfin snapper *Symphorichthys spilurus*	58.0	Indo-Pacific region.

Order	Family	Common Name Scientific Name	Maximum Length cm)	Distribution
		Yellowback fusilier *Caesio xanthonota*	38.0	Tropical Indo-Pacific.
		Yellow banded snapper *Lutjanus adetii*	50.0	Eastern coast of Australia and New Caledonia.
(SURGEONFISHES)				
ACANTHURIFORMES (surgeonfishes and relatives)	ACANTHURIDAE Subfamily: Acanthurinae	Achilles tang *Acanthurus achilles*	28.0	Pacific Ocean.
		Atlantic blue tang *Acanthurus coeruleus*	39.0	Atlantic Ocean.
		Blue eyed tang *Ctenochaetus bintatus*	22.0	Indo-Pacific region.
		Blue lined surgeonfish *Acanthurus nigrosis*	25.0	Indo-Pacific.
		Bristletooth tang *Ctenochaetus striatus*	24.0	Indo-Pacific throughout the region except the Hawaiian, Marquesas and Easter Islands.
		Chevron tang *Ctenochaetus hawaiiensis*	16.0	Throughout parts of the Central Pacific Ocean, namely the Hawaiian Islands.
		Clown tang *Acanthurus lineatus*	38.0	Indian Ocean.
		Convict tang *Acanthurus triostegus*	10.0	Indo- Pacific.
		Desjardini tang *Zebrasoma desjardini*	40.0	Lagoons and reefs in the Indian Ocean.

Order	Family	Common Name Scientific Name	Maximum Length cm)	Distribution
		Doctor fish *Acanthurus chirugus*	39.0	Among rocky outcrops and coral reefs of Atlantic Ocean.
		Dussumieri tang *Acanthurus dussumieri*	54.0	Indo-Pacific.
		Eibli mimic tang *Acanthurus tristis*	25.0	Indian Ocean.
		Gold rim tang *Acanthurus nigricans*	21.3	Reef in the central Indo-Pacific area to the eastern Pacific coast, Hawaii included.
		Kole tang *Ctenochaetus strigosus*	14.6	Reef in Oceania.
		Lavender tang *Acanthurus nigrofuscus*	21.0	Indo-Pacific and Hawaii.
		Lopezi tang *Naso lopezi*	60.0	Western Pacific Ocean.
		Mimic tang *Acanthurus pyroferus*	25.0	Indo- Pacific.
		Naso tang *Naso lituratus*	45.0	Indian Ocean and the Pacific Ocean.
		Orange shoulder tang *Acanthurus olivaceous*	35.0	Indo-West Pacific.
		Powder blue tang *Acanthurus leusternon*	23.0	Indian Ocean.
		Powder brown tang *Acanthurus japonicus*	21.0	Indo-West Pacific.
		Purple tang *Zebrasoma xanthurum*	25.0	Red Sea, Gulf of Aden, Persian Gulf and the Arabian Sea.

Order	Family	Common Name Scientific Name	Maximum Length cm)	Distribution
		Regal/Hippo tang *Paracanthurus hepatus*	30.0	Reefs throughout the Indo-Pacific.
		Sailfin tang *Zebrasoma veliferum*	40.0	Oceania, the Indian Ocean and the South Pacific.
		Scopas tang *Zebrasoma scopas*	40.0	Indo-Pacific region.
		Sohal tang *Acanthurus sohal*	40.0	Red Sea.
		Tennent tang *Acanthurus tennenti*	31.0	Indian Ocean.
		Thompson's surgeonfish *Acanthurus thompsoni*	28.0	Pacific and Indian Oceans.
		Tomini tang *Ctenochaetus tominiensis*	16.0	Central Indo-Pacific region.
		Unicorn tang *Naso unicornis*	70.0	Indo-Pacific.
		Vlamingi tang *Naso vlamingii*	60.0	Indo-Pacific Oceans.
		White freckled surgeon *Acanthurus maculiceps*	40.0	Indo-West Pacific.
		Yellow tang *Zebrasoma flavescens*	20.0	Shallow reefs in the Pacific Ocean.
		Yellowfin surgeon *Acanthurus xanthopterus*	70.0	Near coral reefs in Indo-Pacific.
(TILEFISHES)				
PERCIFORMES	MALACANTHIDAE (tilefishes)	Blue-headed tilefish *Hoplolatilus starcki*	15.5	Central Indo-Pacific.

Order	Family	Common Name Scientific Name	Maximum Length cm)	Distribution
		Purple tilefish *Hoplolatius purpureus*	15.0	Western Central Pacific Ocean.
		Yellow tilefish *Hoplolatilus luteus*	13.0	Indo-west Pacific: Andaman Islands, Indonesia and Solomon Islands.
		Redlined tilefish *Hoplolatilus marcosi*	12.0	Western Central Pacific Ocean.
(TRIGGERFISH) TETRADONTIFORMES	BALSTIDAE (triggerfishes)	Assasi trigger *Rhinecanthus assasi*	30.0	Western Indian Ocean, including the Red Sea and Persian Gulf.
		Blue jaw trigger *Xanthichthys auromarginatus*	30.0	Indo-Pacific.
		Blue line trigger *Pseudobalistes fuscus*	55.0	Tropical Indo-Pacific.
		Bursa trigger *Rhinecanthus verrucosus*	23.0	Indo-Pacific.
		Clown trigger *Balistoides conspicillum*	50.0	Indian Ocean, Western Pacific Ocean and Western Atlantic Ocean.
		Crosshatch trigger *Xanthichthys mento*	29.0	Pacific Ocean.
		Goldenback trigger *Xanthichthys caeruleolineatus*	35.0	Indo-West Pacific.
		Golden heart trigger *Balistes punctatus*	61.0	Eastern Atlantic.

Order	Family	Common Name Scientific Name	Maximum Length cm)	Distribution
		Halfmoon trigger *Sufflamen chrysopterum*	30.0	Tropical Indo-West Pacific area.
		Hawaiian black trigger *Melichthys niger*	50.0	Hawaiian island.
		Lei trigger *Sufflamen bursa*	25.0	Indo-Pacific.
		Indian black trigger *Melichthys indicus*	25.0	Indian Ocean.
		Red toothed triggerfish, Niger trigger *Odonus niger*	50.0	Tropical Indo-Pacific area.
		Picasso trigger *Rhinecanthus aculeatus*	30.0	Indo-Pacific region.
		Pintail trigger *Melichthys vidua*	40.0	Reefs in the Indo-Pacific.
		Queen trigger *Balistes vetula*	60.0	Reefs in the Atlantic Ocean.
		Rectangular trigger *Rhinecanthus rectangulus*	30.0	Reefs in the Indo-Pacific.
		Sargassum trigger *Xanthichthys ringens*	25.0	Western Atlantic.
		Starry trigger *Abalistes stellatus*	60.0	Coasts of the Indian Ocean, the Red Sea, the Persian Gulf and the western edge of the Pacific Ocean.
		Titan trigger *Balistoides viridescens*	75.0	Indo-Pacific except Hawaii.

Order	Family	Common Name Scientific Name	Maximum Length cm)	Distribution
		Undulated trigger *Balistapus undulates*	30.0	Reefs in the Indo-Western Pacific.
		Whitetail trigger *Sufflamen albicaudatum*	22.0	Western Indian Ocean.
(TRUMPETFISHES)				
SYNGNATHIFORMES	AULOSTOMIDAE (trumpetfishes)	Chinese-trumpet fish *Aulostomus chinensis*	80.0	Tropical and subtropical waters from the Indian Ocean and the Pacific Ocean.
WRASSES				
LABRIFORMES (wrasses and relatives)	LABRIDAE (wrasses)	Banana wrasse *Thalassoma lutescens*	30.0	Indian and Pacific Oceans.
		Bicolor cleaner wrasse *Labroides bicolor*	15.0	Indo-Pacific, Indian Ocean and Pacific Ocean.
		Bird wrasse *Gomphosus varius*	28.0	Indo-Pacific.
		Bluehead wrasse *Thalassoma bifasciatum*	28.0	Coral reefs of the tropical waters of the Western Atlantic Ocean.
		Bluestreak cleaner wrasse *Labroides dimidiatus*	14.0	Indo-Pacific.
		Carpenter's flasher wrasse *Paracheilinus carpenteri*	8.0	Western Pacific Ocean.
		Cheeklined maori wrasse *Oxycheilinus diagramma*	36.0	Indian Ocean and the western Pacific Ocean.
		Christmas wrasse *Thalassoma trilobatum*	15.0	Indian Ocean and the Western Pacific Ocean.

Order	Family	Common Name Scientific Name	Maximum Length cm)	Distribution
		Cortez rainbow wrasse *Thalassoma lucasanum*	15.0	Eastern Pacific Ocean.
		Dragon wrasse *Novaculichthys taeniourus*	30.0	Coral reefs and lagoons in the Indo-Pacific region.
		Eightline wrasse *Pseudocheilinus octotaenia*	14.0	Indian and Pacific Oceans.
		Exquisite fairy wrasse *Cirrhilabrus exquisitus*	12.0	Indo-Pacific Oceans.
		Fine-spotted fairy wrasse *Cirrhilabrus punctatus*	10.0	Western Pacific Ocean.
		Flame wrasse *Cirrhilabrus jordani*	10.0	Hawaiian islands .
		Formosa wrasse *Coris formosa*	61.0	Western Indian Ocean, Southern Red Sea, South Africa and east to Sri Lanka.
		Fourline wrasse *Pseudocheilinus tetrataenia*	7.5	Pacific Ocean.
		Goldbar wrasse *Thalossoma hebraicum*	23.0	Western Indian Ocean.
		Greenback fairy wrasse *Cirrhilabrus scottorum*	15.0	Pacific Ocean.
		Harlequin tusk *Choerodon fasciatus*	30.0	Western Pacific Ocean.
		Hawaiian cleaner wrasse *Labroides phthirophagus*	12.0	Hawaiian Islands.
		Hoeven's wrasse *Halichoeres melanurus*	13.0	Western Pacific.

Order	Family	Common Name Scientific Name	Maximum Length cm)	Distribution
		Jansen saddle wrasse *Thalassoma jansenii*	20.0	Indian Ocean.
		Labout's fairy wrasse *Cirrhilabrus laboutei*	8.0	Coral reefs of New Caledonia and Australia.
		Leopard wrasse *Macropharyngodon meleagris*	15.0	Eastern Indian Ocean and the Pacific Ocean.
		Lineatus fairy wrasse *Cirrhilabrus lineatus*	13.0	Coral reefs of New Caledonia and Australia.
		Longfin fairy wrasse *Cirrhilabrus rubriventralis*	8.0	Western Indian Ocean.
		Lyretail wrasse, moon wrasse *Thalassoma lunare*	25.0	Indian Ocean and the Western Pacific Ocean.
		Marble wrasse *Halichoeres hortulanus*	28.0	Indian Ocean to Central Pacific Ocean.
		Mc Cosker's flasher wrasse *Paracheilinus mccoskeri*	15.0	Indian Ocean and the Western Pacific Ocean.
		Multicolor lubbock's wrasse *Cirrhilabrus lubbocki*	8.0	Coral reefs of the Philippines and the island of Sulawesi in Indonesia.
		Multicolor velvet wrasse *Cirrhilabrus cyanopleura*	15.0	Eastern Indian Ocean and the Western Pacific Ocean.
		Mystery wrasse *Pseudocheilinus ocellatus*	10.3	Central Western Pacific Ocean.

Order	Family	Common Name Scientific Name	Maximum Length cm)	Distribution
		Orange-back fairy wrasse *Cirrhilabrus aurantidorsalis*	13.0	Pacific waters of Indonesia.
		Pastel-green wrasse *Halichoeres chloropterus*	20.0	Central western Pacific Ocean.
		Pinkface wrasse *Thalassoma quinquevittatum*	15.0	Indian Ocean and the Western Pacific Ocean.
		Potter's wrasse *Macropharyngodon geoffroyi*	13.0	Hawaiian Islands.
		Red coris wrasse *Coris gaimard*	36.0	Tropical waters of the central Indian Ocean and Western Pacific Ocean.
		Red-head fairy wrasse *Cirrhilabrus solorensis*	13.0	Indonesia and Christmas Island.
		Red velvet wrasse *Cirrhilabrus rubrisquamis*	8.0	Western Indian Ocean.
		Scarlet pin stripe wrasse *Pseudocheilinus evanidus*	8.0	Indian and Pacific Oceans.
		Sixline wrasse *Pseudocheilinus hexataenia*	8.0	Indian Ocean and the Western Pacific Ocean.
		Radiant wrasse *Halichoeres iridis*	15.0	Western Indian Ocean.

Order	Family	Common Name Scientific Name	Maximum Length cm)	Distribution
		Rhomboid fairy wrasse *Cirrhilabrus rhomboidalis*	13.0	Marshall Islands-Kwajalein Atoll.
		Whip fin fairy wrasse *Cirrhilabrus filamentosus*	9.0	Indonesia-Java Sea.
		Yellow wrasse *Halichoeres chrysus*	13.0	Central Indo-Pacific area.
		Yellow & purple wrasse *Halichoeres trispilus*	13.0	Western Indian Ocean.
		Yellow band wrasse *Cirrhilabrus luteovittatus*	13.0	Coral reefs around Pacific Islands.
		Yellowfin fairy wrasse *Cirrhilabrus flavidorsalis*	8.0	Western Pacific Ocean.
		Yellow-flanked fairy wrasse *Cirrhilabrus nyukyuensis*	15.0	Eastern Indian Ocean.

2.11 An Explicit Endevour of some Marine Ornamental Fishes

2.11.1 Blacktip reef shark

Figure 2.80

Carcharhinus melanopterus (Ouoy & Gaimard, 1824)

Common Name(s):

English- Blacktip reef shark.

IUCN Red List status: Near Threatened.

Systematic Position:

Kingdom: ANIMALIA

Phylum: CHORDATA

Grade: CHONDRICHTHYOMORPHI

Class: CHONDRICHTHYES—cartilaginous fishes

Subclass: EUSELACHII (sharks and rays)

Infraclass: ELASMOBRANCHII

Order: CARCHARHINIFORMES—ground sharks

Family: CARCHARHINIDAE—requiem sharks

Genus: *Carcharhinus*

Species: *C. melanopterus*

Scientific Name: *Carcharhinus melanopterus* (Quoy & Gaimard, 1824)

Description:

- Body robustly built with a typically shark-like streamlined form.
- Moderately large and oval eyes and narrow cusped teeth.
- The first dorsal fin is high and second dorsal fin is relatively large. Pectoral fins are large and narrowly falcate, and tapering to points.
- Snout is short, wide and rounded.
- Colour of the body is pale greyish-brown above and white below with white bands on the sides.
- All the fins have black-tips highlighted by lighter coloured borders. This is striking on the first dorsal fin and lower caudal fin lobe. Anterior and posterior dark edging on pectoral fins and upper caudal fin.
- A prominent black tip of first dorsal fin set off absolutely by a light band below it. A dark band on flanks extending to pelvic fins.
- Sexual Dimorphism: Male fishes are little shorter than the female fishes. Males have claspers whereas females lack these organs.

Species Details:

- Distribution:

  Indian Ocean-western; Indian Ocean-eastern; Mediterranean and Black Sea; Pacific eastern central; Pacific-north west; Pacific-western central.

Tropical Indo-West Pacific and Central Pacific, with a range extending from Thailand to China, Japan, the Philippines, New Caledonia and northern Australia.

- Habitat: Commonly found in shallow clear waters on and near coral reefs. Occurs only a few meters deep and is occasionally present in brackish waters. Young ones are found in shallow, sandy flats whereas adults prefer reef edges and reef drop-offs.
- Food and feeding: Predator fish. Prey on teleost fishes, crustaceans, cephalopods and other molluscs. Reported to eat terrestrial and sea snakes.
- Maximum length: 200 cm TL male/unsexed.

Aquarium Aspects:

- Minimum tank size: 5,750 gallons.
- Substrate: Sandy.
- Water conditions: Temperature: 20.0-27.0° C (70.0-80.0° F).

Behaviour:

Social—These have extremely small home ranges and exhibit strong site fidelity, remaining within the same local area for up to several years a time. Seen in small groups though not social. Generally shy. They are harmless unless provoked. They can jump fully out of water.

Reproductive—These fishes are viviparous. Females give birth to 2-5 young ones. Live-bearing females nourish pups in utero via a yolk-sac placenta. Both annual and biennial reproductive cycles have been reported. Parturition and mating generally occur during the summer and early autumn though gestation also varies with location from 8-9 months, 10-11 months and 16 months. Most litters contain 2-4 pups.

Remarks: Blacktip reef shark is one of the most abundant sharks inhabiting the tropical coral reefs of the Indian and Pacific Oceans. They are popular items of public marine aquarium exhibits.

2.11.2 Sand tiger shark

Figure 2.81

Carcharias taurus Rafinesque, 1810

Common Name(s):

English—Sand tiger shark, grey nurse shark, spotted regged tooth shark or blue-nurse sand tiger.

Systematic Position:

Order: LAMNIFORMES--mackerel sharks.

Family: ODONTASPIDIDAE--sand tiger sharks.

Genus: *Carcharias*

Species: *C.taurus*

Scientific Name: *Carcharias taurus* Rafinesque, 1810

Description:

- The body is stout and bulky and its mouth extends beyond the eyes. The eyes are small lacking eyelids.
- The snout is flattened with a conical shape.
- There are three rows of protruding, smooth-edged, sharp-pointed teeth.
- There are two large, broad-based grey dorsal fins set back beyond the pectoral fins.
- Dorsal spines: 0; dorsal soft rays: 0; anal spines :0; anal soft rays :0.
- Caudal fin is elongated with a prominent subterminal notch, a long upper lobe (strongly heterocercal) and a short ventral lobe.
- Body colour pale brown or grey.
- Ventral side paler, with dark spots which are faded in adults.
- Fins plain.
- Sexual Dimorphism: Males have grey claspers with white tips located on the underside of their body. Females lack them.

Species Details:

- Distribution:

  Indo-West Pacific: Red Sea and off the coasts of South Africa, Japan, Korea and Australia. Present in Arafura Sea. Western Atlantic: Gulf of Maine to Argentina. Recorded from Bermuda, south Brazil. Eastern Atlantic: Mediterranean to Cameroon. Northwest Atlantic: Canada.
- Habitat: Found inshore from the surf zone. Inhabit shallow bays to at least 191 m on the outer continental shelves. Found on or near the bottom, in midwater or at the surface.
- Food and feeding: Carnivorous. Feeds on bony fishes, small sharks, rays, squids, crabs and lobsters.
- Maximum length: 330 cm TL male/unsexed.
- Life span: 15 or more years.

Aquarium Aspects:

- Minimum tank size: 5,750 gallons.
- Substrate: Sandy.
- Water conditions: Temperature: 20.0-27.0° C (70.0-80.0° F).

Behaviour:

Social—These sharks are docile, sluggish, and non-aggressive as long as they are not threatened. They prefer to swim singly or in small to large schools. They usually feed at night. They rise to the surface of water to gulp air which they hold in their stomach. This is to keep them buoyant. In winter, they move towards deeper coastal waters.

Reproductive—These tiger sharks are ovophagous. The eggs are fertilized internally in the females. Females have two uterine horns that, during early embryonic development, may have as many as 50 embryos that obtain nutrients from their yolk sacs and possibly consume uterine fluids. When one of the embryos reaches some 10 cm (4 in) in length, it eats all the smaller embryos so that only one large embryo remains in each uterine horn. This process is called intrauterine cannibalism i.e. "embryophagy" or, adelphophagy. Multiple males fertilize a single female, and adelphophagy excludes all but one of them from gaining offspring. These surviving embryos continue to feed on a steady supply of unfertilised eggs. After a lengthy labour, the female gives birth to 1 m (3.3 ft) long, fully independent offspring. The gestation period is approximately eight to twelve months. These sharks give birth only every second or third year (en.wikipedia.org/wiki/sand _ tiger _shark).

Remarks: Sand tiger shark is a very tolerant large shark and very often kept in public aquariums.

2.11.3 Undulate skate

Figure 2.82

Raja undulata Lacepede, 1802

Common Name(s):

English—Undulate ray, undulate skate.

IUCN Red List Status: Endangered.

Systematic Position:

Order: RAJIFORMES--skates

Family: RAJIDAE--skates

Subfamily: RAJINAE--(Hardnose skates)

Genus: *Raja*

Species: *R. undulata*

Scientific Name: *Raja undulata* Lacepede, 1802

Description:

- Body disc-shaped, triangular in front and near-circular in the rear.
- Dermal denticles are developed as spines. In adults, median spines are scattered, in young they are regular.
- The eyes are medium -sized followed by spiracles.
- The mouth is present on the ventral surface and is slightly arched.
- There are five pairs of small gill slits. Tail is as long as body, and has two well-separated dorsal fins near its end. With two spines between them.
- Dorsal surface of the body varies from light brown to citrine or grey, with darker curved bands and numerous small white patches.
- Snout and margins with dark spots.
- The underside is creamy-white with a greyish margin.

- Sexual Dimorphism: The males have one lateral row of spines on each side, whereas the females have three.

Species Details:

- Distribution:

  Native to Mediterranean and Black Sea. Atlantic northeast, Atlantic eastern. Central Eastern Atlantic: Southern Ireland and England to Senegal, including the western Mediterranean and the Canary Islands, Eastern Mediterranean.
- Habitat: Marine, coastal/ supratidal, marine, neritic, marine oceanic, marine intertidal. Inhabit sandy bottoms of shelf- waters. Also found in muddy or detrital bottoms at depths of 50-200 m from coastal regions to the upper continental slope.
- Food and feeding: Carnivorous. Feed on all types of bottom animals. They eat small fish, crustaceans, molluscs and other macrobenthos.
- Maximum length: 100 cm TL male/ unsexed.
- Life span: 21-23 years.

Aquarium Aspects:

- Minimum tank size: 5,750 gallons.
- Substrate: Sandy.
- Water conditions: Temperature: 20.0-27.0° C (70.0-80.0° F).

Behaviour:

Social—They often bury themselves underneath the sediment.

Reproductive-- It is an oviparous cartilaginous fish. Females lay their eggs from March to September. Each egg is a leathery black egg-case with horny corners, known by the common name, "mermaid purse", they are 72-90 mm long and 42-52 mm wide.

Remarks: It is a common skate in public aquarium.

2.11.4 Blackchin Guitarfish

Figure 2.83

Glaucostegus cemiculus (Geoffroy Saint- Hilaire, 1817)

Common Name(s):

English-Blackchin guitarfish.

IUCN Red List Status: Endangered.

Systematic Position:

Order: PRISTIFORMES—guitarfishes and sawfishes

Family: RHINOBATIDAE--guitarfishes

Genus: *Glaucostegus*

Species: *G. cemiculus*

Scientific Name: *Glaucostegus cemiculus* (Geoffroy Saint-Hilaire, 1817)

Description:

- Body, plain brown above and white below.
- A black blotch present on its snout, more noticeable in juveniles than adult fish.
- Eyes small, narrower rostral ridges.
- Shorter front nasal lobe and narrower back nasal flap.
- Short thorns present on the inner margins of the eye, on the shoulders and along the spine and tail.
- Rough skin covered entirely with small denticles. Well-developed thorns, usually along edge of rostral cartilage.
- Sexual Dimorphism: At sexual maturity male fish is 150 cm and female, is about 163 cm long.

Species Details:

- Distribution:

  Atlantic-southeast, Atlantic-northeast, Atlantic-eastern central, Mediterranean and Black Sea. Eastern Atlantic: northern Portugal to Angola.
- Habitat: Marine Neritic, shallow, sandy and muddy stretches of seabed.
- Food and feeding: Invertebrates, shrimps, crabs, other crustaceans, squid and molluscs.
- Maximum length: 180-242 cm TL male/unsexed.

Aquarium Aspects:

- Minimum tank size: 5,750 gallons.
- Substrate: Sandy.
- Water conditions: Temperature: 20.0-27.0° C (70.0-80.0° F).

Behaviour:

Reproductive--Ovoviviparous. The eggs being retained in the uterus of the female and the developing young being nourished first by a yolk sac, and later by uterine secretions. Gestation period is between about five and eight months.

Remarks: A common public marine aquarium fish.

2.11.5 Bowmouth guitarfish — Figure 2.84

Rhina ancylostoma (Muller and Henle, 1841)

Common Name(s):

English—Bowmouth guitarfish, shark ray, mud skate.

IUCN Red List Status: Vulnerable (VU).

Systematic Position:

Order: PRISTIFORMES—guitarfishes and sawfishes

Family: RHINIDAE—bowmouth guitarfishes

Genus: *Rhina*

Species: *R. ancylostoma*

Scientific Name: *Rhina ancylostoma* (Muller and Henle, 1841)

Description:

- Heavily built, body with short, wide and flattened head.
- Broad, rounded snout and jaws with heavily ridged crushing teeth, present on undulating rows.
- Lower jaw has three protruding lobes that fit into the corresponding spaces in the upper jaw.
- Two tall and falcate dorsal fins are present. Broad and triangular pectoral fins have a deep indentation.
- The pelvic fins are much smaller than the pectoral fins and the anal fin is absent.
- The tail is much longer than the body and ends in a large, crescent-shaped caudal fin.
- The entire dorsal surface of the fish has a grainy texture from a dense presence of tiny dermal denticles.
- Along the midline of the back, a thick ridge is present, bearing a band of sharp, robust thorns.
- A pair of thorn-bearing ridges is present in front of the eyes, a second pair running from above the eyes to behind the spiracle and a third pair on the shoulders.
- Dorsal side grey or brownish, ventral white.
- Several white spots present dorsally on fins, body and tail. Black spots are present on head and shoulders.

Species Details:

- Distribution:

  Indo-West Pacific: Persian Gulf, Red Sea and East Africa to Papua New Guinea, north to Japan, south to New South Wales, Australia.
- Habitat: Found in coastal areas and on coral reefs. Inhabit inshore, on sand and mud bottoms, in the water column also.
- Food and feeding: Carnivorous. Feeds on bottom crustaceans, molluscs and bony fishes.
- Maximum length: 300 cm TL male/unsexed.
- Life span: 7 years or more.

Aquarium Aspects:

- Minimum tank size: 5,750 gallons.
- Substrate: Sandy.

- Water conditions: Temperature: 20.0-27.0° C (70.0-80.0° F).

Behaviour:

Social—It is a strong swimmer and is more active at night. It is not territorial.

Reproductive—It is ovoviviparous (aplacental viviparous) with embryos feeding initially on yolk, then receiving additional nourishment from the mother by indirect absorption of uterine fluid enriched with mucous, fat or protein through some specialized structures. Produces litters of 2-11 pups. Size at birth 45 cm.

Remarks: It is a popular fish of public aquariums and generally lives for several years in captivity.

2.11.6 Javanese cownose ray

Figure 2.85

Rhinoptera javanica Muller & Henle, 1841

Common Name(s):

English—Javanese cownose ray, flapnose ray.

IUCN Red List Status: Vulnerable (VU).

Systematic Position:

Order: MYLIOBATIFORMES--stingrays

Family: MYLIOBATIDAE—eagle rays

Subfamily: RHINOPTERINAE (cownose rays)

Genus: *Rhinoptera*

Species: *R. javanica* Muller & Henle, 1841

Description:

- A double-lobed snout and indented forehead.
- The body is flattened, with the pectoral fins broadly expanded and fused with the head and trunk to form a disc.
- This smooth skinned species is characterized by a kite-shaped body disc.
- The body disc is brown on the upper surface and white below.
- The long thin, whip-like tails of cownose rays are strictly demarcated from the body and armed with one or more stings.
- No caudal fin.

Species Details:

- Distribution:

  Indo-West Pacific: off Durban, South Africa and ranging with probably to India, Thailand, Indonesia and southern China. Reported from Persian Gulf, Okinawa, Ryukyu Islands.
- Habitat: Found in bays, estuaries, and near coral reefs, over sand and mud bottoms.
- Food and feeding: Clams, oysters and crustaceans. Adult rays feed on

deep burrowing molluscs and juveniles feed on shallow, non-burrowing bivalves.

- Maximum length: 150 cm WD male/ unsexed.
- Life span: Over 2 years in captivity.

Aquarium Aspects:

- Minimum tank size: 5,750 gallons.
- Substrate: Sandy.
- Water conditions: Temperature: 20.0-27.0° C (70.0-80.0° F).

Behaviour:

Reproductive--Ovoviviparous. Males court by nipping the female's dorsum. The pair usually rests on the substrate. Exhibit ovoviviparity. With embryos feeding initially on yolk then receiving additional nourishment from the mother by indirect absorption of uterine fluid enriched with mucous. Size at birth 60 cm WD.

Remarks: A popular public marine aquarium fish.

2.11.7 Southern stingray

Figure 2.86

Hypanus americanus (Hildebrand & Schroeder, 1928)

Common Name(s):

English—Southern stingray.

IUCN Red List Status: Data Deficient (DD).

Systematic Position:

Order: MYLIOBATIFORMES--stingrays

Family: DASYATIDAE—whiptail stingrays

Genus: *Hypanus*

Species: *H. americanus*

Scientific Name: *Hypanus americanus* (Hildebrand & Schroeder, 1928)

Description:

- The flattened diamond-shaped body has sharp corners.
- Disc has sharp outer corners and irregular row of short spines on upper surface.
- Disc usually uniform dark brown above, greyer in young.
- Upper surface of disc, grey, dark or olive brown.
- Lower surface of disc white or whitish with an edging of grey or brown.
- The slender tail possesses a long, serrated and venomous spine at the base, used for defence.
- The eyes are situated on top of head.
- Sexual Dimorphism: Female stingrays can grow to a disc width of 150 cm and smaller male stingrays reach maximum size at 67 cm.

Species Details:

- Distribution:

  Western Atlantic: New Jersey, USA and northern Gulf of Mexico to southern Brazil, including the Antilles.
- Habitat: Found on sandy bottoms, seagrass beds, lagoons and the reef face. Common in bays and estuaries.
- Food and feeding: Carnivorous. Feeds on worms, bivalves, shrimps, crabs and small fishes.
- Maximum length: 200 cm W.D male/unsexed.
- Life span: 28 years.

Aquarium Aspects:

- Minimum tank size: 4,200 gallons.
- Substrate: Sandy.
- Water conditions: Temperature: 20.0-27.0° C (70.0-80.0° F).

Behaviour:

Social—Remains either singly, in pairs or in aggregations. During the day time buries itself in the sand and forages at night, in seagrass beds. These bottom dwelling, nocturnal predators either spray water from mouths or flap their fins intensely so the substrate is displaced and hidden prey is exposed. Southern stingray exhibits wave-like locomotion using their pectoral fins. Venomous spine present is used for defence. They often visit "cleaning stations" where their external parasites are removed by cleaners like blue head wrasse and Spanish hogfish.

Reproductive—Ovoviviparous. In this aplacental viviparity, the embryos feed initially on yolk, then get nourishment from the mother by direct absorption of uterine fluid or milk, through maternal secretions. This fluid is rich in mucous, fat and protein.

Remarks: They are found in many public aquariums and animal theme parks. Here visitors are allowed to pet the stingray in a touch pool (en.wikipedia.org/wiki/Southern stingray).

2.11.8 Common clownfish Figure 2.87

Amphiprion ocellaris Cuvier, 1830

Common Name(s):

English—Common clownfish, clown anemonefish, false percula clownfish, Nemo fish.

Systematic Position:

Phylum: CHORDATA

Subphylum: CRANIATA

Grade: TELEOSTOMI

Class: OSTEICHTHYES—bony fishes

Subclass: ACTINOPTERYGII—ray-finned fishes

Superorder: ACANTHOPTERYGII

Subseries: OVALENTARIA

Family: POMACENTRIDAE—damselfishes

Subfamily: AMPHIPRIONINAE—(Anemonefishes)

Genus: *Amphiprion*

Species: *A. ocellaris*

Scientific Name: *Amphiprion ocellaris* Cuvier, 1830

Clownfish are grouped into six complexes viz. Percula complex, Maroon complex, Clarkii complex, Saddleback complex, Tomato complex and skunk complex. And there are at present 30 described clownfish species.

Description:

- It is a small fish with oval-shaped body and stocky appearance.
- It is compressed laterally and has a round profile.
- The colouration of the body is generally orange to reddish- brown.
- There are three vertical white stripes present outlined with a fine black line. The first black line passes just behind the eye, the second in the middle of the body widens forward to the head centrally and the third line encircles the caudal peduncle.
- All the fins are also outlined with a fine black line.
- *A. ocellaris* can be distinguished from *A. percula,* in having thinner black outlines, a taller dorsal fin and 11 dorsal fin spines.
- Sexual Dimorphism: The mature female fish is larger in size than the male.

Species Details:

- Distribution:

  Eastern Indian Ocean including Andaman and Nicobar Islands and in the Western Pacific Ocean. Also found in Northern Australia, Southeast Asia and Japan.
- Habitat: Found in small groups on outer reef slopes or in sheltered lagoons at a maximum depth of 15 meters. They inhabit three different species of sea anemones such as Magnificent Anemone (*Heteractis magnifica*), Giant carpet Anemone (*Stichodactyla gigantea*) and *Stichodactyla mertensii.*
- Food and feeding: Omnivorous. Feed on zooplankton, algae and copepods. When in group, the less dominant one eats less than the dominant one.
- Maximum length: 11.0 cm TL male/unsexed.
- Life span: Maximum 12 years. Normal 4-8 years.

Aquarium Aspects:

- Minimum tank size: 10 gallons-20 gallons.

- Suitable for Nano tank: Yes.
- Substrate: Any.
- Lighting: Any.
- Water conditions: Temperature: 75.0-82.0° F (23.3 to 27.8° C); salinity: 1.020-1.025; pH: 7.8-8.4.
- Water movement: Any.
- Water region: Any.
- Water changes: Bi-weekly.

Behaviour:

Social—*A. ocellaris* maintains symbiotic relationship with the sea anemone. The anemone provides protection for the fish with its tentacles, the fish's mucous protection prevents it from being stung by the tentacles. Presence of clownfish helps to lure potential anemone's preys close to the tentacles. And defend the anemone against some reef fishes, which may eat the tentacles.

They are well known for their up and down swimming motion. In captivity, they will often accept corals and other anemones, rock and algae as hosts.

These fish exhibit queue selection and form social hierarchies or social rank. Queue is the term for social groups of *A. ocellaris*. The dominant pair of each queue (social group), reproduces more, compared to the subordinate fishes. There are two types of *A. ocellaris*, settlers and switchers. Settlers prefer shorter queues whereas switchers usually more after settlement.

In captivity, they do well in fish only tanks, full reef tanks, with or without anemones, large or small tanks.

Reproductive—These fish are protandrous hermaphrodites and benthic spawner. All the individuals first develop as males and then later become females (protandrous hermaphrodites). This is also termed as plasticity. The most dominant fish in the group will become female and the next dominant will become male. The female is larger than the male and is more aggressive. Adult fishes will pair up and form a bonded pair. Once they start breeding, they are called a mated pair and will continue to spawn regularly as long as they are alive. About 2000 eggs are laid each time during spawning. Eggs are demersal and adhere to the substrate. Male fish aerate them. These are then guarded and cared until they hatch about 7 to 10 days later.

Remarks: One of the most popular marine aquarium fishes. Bred artificially in Florida for the aquarium trade. As they can be bred and reared in captivity, they are used in research.

2.11.9 Black ocellaris

Figure 2.88

Amphiprion ocellaris Cuvier, 1830

The black ocellaris is a stunning colour morph of *A. ocellaris* and is also referred to as the Darwin ocellaris or the Darwin false percula. This is a naturally occurring variation for the species with a very limited range. Most of the specimens available in the market are captive bred and very few are caught from the wild.

Common Name(s):

English—Black ocellaris, Darwin ocellaris, Darwin false ocellaris, Black and White ocellaris clownfish, black percula, Black percula clown, Black percula clown anemonefish.

IUCN Red List Status: Least Concern.

Systematic Position:

Subfamily: AMPHIPRIONINAE (Anemonefishes)

Genus: *Amphiprion*

Species: *A. ocellaris*

Scientific Name: *Amphiprion ocellaris* Cuvier, 1830

Description:

- Body stocky and is oval in shape.
- It is compressed laterally, with a round profile
- The body is jet-black with three white stripes on each side.
- They are orange coloured when young and become black as they mature. All orange colour disappears as they age and only solid black and white colour remain.
- Sexual Dimorphism: Females are larger than males.
- Colour variations of *A. ocellaris*:

  Snowflake Clownfish, Premium snowflake, Black Ice snowflake clownfish, Black snowflake clownfish, Naked Ocellaris, Midnight Ocellaris, Caramel Ocellaris, Domino Ocellaris, Mis-Bar Ocellaris, Extreme Mis Bar Black Ocellaris.

Species Details:

- Distribution:

  Darwin area of Northern Australia.
- Habitat: Found in association with anemones on the reefs of the coral Sea.
- Food and feeding: Omnivorous. In captivity takes meaty foods and vegetables, high quality marine flake foods, rich in spirulina algae, freeze dried and frozen foods.
- Maximum length: 7.5 cm
- Life span: 20 years.

Aquarium Aspects:

- Minimum tank size: 30 gallons.
- Suitable for Nano Tank: Yes.

- Water condition: Temperature: 72.0-78.0° F; dGH: 8-12; pH: 8.1-8.4; Sg: 1.020-1.025.
- Temperament: Semi-aggressive.
- Reef compatible: Yes.
- Care level: Easy.
- Water region: All.
- Water changes: Bi-weekly.

Behaviour:

Social—Remains either individually or in pairs or small groups within the same anemone such as *Heteractis magnifica* or *Stichodactyla mertensii*.

Reproductive—The pair stays close to each other in the aquarium. They are egg layers and will deposit the eggs on a flat surface, and defend the eggs from other tank mates. The eggs will hatch in 6-11 days depending on the temperature.

Remarks: It is very hardy and is a good candidate for a reef aquarium.

2.11.10 Tomato clownfish

Figure 2.89 A & B

Amphiprion frenatus Brevoort, 1856

Common Name(s):

English—Tomato clownfish, blackback anemonefish, bridled anemonefish, fire clown, red tomato clown.

IUCN Red List Status: Least Concern (LC).

Systematic Position:

Family: POMACENTRIDAE--damselfishes

Subfamily: AMPHIPRIONINAE (Anemonefishes)

Genus: *Amphiprion*

Species: *A. frenatus*

Scientific Name: *Amphiprion frenatus* Brevoort, 1856

Description:

- Body deep and oval- shaped.
- These small-sized fishes have 9-10 dorsal spines, 2 anal spines, 16-18 dorsal soft rays and 13-15 anal soft rays.
- The adult fish is bright orange-red, with a white vertical stripe just behind the eyes, joined over the head. It has a distinctive black outline.
- Juveniles are darker red with two to three white bars.
- Sexual Dimorphism: Females are 1-2 cm larger than males, and turn brown on the sides as they mature. Males retain their vivid orange or reddish colouration throughout their lifetime.

Species Details:

- Distribution:

  Western Pacific: Gulf of Thailand to Southwestern Palau, north to southern Japan, south to Java, Indonesia.
- Habitat: Inhabits lagoon reefs and embayment.
- Food and feeding: They are omnivore. In the wild, they eat filamentous algae, planktonic copepods, and benthic crustaceans such as shrimp, planktonic fish eggs and crustacean larvae. In captivity they feed on frozen *Mysis* and brine shrimp, finely chopped fish and shrimp flesh, any frozen/ thawed prepared foods, flake foods, pellets with spirulina, live foods, some vegetables.
- Feeding frequency: Several feedings per day.
- Maximum length: 14 cm.
- Life span: 17 years.

Aquarium Aspects:

- Minimum tank size: 30 gallons (114 litres).
- Suitable for Nano tank: No.
- Substrate: Any. With live rock plus hiding places.
- Lighting: Any.
- Temperature: 72.0 to 82.0° F (22.2 to 27.8° C).
- Breeding temperature: 79.0° F.
- Specific gravity: 1.023-1.025 SG.
- Range pH: 7.8-8.4.
- Water movement: Any.
- Water region: All.
- Water changes: Bi-weekly.

Behaviour:

Social—They are semi-aggressive, but can also be territorial and aggressive with ages. This clownfish like other clownfish species are known for making sounds. An aggressive clownfish may display, agnostic behaviour, whereas the subordinate one may show appeaser behaviour.

Compatibility with host anemones: The host of Tomato clownfish is the Bubble Tip Anemones *Entamaea quadricolor* in the wild. When kept with an anemone, the Tomato Clownfish remain always with the host.

Breeding—In captivity, these fishes lay around 400 eggs every 10-14 days. These eggs hatch in 7-8 days and their larval period is only 9 days. The fry can be successfully reared in captivity. They are sex switchers. Under the influence of some social cue new born change into juvenile males, and with the opportunity, a dominant fish will become female.

Remarks: A common saltwater aquarium pet.

2.11.11 Orange clownfish

Figure 2.90

Amphiprion percula (Lacepede, 1802)

Common Name(s):

English: Orange clownfish, percula clownfish, clown anemonefish, True percula clownfish.

IUCN Red List Status: Least Concern.

Systematic Position:

Family: POMACENTRIDAE--damselfishes

Subfamily: AMPHIPRIONINAE (Anemonefishes)

Genus: *Amphiprion*

Species: *A. percula*

Scientific Name: *Amphiprion percula* (Lacepede, 1802)

Description:

- True Percula clownfish has an elongated and fairly deep body.
- There are 10 dorsal spines, dorsal soft rays 13-17, anal spines 2, anal soft rays 11-13.
- Body of these fishes have various shades of orange and have 3 white vertical bars on the body. One is over the gill area, the second runs mid-body and usually has a forward projecting bulge, and the third one is at the base of the tailfin.
- All the white bars can vary in shape and are trimmed in a thicker black edge than Ocellaris.
- Sexual Dimorphism: Females are larger than males.
- Various True percula clownfish with designer colours are, True Percula, Misbar, Picasso, Platinum, Snow Onyx. Wild forms with larger amounts of black: Black Crest Percula clownfish, Black Percula clownfish.

Species Details:

- Distribution:

  Western Pacific: Queensland and Melanesia including northern Great Barrier Reef, northern New Guinea, New Britain, Solomon Islands and Vanuatu.
- Habitat: Adults inhabit lagoon and seaward reefs.
- Food and feeding: Omnivorous. In the wild, the foods are algae, very small shrimp, anemone tentacles, planktonic fish eggs, fish larvae, and some polychaete worms. In captivity, they feed on meaty foods, such as *Mysis* and brine shrimp, finely chopped fish and shrimp flesh, flake foods, pellets with spirulina.

- Feeding frequency: Several feedings per day.
- Maximum length: 11.0 cm TL male/ unsexed.
- Life span: 20 years.

Aquarium Aspects:

- Minimum tank size: 10 gallons (38 litres).
- Suitable for Nano tank: Yes.
- Substrate: Any. With typical live rock plus hiding places.
- Lighting: Any.
- Temperature: 74.0 to 82.0° F (23.3 to 27.8° C).
- Breeding temperature: 79.0° F.
- Specific gravity: 1.023-1.025 SG.
- Range pH: 7.8- 8.4.
- Water movement: Any.
- Water region: Top.
- Water changes: Bi-weekly.

Behaviour:

Social—It is semi-aggressive. They do best when kept singly or as a pair.

Host anemones: The true percula clownfish, is associated with the following host anemones in the wild: 1. Sebae anemone (*Heteractis crispa*), 2. Magnificent sea anemone/Ritteri anemone (*Heteractis magnifica*), 3. Giant carpet anemone (*Stichodactyla gigantea*), 4. Marten's carpet anemone (*Stichodactyla mertensi*).

Reproductive—Each group of the True percula clownfish consists of a breeding pair and 0-4 non breeders. There is a size-based hierarchy within each group. The male is second largest and the non-breeders are progressively smaller according to the descending hierarchy. In case, the female dies, the male changes sex and becomes the breeding female, and the largest non-breeder becomes the breeding male. This fish is oviparous, forms distinct pair during breeding. The female when ready to lay eggs, nips at the anemone, so it retracts and exposes the spawning sight. She then lays her eggs. Eggs number between 67 to 649 with an average being 331. Eggs are demersal and get attached to the substrate. Male fishes guard and aerate the eggs. These fishes have been bred in captivity and the fry are successfully reared.

Remarks: This is a highly prized aquarium fish. The Percula Clown has been successfully bred in captivity and a number of "designer" clownfish combinations have emerged.

2.11.12 Emperor angelfish

Figure 2.91

Pomacanthus imperator (Bloch, 1787)

Common Name (s):

English—Emperor angelfish, Imperial angelfish.

IUCN Red List Status: Least Concern (LC).

Systematic Position:

Order: PERCIFORMES--perches

Family: POMACANTHIDAE--angelfishes

Genus: *Pomacanthus*

Species: *P. imperator*

Scientific Name: *Pomacanthus imperator* (Bloch, 1787)

Description:

- The body is deep, strongly compressed, slightly elongated with strong spine at angle of preopercle.
- These fishes have yellow and blue horizontal lines throughout their body that start from just behind the gills to the area just before the tail fin.
- From the top of the dorsal fin to the anal fin it is blue and can have some blue striping.
- The dorsal fin is trimmed white, and the tail fin is yellow.
- The mouth and snout area are white, and sharply cut off by a black mask that covers the eyes. The mask is black, trimmed in blue. There is a yellow vertical bar that extends from the yellow in the head downward. This divides the black mask from a thicker vertical black band that starts about ¼ of the way down the body and continues down the pelvic fins. The black band occupies the same area of vertical space as the pectoral fins and gills.
- Dorsal fin spines: 13-14; dorsal fin soft rays: 17-21; anal spines: 3; anal soft rays:18-21.
- Juveniles are dark blue with electric blue and white rings. It takes about 24 to 30 months for the juveniles to achieve the adult colouring.
- Sexual Dimorphism: The blue in front of the eye mask is intense in male Emperor angels and grey blue in females.

Species Details:

- Distribution:

  Indo-Pacific: Red Sea and East Africa to the Hawaiian, Line and Tuamotu Islands, north to southern Japan and the Ogasawara Islands, south to the Great Barrier Reef, New Caledonia, and the Austral Islands.
- Habitat: Adult fishes are benthopelagic and inhabit ledges and caves in areas of rich coral growth on clear lagoon, channel, or seaward reefs. Juveniles are found under ledges, or in holes of outer lagoon patch reefs or semi-protected areas of exposed channels and outer reef flats. Subadults are found in the reef front holes and surge channels.
- Food and feeding: This fish is an omnivore. In the wild they feed on a wide variety of sponges and encrusting organisms along with small amounts of algae, tunicates, hydroids and bryozoans. In captivity, the fed should be

vegetable- based foods as well as meaty foods like chopped squid, scallop, and shrimp. Prepared foods may be marine sponge and tunicates, colour enhancing foods, mussels, shrimp, squid and spinach. Flakes and pellets with spirulina and sponge may also be given. Live foods, vegetable foods, meaty foods are also given.

- Feeding frequency: Several feedings per day.
- Maximum length: 40.0 cm SL male/ unsexed.
- Life span: 21 years.

Aquarium Aspects:

- Minimum tank size: 125 gallons (473 litres).
- Suitable for Nano tank: No.
- Substrate: Any with live rocks and hiding places.
- Lighting: Any.
- Temperature: 72.0 to 82.0° F (22.2 to 27.8° C).
- Specific gravity: 1.023-1.025 SG.
- Range pH: 8.1-8.4.
- Water movement: Any.
- Water region: All.
- Water changes: Bi-weekly.

Behaviour:

Social—They are in general peaceful but can be aggressive towards other Emperor angelfish, and other large angelfish species. In the wild, they are found singly, in pairs or in a harem of one male and two or more females.

Breeding—The pair of Emperor angelfish will spawn once a year. Males have 2 or more females, with each female having their own territory. The larvae float as planktons for a few weeks before turning into fry. The males produce 'grunt" while descending from spawning.

Remarks: This unusually beautiful fish is one of the most recognized and prized seawater ornamental fish.

2.11.13 Blacktail angelfish

Figure 2.92

Centropyge eibli Klausewitz, 1963

Common Name(s):

English—Blacktail angelfish, red stripe angelfish, orangelined angelfish, eibli dwarf angel.

IUCN Red list Status: Least Concern (LC).

Systematic Position:

Family: POMACANTHIDAE—angelfishes

Genus: *Centropyge*

Species: *C.eibli*

Scientific Name: *Centropyge eibli* Klausewitz, 1963

Description:

- Body is small, strongly compressed, elongated and oval shaped with rounded fins.
- There is a strong spine at the angle of preopercle.
- Dorsal fin continuous with 9-15 spines and 15-37 soft rays.
- This fish has a pearlescent, light grey to greenish tan body with 8-10 vertical narrow scribbled orange stripes that run from the top of the fish to the belly and a few in between that are interrupted dashes.
- The eyes have orange rings and the face and head are little duskier.
- The edges of the dorsal fin are typically black, with the back edges of the dorsal, tail and anal fins being edged in blue.
- The dorsal fin and pelvic fins may be greenish to grey or yellow colouring and the pectoral fins are yellow to yellowish grey.
- Juveniles are with fewer stripes on the side.

Species Details:

- Distribution:

  Eastern Indian Ocean: Sri Lanka to Western Australia, including Indonesia and the Andaman Islands.
- Habitat: Inhabits rocky areas of seaward reefs as well as rich coral areas.
- Food and feeding: Omnivorous. In the wild, it takes benthic algae, benthic weeds and hard coral polyps. In the aquarium, they can be fed on meaty food, dried flakes, mysid shrimps, brine shrimps, prepared food with marine algae and frozen prepared diets for sponge and algae eaters. Formula II and Angel formula may also be given. They may also be provided with Green algae like Crunchy Hair Grass of the Boodlea genera, several Green Hair algae, Sea Lettuces like *Ulva ridida* and *Ulva lactuca*.
- Feeding frequency: Several feedings per day.
- Maximum length: 15.0 cm TL male/unsexed.
- Life span: 5 years.

Aquarium Aspects:

- Minimum tank size: 55 gallons (208 liters).
- Suitable for Nano tank: Sometimes.
- Substrate: Any.
- Lighting: Normal to moderate.
- Temperature: 72.0-82.0° F (22.2 to 27.8° C).

- Breeding temperature: 80.0 F -82.0° F (27-28° C).
- Specific gravity: 1.023-1.025.
- Range pH: 8.0-8.4.
- Water movement: Any.
- Water region: Bottom.
- Water changes: Bi-weekly.

Behaviour:

Social—It is a semi-aggressive fish.

Reproductive—This fish is born as female, as they grow in a group, the larger and more dominant fish will become male. The male forms a harem with one to four smaller mature females and up to nine juveniles. They are pelagic spawner. Each pair will rise up several feet above the reef and release the eggs and sperms together directly into the water column. The eggs are fertilized and will rise to the plankton rich surface. Within a day, the newly fertilized eggs will hatch.

Remarks: It is a very pretty aquarium fish and is in much demand among the hobbyists.

2.11.14 Bicolor anthias

Figure 2.93

Pseudoanthias bicolor (Randall, 1979)

Common Name(s):

English—Bicolor anthias, Fancy Bass anthias.

IUCN Red List Status: Least Concern.

Systematic Position:

Order: PERCIFORMES

Family: SERRANIDAE—sea basses

Subfamily: ANTHIINAE

Genus: *Pseudanthias*

Species: *P. bicolor*

Scientific Name: *Pseudanthias bicolor* (Randall, 1979)

Description:

- The upper half of the body is yellow or orange, the lower half can be white pinkish, pink or purple.
- Edges of the fins are deep blue.
- The dorsal fin of the male is edged in purple and the first ray is elongated.
- The males have a more intense coloration with a violet body and yellow running along the back through the upper caudal lobe.
- The females tend to be lavender in colour with a yellow back and caudal fin.

- Sexual Dimorphism: Male has two yellow-tipped filaments heading the dorsal fin that are used for display.

Species Details:

- Distribution:

  Indo-Pacific: Mauritius to the Hawaiian and Line islands, north to the Ryukyu Islands, south to the Loyalty Islands; Marshall and Caroline islands in Micronesia.
- Habitat: Found in deep coastal to outer reef slopes, in current prone areas. Benthopelagic. It occurs in small groups above coral outcrops or crevices or ledges.
- Food and feeding: In the wild, feed primarily on zooplanktons. In the aquarium, take cyclops, artemia, frozen *Mysis*, enriched frozen brine shrimp, high quality flakes.
- Feeding frequency: Small quantities throughout the day.
- Maximum length: 13.0 cm TL male/unsexed.

Aquarium Aspects:

- Minimum tank size: 55-125 gallons.
- Substrate: With lots of hiding places.
- Water conditions: Temperature: 72.0-78.0° F; specific gravity: 1.020-1.027; pH: 8.1-8.4.
- Water region: Middle.
- Reef compatibility: Yes.

Behaviour:

Social: Semi-aggressive. Form massive schools among the reefs. Does best when kept with its own species in an aquarium. In a group with one male and multiple females.

Reproductive—Hermaphroditic. When a male is needed, a female changes sex and takes on the role.

Remarks: It is one of the most sought after and popular fish to keep in a reef tank. They are very popular because of their stunning colours and their propensity to school.

2.11.15 Lyretail anthias

Figure 2.94

Pseudanthias squamipinnis (Peters, 1885)

Common Name(s):

English—Sea goldie, lyretail coralfish, lyretail anthias, scalefin anthias.

IUCN Red List Status: Least Concern (LC).

Systematic Position:

Family: SERRANIDAE--sea basses

Subfamily: ANTHIINAE

Genus: *Pseudanthias*

Species: *P. squamipinnis*

Scientific Name: *Pseudanthias squamipinnis* (Peters, 1885)

Description:

- This fish is with dorsal spines:10; dorsal soft rays :15-17, anal spines :3, anal soft rays: 6-7.
- In males, third ray of the dorsal fin is elongated. A red patch, on the pectoral fin and elongated margins of the tail.
- Sexual Dimorphism: Males have bright red colouration of varying hues, sometimes with a splash of blue along the fins. Males usually have an elongated dorsal spine as well. Female fish are more orange in colouration, with a blue ring around their eyes as well as a blue stripe running horizontally from their eye to their pectoral fin. Males are 15 cm and females are 7 cm in length.

Species details:

- Distribution:

  Indo-West Pacific: Red Sea and Natal, South Africa to Niue, north to Japan, south to Australia.
- Habitat: Adult fishes inhabit coral outcrops or patch reefs of clear lagoons, channels or outer reef slopes.
- Food and feeding: In the wild, they feed on zooplankton. In aquarium, they should be fed on high protein content frozen/ dried live foods, frozen *Mysis*, shrimp, enriched frozen brine shrimp, high quality flake foods.
- Feeding frequency: Several time throughout the day.
- Maximum length: 15.0 cm TL male/ unsexed.

Aquarium Aspects:

- Minimum tank size: 125 gallons.
- Water conditions: Temperature: 72.0-78.0° F; dGH-8-12; pH-8.1-8.4; sg-1.020-1.025.
- Substrate: Coral filled reef.

Behaviour:

Social—It is semi-aggressive. This species is the hardiest and is the most territorial among the anthias. They are best kept in a group and housed in species-specific aquarium.

Reproductive—They are protogynous hermaphrodite, a male retains a harem of five to 10 females, but when the male dies, one of the females will undergo sex reversal and take the place of missing male. Spawning occurs at sunset. In the Red Sea, between December and February. It is a pelagic spawner, a monandric species and females are induced to change sex by removal of males from social groups. Sex-reversal is completed in 2-4 weeks after male removal in the area, or 170-280 days, in all female group. Successful males spawn with successive females. Males maintain harem and are territorial.

2.11.16 Dogface pufferfish

Figure 2.95

Arothron nigropunctatus (Bloch & Schneider, 1801)

Common Name(s):

English—Blackspotted puffer, dogface pufferfish, dog-faced puffer, hush puppy puffer.

IUCN Red List Status: Least Concern (LC).

Systematic Position:

Order: TETRADONTIFORMES--plectognaths

Family: TETRADONTIDAE—puffers

Subfamily: TETRADONTINAE

Genus: *Arothron*

Species: *A. nigropunctatus*

Scientific Name: *Arothron nigropunctatus* (Bloch & Schneider, 1801)

Description:

- Body broadly rounded, somewhat elongated, with a head and snout which looks like a dog.
- Body is covered with prickles and with large black spots.
- There are no dorsal spines, dorsal soft rays: 10-11; anal spines: 0; anal soft rays: 10-12.
- There is no pelvic fin and no lateral line.
- This snout is short with two pairs of nostrils and its mouth is terminal with four strong teeth.
- It comes in a variety of colour, although most commonly blue or grey, with brown on the chin and a white band just below the eyes.
- Sexual Dimorphism: Unknown.

Species Details:

- Distribution:

  Indo-Pacific: East Africa to Micronesia and Samoa, north to southern Japan, south to New South Wales.
- Habitat: Found in coastal to outer reef crest and slopes with rich invertebrate growth. They thrive along reefs, lagoon patch reefs and reef faces.
- Food and feeding: In the wild, they are predatory. Feed on corals (generally *Acropora* tips), crustaceans, molluscs, sponges, tunicates and algae. In

captivity they feed on a large variety of all kinds of live and frozen meaty foods. Live foods include snails, crabs, crayfish, shrimp and earthworms. Frozen foods include prawn, crabs, bloodworms, blackworms, silversides and mussels. Pellets and flakes are not suitable.

- Feeding frequency: Small amounts several times a day.
- Maximum length: 33.0 cm TL male/unsexed.

Aquarium Aspects:

- Minimum tank size: 100 gallons.
- Light: No special requirement.
- Water movement: No special requirements.
- Water region: No special requirements.

Behaviour:

Social—This fish is not aggressive. It is a diurnal, solitary and territorial fish.

Reproductive—It is an oviparous fish.

Remarks: They are favourite with aquarists. These fishes make a special pet that can recognize their care givers and show excitement when given attention. As most other puffers, blackspotted puffers are highly poisonous. These creatures are the second most poisonous creature in the world. It can inflate their bodies by swallowing air or water.

2.11.17 Threadfin butterflyfish

Figure 2.96

Chaetodon auriga Forsskal, 1775

Common Name(s):

English—Threadfin butterflyfish.

IUCN Red List Category: Least Concern.

Systematic Position:

Order: PERCIFORMES

Family: CHAETODONTIDAE

Genus: *Chaetodon*

Subgenus: *Rabdophorus*

Species: *C. auriga*

Scientific Name: *Chaetodon auriga* Forsskal, 1775

Two sub-species: *Chaetodon auriga auriga* occurs in the Red Sea population lacks the dorsal eye spot.

Chaetodon auriga setifer is the spotted population occurring outside the Red Sea.

Description:

- Body is white with chevron markings on the side.

- Rear edge of the dorsal fin has a prominent black spot and a prominent black vertical band runs through the eye.
- The rear of soft dorsal fin has a trailing filament.
- It has a belly patch of descending oblique dark lines as well as bright yellow fins.

Species Details:

- Distribution:

  Indian Ocean-western, Pacific-eastern central, northwest, southeast, Atlantic-southeast, Indian Ocean-eastern, Pacific-southwest, Pacific-western central. Red Sea and East Africa to the Hawaiian, Marquesan, and Ducie islands, north to southern Japan, south to Lord Howe and Rapa islands.
- Habitat: From rich coral reefs to weedy and rubble covered areas. Benthopelagic. May be found in singly, in pairs and in aggregations.
- Omnivorous species. In the wild, they feed chiefly by tearing pieces from annelid worms, coral polyps and sea anemones, and algae. In the aquarium, it accepts fresh, frozen, and freeze- dried food, vegetables, shrimps, plankton, algae- based food.
- Maximum length: 23.0 cm TL male/unsexed.

Aquarium Aspects:

- Minimum tank size: 15-50 gallon/60-200 litres.
- Temperature: 72.0-78.0°F/ 22.0-25.0°C.
- pH: 8.2-8.4.
- Specific gravity: 1.021 and 1.023.
- Aquarium care: Fairly hardy.

Behaviour:

Social: They are peaceful in nature.

Reproductive--Distinct pairing. Monogamous mating as both obligate and social. It is an egg-laying species that scatter the eggs. During spawning, the female will release her eggs in mid-water and the male will fertilize them. Tidal movements and lunar cycle may trigger the spawning. The larvae of this fish, go as in an armoured stage, when they drift around the ocean as free-floating plankton.

Remarks: A popular marine aquarium fish.

2.11.18 Addis butterflyfish

Figure 2.97

Chaetodon semilarvatus Cuvier, 1831

Common Name(s):

English—Addis butterflyfish, Bluecheek butterflyfish, Golden butterflyfish, Redlined butterflyfish, Yellow butterflyfish.

IUCN Red List Status: Least Concern.

Systematic Position:

Order: PERCIFORMES

Family: CHAETODONTIDAE

Genus: *Chaetodon*

Species: *C. semilarvatus*

Scientific Name: *Chaetodon semilarvatus* Cuvier, 1831

Description:

- Body is disc-like and laterally compressed and has a protruding snout tipped with a small mouth.
- Bright yellow body, with thin slate blue vertical lines on the sides.
- A slate blue cheek patch in lieu of the usual black eye stripe of most chaetodon.
- The dorsal fin is continuous and it has a rounded tail fin.
- The fins are yellow except the pectoral ones which are translucent, the dorsal and anal fins have a narrow blue sub-marginal line, and the caudal fin is translucent posteriorly.
- Sexual Dimorphism: No sexual dimorphism is noticed in this species.

Species Details:

- Distribution:

  Endemic to the Western Indian Ocean: Red Sea and Gulf of Aden, northeast to the southern coast of Oman. Native to Djibouti; Egypt; Eritrea; Israel; Jordan; Saudi Arabia; Somalia; Sudan; Yemen.
- Habitat: Found at depths of between 3 to 20 m, in areas of rich coral growth in relatively shallow water. Take rest beneath *Acropora* plate corals during the day, becoming active in late afternoon. This species is an obligate corallivore.
- Food and feeding: They are omnivores. In the wild they eat live coral polyps, sessile invertebrates and zooplankton. Prefer meaty foods, dried flakes, shrimps and tablets, pellets, live foods and vegetable food.
- Feeding frequency: Several feedings per day.
- Maximum length: 23.0 cm TL male/unsexed.

Aquarium Aspects:

- Minimum tank size: 100 gallons (379 L).
- Suitable for Nano Tank: No.
- Substrate type: Sand and coral mixed.
- Lighting needs: Moderate to normal lighting.
- Temperature: 75.0 to 79.0° F.

- pH: 8.1-8.4.
- Water movement: Weak.
- Water region: All.

Behaviour:

Social--This species is not aggressive and is peaceful so it is best to select tank mates that are not overly territorial or aggressive.

Reproductive--In the wild, this species is pelagic spawner that release many tiny eggs into the planktonic water column where they float, until they hatch. There is no report of successful breeding of the species in captivity.

Remarks: This is a very popular butterflyfish with large size, vibrant colours and graceful swimming. This pretty marine fish is much sought after in the aquarium trade and commands a relatively high price.

2.12.19 Pennant coralfish

Figure 2.98

Heniochus acuminatus (Linnaeus, 1758)

Common Name(s):

English—Pennant coralfish, longfin bannerfish, reef bannerfish.

IUCN Red List Status: Least Concern.

Systematic Position:

Order: PERCIFORMES

Family: CHAETODONTIDAE

Genus: *Heniochus*

Species: *H. acuminatus*

Scientific Name: *Heniochus acuminatus* (Linnaeus, 1758)

Description:

- This fish has laterally compressed body.
- The snout is little stretched with a small terminal protractile mouth.
- The colour of the body is white with two large black diagonal bands.
- The dorsal, caudal and pectoral fins are yellow.
- The head is white, the eyes are black and linked together by a black band.
- The snout is spotted with black.
- The first ray of its dorsal fin extends in a long white filament.
- The juvenile does not have any white area after the second black stripe like adults.
- Sexual Dimorphism: There is no morphological difference between male and female.

Species Details:

- Distribution:

It is widespread throughout the tropical and subtropical waters of the Indo-Pacific from the eastern coast of Africa, Red Sea to Polynesia and from south Japan to the south of the Great Barrier Reef.

- Habitat: Marine neritic. Relatively deep waters from protected lagoon, channels or outer reef slopes from 15 to 75 meters deep.
- Food and feeding: In the wild, feeds on zooplankton in the water column and sometimes benthic invertebrates. Juveniles feed by cleaning other fishes. In the aquarium, takes, flake food, pellet, live foods, vegetable foods.
- Feeding frequency: Several feedings per day.
- Maximum length: 25.0 cm TL male/ unsexed.
- Life span: 5 years.

Aquarium Aspects:

- Minimum tank size: 55 gallons.
- Suitable for Nano tank: No.
- Substrate type: Sand and coral mixed.
- Lighting needs: Moderate to normal lighting.
- Temperature: 72.0 to 78.0° F.
- pH: 8.1-8.4.
- Diet type: Omnivore.
- Aquarium hardiness: Moderately hardy.
- Water changes: Bi-weekly.

Behaviour:

Social- Peaceful fish. Adults are often seen in pairs, though sometimes seen singly. Juveniles are solitary.

Reproductive- These are pelagic spawner and release many tiny eggs into the water with planktons, where they float until they hatch. There is no report yet of their breeding in captivity.

Remarks: This is a very popular butterflyfish and an excellent choice for the aquarists with some experience.

DAMSELFISH

2.11.20 Sergeant major damselfish

Figure 2.99

Abudefduf saxatilis (Linnaeus, 1758)

Common Name(s):

English—Sergeant major damselfish.

IUCN Red List Status: Least Concern (LC).

Systematic Position:

Superorder: ACANTHOPTERYGII

Subseries: OVALENTARIA

Family: POMACENTRIDAE--damselfishes

Subfamily: POMACENTRINAE

Genus: *Abudefduf*

Species: *A. saxatilis*

Scientific Name: *Abudefduf saxatilis* (Linnaeus, 1758)

Description:

- A very deep-bodied fish, white, silvery to greenish grey colour, with a yellow top.
- It has 5 vertical stripes which are black.
- A faint sixth stripe may be present on the caudal peduncle.
- There is a dark spot around its pectoral fin.
- Sexual Dimorphism: Adult males have a more bluish colouration and its stripes are less visible. During spawning, the male exhibits a white patch on the face.

Species Details:

- Distribution:

  Atlantic Ocean: Canada to Rhode Island, USA to Uruguay in the Western Atlantic, abundant on Caribbean reefs; around islands of the mid- Atlantic; Cape Verde, and along the tropical coast of western Africa south to Angola.
- Habitat: Juveniles are common in tide pools while adults are found over shallow reef tops.
- Food and feeding: Omnivore. In the wild, feed on all types of foods including small fish, fish eggs, various life stages of invertebrates, many types of algae, zoanthids. In captivity, any type of prepared meaty or vegetable foods that are freeze dried, frozen, flake or pellet are taken as food.
- Feeding frequency: Several feedings per day.
- Size of fish: 14.99 cm.
- Life span: 15 years.

Aquarium Aspects:

Minimum tank size: 75 gallons (284 litres).

Suitable for Nano tank: No.

Live-rock requirements: Typical plus hiding places.

Substrate: Any.

Lighting: Any.

Temperature: 72.0-to 84.0°F (22.2 to 28.9°C).

Specific gravity: 1.023-1.025 SG.

Range pH: 8.1-8.4.

Water movement: Any.

Water region: All.

Water changes: Bi-weekly.

Behaviour:

Social: They are very aggressive and this aggression increases as they get older. They can be kept singly or as a male/female pair. They can also be kept as a school of 5 or more but they should be the only damsels in the tank.

Reproductive: Excellent water parameters and large non-predatory aquarium is needed for successful breeding. The eggs and larvae are much smaller and the fry are difficult to rear. During the spawning, both male and female will develop deep bluish colour, almost hiding the black vertical bars.

Remarks: This fish is an easily available, cost-friendly and hardy pet of the aquarium.

2.11.21 Yellowtail damselfish

Figure 2.100

Chrysiptera parasema (Fowler, 1918)

Common Name(s):

English—Yellowtail damselfish, yellowtail blue damsel, gold tail demoiselle.

IUCN Red List Status: Not Evaluated (NE).

Systematic Position:

Subseries: OVALENTARIA

Family: POMACENTRIDAE--damselfishes

Subfamily: POMACENTRINAE

Genus: *Chrysiptera*

Species: *C. parasema*

Scientific Name: *Chrysiptera parasema* (Fowler, 1918)

Description:

- Small and deep bodied fish, laterally compressed and with spiny dorsal fin.
- The body of this fish is blue with a distinctive yellow caudal fin.
- The yellow colour may extend up to the back of the anal and dorsal fins, especially in species from Papua New Guinea.
- The pelvic or pectoral fins may also be yellow.
- Sexual Dimorphism: Sexual dimorphism is unknown, though males may be larger.

Species Details:

- Distribution:

  Western Pacific, in the Solomon Islands, northern Papua New Guinea, Philippines and Ryukyu Islands.
- Habitat: Inhabits densely populated coral groupings in sheltered lagoons and inshore reefs. They prefer to remain at a depth between 3-52 feet (1-16 m).
- Food and feeding: They are omnivores. In the wild, they feed on zooplankton, other planktonic foods and algae. In captivity, they should be provided with foods plenty of proteins, some vegetables, mysis, shrimp, vitamin-enriched brine shrimp, cyclops, finely shredded frozen seafoods. Freeze dried, frozen, sinking pellets, flake or fresh foods and colour enhancing foods may be given.
- Feeding frequency: Several feedings per day.
- Maximum length: 6.99 cm.
- Life span: 6-15 years.

Aquarium Aspects:

- Minimum tank size: 20 gallons (76 litres).
- Suitable for Nano tank: Yes.
- Substrate: Sand, live rocks with hiding places.
- Lighting: Any.
- Temperature: 74.0 to 84.0° F (23.3 to 28.9° C).
- Breeding temperature: 79.0 to 82.0° F (26.0-28.0° C).
- Specific gravity: 1.023-1.025 SG.
- Range pH: 8.2-8.4.
- Water movement: Any.
- Water region: Bottom.
- Water change: Bi-weekly.

Behaviour:

Social—Semi-aggressive though they are one of the less aggressive of the genus. They contain chromatophores and can change colour in response to external stimuli. When in stress, they generally turn darker in colour.

Reproductive—This fish readily spawns in captivity. Females usually lay around 300 eggs, which will hatch within 96 hours. The eggs and larvae, due to their significantly small size, are difficult to rear. The female chooses a fit and healthy male, and inspecting the number of eggs in the nest of the male, she chooses her partner. She chooses the best candidate. After spawning, the male will stay and protect the eggs until they have hatched, which may take up to 4 days in the wild, depending on the water temperature. The larval stage can last 10-15 days.

Remarks: This beautiful, small, active and very hardy popular fish makes a great choice for beginners though enjoyed by all aquarists.

2.11.22 Green moray eel

Figure 2.101

Gymnothorax funebris Ranzani, 1839

Common English Name (s):

English--Green moray eel, green moray.

IUCN Red List Status: Least Concern.

Systematic Position:

Order: ANGUILLIFORMES--eels

Family: MURAENIDAE--moray eels

Subfamily: MURAENINAE

Genus: *Gymnothorax*

Species: *G. funebris*

Scientific Name: *Gymnothorax funebris* Ranzani, 1839

Description:

- Very large, muscular scale-less body, is laterally compressed.
- The dorsal fin begins just behind its head, extends along the length of its body and is fused with the caudal and anal fins.
- It lacks both pelvic and pectoral fins.
- Body is uniformly dark brown in colour, the apparently green colour comes from a yellow green mucous that covers its skin to help protect it from parasites and bacteria.
- It has a row of very rear-hooked teeth and opens and closes its mouth continually.
- This keeps the water circulating over its gills.
- It has pharyngeal teeth that is used to catch prey.

Species Details:

- Distribution:

  Western Atlantic: New Jersey (USA), Bermuda, and northern Gulf of Mexico to Brazil.
- Habitat: This species is benthic and solitary. Inhabits rocky shorelines, reefs and mangroves. Generally found shallower than 30 m. It hides among cracks and crevices, along rocky shorelines, and in coral reefs.
- Food and feeding: It is carnivorous and is sedentary predator. This fish feeds mostly at night on fish, crabs, shrimp, octopus, and squid.
- Maximum length: 250 cm TL male/unsexed.
- Life span: 10-30 years.

Aquarium Aspects:

- Minimum tank size: 30 gallons.
- Tank set up, substrate: Bare bottomed tank, sand or fine gravel, corals and crevices.
- Water conditions: Temperature: 74.0-79.0° F; pH: 7.6-8.4; salinity: 1.022-1.025.

Behaviour:

Social-They are semi-aggressive, though calm and curious and are very protective of their territory. It is nocturnal predator. During the day, it hides in a hole in the rocks. As they have very poor vision, they use their sense of smell to find its prey. They sit and wait for the prey instead of going hunting.

Breeding-The green moray's breeding season is from July to September. The fertilized eggs float on top of the water until they hatch in to larvae called leptocephali. Leptocephali are transparent, ribbon-like larvae. The larvae float around with the plankton for 8 months while they develop into a young adult eel (elver).

Remarks: They are common in large commercial aquarium.

2.11.23 Zebra moray eel Figure 2.102

Gymnomuraena zebra (Shaw, 1797)

Common Name(s):

English—Zebra moray eel, zebra moray.

IUCN Red List Status: Least Concern (LC).

Systematic Position:

Order: ANGUILLIFORMES

Family: MURAENIDAE—moray eels

Subfamily: MURAENINAE

Genus: *Gymnomuraena*

Species: *G. zebra*

Scientific Name: *Gymnomuraena zebra* (Shaw, 1797)

Description:

- It has a blunt rounded short snout.
- Zebra striping i.e., narrow white bands are present on a chocolate or golden black background.
- Sexual Dimorphism: Unknown.

Species Details:

- Indo-Pacific and the tropical eastern Pacific, they cover a wide range from the Red Sea and east coast of Africa across the Indian Ocean to the Philippines and north to the Hawaiian Islands, also found in the Gulf of California off Panama, Mexico and Galapagos Islands.

- Habitat: They inhabit seaward reefs at depths of 20 feet to 145 feet (6-44 meters), dwelling close to the bottom among corals and rubble.
- Food and feeding: They are carnivorous. With blunt teeth suitable for grinding, they primarily eat crabs, crustaceans, snails, urchins and other molluscs. In captivity they may be given, crab meat, shrimp, clams, squid, scallops and fish flesh.
- Maximum length: 150 cm TL male/unsexed.

Aquarium Aspects:

- Minimum tank size: 75 gallons (284 litres).
- Light: No special requirement.
- Temperature: 74.0-79.0° F.
- Water movement: All.
- Water region: Remains at the bottom of the tank and spends most of their time under a rock in cave.
- Water changes: 20% water changes in a month.

Behaviour:

Social—It is a peaceful and sociable eel. A reef aquarium or an aquarium with lots of rock is suitable for this fish. It is a community fish but cannot be kept with ornamental crustaceans.

Reproductive—Captive breeding is probably not yet possible. In the wild, the female will spawn either as a pair or with several smaller males. At dusk eggs are released and they are pelagic. The larval stage lasts for 6-10 months.

Remarks: It is an easily available eel and is kept in public marine aquarium.

2.11.24 Spotted garden eel

Figure 2.103

Heteroconger hassi (Klausewitz & Eibl-Eibesfeldt, 1959)

Common Name(s):

English—Spotted garden eel.

IUCN Red List Status: Least Concern (LC).

Systematic Position:

Order: ANGUILLIFORMES—eels

Family: CONGRIDAE—conger eels

Subfamily: HETEROCONGRINAE—garden eels

Genus: *Heteroconger*

Species: *H. hassi*

Scientific Name: *Heteroconger hassi* (Klausewitz & Eibl-Eibesfeldt, 1959)

Description:

- Body very elongate and slender.

- Mouth short and lower jaw projecting beyond upper.
- They remain in the burrows in large colonies with their tail down, and the body relatively straight, and the head up.
- Dorsal and anal fin rays unsegmented.
- Pectoral fins minute.
- Two large black spots are present.
- Body diameter to about 14 mm.
- Dorsal fin spines: 0; dorsal fin soft rays: 449-540; anal spines: 0; anal soft rays: 273-334.

Species Details:

- Distribution:

  Indo-Pacific: Red Sea and East Africa to the Society Islands, north to the Ryukyu and the Ogasawara islands, south to north-western Australia and New Caledonia, throughout Micronesia.
- Habitat: Found on sandy slopes below 15 m.
- Food and feeding: Planktivorous in nature. They eat tiny planktonic animals (zooplanktons), that float by them in the water current.
- Maximum length: 40 cm TL male/unsexed.

Aquarium Aspects:

- Minimum tank size: 125 gallons.
- Substrate: Sand with live rock.
- Water conditions: Temperature: 72.0-80.0° F; dKH: 8-12; pH: 8.1-8.4; sg: 1.020-1.025.

Behaviour:

Social—Burrow into the sandy sea bottom to make their home. They secrete slime from their skin to cement the burrow walls. They will escape from enemies by diving tail-first into the burrow, which they rarely leave. They are reported to have good eye sight.

Reproductive—They are pelagic spawner, after mating, the fertilized eggs are released into the water current. The eggs float in the open water near the surface until they hatch (Randall et al., 1990). When they grow enough, they swim down and make their burrows. The burrows come closer together during breeding season until contact between partners is possible (en.wikipedia.org>wiki>Spotted_ garden _ eel).

Remarks: These eels are kept in public marine aquaria.

GRUNTS

2.11.25 Pork fish

Figure 2.104

Anisotremus virginicus (Linnaeus, 1758)

Common Name (s):

English--Porkfish.

IUCN Red List Status: Least Concern (LC).

Systematic Position:

Order: PERCIFORMES

Family: HAEMULIDAE--grunts

Genus: *Anisotremus*

Species: *A. virginicus*

Scientific Name: *Anisotremus virginicus* (Linnaeus, 1758)

Description:

- It is a deep-bodied fish with a blunt snout, thick lips and a small mouth positioned low on the head.
- Dorsal fin is continuous. Dorsal fin spines: 12; dorsal fin soft rays: 16-18; anal fin spines: 3; anal fin soft rays: 9.
- Body striped golden-yellow and silvery blue.
- Fins are yellow, 2 black bars, one from nape through eye to mouth, another from origin of dorsal fin to base of pectoral fin.
- Only grunt fish in the Caribbean with two vertical black bars and yellow stripes.
- Young porkfish do not have black bars but have a large dark spot near the base of caudal fin.
- Sexual Dimorphism: Unknown.

Species Details:

- Distribution:

  Western Atlantic: USA to Brazil including the Gulf of Mexico and the Caribbean Sea.
- Habitat: Inhabits reefs and rocky bottoms, depth of 2-20 m and corals.
- Food and feeding: Nocturnal predators, targeting benthic prey items. Feeds at night on molluscs, echinoderms, annelids and crustaceans. Youngs pick parasites from bodies of larger fishes.
- Maximum length: 40.6 cm TL male/unsexed.

Aquarium Aspects:

- Substrate: Live rock and hardy corals.
- Water conditions: Temperature: 72.0-78.0°F; pH: 8.1-8.4; dKH: 8-12.

Behaviour:

Social--Produce grunt like sounds via their pharyngeal teeth and swim bladder.

Reproductive--Distinct pairing takes place between a supermale and a large group of females during breeding. The supermale is larger than other males and has distinct colours and patterns on its skin. When a supermale dies a larger female becomes supermale.

Remarks: It is a popular fish in public marine aquarium.

2.11.26 Spanish hogfish

Figure 2.105

Bodianus rufus (Linnaeus, 1758)

Common Name(s):

English—Spanish hogfish.

IUCN Red List Status: Least Concern (LC).

Systematic Position:

Order: LABRIFORMES—wrasses and relatives

Family: LABRIDAE--wrasses

Genus: *Bodianus*

Species: *B. rufus*

Scientific Name: *Bodianus rufus* (Linnaeus, 1758)

Description:

- Body elongate, laterally compressed.
- Mouth protractile, elongated, snout-like.
- Dorsal fin spines: 12; dorsal fin soft rays: 9-11; anal fin spines: 3; anal fin rays: 11-13.
- The upper front of head, part of body and front of dorsal fin is blue, if from shallow water, red if from deep, rest of body and tail is yellow.
- Dorsal, anal and pelvic fins mostly blue or red-yellowish at rear.
- Sexual Dimorphism: No notable sexual dimorphism.

Distribution:

- Western Atlantic: Bermuda, southern Florida (USA) and throughout the Gulf of Mexico and Caribbean Sea to southern Brazil.
- Habitat: Adults stay in rocky or coral reefs.
- Food and feeding: Carnivorous. Feed on marine invertebrates such as molluscs, crustaceans, brittle stars and sea urchins. In captivity, they may be given, *Mysis*, brine shrimp frozen or live, other vitamin enriched frozen meaty food. Juveniles are cleaner fish, pick parasites from larger fishes.
- Maximum length: 40.0 cm TL male/ unsexed.

Aquarium Aspects:

- Minimum tank size: 150 gallons.
- Temperature: 72.0-78.0° F.
- Range pH: 8.1-8.4.

- Specific gravity: 1.020-1.025.

Behaviour:

Social—They are diurnal, i.e., active during the day time. It is aggressive towards smaller, peaceful fishes. Spanish hogfish is a very hardy tropical reef fish.

Reproductive—It is a protogynous (female-first) hermaphrodite. Females can undergo transformation to become a sexually active male. Sex reversal is completed in 7-10 days. Male fish forms permanent harem with a single male and several smaller females. When the male fish dies, or is otherwise removed, from the harem, the largest and the most dominant female is able to change her sex and form the harem. This species is oviparous and form distinct pair during breeding. Fertilization is external and occurs all year long. This fish does not exhibit parental care of any kind. Eggs are pelagic. Larvae hatch within 24 hours of fertilization.

Remarks: It is a favourite pet in the aquarium trade.

2.11.27 Golden trevally

Figure 2.106

Gnathanodon speciosus (Forsskal, 1775)

Common Name (s):

English-Golden trevally, golden kingfish, banded trevally, king trevally.

IUCN Red List Status: Least Concern (LC).

Systematic Position:

Order: CARANGIFORMES--jacks

Family: CARANGIDAE--jacks and pompanos

Subfamily: CARANGINAE

Genus: *Gnathanodon*

Species: *G. speciosus*

Scientific Name: *Gnathanodon speciosus* (Forsskal, 1775)

Description:

- It is a relatively large fish with compressed, oblong body.
- The dorsal profile slightly more convex than the ventral profile, particularly at the anterior side.
- The mouth is highly protractile and fleshy.
- The dorsal spines: 8; dorsal soft rays: 18-20; anal spines: 3; anal soft rays: 15-17.
- Juveniles of this fish are of bright golden yellow colour over their entire body and all fins, with 7 to 11 black vertical cross bars all over the body.
- These bars generally alternate between broad and narrow.
- The caudal fin lobes have dark tips and there is a prominent black edge to the operculum.
- As the fish grows old, the body colour becomes more silver to silver golden

and the cross bars fade or disappear, often replaced by dark blotches.

- The fins remain yellow, often with greenish tinges.
- The dark edge of the operculum also fades with edge.

Species Details:

- Distribution:

  Indo-Pacific: from Kwazulu-Natal, South Africa, north to Mie Prefecture, Japan and east to southwestern coast of Baja California Sur, Mexico and Gulf of California to Ecuador.
- Habitat: Adults inhabit deep lagoon and seaward reefs. Small juveniles occur among the tentacles of jellyfish.
- Food and feeding: Carnivorous. Feed on crustaceans, such as shrimps, crabs, and amphipods, molluscs and small fish.
- Maximum length: 120 cm TL male/ unsexed.

Aquarium Aspects:

- Minimum tank size: 300 gallons.
- Substrate: Coral or rocks, may be plants.
- Light: High.
- Water conditions: Temperature: 72.0-78.0° F; pH: 8.1-8.4; sg: 1.020-1.025.

Behaviour:

Social—It remains either as a solitary individual or in small schools. Young ones form larger schools and follow other larger fishes. This diurnal foraging carnivore, sucks prey out of both reef and algae-dominant habitats, and form sandy substrate.

Reproductive—Spawning, in the Indian Ocean, in the southern Persian Gulf, occurs in April and May. In Hawaii, spawning occurs from late February to early October.

Remarks: Golden trevally are kept in marine aquaria. Culture of this fish for aquarium trade has been achieved in Singapore.

2.11.28 Snubnose pompano

Figure 2.107

Trachinotus blochii (Lacepede, 1801)

Common Name(s):

English-Snubnose pompano, Asian pompano, Golden pompano, Dart, moonfish.

IUCN Red List Status: Least Concern.

Systematic Position

Order: CARANGIFORMES--jacks

Family: CARANGIDAE--jacks and pompanos

Subfamily: TRACHINOTINAE

Genus: *Trachinotus*

Species: *T. blochii*

Scientific Name: *Trachinotus blochii* (Lacepede, 1801)

Description:

- Body ovate in young to sub-ovate in large adults and is compressed laterally.
- Profile of snout broadly rounded, in adults becoming nearly straight to interorbital region.
- Both jaws with bands of small villiform teeth.
- Two separate dorsal fins.
- Head and body generally silvery, blue grey above, paler below.
- Large adults sometime with most of body golden orange especially snout and lower half of body.
- Second dorsal fin dark, lobe of fin dusky orange, anal fin dusky dirty orange, lobe with a brownish anterior margin.
- Caudal fin dark to dirty orange with leading edges of fin darkest. Fins are golden yellow with dusky tips, pelvic fins white to dirty orange, pectoral fins dark.

Species Details:

- Distribution:

  Indo-Pacific: Red Sea and East Africa to the Marshall Islands and Samoa, north to southern Japan, south to Australia.
- Habitat: Juveniles inhabit sandy shorelines and shallow sandy or muddy bays near river mouths while adults move out in schools to clear seaward coral and rock reefs. Depth range 1-7 m.
- Food and feeding: Benthic feeder. Feeds on small shrimps, crabs and bivalves.
- Maximum length: 50.0 cm TL male/unsexed.

Aquarium Aspects:

- Minimum tank size: 2000 litres.
- Substrate: Coarse sand and corals.
- Light: Moderate.
- Temperature: 25° C.

Behaviour:

Social-Found singly or in small schools over clear water seaward coral and rocky reefs.

Reproductive—Florida Pompano spawn in offshore waters between early spring to middle autumn, from April to June. The pelagic eggs hatch in 24 hours. The larvae undergo metamorphosis after 15 days. The estimated fecundity is 133000 to 800000 eggs per female each season (www.fao.org/fishery/culturedspecies/Trachinotus_spp/en).

Remarks: Common in public aquaria.

2.11.29 Red lionfish Figure 2.108

Pterois volitans (Linnaeus, 1758)

Common Name(s):

English—Red lionfish, zebrafish, fire fish, turkey fish, red lionfish, scorpion volitans, devil fire fish.

IUCN Red List Status: Least Concern.

Systematic Position:

Order: SCORPAENIFORMES—mail-cheeked fishes

Family: SCORPAENIDAE--scorpionfishes

Genus: *Pterois*

Species: *P. volitans*

Scientific Name: *Pterois volitans* (Linnaeus, 1758)

Description:

- The body is almond-shaped with brown or maroon and white zebra stripes or bands covering the head and body.
- They have fleshy tentacles above their eyes and below the mouth.
- The colour of the fish varies according to their habitat.
- The number of dorsal spines :13; dorsal soft rays: 10-11; anal spines: 3; anal soft rays: 6-8.
- Pectoral fins are fan-like.
- Long, separated dorsal spines.
- The soft dorsal, anal and caudal fin are spotted.
- Venom is delivered by the spines and it is produced by glands at the base of the spines.
- Sexual Dimorphism: Unknown.

Species Details:

- Distribution:

  Pacific-Ocean: Cocos-Keeling Islands and Western Australia in the Oeno island Pitcairn eastern Indian Ocean to the Marquesas and Oeno (Pitcairn group), north to southern Japan and southern Kores south to Lord Howe Island, northern New Zealand, and the Austral Islands.
- Habitat: Found in lagoons and seaward reefs from turbid inshore areas to depths of 50 m.
- Food and feeding: Carnivorous. Eat small fish and invertebrates. Solitary predator of small fishes, shrimps and crabs.

- Maximum length: 38.0 cm TL male/ unsexed.
- Average life span in the wild: Up to 15 years.

Aquarium Aspects:

- Minimum tank size: 75 gallons.
- Temperature: 74.0-79.0° F.
- Lighting: No special requirements.
- Water medium: No special requirement.
- Water region: All.

Behaviour

Social--Peaceful and sociable. It is a community fish but may eat small tank-mates. They are probably nocturnal hunters. During the day, they retreat among the rocks and corals. They move by slowly undulating the soft rays of the dorsal and anal fins. The venom of the red lionfish, delivered via an array of up to 18 needle-like dorsal fins, is purely defensive. It relies on camouflage and lightning-fast reflexes to capture prey, mainly fish and shrimp. A sting from a lionfish is extremely painful to humans and can cause nausea and breathing difficulties, but is rarely fatal.

Reproductive--They are mainly a solitary species and courting is the only time they aggregate, generally one male, with several females. They are gonochoristic, only showing sexual dimorphism, during reproduction. After courtship, the female releases two egg masses, fertilized by the male before floating to the surface. The embryos secrete an adhesive mucous allowing them to attach to nearly intertidal rocks and corals before hatching. During one mating season, females can lay up to 30,000 eggs.

Remarks: It is an invasive species in the Atlantic in U.S, Southeast and Caribbean coastal waters. Lionfish are still very popular and common aquarium fish.

MOONFISHES

2.11.30 Silver moony

Figure 2.109

Monodactylus argenteus (Linnaeus, 1758)

Common Name(s):

English—Silver moonfish, silver moony, butter bream, diamond fish, Mono.

IUCN Red List Status: Least Concern.

Systematic Position:

Order: PERCIFORMES

Family: MONODACTYLIDAE--moonfishes

Genus: *Monodactylus*

Species: *M. argenteus*

Scientific Name: *Monodactylus argenteus* (Linnaeus, 1758)

Description:

- Body disc shaped and flattened.
- Eyes large and mouth is small.
- Dorsal fin spine:7-8; dorsal fin soft rays: 27-31; anal spines: 3; anal spine soft rays: 27-32.
- It is bright shiny silver with yellowish edges to the fins.
- The dorsal and anal fins have black tips.
- Juveniles have more yellow colouration and are distinguished by two vertical black bands.
- Fins yellow except pectorals are translucent.

Species Details:

- Distribution:

  Indo-West Pacific: Red Sea and East Africa, including the Persian Gulf to Samoa, north to the Yaeyama islands.
- Habitat: Inhabits bays, mangrove estuaries, tidal creeks, and lower reaches of freshwater streams, sometimes in silty coastal reefs.
- Food and feeding: Omnivorous. Feeds on plankton and detritus. In aquarium they may be given dried food, fresh or frozen brine shrimp, blood worms, algae, blanched lettuce, spinach and boiled peas.

Aquarium Aspects:

- Minimum tank size: 55 gallons (208 litres).
- Suitable for Nano tank: No.
- Substrate: Mixed type-sand and gravel.
- Lighting: Moderate -normal.
- Temperature: 75.0-82.0° F (23.9-27.8° C).
- Range pH: 7.2-8.5.
- Hardness range: 8-12 dGH.
- Water medium: Moderate.
- Water region: All.
- Water changes: Weekly.

Behaviour:

Social—Peaceful fish. Adults remain in schools and within their school they develop a hierarchy. Young fishes remain either solitary or in small aggregations. Highly territorial in nature and may eat smaller fishes.

Reproductive—There is no report of captive breeding of this fish.

Remarks: This beautiful silvery fish is one of the favourites of marine aquarium.

2.11.31 Messmate pipefish

Figure 2.110

Corythoichthys haematopterus (Bleeker, 1851)

Common Name(s):

English- Messmate pipefish, dragon-faced pipefish, bloodspot pipefish, reef pipefish, reef-top pipefish and yellow-streaked pipefish.

IUCN Red List: Least Concern (LC).

Systematic Position:

Order: SYNGNATHIFORMES—pipefishes and seahorses

Family: SYNGNATHIDAE—pipefishes and seahorses

Subfamily: SYNGNATHINAE (pipefishes)

Genus: *Corythoichthys*

Species: *C. haematopterus*

Scientific Name: *Corythoichthys haematopterus* (Bleeker, 1851)

Description:

- Long, slender body with a tiny, tubular mouth and a small pink tail.
- Fine lines on the head.
- Body is encased in a series of bony rings.
- Series of spots over the back and along trunk.
- Caudal fin mainly pink.
- The body is cream coloured with highly variable black, pink and yellow markings.
- Each individual has a unique pattern, and some can be very striking and colourful.

Species Details:

- Distribution:

  Indo-Pacific: East Africa to Vanuatu, north to southern Japan southward to Australia.
- Habitat: Marine, reef-associated, depth range 0-21 m. Adults inhabit shallow protected areas of rubble and sand, usually semi-silty zones.
- Food and feeding: Carnivorous. In captivity, it may be given live copepods, live baby brine shrimp, frozen cyclops, small *Mysis*, Nutra mar Tigrio bottled live copepods and vitamin-enriched live, baby brine shrimp, also frozen cyclops, small *Mysis* and Nutra mar ova.
- Maximum length: 19.8 cm SL male/ unsexed.

Aquarium Aspects:

- Minimum tank size: 50 gallons.
- Water conditions: Temperature: 72.0-78.0°F; pH: 8.1-8.4; specific gravity: 1.020-1.025.
- Substrate: Plenty of live rocks.

Behaviour:

Social- Peaceful and social fish. It is a poor swimmer and prefers to travel by sliding its body over substrate and live rock similar to the way a snake moves. It is among the hardiest pipefish. It exhibits bioluminescence.

Reproductive- Ovoviviparous. Males carry the eggs in a brood pouch, which is found under the tail. Males may be brooding at 9.0 cm SL. Monogamous mating is observed as both obligate and genetic.

Remarks: It is a favourite pet in reef aquarium. This species consumes "Red bugs" infecting *Acropora* corals. So, this pipefish may be installed in the reef aquarium to save *Acropora* corals from Red bug infestation.

RABBITFISHES

2.11.32 Bicoloured foxface

Figure 2.111

Siganus uspi Gawel & Woodland, 1974

Common Name(s):

English—Bicoloured foxface, Fiji foxface.

IUCN Red List Status: Near Threatened.

Systematic Position:

Order: PERCIFORMES

Family: SIGANIDAE--rabbitfishes

Genus: *Siganus*

Species: *S. uspi*

Scientific Name: *Siganus uspi* Gawel & Woodland, 1974

Description:

- Body has a unique colouration. The front two-third part of the body is dark brown and the posterior one-third is yellow in colour.
- Dorsal fin spines:13; dorsal fin soft rays: 10; anal fin spines: 7; anal soft rays: 9.
- Snout slightly shorter and less tubular, less concave above the nostrils.
- Caudal fin slightly incised.
- Spines stout, pungent and venomous.
- Colour dark purplish brown except pectoral fins, soft dorsal fin and anal fins, caudal fin and posterior part of body.
- Demarcation of two colours is distinct.

Species Details:

- Distribution:

  Western Pacific: endemic to Fiji.
- Habitat: Found among hard corals at the faces of drop-offs at reef edges, or in deep pools inside reef crests. Occurs at the depth of 3-30 m.

- Food and feeding: Omnivorous but require large amount of algae-based food in the diet. Feed on sea weeds, soft corals, large polyp stone coral, small crustaceans (Krill, *Mysis*, *Artemia*), microalgae and macroalgae.
- Maximum length: 24.0 cm TL male/unsexed.

Aquarium Aspects:

- Minimum tank size: 125 gallons.
- Water conditions: Temperature: 72.0-78.0° F; pH: 8.1-8.4; specific gravity: 1.020-1.025.
- Hardness: Average.
- Aquarium conditions: They should be kept in a well- run aquarium where they can graze algae from rocks and stones. Requires plenty of space for swimming.

Behaviour

Social: Mostly peaceful but might be aggressive towards similar species. This species nibbles soft coral and LPS when there is insufficient food. This species changes colour when afraid. They generally become pale or brownish.

Reproductive—It is an egg layer fish.

Remarks: It is an excellent choice for any marine aquarium.

2.11.33 Foxface rabbitfish

Figure 2.112

Siganus vulpinus (Schlegel & Muller, 1845)

Common Name(s)

English—Foxface, Foxface rabbitfish.

IUCN Red List Status: Least Concern (LC).

Systematic Position:

Order: PERCIFORMES

Family: SIGANIDAE--rabbitfishes

Genus: *Siganus*

Species: *S. vulpinus*

Scientific Name: *Siganus vulpinus* (Schlegel & Muller, 1845)

Description:

- Spines are stout, pungent and venomous.
- Caudal fin only slightly incised.
- A dark chocolate ocular band running in an arc from base of 1st dorsal spine to chin.
- On head, a narrow white blaze is present on midline.
- Thorax dark chocolate.
- On trunk a broad white zone is present behind ocular bar.

- Rest of the body is yellow-orange.
- A large black spot below the aft of the dorsal fin.

Species Details:

- Distribution:

  Western Pacific: western Philippines, Indonesia, New Guinea, Great Barrier Reef, Vanuatu, New Caledonia, Caroline Islands, Marshall Islands, Nauru and Kiribati, Tonga.
- Habitat: Occurs in coral-rich areas of lagoon and seaward reefs. Found among staghorn corals. Depth range 1-30 m.
- Food and feeding: Omnivorous, eating mostly algae and zooplankton. It may nip at coral, such as Zoanthids. In captivity, they take *Mysis*, shrimp, dried seaweed, marine flake food containing algae.
- Maximum length: 25.0 cm SL male/unsexed.

Aquarium Aspects:

- Minimum tank size: 70 gallons (265 litres).
- Water conditions: Temperature: 72.0-78.0°F (22.0-25.0°C); range pH: 8.1-8.4; specific gravity: 1.020-1.025.

Behaviour:

Social: Sometime territorial and is capable of inflicting painful stings. Occurs singly or in pairs, but juveniles and subadults occasionally occur in large schools among *Acropora* where they feed on algae growing on the dead bases of the coral branches.

Reproductive: Once they form pair, it lasts until one of the partner dies. It is an egg laying species. Breeding is possible in captivity.

Remarks: It is popular in marine aquarium trade. It is popular with aquarists as it eats the macroalgae that generally overgrow the rockwork in home aquaria. Within a matter of days, it removes the algae and clears the aquarium.

2.11.34 Eschmeyer's scorpionfish

Figure 2.113

Rhinopias eschmeyeri Conde', 1977

Common Name (s):

English—Eschmeyer's scorpionfish, paddle-flap scorpionfish.

IUCN Red List Status: Least Concern (LC).

Systematic Position:

Order: SCORPAENIFORMES—mail-cheeked fishes

Family: SCORPAENIDAE—scorpionfishes (rockfishes)

Subfamily: SCORPAENINAE

Genus: *Rhinopias*

Species: *R. eschmeyeri*

Scientific Name: *Rhinopias eschmeyeri* Conde', 1977

Description:

- Body deep and laterally compressed.
- Have an upturned mouth. The eyes are on top of the head and have a special kind of eyebrows.
- The mouth includes strange protrusions. The body is relatively narrow but sometimes also contains this same type of protrusions.
- Dorsal fin spines: 12; dorsal fin soft rays: 8-9; anal fin spines:3; anal fin soft rays:5.
- Have two tentacles on the underside of the lower jaw.
- Lack tentacles on the frontal below the eyes in anterior.
- Lack distinct tentacles on the lateral surface of the body above the lateral line.
- Have short tentacles without distinct branches along distal margins on the supraocular and posterior lacrimal spines.
- Body, head, fins and tentacles without distinct pigmentation or markings.
- Colour of the body may be red, pink, white or purple.

Species Details:

- Distribution:

  Indo-West Pacific.
- Habitat: Found in coral reefs and rubble bottoms in 18-55 m.
- Food and feeding: Carnivorous. Feeds on crustaceans, cephalopods and small fishes.
- Maximum length: 23.0 cm TL male/unsexed.
- Life span: 20 years.

Aquarium Aspects:

- Minimum tank size: 20-30 gallons.
- Suitable for Nano tank: Yes.

Behaviour:

Social—They remain solitary, living in caves or reefs. They camouflaged themselves to blend with the surroundings. Instead of swimming they move by hobbling along the bottom with their pectoral and pelvic fins. When hunting, they remain motionless and wait for their prey to approach.

Reproductive—They are oviparous, laying eggs. The eggs are laid as a clear or greenish gelatinous mass that floats near the surface. The eggs hatch within five days.

Remarks: It is a popular aquarium fish, and is one of the most searched fish in the world. This rare species is described as the "Holy Grail" of underwater photography. It tops the list of weird and strange creatures.

2.11.35 Dwarf seahorse

Figure 2.114

Hippocampus zosterae Jordan & Gilbert, 1882

Common Name (s):

English-Dwarf seahorse, sea pony.

IUCN Red List Status: Data Deficient.

Systematic Position:

Order: SYNGNATHIFORMES—pipefishes and seahorses

Family: SYNGNATHIDAE--pipefishes and seahorses

Subfamily: HIPPOCAMPINAE (seahorses)

Genus: *Hippocampus*

Species: *H. zosterae*

Scientific Name: *Hippocampus zosterae* Jordan & Gilbert, 1882

Description:

- The head is angled at right angles to its body.
- It has a short snout.
- Dorsal fin spines: 0; Dorsal fin soft rays: 12.
- The dorsal fin is used for propelling the dwarf seahorse and pectoral fins help in steering.
- The skin may be mottled, have dark spots and covered in tiny warts.
- It can change colour.
- They have 9-10 bony rings around their trunk and 31-32 rings around their tails.
- It is most often white in colour but the colour may range from tan, brown, yellow and green.
- In the wild, this fish has small skin growths called cirri, that resemble algae.
- Sexual Dimorphism: The male seahorse has brood pouch; the female is without brood pouch.

Species Details:

- Distribution:

  Western Atlantic Ocean, in southern Florida, Bermuda, Bahamas and the Gulf of Mexico.
- Habitat: They inhabit shallow waters populated with seagrasses especially *Zostera* and other seagrasses.
- Food and feeding: Carnivorous. They feed on small crustaceans and small fishes. They are "ambush predators" i.e., using their long snout as a pipette to suck in their food. In captivity, they may be given live brine shrimp.

- Feeding frequency: Twice a day.
- Maximum length: 5.0 cm TL male/unsexed.
- Life span: 1 year.

Aquarium Aspects:

- Minimum tank size: 5-10 gallons.
- Substrate: Preferably bare-bottom.
- Light: No special requirement.
- Water conditions: Temperature: 70.0-74.0° F; pH: 8.0-8.3; nitrate:>20 ppm.
- Water changes: 25-50% weekly water changes.

Behaviour:

Social—Diurnal, male more site faithful than females.

Reproductive—The breeding season is from February to November. They have a four- phase courtship ritual involving colour changes and vibrations while attached to a holdfast. Dwarf seahorses are ovoviviparous, and eggs produced by the females are reared in the male's brood pouch. About 55 eggs are produced. The eggs take about 11 days to hatch into miniature seahorses. Breeding and rearing are possible in captivity.

Remarks: Often seen in public aquaria.

2.11.36 Slender seahorse

Figure 2.115

Hippocampus reidi Ginsburg, 1933

Common Name(s):

English—Longsnout seahorse, slender seahorse.

IUCN Red List Status: Data Deficient.

Systematic Position:

Order: SYNGNATHIFORMES

Family: SYNGNATHIDAE

Subfamily: HIPPOCAMPINAE

Genus: *Hippocampus*

Species: *H.reidi*

Scientific Name: *Hippocampus reidi* Ginsburg, 1933

Description:

- Body elongate, slender and encased in a series of bony rings.
- Mouth small, at the end of tube-shaped snout.
- A unique lobate gill-filaments, lophobranch present.
- Both males and females have brown or white lines and spots placed sporadically upon their body.

- During the courtship period, these spots may also change into a pink or white colour.
- Dorsal spines 0, dorsal soft rays, 16-19.
- On top of their head is a coronet (crown-shaped piece of skin located at the top of the head) that is low and convoluted.
- Their skin stretches over bony rings visible on their body. They have 11 rings on their trunk and 31-39 rings on their tail.
- Sexual Dimorphism: Males are usually orange, while the females are yellow. Male seahorse has brood pouch whereas in female it is absent.

Species Details:

- Distribution:

  Western Atlantic: North Carolina in USA, Bermuda and Bahama to Santa Catarina, Brazil.
- Habitat: Marine, brackish; reef-associated; non-migratory; depth range 0-55m. Usually attached to gorgonians or seagrasses but may occur in floating Sargassum or swimming freely in midwater.
- Food and feeding: They eat ghost shrimp, grass shrimp, opossum shrimp, amphipods and other small crustacean plankton and plants, using their long snout with a pipette-like motion to suck in their food. They feed during the day and rest at night by attaching to structures in the water such as mangroves or seagrasses.
- Maximum length: 17.5 cm OT male/ unsexed.
- Life span: 1-4 years.

Aquarium Aspects:

- Minimum tank size: 55 gallons.
- Water conditions: Temperature: 72.0-78.0° F; pH: 8.0-8.4; specific gravity: 1.021-1.025.

Behaviour:

Social—It can change colour quickly for camouflage with the surroundings. A poor swimmer, ambushes small prey from its secure hiding place.

Reproductive–Ovoviviparous. Breeding season more than 8 months in laboratory. Gestation period 2 weeks depending upon water temperature. They are sexually mature when they are about 3 inches long. They show a dramatic courtship ritual in which the male may change colour and inflate his pouch and the male and female perform a dance around each other. When courtship is complete, the female deposits her eggs in the male's brood pouch, where they are fertilized. About 1,600 eggs are there about 1.2 mm in diameter. It takes 2 weeks for the eggs to hatch, when the babies are about 5-14 mm (2 inches), they are born. They look like miniature versions of their parents.

Remarks: It is an important species in the aquarium trade. This seahorse is one of the most exported Brazilian marine aquarium fish species. Seahorse trade has become a multi-billion $ industry across the globe today leaving the wild population as endangered. A successful captive breeding protocol has been developed for their conservation while creating high income generating opportunities (Hettiarachchi and Ediringhe, 2017).

2.11.37 Long-snouted seahorse

Figure 2.116

Hippocampus guttulatus Cuvier, 1829

Common Name(s):

English-Long- snouted seahorse.

IUCN Red List: Data Deficient (DD).

Systematic Position:

Order: SYNGNATHIFORMES

Family: SYNGNATHIDAE

Genus: *Hippocampus*

Species: *H.guttulatus* Cuvier, 1829

Description:

- A small sized fish with slender body, long snout and prehensile tail.
- The head and dorsal ridge have often some more or less long and numerous dermal filaments which may be simple or bifid.
- Dorsal fin spines:0; dorsal fin soft rays: 17-20; rings: 35-40.
- Colour of the body varies from dark green to brown with prominent white spots on the body.
- Spines medium to well-developed with blunt tips.
- Prominent rounded eye spines, often with a mane of thick skin fronds on neck and head.
- Coronet small but distinct, with 5 rounded knobs or blunt points.
- Horizontal plate in front of coronet as high as coronet itself with a more or less prominent spine at its front edge.
- Sexual Dimorphism: Male fish with brood pouch and female fish without it.

Species Details:

- Distribution:

  Eastern Atlantic: British Isles and the Netherlands to Morocco, Canary Islands, Madeira and the Azores, including the Mediterranean.
- Habitat: Occurs mostly in shallow inshore waters including littoral lagoons among algae and eel grass, or among rocks and in gravel bottoms. Juveniles are found in vegetated habitats.
- Food and feeding: Carnivorous. Feeds on small crustaceans, larvae, fish eggs and other planktonic organisms.

- Maximum length: 21.5 SL male/ unsexed; 18.0 cm OT female.
- Life span: 4-7 years.

Behaviour:

Social—Adult fish maintains a very restricted home ranges over years and do not move much.

Reproductive: It is ovoviviparous and the male broods the eggs, deposited by the female, in its ventral brood pouch. This has villi rich in capillaries that surround each fertilized egg creating a sort of placenta supplying the embryo when fully grown pups will be expelled from the pocket and evolve in complete autonomy. Gestation usually lasts for 21 days and brood sizes up to 581. Newly hatched young ones are planktonic for about 8 weeks.

Remarks: It is favourite marine ornamental species.

2.11.38 Striped razorfish

Figure 2.117

Aeoliscus strigatus (Gunther, 1861)

Common Name(s): Striped razorfish, Razorfish, jointed razorfish, coral shrimpfish.

IUCN Red Data List: Data Deficient (DD).

Systematic Position:

Class: OSTEICHTHYES

Order: SYNGNATHIFORMES—pipefishes and seahorses

Family: CENTRISCIDAE--shrimpfishes

Genus: *Aeoliscus*

Species: *A. strigatus*

Scientific Name: *Aeoliscus strigatus* (Gunther, 1861)

Description:

- Extremely compressed, razor-like body which ends in a long fine beak with sharp ventral edge.
- The body of this fish is almost entirely encased by thin bony plates that are expansions of the vertebral column.
- First dorsal spine long and sharp at extreme end of body followed by two short spines.
- Soft dorsal fin and caudal fin is displaced ventrally.
- No lateral line.
- Swimming is in a vertical position with snout down.
- The fins are considerably reduced and transparent.
- A black to brown median band runs the length of the fish crossing also the eyes. Though the colour of the body is variable with the habitat.
- Sexual Dimorphism: There is no known sexual dimorphism.

Species Details:

- Distribution:

  Indo-West Pacific: Tanzania and Seychelles; then from southern Japan to New South Wales, Australia eastward to Vanuatu.
- Habitat: Marine, reef-associated, depth range 2-42 m.
- Food and feeding: Carnivorous. Eats mainly brine shrimps and small invertebrates, minute crustaceans.

Aquarium Aspects:

- Minimum tank size: 50 gallons.
- Water conditions: Temperature: 72.0-78.0° F; dKH: 8-12; pH: 8.1-8.4; sg: 1.020-1.026.

Behaviour

Social- In the wild, they hide in the spines of sea urchins. Form schools among the spines of *Diadema* or staghorn corals.

Reproductive- This fish is oviparous and the eggs and larvae are pelagic, the juveniles settle when they attain 20 millimetres in length.

Remarks: This unique fish is often seen in public aquaria.

2.11.39 Longspine snipefish

Figure 2.118

Macroramphosus scolopax (Linnaeus, 1758)

Common Name(s):

English—Longspine Snipefish, Snipefish, Trumpet fish, Bellowfish.

IUCN Status: Least Concern.

Systematic Position: Figre

Order: SYNGNATHIFORMES

Family: MACRORAMPHOSIDAE—snipefishes

Genus: *Macroramphosus*

Species: *M. scolopex*

Scientific Name: *Macroramphosus scolopax* (Linnaeus, 1758)

Description:

- Body is compressed and reddish pink dorsally and silvery bellies.
- With bony plates on each side of back.
- No barbels on jaw.
- Eyes large, long tubular snout and a slender spine protruding dorsally.
- Dorsal spines 4-8. Second spine of first dorsal fin very strong.
- Mouth is small and is placed at the end of the long tubular snout.
- Larvae and juveniles are silvery in colour. The back is bluish black. Adults are pinkish or reddish on upper sides.

Species details:

- Distribution:

  In the western Atlantic off the east coast of the USA and the Greater Antilles, northern South America and Argentina. It also occurs in the east Atlantic, the Mediterranean and the Indo-West Pacific.
- Habitat: Benthopelagic, subtropical species typically found on the lower continental shelf over sand substrates.
- Food and feeding: Carnivore, feeds on crustacean zooplankton such as copepods and ostracods, as well as benthic invertebrates. The tooth less mouth at the end of the long snout sucks up small bottom-dwelling invertebrates. Juveniles feed on pelagic copepods.
- Maximum length: 20.0 cm TL male/ unsexed.

Aquarium Aspects:

- Minimum tank size: 30 gallons (113 litres).
- Water condition: Temperature: 23.0-26.0°C (73.4-78.8° F); pH: 8.0-8.3; specific gravity: 1.020-1.025.

Behaviour:

Social—They are gregarious, form large schools when search for food. They generally swim in vertical position, with head down.

Reproductive—Oviparous in nature with dimorphic individuals. During the mating process, they change colours. Their planktonic eggs and larvae are pelagic.

Remarks: A unique aquarium fish.

2.11.40 Striped squirrelfish

Figure 2.119

Sargocentron xantherythrum (Jordan & Evermann, 1903)

Common Name (s):

English-Striped squirrelfish, Hawaiian squirrelfish.

IUCN Red List Status: Least Concern (LC).

Systematic Position:

Series: BERYCIDA

Order: HOLOCENTRIFORMES--squirrelfishes

Family: HOLOCENTRIDAE--squirrelfishes

Subfamily: HOLOCENTRINAE—squirrelfishes

Genus: *Sargocentron*

Species: *S. xantherythrum*

Scientific Name: *Sargocentron xantherythrum* (Jordan & Evermann, 1903)

Description:

- The body is bright red with 10 silvery white stripes alternating.
- Head red, and operculum with silvery reflections. There are two diagonal white lines on head.

- A white streak from above the edge of upper jaw to the corner of preopercle.
- A narrow white band on posterior edge of preopercle.
- Dorsal spines: 11; dorsal soft rays: 13-14; anal spines 4; anal soft rays: 9.
- Characteristic large eyes are present.
- Sexual Dimorphism: There are no noticeable morphological differences among the sexes.

Species Details:

- Distribution:

  Eastern Central Pacific: Unknown from the Hawaiian Islands and Johnston Atoll.
- Habitat: Found among seaward reefs below the surge zone, common near caves and ledges.
- Food and feeding: Carnivorous, feeds mainly on crustaceans, shrimp, crab, marine worms, brittle and serpent stars, motile invertebrates.
- Maximum length: 17.0 cm TL male/unsexed.
- Life span: 2-4 years.

Aquarium Aspects:

- Minimum tank size: 50-70 gallons.
- Tank level: Middle.
- Aquarium conditions: Temperature: 72.0-78.0° F; pH: 8.1-8.4; hardness: 8-12 dGH.
- Lighting: Low
- Substrate: Lots of hiding places required. Live rocks, crevice-type formations.

Behaviour:

Social—Peaceful and a nocturnal schooling fish. It can make a squirrel-like noise with its swim-bladder. They do best when kept in large groups of six or more.

Reproductive—Reproduction is by external fertilization. They are pelagic, or egg scatterers. They do not guard their offspring and at the water column carry the eggs in open waters. There is little information about their captive breeding.

Remarks: It is an attractive aquarium fish.

TANGS

2.11.41 Achilles tang Figure 2.120

Acanthurus achilles Shaw, 1803

Common Name(s):

English: Achilles tang.

IUCN Red List Status: Least Concern (LC).

Systematic Position:

Order: ACANTHURIFORMES—surgeonfishes and relatives

Family: ACANTHURIDAE—surgeonfishes

Subfamily: ACANTHURINAE

Genus: *Acanthurus*

Species: *A. achilles*

Scientific Name: *Acanthurus achilles* Shaw, 1803

Description:

- Dark brown, nearly black in colour with striking orange and white lining along the fins and tail of the tang.
- A prominent orange drop shape develops on the caudal area, terminating into a sharp spine.
- Erectile spine (sharp and forward pointing) on each side of caudal peduncle which folds down into a groove.
- Mouth small, snout noticeably extended.
- Light blue ring around chin. A spot of same colour is present on gill cover at angle of gill opening.
- Soft part of the dorsal fin has longer base than spinous part.
- Young ones lack the large orange spot on caudal area.
- Spine on caudal peduncle may be venomous.
- Sexual Dimorphism: Unknown.

Species Details:

- Distribution:

  Western Pacific: oceanic islands of Oceania to the Hawaiian and Pitcairn islands.
- Habitat: Found in clear seaward reefs, usually in groups. Benthopelagic. Depth range 0-10 m, usually 0-4 m.
- Food and feeding: It is herbivorous, eating mostly benthic algae. In captivity accepts frozen and meaty foods, such as brine shrimp and Mysis shrimp.
- Feeding frequency: Several feedings a day in small amounts.
- Maximum length: 25.0 cm TL male/unsexed.
- Life span: 7 years or longer.

Aquarium Aspects:

- Minimum tank size: 120 gallons.
- Water movement: In good water movement.

- Water region: All.
- Water conditions: Temperature: 78.0- 82.0° F (26.0-28.0° C); pH: 8-8.4; sg: 1.022-1.026.

Behaviour:

Social: It is generally found singly, form territories in shallow water areas of the reef. In aquarium, also form territories.

Reproductive: Monogamous. Mating is observed as both facultative and social.

Remarks: It is fairly expensive though readily available.

2.11.42 Blue tang surgeonfish Figure 2.121

Acanthurus coeruleus Bloch & Schneider, 1801

Common Name (s):

English-Blue tang, blue tang surgeonfish, Atlantic blue tang, blue barber, blue doctor, blue doctor fish.

IUCN Red List Status: Least Concern (LC).

Systematic Position:

Order: ACANTHURIFORMES

Family: ACANTHURIDAE

Subfamily: ACANTHURINAE

Genus: *Acanthurus*

Species: *A. coeruleus*

Scientific Name: *Acanthurus coeruleus* Bloch & Schneider, 1801

Description:

- A deep-bodied surgeonfish with a conspicuous yellow caudal spine.
- Dorsal fin spine: 9; dorsal fin soft ray: 26-28; anal fin spine: 3; anal fin soft ray: 24-26.
- Colour of the body varies from light to dark blue.
- The dorsal, anal and caudal fins are golden blue.
- The colour of this species changes during growth from a yellow juvenile, yellow tailed blue subadult and the blue adult phase.
- This fish exhibits biofluorescence.
- Sexual Dimorphism: Unknown

Species Details:

- Distribution:

  Western Atlantic: New York, USA and Bermuda to the Gulf of Mexico and Brazil; Eastern Atlantic: Ascension Island.
- Habitat: Inhabits coral reefs, inshore grassy or rocky areas. Forms small groups. Mainly diurnal. Depth range 2-40 m, usually 2-18 m.

- Food and feeding: In the sea, they eat mainly zooplankton. In aquarium, brine shrimp, *Mysis*, Grindal worms and flake food, live, frozen and prepared formula foods.
- Maximum length: 39.0 cm TL male/ unsexed.

Aquarium Aspects:

- Minimum tank size: 75-100 gallons.
- Water movement: No special requirements.
- Water region: Swims all over the aquarium.
- Water conditions: Temperature: 75.0- 82° F (25.0-28.0° C); pH: 8.1-8.4; sg: 1.020-1.025; dKH: 8-12.
- Light: No special requirements.
- Feeding: Small amounts several times a day.

Behaviour:

Social: Adults form solitary groups. They are territorial, schooling and wandering in nature. Mainly diurnal. Juveniles live in groups in current rich outer reef terraces. These young ones hold cleaning stations and engage in cleaning parasites and other material from the body of other fishes. This tang exhibits biofluorescence, which may help in intraspecific communication and camouflage.

Reproductive: They are gonochoristic, each fish is either male or female.

Remarks: Not very expensive and is readily available. They are common in aquarium trade.

2.11.43 Clown tang

Figure 2.122

Acanthurus lineatus (Linnaeus, 1758)

Common Name (s):

English-Clown tang, lined surgeonfish, blue banded surgeonfish, blue-lined surgeonfish, clown surgeonfish, pyjama tang, striped surgeonfish, zebra surgeonfish.

IUCN Red List Status: Least Concern (LC).

Systematic Position:

Order: ACANTHURIFORMES

Suborder: ACANTHUROIDAE

Family: ACANTHURIDAE

Subfamily: ACANTHURINAE

Genus: *Acanthurus*

Species: *A.lineatus*

Scientific Name: *Acanthurus lineatus* (Linnaeus, 1758)

Description:

- Body deep and compressed. Body of juveniles relatively deeper than that of adults.

- Mouth small, teeth spatulate, close-set.
- A lancet-like spine on each side of caudal peduncle which folds into a deep horizontal groove. This spine is long and venomous.
- Dorsal fin spines: 9; dorsal fin soft rays: 27-30; anal fin spines: 3; anal fin soft rays: 25-28.
- Colour of the body is with ¾ part with alternating black-edged blue and yellow stripes. Those on the head are mainly diagonal.
- Lower fourth portion is pale lavender to bluish white.
- Dorsal fin finely striped pale blue and yellowish.
- Anal fin grey with a yellow basal band and light blue margin.
- Caudal fin blackish with a large grey crescent centre-posteriorly, edged at the front in bluish white and at the posterior margin with black.
- Pectoral fins are pale except at the base where they are coloured like the body.
- Pelvic fins are yellow-orange with a white lateral margin and a blackish sub-marginal line.
- Sexual Dimorphism: Males are larger than females.

Species Details:

- Distribution:

  Indian Ocean from East Africa to the western Pacific Ocean to the Great Barrier Reef, Japan, Polynesia and Hawaii.
- Habitat: Inhabits reefs, in marine waters a few meters deep. It is benthopelagic.
- Food and feeding: It is herbivore in food habit. Grazes on filamentous algae. It grazes during the day. In captivity, dried seaweed, dried algae with a lettuce clip, algae (Nori), brine, *Mysis* shrimp may be given. Flakes, pellet foods, and live foods, dried nori may be given.
- Feeding frequency: Small amounts, several times a day.
- Maximum length: 36 cm.
- Life span: 10 years or more.

Aquarium Aspects:

- Minimum tank size: 75 gallons.
- Water movement: Moderate.
- Water region: All levels.
- Light: In areas with sunlight.
- Water conditions: Temperature: 75.0-83.0° F (24.0-28.0° C); pH: 8-8.4: specific gravity: 1.020-1.025.
- Care level: Moderate to difficult.

Behaviour:

Social- They may become more territorial and extremely aggressive when grow older. Live singly or in small groups with a territorial male and several females. The adults may also school and they gather during spawning. The juvenile is solitary.

Reproductive-They are egg scatterers. They form spawning aggregations and pair off to spawn. No report of breeding in captivity.

Remarks: This species is of great ornamental value.

2.11.44 Desjardini tang

Figure 2.123

Zebrasoma desjardinii (Bennett, 1836)

Common Name (s):

English--Desjardini tang, Red Sea sailfin tang, Desjardin's sailfin tang.

IUCN Red List Status: Not Listed.

Systematic Position:

Family: Acanthuridae

Subfamily: Acanthurinae

Genus: *Zebrasoma*

Species: *Z. desjardini*

Scientific Name: *Zebrasoma desjardini* (Bennett, 1836)

Description:

- The body is oval or disc-shaped.
- Dorsal and anal fins are erectile and much elevated.
- In the middle of the caudal peduncle, there is a defensive dark spine surrounded by a blue zone.
- This spine is hinged and unfolds to 80°.
- The upper side of the body alternates orange and dark blue vertical bands, with a larger blue band on the eyes.
- Ventral part is spotted and several white spots are present on the head.
- A pattern of horizontal alternate orange and blue bands are present on the dorsal and anal fins.
- The caudal fin shows white spots and lines.
- Sexual Dimorphism: Mature males are larger than mature females.

Species Details:

- Distribution:

  Indian Ocean from the northern Red Sea to Kwazulu-Natal Province in South Africa and as far east as India and Java.
- Habitat: Inhabits lagoons and reefs in a tropical climate. They generally

live at water depths of 2-30 m or more.

- Food and feeding: They are primarily herbivores. In the wild, they feed on filamentous algae scraping from hard surface. Often feed on jelly fish and comb jellies in the sea. In the aquarium, they feed on vegetable matter, meaty foods. They may be given lots of marine algae, prepared frozen formulas containing algae or spirulina, frozen brine and mysid shrimp and flake foods. Japanese Nori or other seaweed can be adhered to the sides of the tank with a vegetable clip. Sometimes providing live rock with micro and macro organisms may be good. Culture of macroalgae like *Chaetomorphia* in the tank may also be done. Dried pellets may be soaked with liquid vitamins, vitamins may be added to the food, or a liquid vitamin may be added into the water. Boiled or frozen zucchini, broccoli, spinach and leaf lettuce are also given.
- Feeding frequency: Three times a day in smaller amounts.
- Maximum length: 40.0 cm TL male/unsexed.

Aquarium Aspects:

- Minimum tank size: 75 gallons for a young fish and 125 gallons for an adult.
- Light: In nature it inhabits sunlit areas. In aquarium, normal to very bright light may be provided.
- Water conditions: Temperature: 74.0-82.0° F (23.0-28.0° C); pH: 8.1-8.4; specific gravity: 1.020-1.025.
- Water movement: Good water movement.
- Water region: Open water. At night it sleeps in crevices.

Behaviour:

Social-Adult fishes are found in pairs, whereas juveniles are solitary. If disturbed, these fishes display their large ventral and dorsal fins. They can remain well with more aggressive fishes also.

Reproductive-They are pair spawner. But regular spawning and rearing of young in captivity has not yet been reported.

Remarks: An attractive tang, common in public marine aquaria.

2.11.45 Mimic tang

Figure 2.124

Acanthurus pyroferus Kittlitz, 1834

Common Name (s):

English-Mimic tang, mimic surgeonfish, chocolate surgeonfish, orange-gilled surgeonfish, Pacific mimic surgeon, yellow spot surgeon.

IUCN Red List Status: Least Concern (LC).

Systematic Position:

Family: ACANTHURIDAE

Subfamily: ACANTHURINAE

Genus: *Acanthurus*

Species: *A. pyroferus*

Scientific Name: *Acanthurus pyroferus* Kittlitz, 1834

Description:

- The body is disc-like with a spine on either side of the caudal peduncle.
- The colour of the body is yellow to dark brown or purplish black with an orange patch just above the pectoral fin base.
- A broad dark band is present starting at the top of the gill opening and arcing down to below the chin.
- There is a narrow white band encircling the chin.
- A broad yellow band is present on the end of the caudal fin.
- As juveniles, they are mimics of several of the dwarf angelfish and have body colouration similar to these fishes.
- Juveniles can exhibit 3 different colour patterns mimicking *Centropyge* species.
- Caudal fin is rounded in juveniles.
- Dorsal fin spines: 8; dorsal fin soft rays: 27-30; anal fin spines: 3; anal fin soft rays: 24-28.
- Sexual Dimorphism: Unknown.

Species Details:

- Distribution:

  Indo-Pacific: Seychelles to the Marquesas and Tuamotus islands, north to southern Japan, south to Great Barrier Reef and New Caledonia.
- Habitat: Inhabits lagoons and seaward reefs. Found in areas of mixed coral, rock or sand at the base of reefs or ledges, on silty reefs.
- Food and feeding: These fishes are primarily herbivore. Feeds on algae. In the wild, they graze on benthic algae. In captivity, they should be provided with vegetable matter, meaty foods, lots of algae, prepared frozen formulas containing algae or spirulina, frozen brine and *Mysis* shrimp and flake foods. Japanese Nori or other seaweed may be attached to the sides of the aquarium with vegetable clips. Live rock with micro and macro organisms may also be provided. Macroalgae such as *Chaetomorphia* may be cultured in the tank. Vitamin supplement (including Vitamin C) should be provided.
- Feeding frequency: At least 3 times a day.
- Maximum length: 29.0 cm TL /unsexed.

Aquarium Aspects:

- Minimum tank size: 75-125 gallons.

- Lighting: Normal lighting to very bright light with some dimly lit space.
- Water conditions: Temperature: 78.0-82.0° F (26.0-28.0° C); pH: 8.1-8.4; Specific gravity: 1.020-1.025.
- Water movement: Good water movement.
- Water region: Open water. It sleeps in crevices at night.

Behaviour:

Social--Peaceful. They are not overly aggressive and can remain with most tank mates.

Reproductive—This tang has not yet been bred in captivity.

Remarks: It is a favourite pet in aquarium trade.

2.11.46 Regal tang

Figure 2.125

Paracanthurus hepatus (Linnaeus, 1766)

Common Name(s):

English-Regal tang, palette surgeonfish, royal blue tang, hippo tang, blue hippo tang, flagtail surgeonfish, Pacific regal blue tang and blue surgeonfish.

IUCN Red List Status: Least Concern (LC).

Systematic Position:

Family: ACANTHURIDAE

Subfamily: ACANTHURINAE

Genus: *Paracanthurus*

Species: *P. hepatus*

Scientific Name: *Paracanthurus hepatus* (Linnaeus, 1766).

Description:

- Body is rather flat, with a circular body shape.
- A pointed snout-like nose is present.
- The body colouration is royal blue, yellow tail and black "palette" design.
- The lower body is yellow in the west-central Indian Ocean.
- Dorsal fin spines: 9; dorsal fin soft rays: 19-20; anal fin spines: 3; anal fin soft rays: 18-19.
- Sexual Dimorphism: Unknown.

Species Details:

- Distribution:

  Indo- Pacific: East Africa, including the Mascarene Islands to Kiribati, north to southern Japan, south to the southern Great Barrier Reef, New Caledonia and Samoa.
- Habitat: Inhabits clear, current-swept terraces of seaward reefs. Found 1 or 2 meters above the bottom. Juveniles and substrates are found in

groups near isolated *Pocillopora eydouxi* coral heads. If disturbed, they hid themselves tightly among the branches. They are benthopelagic.

- Food and feeding: They feed on zooplankton and sometimes on algae. In captivity, they should be fed a good variety of live, frozen, and prepared formula foods. Brine shrimps, Mysis, grinded worms and flake foods may be given.
- Feeding frequency: Several times a day in small amounts.
- Maximum length: 31.0 cm TL male/unsexed.

Aquarium Aspects:

- Minimum tank size: 75 gallons.
- Light: Normal.
- Water movement: No special requirement.
- Water region: All.
- Water conditions: Temperature: 75.0 and 79.0°C (72.0-78.0° F); pH: 8.1-8.4; sg: 1.020-1.025; dKH: 8-12.

Behaviour:

Social--Adults are solitary. Juveniles remain in groups in the current rich outer reef terrace. When disturbed, juveniles take refuge in stalks of the coral *Pocillopora eydouxi*.

Reproductive-- Captive breeding is not fully achieved. Males court females quite aggressively and there is a quick upward spawning rush toward the surface of the water during which eggs and sperms are released. The eggs are small and pelagic, containing a single droplet of oil for floatation. The fertilized eggs hatch in twenty-four hours into small translucent larvae with silvery abdomens and rudimentary caudal spines. Adults also become semi-transparent when disturbed.

Remarks: A popular fish in marine aquaria.

2.11.47 Sailfin tang

Figure 2.126

Zebrasoma veliferum (Bloch, 1795)

Common Name(s):

English-Sailfin Tang.

IUCN Red List Status: Least Concern (LC).

Systematic Position:

Family: ACANTHURIDAE

Subfamily: ACANTHURINAE

Genus: *Zebrasoma*

Species: *Z. veliferum*

Scientific Name: *Zebrasoma veliferum* (Bloch, 1795)

Description:

- Disc-shaped body with giant dorsal and anal fins that are approximately the same size as the body.
- Dorsal fin very elevated, the longest ray 2.1 to 2.5 times in SL.
- No brush-like patch of setae posteriorly on side of body.
- Possesses fewer and larger pharyngeal teeth.
- The body of the fish has beautiful striped patterning, broad and pale-yellow bands alternating with darker bands. These bands extend into the dorsal and anal fins.
- The dark bands are with some yellow markings, dots and stripes.
- The head is white with yellow spots. There is a darkest band running through the eye and a second very dark band just behind it. These bands have yellow dots and lines in them.
- Small juveniles have alternating yellow and black bars.
- On each side of the caudal peduncle there is a single spine, when not in use the spine is folded down into a groove.
- Dorsal fin spines: 4-5; dorsal fin soft rays: 29-33; anal fin spines: 3; anal soft rays: 23-26.
- Sexual Dimorphism: No significant difference though the males of the species grow larger than the females.

Species Details:

- Distribution:

  Western Indian Ocean: Mozambique Pacific Ocean: Indonesia to the Hawaiian and Tuamoto islands, north to southern Japan, south to the southern Great Barrier Reef, Caledonia and Rapa Island.
- Habitat: Found in lagoon and seaward reefs from the lower surge zone to a depth of at least 30 m. Young solitary ones found among rocks or coral of shallow, protected sometimes turbid reefs. Benthopelagic.
- Food and feeding: Herbivores. Feeds on leafy macroalgae. They are primarily herbivores. In captivity, they may be given vegetable matter, meaty foods, marine algae, prepared frozen formulas containing algae or spirulina, frozen brine and Mysid shrimp, and flake foods. Japanese Nori or other sea weed may be attached to the aquarium glass with a vegetable clip. Live rock with micro and macro organisms may be kept in the aquarium. *Chaetomorphia*, the macroalgae may be cultured in the tank. Vitamin supplement (including vitamin C) may be given.
- Feeding frequency: Three times a day in smaller amounts.
- Maximum length: 40.0 cm SL male/ unsexed.

Aquarium Aspects:

- Minimum tank size: 125 gallons.

- Water conditions: Temperature: 72.0-82.0° F (23.0-28.0° C); pH: 8.1-8.4; hardness: 8-12 dGH.
- Light: Normal lighting to very bright light.
- Water movement: Good water movement.
- Water level: All areas.

Behaviour:

Social—It is mild tempered and gets along with most fishes except others of their own genus. They may be territorial also.

Reproductive—They have not been bred in captivity.

Remarks: It is a favourite in marine aquarium.

2.11. 48 Sohal tang

Acanthurus sohal (Forsskal, 1775)

Common Name(s):

English: Sohal surgeonfish, sohal tang.

IUCN Red List Status: Least Concern.

Systematic Position:

Order: ACANTHURIFORMES

Family: ACANTHURIDAE

Genus: *Acanthurus*

Species: *A. sohal*

Scientific Name: *Acanthurus sohal* (Forsskal, 1775)

Description:

- Laterally compressed and ovoid shaped body.
- It has contrasting and sometimes vermiculated stripes alternate from light to dark. Head is with vermiculated (wavy) stripes.
- The dorsal, anal and pelvic fins are black while the pectoral and caudal fins are yellowish.
- All the fin margins are a striking neon blue.
- There is a patch of orange post gill plate.
- Scalpel like spine is located on the caudal peduncle which is highlighted in orange.
- Sexual Dimorphism: Not noticed.

Species Details:

- Distribution:

  Western Pacific Ocean-Red Sea to the Persian Gulf.
- Habitat: In seaward edges of reefs exposed to surge. Depth range 0-7m.

- Food and feeding: Herbivore. Prefers algae-based foods, such as frozen herbivore formulations, herbivore pellets, and dried marine algae. It also grazes on various algae growing naturally within the system. Various kinds of algae mainly Sargassum, and fine filamentous green algae.
- Feeding frequency: Several times in a day in small amounts.
- Maximum length: 40.0 cm TL male/ unsexed.

Aquarium Aspects:

- Minimum tank size: 200 gallons.
- Water movement: No special requirements.
- Water region: Swims all over the aquarium.
- Water conditions: Temperature: 72.0-78.0° C; dGH: 8-12; pH: 8.1-8.4.
- Light: No special requirements.

Behaviour:

Social-- An already established tang gets territorial if another tang is added in the aquarium. So, it may be rated as aggressive and territorial.

Reproductive—No report of breeding in captivity.

Remarks: A popular marine aquarium fish.

2.11.49 Doubleband surgeonfish Figure 2.127

Acanthurus tennenti Gunther, 1861

Common Name(s):

English--Tennenti tang, Doubleband surgeonfish, lieutenant tang, Doctor fish, lancetfish, thorn tail.

IUCN Red List Status: Least Concern (LC).

Systematic Position:

Family: ACANTHURIDAE

Subfamily: ACANTHURINAE

Genus: *Acanthurus*

Species: *A. tennenti*

Scientific Name: *Acanthurus tennenti* Gunther, 1861

Description:

- Body brown to grey in colour.
- There is a black horseshoe mark on shoulder region breaking into 2 horizontal curved bands in larger specimens.
- Caudal spine surrounded by a black area which is then surrounded by a bluish white area.
- Caudal fin with a white posterior border.

- Dorsal fin spines: 9; dorsal fin soft rays: 23-24; anal fin spines: 3; anal fin soft rays: 22-23.

Species Details:

- Distribution:
- Indian Ocean: East Africa to Sri Lanka and the Lesser Sunda Islands of southern Indonesia.
- Habitat: Inhabits lagoon and seaward reefs, 1 to over 20 m depth. Also found in slopes from silty inshore to outer reefs.
- Food and feeding: Herbivorous. Chiefly feed on hair algae, grazes on benthic algae, detritus, diatoms, frozen brine and mysid shrimp and flake foods. It is recommended to feed marine algae (Nori), Spirulina flakes, frozen *Mysis* shrimp, high quality herbivore flakes with vitamin supplement like selcon or a garlic supplement.
- Feeding frequency: Three times a day.
- Maximum length: 31.0 cm TL male/unsexed.

Aquarium Aspects:

- Minimum tank size: 150 gallons.
- Substrate: Plenty of live rocks that encourage the growth of algae.
- Water conditions: Temperature: 72.0-78.0° F; pH: 8.1-8.4; dKH: 8-12; specific gravity: 1.020-1.025.
- Care level: Moderate.

Behaviour:

Social-Relatively peaceful and tolerates other types of tangs. Found singly or in schools in silty waters of lagoons and seaward reefs. Temperament is semi-aggressive.

Remarks: It is suitable for reef and community aquariums.

2.11.50 Bignose unicornfish Figure 2.128

Naso vlamingii (Valenciennes, 1835)

Common Name(s):

English--Bignose unicorn fish, scribbled unicornfish.

IUCN Red List Status: Least Concern (LC).

Systematic Position:

Order: ACANTHURIFORMES

Family: ACANTHURIDAE

Subfamily: NASINAE

Genus: *Naso*

Species: *N. vlamingii*

Scientific Name: *Naso vlamingii* (Valenciennes, 1835)

Description:

- Body is deep and elongated with a narrower caudal peduncle.
- Adult fish has a convexly rounded prominent snout.
- Unusually tall dorsal and anal fins are present.
- A broad blue band present from eye to front of rostral protuberance.
- Vertical blue lines are present on sides and small blue spots present dorsally and ventrally.
- Dorsal fin spines: 6; dorsal soft rays: 26-27; anal spines: 3; anal soft rays: 27-29.
- Sexual Dimorphism: The male fish is larger in size than the female fish.

Species Details:

- Distribution:

  Indo-pan-Pacific: East Africa to the Galapagos Is. North to southern Japan, south to the southern Great Barrier Reef and New Caledonia, Tuamotus, throughout Micronesia.
- Habitat: Inhabits deep lagoon and seaward reefs.
- Food and feeding: Primarily herbivores. In the wild they feed mainly on leafy brown algae, may also eat small crustaceans. In tank, lots of algae, large chunk types of prepared frozen formulas containing algae or spirulina and flakes. Also feeds on zooplankton, shrimp, Mysis and fish pieces.
- Feeding frequency: Three times a day in small amounts.
- Maximum length: 60.0 cm TL male/unsexed.
- Life span: 30-45 years.

Aquarium Aspects:

- Minimum tank size: 50-180 gallons.
- Water movement: Good water movement, lots of oxygen, strong water movement at least in one area of the tank.
- Water region: Open water.
- Suitable for Nano tank: No.
- Light: In nature, it is found in sunlit zones. In the aquarium, normal lighting condition. Bright light with dimly lit space is also suitable.
- Water condition: Temperature between 75.0-82.0° F (24.0-28.0° C); pH: 8.1-8.4; specific gravity: 1.023-1.025.

Behaviour:

Social-They are peaceful and suitable for community tank. They may be territorial.

Reproductive- No report of breeding in captivity. Males show brilliant colour during courtship. The pair rises to the surface to release eggs and sperms. Eggs are pelagic.

Remarks: It is one of the largest fish to be sold in the aquarium market.

2.11.51 Yellow tang

Figure 2.129

Zebrasoma flavescens (Bennett, 1828)

Common Name(s):

English—Yellow Tang, Lemon Sailfin, Somber Surgeonfish, Yellow Sailfin Tang.

IUCN Red List Status: Least Concern.

Systematic Position:

Order: ACANTHURIFORMES

Family: ACANTHURIDAE

Subfamily: ACANTHURINAE

Genus: *Zebrasoma*

Species: *Z. flavescens*

Scientific Name: *Zebrasoma flavescens* (Bennett, 1828)

Description:

- Body very deep, snout moderately protruding.
- Mouth small, teeth spatulate, close-set, the edges are denticulate.
- Like other surgeon fishes, scalpel-sharp spines are present on either side of the tail.
- They are bright yellow in colour, this colour fades a little at night.
- Sexual Dimorphism: Adult males are larger than females. Males change colour during mating season and have a "shimmering" behaviour.

Species Details:

- Distribution:

  Central and western Pacific from the Hawaiian Islands and Johnston Island to the Marshall Islands, Wake, Mariana Islands, Philippines, Ogasawara and Ryukyu Islands, Japan and southern Taiwan.
- Habitat: Found in inshore coral reefs, protected bay and lagoons, from below the surge one to above 46 m.
- Food and feeding: Herbivorous. In the wild, this species feed on turf algae and other marine plant materials. In captivity, dried and frozen herbivore food, Zucchini, broccoli and lettuce may be given.
- Maximum length: 20.0 cm TL male/ unsexed.

Aquarium Aspects:

- Minimum tank size: 55-100 gallons.

- Water conditions: Temperature: 24.0-28.0° C; pH: 8.1-8.4; specific gravity: 1.020-1.025.

Behaviour:

Social—They may move alone or in loose schools. It may be aggressive towards other Yellow tangs and other surgeon fish, if they are not introduced at the same time.

Reproductive—It is a group spawner. In the wild, they spawn during the full moon.

Remarks: It is one of the most popular and very commonly kept aquarium fish. They are prone to the fish disease "ich" and may damage coral in the reef tank.

2.11 52 Yellowfin tang Figure 2.130

Acanthurus xanthopterus Valenciennes, 1835

Common Name (s):

English-Yellowfin tang, yellowfin surgeonfish, Cuvier's surgeonfish.

IUCN Red List Status: Least Concern (LC).

Systematic Position:

Order: ACANTHURIFORMES

Family: ACANTHURIDA1

Genus: *Acanthurus*

Species: *A.xanthopterus*

Scientific Name: *Acanthurus xanthopterus* Valenciennes, 1835

Description:

- Body is purplish grey. A dull yellow portion is present in front of eye, a lesser extension is present posterior to lower part of eye.
- The outer 1/3 part of pectoral fin is yellow, extreme distal part is hyaline.
- Dorsal and anal fins yellowish grey basally, dull yellow distally.
- Caudal fin is whitish at the base.
- Caudal spine is small.
- Dorsal spines: 8-9; dorsal soft rays: 25-27; anal spines: 3; anal soft rays: 23-25.

Species Details:

- Distribution:

  Indo-Pacific: coast of East Africa to the Hawaiian Islands and French Polynesia, north to southern Japan, south to the Great Barrier Reef and New Caledonia. Eastern Pacific: lower Gulf of California and Clipperton Island to Panama and the Galapagos Islands.
- Habitat: Inhabits various reefs, sand slopes and lagoon. Juveniles inhabit

shallow, protected, turbid, inshore waters whereas adults are found in deeper areas of protected bays and lagoons and also in outer reef areas. Benthopelagic.

- Food and feeding: Feed on diatoms, detritus, seaweed film of sand, filamentous algae, hydroids, and pieces of fish. In captivity, they are to be offered plenty of marine based seaweed and algae, sea veggies, seaweed salad and Ocean Nutrition.
- Feeding frequency: 2-3 times daily.
- Maximum length: 70 cm TL male/unsexed.
- Life span: 5-8 years.

Aquarium Aspects:

- Minimum tank size: 360 gallons.
- Water condition: Temperature: 72-78° F (22.2-25.6° C); dKH: 8-12; pH: 8.1-8.4; sg: 1.020-1.025.
- Substrate: Sandy, prefers caves for hiding. Live rocks are necessary.

Behaviour:

Social-Schooling species. Semi-aggressive. Cannot be kept with other fish of the same genus or similar tangs. Juveniles can be kept in groups. They do not cause harm to corals or invertebrates.

Remarks: It is ideal for grand show displays and in public aquariums.

2.11.53 Red toothed triggerfish

Figure 2.131

Odonus niger (Ruppell, 1836)

Common Name (s):

English—Redtoothed triggerfish.

IUCN Red List Status: Not Evaluated.

Systematic Position:

Order: TETRADONTIFORMES--plectognaths

Family: BALISTIDAE--triggerfishes

Genus: *Odonus*

Species: *O. niger*

Scientific Name: *Odonus niger* (Ruppell, 1836)

Description:

- The mouth has the appearance of grinning and there are tiny red teeth that are needle sharp with two long red teeth in the upper jaw.
- The tail is lyre-shaped.
- The pectoral fins are quite small. They steer mostly with their dorsal and anal fins.

- Body deep purple with bluish-green markings on their heads and glowing light blue margins on the tail lobes and fins.
- They can change their colour depending on their mood, food, feeding and water quality from purple to blue and bluish-green.
- Dorsal fin spines: 3; dorsal fin soft rays: 33-36; anal fin spines: 0; anal fin soft rays:28-31.
- Triggerfish have the characteristic feature of locking and unlocking mechanism of their first dorsal fin spines. The dorsal fin is positioned in a flat groove on the top of the body. This can be locked in the upright position in order to wedge the fish into some crevice in which the fish has taken shelter.

Species Details:

- Distribution:

  Indo-Pacific: Red Sea south to Durban, South Africa and east to Marquesas and Society Islands, north to southern Japan, south to the southern Great Barrier Reef in Australia and New Caledonia.
- Habitat: Inhabits reef channels or along slopes with strong currents. Also found in current-swept seaward coral reefs. Juveniles are present with isolated patches of rubble or crevices with holes.
- Food and feeding: Omnivorous. Feed on zooplankton in large shoals. Mainly planktivorous. They also eat different types of meaty foods like krill, clams, squid, urchins, fishes. In captivity, dried, frozen, flaked marine fish food may be given.
- Feeding frequency: 2-3 times a day.
- Maximum length: 50.0 cm TL male/unsexed.

Aquarium Aspects:

- Minimum tank size: 180 gallons.
- Water conditions: Temperature: 72.0-79.0° F; pH: 8.2-8.4; dKH: 8-12; sg: 1.020-1.025.
- Light: Any.
- Water movement: Any.
- Water region: All.

Behaviour:

Social—They are friendly and peaceful towards their tankmates, may be considered as a community fish. But may become aggressive sometime towards other fishes.

Reproductive—Spawning in this fish is timed with lunar cycles, tides and time of change- over of tides. Distinct pairing. They are demersal spawner. Eggs are scattered and attached to sand particles. Only females show "tending" care of the eggs but both the parents keep guard. The males set up their territories, and the pair meet at their mating ground. They build nest to lay their eggs in. They do not breed in captivity.

Remarks: Often seen in public marine aquarium.

2.11.54 Titan triggerfish

Figure 2.132

Balistoides viridescens (Bloch & Schneider, 1801)

Common Name(s):

English—Titan triggerfish, giant triggerfish, moustache triggerfish.

IUCN Red List Status: Not Evaluated.

Systematic Position:

Order: TETRADONTIFORMES

Family: BALISTIDAE

Genus: *Balistoides*

Species: *B. viridescens*

Scientific Name: *Balistoides viridescens* (Bloch & Scneider, 1801)

Description:

- Body laterally compressed.
- It has a deep grove before eye.
- An area without scale is present around the lips, which continues and gets narrow in the posterior to the corner of mouth.
- Small, forward-curving spines are present in about five rows on side of and a short distance anterior to caudal peduncle.
- Caudal peduncle is compressed.
- Dorsal fin spines: 3; dorsal fin soft rays: 24-26; anal fin spines :0; anal fin soft rays: 22-24.

Species Details:

- Distribution:

  Indo-Pacific: Red Sea south to Delagoa Bay, Mozambique and east to the Line and Tuamotus islands, north to southern Japan, south to New Caledonia.
- Habitat: Inhabits lagoons, seaward reefs and sheltered inner reef slopes. Juveniles are found in isolated patches of branching coral or rubble of shallow sandy protected areas. Adults are generally found either singly or in pairs on the slopes of deep lagoon or seaward reefs.
- Food and feeding: Carnivorous. Feed on sea urchins, coral, crabs and other crustaceans, molluscs and tube worms.
- Feeding frequency: Several feedings a day in small amounts.
- Maximum length: 75.0 cm TL male/unsexed.

Aquarium Aspects:

- Minimum tank size: 180 gallons.

- Water conditions: Temperature: 72.0-79.0° F; pH: 8.0-8.4; dKH: 8-12; sg: 1.021-1.025.
- Light: Any.
- Water movement: No special requirement.
- Water region: All.

Behaviour:

Social—It is diurnal and solitary in nature. It becomes aggressive to other fish who enter their territory. It often bites pieces of branching coral. Often feeds by turning over rocks and stirring up sand.

Reproductive—This fish, builds nest on a sandy patch. The nest is cone shaped and it is the place to keep the eggs. The female fish guards the nest.

Remarks: It is often kept in public marine aquaria.

2.11.55 Trumpet fish

Figure 2.133

Aulostomus chinensis (Linnaeus, 1766)

Common Name(s):

English—Chinese trumpetfish, Trumpet fish.

IUCN Red List Status: Least Concern.

Systematic Position:

Class: ACTINOPTERYGII

Order: SYNGNATHIFORMES—pipefishes and seahorses

Family: AULOSTOMIDAE--trumpetfishes

Genus: *Aulostomus*

Species: *A. chinensis*

Scientific Name: *Aulostomus chinensis* (Linnaeus, 1766)

Description:

- Body of this fish is elongated, and compressed laterally, with a long, tubular snout and which has a small barbel at its inferior extremity.
- The protrusible mouth is used for catching prey.
- The first dorsal fin is with a set of 8-12 isolated spines. The second dorsal fin is shaped like the anal fin and is found directly above it.
- The pelvic fins are present in the middle of the body and are small, with one basal black spot.
- Fleshy barbel is present at the tip of lower jaw.
- Caudal fin is rounded.
- Dorsal fin spine: 8-12; Dorsal fin soft rays: 24-27; anal spines: 0; anal soft rays: 26-29.
- The colour of the body is either uniform or mottled in a range of grey,

brown or dark green. A uniformly bright yellow colour may also be present. The rest of the body is black with white dots. Two black spots are present on the tail.

- The long caudal peduncle and the caudal fin is yellowish in spite of the body colour.
- Caudal fin usually has two round black spots.

Species Details:

- Distribution:

  Indo-Pacific: East Africa to Hawaii and the Easter Island, north to the southern Japan south to Lord Howe Island. Eastern Central Pacific: Panama, Revillagigedo Islands, Clipperton Island, Cocos Island, and Malpelo Island.
- Habitat: Found in clear, shallow water in rocky and coral areas of protected and seaward reefs. Benthopelagic. Depth range 3-122 m.
- Food and feeding: Feeds on small fishes and shrimps.
- Maximum length: 80.0 cm TL male/ unsexed.

Aquarium Aspects:

- Minimum tank size: 396 gallons.
- Water conditions: Temperature: 72.0-79° F; pH:8.0-8.4; sg: 1.021-1.025.

Behaviour:

Social- It is diurnal, slow-moving benthopelagic and solitary. For catching prey, it either lies in wait for a potential prey close to hard coral, black coral bush, or gorgonian, or it stays close to some big fishes and then approaches the suspecting prey.

Reproductive-Oviparous with external fertilization and gonochorous. Eggs and larvae produced are pelagic. Juveniles develop in the epipelagic zone then move to seagrass and coral reef environments.

Remarks: This unusual looking fish is a favourite marine aquarium fish.

Figure 2.80 Blacktip reef shark

Figure 2.81 Sand tiger shark

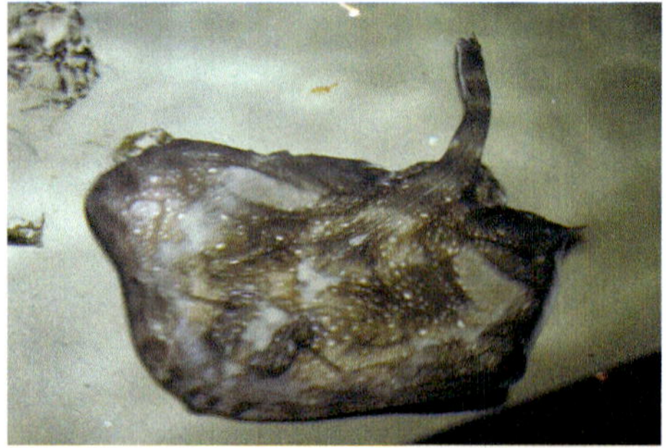
Figure 2.82 Undulate skate

Figure 2.83 Blackchin guitarfish

Figure 2.84 Bowmouth guitarfish

Figure 2.85 Javanese cownose ray

Figure 2.86 Southern sting ray

Figure 2.87 Common clownfish

Figure 2.88 Black ocellaris

Figure 2.89 A Tomato clownfish

Figure 2.89 B Tomato clownfish

Figure 2.90 Orange clownfish

Figure 2.91 Emperor angelfish

Figure 2.92 Blacktail angelfish

Figure 2.93 Bicolour anthias

Figure 2.94 Lyretail anthias

Figure 2.95 Dogface pufferfish

Figure 2.96 *Chaetodon auriga*

Figure 2.97 *Chaetodon semilarvatus*

Figure 2.98 Pennant coralfish

Figure 2.99 Sergeant major damselfish

Figure 2.100 Yellowtail damselfish

Figure 2.101 Green moray eel

Figure 2.102 Zebra moray eel

Figure 2.103 Spotted garden eel

Figure 2.104 Pork fish

Figure 2.105 Spanish hogfish

Figure 2.106 Golden trevally

Figure 2.107 Snubnose pompano

Figure 2.108 Red lionfish

Figure 2.109 Silver moony

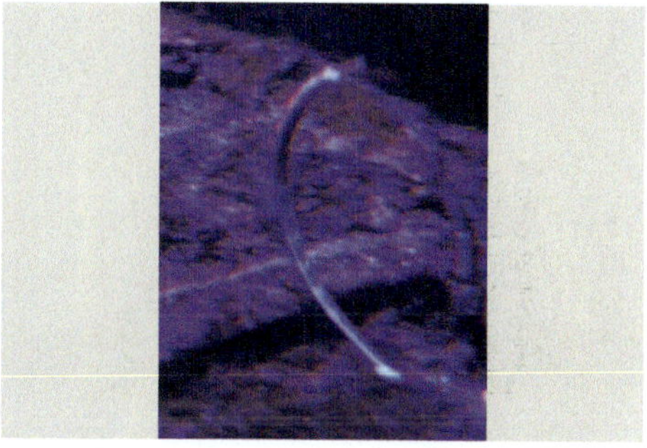

Figure 2.110 Messmate pipefish

Figure 2.111 Bi-coloured foxface

Figure 2.112 Foxface rabbitfish

Figure 2.113 Eschmeyer's scorpionfish

Figure 2.114 Dwarf seahorse

Figure 2.115 Slender seahorse

Figure 2.116 Long snout seahorse

Figure 2.117 Striped razorfish

Figure 2.118 Longspine snipefish

Figure 2.119 Striped squirrelfish

Figure 2.120 Achilles tang

Figure 2.121 Blue tang surgeonfish

Figure 2.122 Clown tang

Figure 2.123 Desjardini tang

Figure 2.124 Mimic tang

Figure 2.125 Regal tang

Figure 2.126 Sailfin tang

Figure 2.127 Doubleband surgeonfish

Figure 2.128 Bignose Unicornfish

Figure 2.129 Yellow tang

Figure 2.130 Yellowfin tang

Figure 2.131 Red toothed triggerfish

Figure 2.132 Titan triggerfish

Figure 2.133 Chinese trumpet fish

2.12 References

Alfred J.R.B. and Nandi N.C. Wetlands: Freshwater. In: *Ecosystems of India*, Edited by Alfred JRB, Das AK, Sanyal AK. ENVS-Zool. Surv. India. 2001, pp. 165-193.

Barman, J. Baruah U.K., Goswami U.C. Indigenous Techniques of catching the mud eel, *Monopterus cuchia* (Ham) in Goalpara district, Asssam. Indian Journal of Traditional Knowledge, 2013. 12(1):109-115.

Chaudhuri S. *Economic Zoology*. New Central Book Agency (P) Ltd. 2017, pp. 1-693.

De D. K. Estuarine Fisheries. *Handbook of Fisheries and Aquaculture*. Published by Directorate of Knowledge management in Agriculture. ICAR. New Delhi. 2011, pp. 208-237.

FAO. 2007-2020. Fisheries and Aquaculture Fact Sheets. In FAO Fisheries Division (online). Rome. Updated 22 May 2013. www.fao.org/fishery/culturedspecies/Trachinotus_spp/en. Cited 17 August 2020.

Hettiarachchi S. and Ediringhe U. Captive breeding of *Hippocampus reidi* Ginsburg, 1933 (Longsnout seahorse) in Sri Lanka under artificial conditions. Tropical Agricultural Research, 29 (1). 2017, p.77.

Jayaram K.C. *The Freshwater Fishes of India, Pakistan, Bangladesh, Burma and Sri Lanka-a handbook*. Edited by the Director, Zoological Survey of India. 1981, pp. 1-475.

Kapoor D., Dayal R. and Ponniah A.G. *Fish Biodiversity of India*. Natinal Bureau of Fish Genetic Resources Lucknow, India. 2002, 775p.

Karmakar A.K. and Das A. *Endemic Freshwater Fishes of India*. Rec. zool. Surv. India. Occ. Paper No.230. Published by Director, Zool. Surv. India, Kolkata. 2005, pp. 1-125.

Menon A.G.K. *Check-list freshwater fishes of India*. Rec. Zool. Surv. India, Misc. Publ., Occas. Paper No. 175. 1999, pp. 1-366.

Mercy T.V.A, Gopalakrishnan A., Kapoor D. and Lakra W.S. *Ornamental Fishes of the Western Ghats of India*. National Bureau of Fish Genetic Resources. CMFRI Digital Repository.2007. http: //eprints.cmfri.org.in/11778/1/.

Murty V.S. *Marine ornamental fishes of India. Proceedings of the seminar on Fisheries. A Multimillion Dollar Industry*, Madras.1996, pp. 23-34.

Murty V.S. and Philipose K.K. and Koya K.P.S. *Marine Ornamental Fishes*. In Proceedings of the Workshop National Aquaculture Week. The Aquaculture Foundation of India, Chennai. 1998, pp. 154-158.

Nandi N.C. Wetlands: Brackish water. In: *Ecosystems of India*. Edited by Alfred JRB, Das AK, Sanyal AK. ENVIS-Zool. Surv. India. 2001, pp. 195-217.

Odum E.P. *Fundamentals of Ecology*. W.B. Saunders Company. 1971, pp. 1-574.

Ramakrishna and Venkataraman K. Marine. In: *Ecosystems of India*. Edited by Alfred JRB, Das AK, Sanyal AK. ENVIS-Zool. Surv. India. 2001, pp. 291-315.

Randall J.E., Allen G.R. and Steene R.C. *Fishes of the Great Barrier Reef and Coral Sea*. University of Hawaii Press, Honolulu, Hawaii. 1990, pp. 1-506.

Rao C.A. Estuaries. In *Ecosystems of India*. Edited by Alfred JRB, Das AK, Sanyal AK. ENVIS-Zool. Surv. India. 2001, pp. 219-237.

Sen T.K. Freshwater Fish. *State Fauna Series 3: Fauna of West Bengal*, Part 2. 1992, pp. 101-242.

Silas E.G., Gopalakrishnan A., Ramachandran A., Anna Mercy T.V., Sarkar K., Pushpangadan K.R., Anil Kumar P., Ram Mohan M.K. and Anikuttan K.K. *Guidelines for green certification of freshwater ornamental fish*. The Marine Products Export Development Authority, Kochi, India. 2011, pp.1-106.

Sinha A. *Ornamental Fish of India*. Published by Director, Central Institute of Fisheries Education. ICAR. 2008, pp. 1-254.

Swain S.K., Madhu K., Madhu R., and Gopakumar G. Ornamental fish breeding and culture. In: *Handbook of Fisheries and Aquaculture*. Published by Directorate of Knowledge management in Agriculture. ICAR. New Delhi. 2011, pp. 500-532.

Swain S.K., Sarangi N., Ayyappan S. *Ornamental Fish Farming*. Published by Directorate of Information and Publications of Agriculture. ICAR. New Delhi. 2010, pp. 1-145.

Talwar P.K. and Jhingran A.G. 1991. *Inland Fishes of India and Adjacent Countries*. Vol. 1. 1991, pp. 1-1158.

Varghese M., Manisseri M.K., Ramamoorthy N. and Geetha P.M. *Coral reef fishes of Gulf of Mannar, S.E. India*. 2011. Research Gate. www.researchgate, net/ publication 277789788_coral.

Venkataraman K. and Alfred J.R.B. 2001. Coral Reefs. In: *Ecosystems of India*. Edited by Alfred JRB, Das AK, Sanyal AK. ENVIS-Zool. Surv. India. 2001, pp. 261-290.

Sand tiger shark. Wikipedia the free encyclopedia. 8 August 2020. en.wikipedia. org/wiki/sand_ tiger_ shark. Accessed 17 August 2020.

Spotted garden eel. Wikipedia the free encyclopedia. 1 August 2020. en.wikipedia. org> wiki > Spotted_ garden_ eel. Accessed 17 August 2020.

www.researchgate.net> publication> 331867774_ Captive_ breeding _ of _ *Hippocampus reidi*.

3

Anecdote of Ornamental Fish Keeping

3.1 History of Keeping Ornamental Fishes as Pets

3.1.1 The inception

According to archaeological evidences, the fishkeeping began with the Sumerians, during 2500 BCE, and with the Babylonians, in 500 BCE ("Aquarium" 2018). Ancient Egyptian art depicts, that the sacred fish Nile Perch *Oxyrhynchus,* during the period of 500 BCE, were kept in rectangular temple ponds, but this was not for the purpose of decoration ("Fishkeeping", n.d.). The Indus Valley discoveries revealed that, the fish was one of the religious symbols of the Mohenjo Daro, and during that period, (2600-1900 BCE) the most popular of all forms of God was the fish (Karmakar, 1943).

3.1.2 Beginning of keeping fishes for decorative purposes and further developments

In the Roman Empire, during 50 AD, sea fishes were kept in small marble tanks, and one wall of the marble tanks was replaced with glass panes.

Records show that the Chinese, during the Sung dynasty in China in 1000 AD, kept carp only for decorative purposes and began breeding them selectively. The Chinese people were barred from eating them during 968-975. Emperor Hiau-Tsung in 1136 encouraged breeding and keeping of goldfish in a more controlled environment. In 1369, during the reign of Hongwu Emperors of China, large porcelain tubs were used for maintaining goldfish. Goldfish, from China, came to Japan, during 1500 AD.

By 1510, goldfish became common among all people and no longer a luxury for elite. Successful breeding techniques were kept secret. The very first book on goldfish, "Essay about the Goldfish" was written in China in 1596. As goldfish had arrived in Japan in 1616, the breeding of this fish was mastered, consequently. The Japanese became the largest exporters of goldfish eventually. Goldfish reached in Portugal, Europe in 1691. From there goldfish appeared in England in 1728. This fish was popular throughout Europe's privileged ruling class. Holland was the first country to breed the goldfish in Europe in 1780.

Robert Warrington in 1805 discovered the "tank water cycle principle". This helped in establishing first indoor tanks. In 1832, Jeanne Villepreux-Power, a marine biologist, created aquaria for experimenting with aquatic organisms. In 1838, Felix Dujardin used a saltwater aquarium. Dr Nathaniel Bagshaw Ward, placed aquatic plants and toy fish in aquarium, later on he, housed live animals.

Anne Thynne in 1846, established the first balanced marine aquarium in London. In 1850, English chemist Robert Warrington fully developed the aquarium principles and explained the importance of plants in aquarium in providing oxygen to support animals. Tank designs and techniques for maintaining water quality were developed by Warrington. Edward Edwards developed aquaria in 1858 with water slowly circulating to a reservoir beneath.

In 1853, Philip Henry Gosse helped in establishing the first public aquarium in the London Zoo, which came to be known as the Fish House. In 1854, Gosse in his book "The Aquarium: An unveiling of the wonders of deep sea" discussed mainly the saltwater aquarium. During the 1870s, some of the first aquarist societies appeared in Germany. In 1893, the first aquarist society was founded in New York City in the United States. The world's first aquarium magazine 'The New York Aquarium Journal', was published in 1876.

In 1908, the first mechanical aquarium air pump was invented. This was powered by running water instead of electricity. And with the availability of electric supply, artificial lighting, aeration, filtration, heating arrangements of the water were possible. After World War I, artificial lighting, aeration, filtration and heaters were made to enter the aquariums. In the 1950s, plastic shipping-bags were introduced making it easier to ship fish. The development of tar and silicone sealant allow, in 1960, the first all-glass aquaria made by Martin Horowitz in Los Angeles. This allowed better water-proofing and better versatility in fish keeping. Acrylic aquariums first became available to the public in the 1970s. A wide variety of tanks were possible with the use of acrylic, which are more lightweight, more crack resistant and can be made into different shapes.

In the United States, since 1996, aquarium keeping is the second most popular hobby after stamp collection. It has been reported that, in 1999, about 9.6 million US households owned an aquarium. According to the National Pet Owners Survey's report (2005-06), the Americans own approximately 139 million freshwater fish and 9.6 million saltwater fish as pets. About 3.9 million households in the UK, own fish as ornamental pets (Customer Profile Survey Reports, 2011-2012). In India, households, 7.95% in Chennai, 7.32% in Kolkata, 3.60% in Mumbai and 1.48% in Delhi, keep aquariums.

3.2 The Ornamental Fish Abode—Aquatic Set-ups

3.2.1 Aquarium

Philip Henry Gosse, in 1854, in his book, for the first time used the term "aquarium". He did so, by joining the Latin word aqua meaning water, with the suffix -arium meaning "a place for relating to".

An aquarium (*pl.* aquariums or aquaria) can be simply defined, as a container of any size, with at least one transparent side, in which water dwelling animals and plants are kept and displayed (Figure 3.1). Glass or high strength acrylic aquaria are used by the fish-keepers to keep fish, invertebrates, amphibians, aquatic reptiles and aquatic plants. Fish tanks cuboid-shaped and fish bowls which are bowl shaped are in use. Size ranges from a small glass bowl to immense public aquaria.

Various equipment used, are to maintain water quality and other characteristics suitable for the aquarium's environment. During the last half of 19th century, a variety of aquarium designs were explored, such as hanging aquarium on a wall, using it as a part of window, as a great corner piece, as room divider, as wall unit with fancy shapes, as a focal point for rooms, or combining it with a bird-cage. Aquarium designers make the aquariums as a substitute for tea/coffee table, providing a particular theme to the room. Toilet aquarium, wash basin aquarium, clock aquarium, lamp aquarium, TV set, computer and

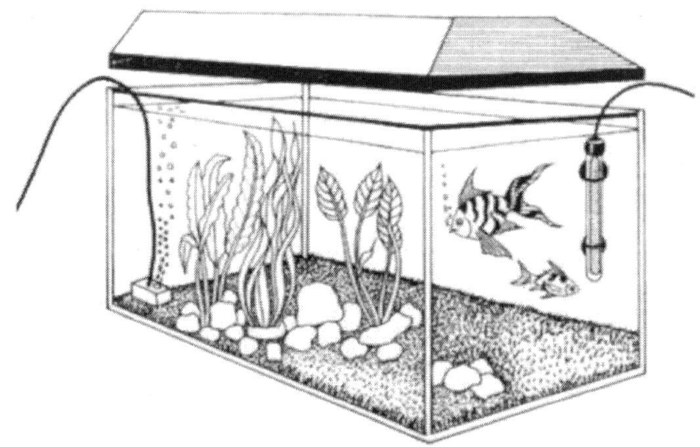

Figure 3.1 An aquarium.

CPU aquariums are also available in the market. Cuboid, hexagonal, angled to fit in a corner (L-shaped) and bow- front (the front side curves outwards) are the shapes of the aquarium preferred by hobbyists. Fish bowls are either spherical or round-shaped. Bowls for keeping goldfish, in particular are available as, waterfall glass goldfish bowl, glass goldfish bowls by CYS EXCEL, drum glass fish bowls, biOrb classic fish bowl, and hanging fish bowl.

3.2.1.1 Materials used for making aquariums

Most aquariums consist of glass panes bonded together by 100% silicone sealant with plastic frames attached to the upper and lower edges of the aquarium. Developmental history related to aquarium materials are:

- The first modern aquarium made of glass was developed in the 19th century by Robert Warrington.
- During the Victorian age, glass aquariums were with slate or steel bottoms, for allowing these to be heated underneath. These aquariums had the glass panels attached with metal frames and sealed with putty.
- The modern, silicone-sealed replaced metal-framed aquariums.
- Acrylic aquariums first became available to the public in the 1970s. Laminated glass is sometimes used, which combines the advantages of both glass and acrylic.

- Glass aquaria are always a popular choice for home and hobbyist aquarists. Silicone sealant is a long-term water-tight seal and so it eliminated the need for a structural frame.
- Fish bowls are either made up of plastic, acrylic or glass.

3.2.1.2 Classification of aquariums

Aquariums may be classified based on criteria such as salinity, temperature, type of fish and ecosystem (Table 3.1).

Table 3.1: Classification of aquariums

Criteria		Type of Aquariums	
Salinity	Freshwater	Brackish water	Marine
Temperature	Tropical	Temperate	Cold water
Type of fish	Community	Aggressive	Solitary
Ecosystem	Freshwater	Brackish water	Marine

3.3 Ideas and Implementations Related to Aquariums

3.3.1 Aquahouse

Aquahouse is the abode where polyhouse technology with appropriate modifications has been extended to culture aquatic organisms under unfavourable weather conditions. Aquahouse technology has been well adopted in developed countries like Germany, the Netherlands, Belgium, Japan, Israel, the USA, etc., for production of aquacrops (Chakrabarti and Sharma, 2008). Aquahouse are of two types, earthen pond based and land based. In the first type, the whole water body is covered. Whereas in the second type, aquacrop is grown in a housing facility, with tanks and water purification plant inside (Figure 3.2 A, B, C). Climatic conditions are fully controlled.

3.3.1.1 In aquarium fish culture

For commercial scale aquarium fish culture, aquahouse provides ideal conditions. Environmental factors are under control here so maintaining brooders and growing young ones are much easier than outdoor culture. Aquahouse fish culture is easily manageable as everything is inside the housing facility and so women, physically handicapped people and aged persons can also be employed.

3.3.2 Aquascaping

Aquascaping is the creation of an exquisite garden under water, by designing and arranging aquatic plants, rocks, stones, cave-work or driftwood and fish within an aquarium (Figure 3.3). The purpose of aquascaping is to create an artful underwater landscape. Though technical aspects of tank maintenance and care of plants and fishes are to be taken care of.

In this closed system of aquarium, filtration, maintaining carbon dioxide at levels sufficient to support photosynthesis, underwater substrate and fertilization, lighting and algae control are considered carefully.

Aquascape designs

- Dutch style: Here multiple types of plants with diverse leaf colours, sizes and textures are displayed.
- Japanese style: Relatively few species of plants carefully selected stones or driftwood are placed.
- Iwagumi style: It is a layout where stones play a leading role. Rocks provide the bony structure of the aquascape, where three stones are mainly used.
- Jungle style: In this, there is little open space, no landscape materials are present, and plants are given an untrimmed, natural look.

3.3.3 Biotope

Biotope aquariums are designed to replicate exactly a particular aquatic habitat at a particular geographic location (Figure 3.4). Plants and fishes present should represent the particular habit that is present in nature. Some of the most popular biotopes are the freshwater habitats of the Amazon and Rio Negro rivers, saltwater coral reefs of Australia, the Red Sea, the Caribbean. The African rift lake environments of Lake Malawi and Lake Tanganyika.

3.3.4 Vivarium

A vivarium (Latin: place of life) is an enclosed place where animals and plants are kept and raised for observation or research (Figure 3.5). Sometimes, a portion of ecosystem is simulated on a small scale with environmental conditions being controlled. A vivarium may be small enough to be placed on a table or may be a large structure to be placed outdoors.

Different types of vivarium

There are various types of vivarium recognised as set-ups.

- Aquarium: It is simulating an aquatic habitat, and only the submerged area of lakes, rivers and oceans. Aquatic plants and aquatic animals of the particular habitat are kept.
- Oceanarium: Large ocean dwelling fish and mammals such as sharks and dolphins are displayed here.
- Paludarium: This is a semi-aquatic enclosure.
- Insectarium: Having insects, arachnids and other arthropods.
- Formicarium: Having different species of ants.
- Terrarium: Displaying a dry habitat, such as desert or savannah.

3.3.5 Paludarium

It is an aquarium where aquatic and terrestrial environments are represented. This semi-aquatic enclosure simulates a rain forest (Figure 3.6), swamp, riverbanks, bogs, beach even jungles. It is also considered as an aquarium connected with a terrarium, with both submerged area and shore. Certain environmental factors like temperature, light, humidity etc., are to be controlled as required.

3.3.6 Oceanarium

An oceanarium represents either an ocean habitat, with marine animals (Figure 3.7), especially large ocean dwellers such as sharks, teleost fish, turtles or marine mammals.

3.3.6.1 World's largest oceanarium

Chimelong Ocean Kingdom a theme park, in Hengqin, Zhuhai, is the world's largest oceanarium.

3.3.6.2 Marine mammal parks

The world's first oceanarium is the "Marine land of Florida", one of the first theme parks in Florida, United States, started in 1938. There are marine mammal parks now in different countries of the world.

3.3.6.3 Marine Public Aquariums

Natural marine environments are created in the modern marine public aquariums, as in London Sea Life Aquarium (Figure 3.8), situated on the South Bank of the River Thames in Central London. There is a four-story cylindrical tank with lots of marine animals in the New England Aquarium, in Boston. In the National Aquarium in Baltimore, there is a walkway spirals up through the centre of two gigantic cylindrical tanks. The Monterey Bay Aquarium, in California is designed to display local environments.

3.3.7 Micro Landschaft

A micro landschaft is a small landscape which includes aquariums, terrariums, bosai, and even small gardens and ponds, etc. The term may be used to describe a variety of terrariums, small planters, units with hanging planters, ant farms (formicaria), small reptile habitats and even small aquariums, with betta, koi or goldfish (Figure 3.9). Micro landschaft overall represents man-made objects and habitats.

3.3.8 Reef Aquarium

A reef aquarium is a marine aquarium in which live corals, other marine invertebrates and fish community, associated with the coral environment are present (Figure 3.10).

- Components in a reef aquarium

  In addition to the biotic and abiotic components, some essential material components are needed for a reef aquarium. These are:

  1. Display tank: The primary tank made up of glass or acrylic and with holes drilled. In this, the livestock are kept and displayed.
  2. Stand: The display tank is placed on the stand.
  3. Sump: An accessory tank in which mechanical equipment is kept. A remote sump allows for a clutter-free display tank.

4. Refugium: It is an accessory tank used for the cultivation of macroalgae and microflora/fauna.
5. Lighting: Most popular aquarium lights are metal halide lamps, with very high output or VHO, compact fluorescent and T5 high output lighting systems. Recently light emitting diodes (LEDs) are used as lighting systems. Colour temperature, colour rendering index (CRI), Photosynthetically Active Radiation (PAR), and lumens are used.
6. Canopy: It houses the light fixtures and provides access to the tank for feeding and maintenance.
7. Filtration and water movement: Different types of filtration and water movement strategies are utilized.
8. Water chemistry: Optional water parameters are- Salinity: 1.022---1.025 dg; Temperature: 24—27°C; Ammonia: 0 ppm; Nitrite: 0 ppm; Nitrate: 0—10 ppm; Phosphate: 0 ppm; pH: 8.2—8.6; Calcium: 400—450 ppm; Alkalinity: 7—12 dKH.
9. Livestock: Fish-Clownfish, Tangs, Gobies etc.; Invertebrates- Plating Montipora coral, Frogspawn coral, Hammer coral, Pulse coral, Brain coral, Button polyp, Star polyp, Cleaner shrimp etc.

3.3.9 Nano Reef

A Nano reef is a type of marine aquarium with a capacity of less than 140 litres (37 US gallons). The difference between a nano reef and a regular reef may be the capacity limit of 140 litres. Nano reefs have become quite popular nowadays due to their smaller size, maintainability and lower costs. Nano reefs as complete kits contain the tank, stand, power compact T5, T8, PL lamps or metal halide lighting, protein skimmer, UV steriliser, 3 or more stage filtration, a heater and a water pump or powerhead. More powerful protein skimmer and better lighting systems are also used.

3.3.10 Pico Reef

Pico reef (Figure 3.11) is the smallest of Nano reef aquarium. Capacity of these tanks may be 10 litres (2.5 gallons) or less. In the smallest of Pico reefs, instead of a fish, only live rock, jellyfish, hardy corals, and small invertebrates such as hermit crabs and marine snails are kept. In this reef, allelopathy is present as the corals compete for space by chemical and physical means. These tiny aquariums require more attention and care regarding water changes and water chemistry.

3.3.11 Kreisel

A Kreisel tank is an aquarium, which is shaped like a horizontal cylindrical structure. Delicate animals such as jellyfish (Figure 3.12) and new born seahorses are kept here. There are different types of Kreisel tanks available. True Kreisels are circular tanks with a circular-submerged lid. Pseudo-Kreisels are U or semi-circular in shape, without a lid. Stetch Kreisels are a "double gyre" Kreisel design.

Figure 3.2 A Land based aquahouse.

Figure 3.2 B Inside an aquahouse facility.

Figure 3.2 C Water purification plant in aquahouse.

Figure 3.3 Aquascaping.

Figure 3.4 Biotope representing the Amazon river.

Figure 3.5 Vivarium of marine habitat.

Figure 3.6 Paludarium representing rain forest.

Figure 3.7 Oceanarium with fishes and corals.

Figure 3.8 London Sea Life Aquarium.

Figure 3.9 Micro landschaft with goldfish.

Figure 3.10 Reef aquarium with fish and corals.

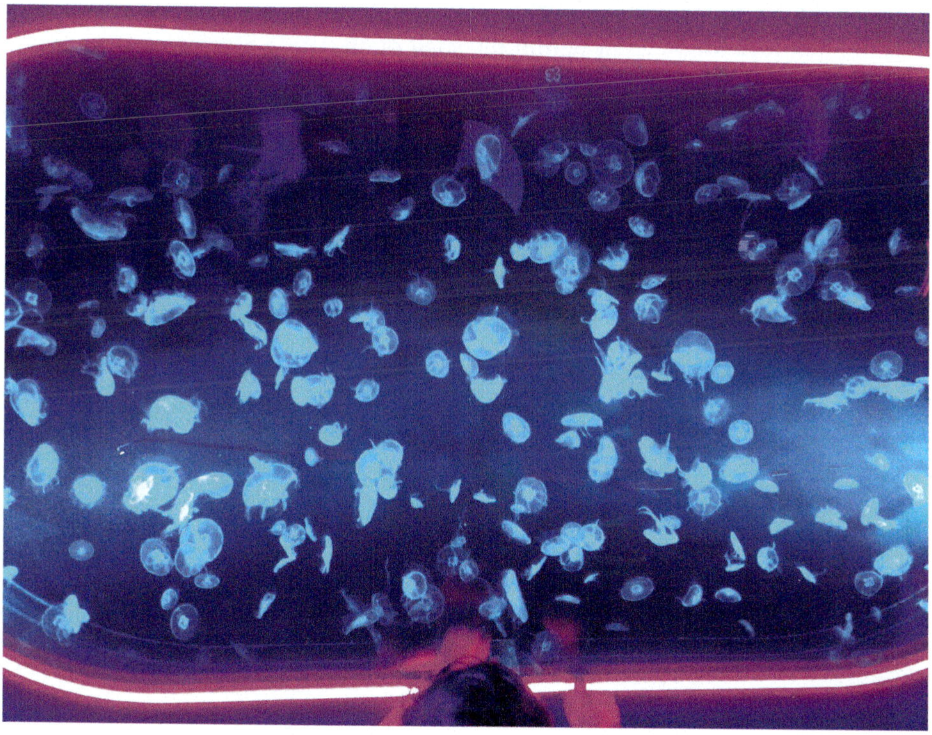

Figure 3.11 Pico reef.

Figure 3.12 Kreisel tank with jellyfsih.

3.3.12 Macquarium

The term Macquarium was coined by computer writer Andy Ihnatko in 1992, and it is an aquarium made to sit within the shell of an Apple Macintosh Computer. Macquariums have been built by Ihnatko himself and others. In the early 1990s, several Mac models, in the form factor as the Macintosh 128K, Macintosh 512 K and Macintosh Plus, were becoming obsolete and Ihnatko believed that turning this into an aquarium might be "the final upgrade" and also an affordable way to have a colour Compact Mac. He designed it without a visible water line and which gave an illusion of a really good screensaver. Moreover, it allowed the external case of the donor Mac to remain intact.

Ihnatko's designed Macquarium is a slant-front tank, made of glass, had a nominal capacity of approximately 10 litres (2.5 gallons). Other designs have utilized acrylic glass or Lexan. Macquarium is considered a form of Pico aquarium, because of its small capacity and which requires a higher level of diligence to maintain proper water chemistry and cleanliness.

Goldfishes (2-3) are often stocked in the macquariums as they do not require tank heaters and are cheap. Siamese Fighting Fish and small shrimp are better for stocking.

Macintosh TV, the Apple Lisa and the Power Mac G4 Cube are used to make macquariums. A maquarium in 1995, based on a Mac LC 575 appeared in a Macintosh magazine titled, "Macquarium 95".

3.3.13 Virtual Aquarium

A virtual aquarium is a computer program, which uses 3D graphics to reproduce an aquarium on a computer. The fish in the aquarium, may appear to swim both in front and behind them, as the objects on the floor of the tank are mapped in simple planes. A relatively simple 3D map of the general shape of objects are used to allow light and ripples on the surface of the water to cast realistic shadows. Bubbles and water noises are common. Sea star, jellyfish, seahorses, sea turtles and different fishes are selected. Many companies produce virtual aquarium software and offer types of fish and other objects for sale via internet download.

3.3.14 Public Aquarium

Public aquariums are characterized by large tanks or smaller tanks housing living aquatic animals and plant specimens for public viewing. These aquatic counterparts of zoo, have become very popular since their inception and so their number has increased a lot during the last few years. The first public aquarium was opened in London Zoo in May 1853.It was named the Fish House. Later, several public aquaria were established in cities of developed Nations.

3.3.14.1 Modern Public Aquariums

Modern Public Aquariums can hold millions of litres of water and can house large aquatic animals such as dolphins, sharks, beluga whales, even otters, seals and penguins. Front walls of these aquariums are made up of thick, clear acrylic glass windows. Most of these public aquariums are located close to the ocean, for steady supply of sea water. Some public aquariums use freshwater from the city water system and later on add salt to it, as required.

Activities

Public aquariums are places for marine research, conservation programmes and marine animal rehabilitation.

3.3.14.2 Famous Public Aquariums of the world

Some of the huge and famous public aquariums are as follows:

1. Chimelong Ocean Kingdom situated in Hengqin, Zhuhai, People's Republic of China is the world's largest oceanarium with a total of 10.72 million UK gallons of water.
2. World's second largest oceanarium, is the Marine Life Park located within Resorts, World Sentosa situated in southern Singapore. With 9.9 million UK gallons of water and more than 100,000 animals, of more than 800 species, this Marine Life Park inspires research and education with an aim to protect oceans.
3. Georgia Aquarium (capacity 6.3 million gallons). Situated in Atlanta, this is the third largest aquarium in the world, housing more than 100,000 sea creatures including whale sharks.
4. Dubai Mall Aquarium (capacity 2.64 million gallons). In this public aquarium in Dubai more than 33,000 animals including over 400 sharks and rays are displayed. It is the world's largest Acrylic panel Aquarium.
5. Okinawa Churaumi Aquarium (capacity 1.98 million gallon). It is situated in Ocean Expo Park in Japan. Whale sharks and manta rays are kept along with many other fish species from Kuroshio sea.

6. L' Oceanographic (capacity 1.85 million gallon). It is in the city of Valencia, Spain.
7. Turkua Zoo (capacity 1.32 million gallon). Situated in Turkey, this aquarium features a rainforest, flooded forest and tropical sea zones. About 10,000 sea creatures are present.
8. Monterey Bay Aquarium (capacity 1.2 million gallons). Located on Connery row in Monterry, California.
9. U Shaka Marine World (capacity 1 million gallon). It is a theme park located in Durban, South Africa.
10. Shanghai Ocean Aquarium (capacity 1 million gallon). This aquarium has world's largest underwater tunnel at the depth of 156 meters. It shows China zone, Antarctic zone and the Australia.
11. Aquarium of Genoa (capacity 1 million gallon). The Aquarium of Genoa in Italy is one of the largest aquariums in Europe. The aquarium's 70 tanks reproduce marine and terrestrial habitats from throughout the world and provide a home for more than 6000 animals. Natural environments from different areas as the Mediterranean, the Red Sea and the Indian ocean are reproduced in some tanks. Sharks, dolphins and seals are also housed here.
12. Aquarium of Western Australia (capacity 0.8 million gallons). Located in the coastal suburb of Perth, the Aquarium of Western Australia or AQWA contains Australia's largest aquarium tanks. Main tank is 40 m (130 ft) long and 20 m (66 ft) wide and holds 3,000,000 litres (793000 gallons) of sea water.

3.3.14.3 Public Aquariums in India

Specimen-rich and public-friendly aquariums till now 23 in number ("List of aquaria in India", n.d.) are situated, in different states/ Union Territories of India (Table. 3.2).

Table 3.2: Public aquariums in India

Names of aquariums	City	State/ Union Territory
Bagh-e- Bahu Aquarium	Jammu	Jammu & Kashmir
Bangalore Aquarium (Figure 3. 13)	Bangalore	Karnataka
Calcutta Aquarium	Kolkata	West Bengal
CIFA Aquarium	Bhubaneswar	Odisha
Dr. A.M Michael Aquarium	Panagad, Kochi	Kerala
District Tourism Promotion Council (DTPC) Aquarium	Kozhikode	Kerala
Jagdish Chandra Bose Aquarium	Surat	Gujarat
Jawahar Aquarium	Mussoorie	Uttarakhand
Kaukaria Aquarium	Ahmedabad	Gujarat

Names of aquariums	City	State/ Union Territory
Kollam Aquarium	Kollam	Kerala
Lal Bagh Aquarium	Bangalore	Karnataka
Machhli Ghar	Bhopal	M.P
MARC Aquarium (Figure 3.14)	Digha	West Bengal
Marine Biological Research Station	Ratnagiri	Maharashtra
Marine Life Aquarium	Chennai	Tamil Nadu
Matsyadarsini Aquarium	Visakhapatnam	Andhra Pradesh
Nandankanan Zoo Aquarium	Bhubaneshwar	Odisha
Sanjay Gandhi Jaivik Udyan	Patna	Bihar
Star Aquarium	Karunagappalli	Kerala
Taraporewala Aquarium (Figure 3.15)	Mumbai	Maharashtra
Travancore Royal Aquarium	Trivandrum	Kerala
ICAR-NBFGR Ganga Aquarium	Lucknow	Uttar Pradesh
Varkala Aquarium	Trivandrum	Kerala

Figure 3.13 Bangalore (Government) Aquarium, India.

Figure 3.14 Marine Aquarium, ZSI, West Bengal, India.

Figure 3.15 Taraporewala Aquarium, Mumbai, India.

3.4 References

Chakraborti R. and Sharma J.G. *Aquahouse, New Dimension of sustainable Aquaculture.* Published by Directorate of Information and Publications of Agriculture. ICAR. 2008 pp. 1-54.

Gosse P.H. *The Aquarium: an unveiling of the wonders of the deep sea.* London. 1854,pp I-vii. 1-303.

Karmakar A.P. *The Fish in Indian folklore and the age of the Atharvaveda.* Annals of the Bhandarkar Oriental Research Institute. 1943, 24 (3/4): 191-206.

Fishkeeping. Wikipedia. The Wikimedia Foundation. 3rd August 2020. En.wikipedia.org. > wiki> Fishkeeping. Accessed 16 August, 2020.

List of aquaria in India. Wikipedia: The Free Encyclopedia, Wikimedia, 15 August 2020. En.wikipedia.org/wiki/ List _ of _aquaria_ in _ India. Accessed 16 August 2020.

The Editors of Encyclopaedia Britannica. Aquarium. *Encyclopaedia Britannica.* February 28, 2018. Accessed August 15, 2020. https:// www.britannica.com/ science/aquarium.

4

Fabrication and Setting Up of Aquarium

Fabrication and aquarium setting are the initial manoeuvres followed in ornamental fish keeping. Depending on the location and type of fish, size, shape and volume of the aquaria are to be decided.

4.1 Fabrication of an Aquarium

Fabrication of an aquarium is the art of combining or assembling parts to make an aquarium.

4.1.1 Types of fish tanks

Aquaria are made up of different materials like glass (Figure 4.1 A), concrete (Figure 4.1 B), wood, fibre glass, acrylic, and plastic sheet (Figure 4.1 C) etc., depending on the location, type of fish, cost and durability.

Figure 4.1 A Glass tanks.

Figure 4.1 B Concrete tanks.

Figure 4.1 C Plastic tanks.

4.1.2 Materials required for construction of an all-glass aquarium

Glass panes (4-5 in numbers) of particular sizes, silicone gel and squeezing gum are required.

4.1.3 Glass tanks

Aquariums made up of glass are the most common one. The **acrylic tank**s are costlier compared to any other glass aquaria and though they are breakage -proof, they are not scratch-proof. Regarding glass tanks, these are either **all glass tank**s or **metal framed tank**s. In all glass tanks, glass walls are fitted together edge-to-edge using silicone rubber adhesive. In the metal (steel or iron) framed tanks glass panels are held in place with putty. Metal-framed tanks are popular now-a-days for their slim appearance.

4.1.3.1 Shape of aquariums

The shape of the fish tanks may be circular, square, rectangular, oval, hexagonal or octagonal. Rectangular tanks are widely used as these provide enough space for free swimming of the fishes.

4.1.3.2 All glass aquarium tank

Usually glass sheets of thickness of 5-10 mm are cut to make all **glass tank**s. For rectangular aquarium, 5 pieces of specific glass-cuts are required. Back-glass may be replaced with a mirror. The cut glass walls are fitted together with synthetic rubbery sealant called silicone gum. Side walls of the tank are to be arranged according to the desired shape. The glass walls are tied with a rope in order to keep the correct shape of the tank. The bottom sheet of the tank is then placed below the arranged sidewalls. A thermocol sheet of suitable size is placed in between the tank and the place on which tank is placed. A firm wooden or slotted iron stand is used for keeping the aquarium.

4.1.3.3 Steps for construction of an all glass fish tank (Figures 4.2 A-E):

- After deciding the size of the aquarium, dimensional cuttings of panel glasses are to be made.
- Rough glass edges are smoothened by rubbing with rubbing stone or a chisel fine file.
- To remove any glass powder and other foreign particles, glass should be thoroughly washed with soap and water and then the clean panel should get dried up.
- Large table or a working bench is to be arranged for the fabrication purpose.
- Thermocol sheet is to be placed on the surface.
- The bottom glass is placed on the thermocol.
- A thick string of silicone gel is applied along the edge of back surface (Figure 4.2 A).
- The back panel is to be raised (Figure 4.2 B).

- Gel is applied to the two side edges of bottom glass panel and also edges of fixed back glass pane.
- Side glasses are to be placed gradually and joined with other already fixed panes (Figures 4.2 C, D, E).
- The gel is to be smoothen.
- The aquarium tank is to be left as such for a day.
- The newly constructed aquarium is to be checked for any leakage by pouring water in the tank.
- After rinsing with water, the aquarium is ready for use.

4.1.3.4 Other parts of an aquarium

- Glass cover—A glass cover is placed on the all glass aquarium to prevent escape and foreign objects falling inside the tank. The glass cover should be designed as such to permit free exchange of air.
- Hood—A hood or tank cover is made of glass, wood or tin plate. Such hoods act as "lid" for the tank and helps in preventing accumulation of dust particles, to install lamps, with provision for feeding fish. Hoods or tank covers should match the size of the tank.
- Aquarium stand—It is made up of wood or metal and is required to keep aquarium tank in position. Aquarium stand should be capable of carrying out the weight of the tank. The height of the stand is also decided as necessary.

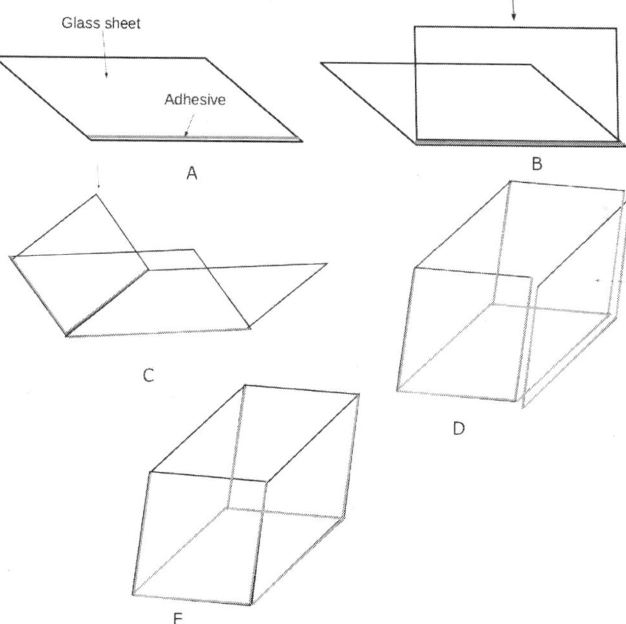

Figure 4.2 A, B, C, D and E. Construction of glass aquarium.

4.1.3.5 Cost of aquariums

Approximate cost of an aquarium depends on the size, along with the cut glass size, numbers and thickness of the glass (Table 4.1).

Table.4.1. Aquarium size, cut-glass size, numbers, thickness and approximate price of aquarium (after Swain et al., 2010)

Aquarium size (cm)	Size of cut-glass =Numbers required	Thickness (mm)	Approximate cost (Rs)
60x30x30	60x30=3; 30x29=2	5	500
75x 30x 30	76x30=3; 30x29=2	5	550
76x30x38	76x30=1; 76x38=2; 38x29=2	5	650
76x38x46	76x38=1; 76x46=2; 46x36.25=2	6	800
100x38x45	100x38=1; 100x53=2; 53x36=2	8	2000
120x45x45	120x45=3; 45x43=2	8	2500
120x45x53	120x53=2; 120x45=1; 53x43=2	10	3000
150x61x61	150x61=3; 61x58=2	12	6500
200x61x61	200x61=3; 61x57=2	19	70,000
250x61x61	250x61=3; 61x55=2	25	1,12,000

4.2 Setting Up of Aquarium

Setting up of an aquarium is known as aquascaping. Aquascaping of an aquarium is done with the objectives of making the tank more attractive to the viewer and for simulating natural ecosystems by keeping colourful fishes, aquatic animals, aquatic plants, rocks and gravels.

4.2.1 Steps involved in setting up of an aquarium:

- Location: Aquarium should be placed at a convenient and attractive place which is away from direct sunlight and not in close proximity of an open fire.
- Cleaning of the tank: Soaps and detergents should not be used in cleaning inside surfaces of the tank. Only lukewarm water with a dish-scrub pad is to be used. A light solution of herbal shampoo may be used.
- Cleaning gravel and chips: Gravel, chips and coloured stones are to be washed thoroughly with water without using any soap or detergent.
- Water depth: In an aquarium, minimum water depth to be maintained is 30 cm.
- Coloured stones, gravel and chips are to be spread in the tank. Under gravel filters may also be placed. Rocks or driftwoods can be added.
- Adding water: While filling the tank with water care should be taken

not to disturb the already placed materials inside the tank. So, water is generally poured over a piece of clean paper covering required portion.

- Water: Clean and pathogen free water should be utilized for the tank. Chlorinated water should be dechlorinated first and then used.
- Cover glass or plastic sheet: To cover the top of the tank, cover glass or plastic sheet are used.
- Hood and lighting: Most hoods are made up of acrylic, fiberglass, reinforced plastic, aluminium or waterproof ply attached with either a fluorescent or general tube light. Low wattage bulbs should be used with incandescent features of minimizing water temperature rise and saving electricity.
- Plants: Plants should be thoroughly cleaned and disinfected before planting in the tank. The roots of the plants should be pushed deep into the gravel for proper anchorage. Taller plants are placed in the back, broader leaved plants in the middle and smaller plants are kept in front. Plantations are done when the tank is only half full.
- Fishes: Healthy, colourful, companionable and compatible fishes are displayed in the tank. Fishes are to be stocked according to the recommended stocking density.
- Introducing fishes in the aquarium: The newly brought fishes should not be released into the aquarium tank immediately. The polythene bags containing the fishes are placed in the tank water for about 15 minutes. Water inside the polythene bag gets mixed with tank water a little and the fishes get acclimatized to the new water conditions. Fishes should not be fed immediately.
- Number of fishes that can be stocked in an aquarium:

A simple calculation is done to ascertain the number of fishes that can be stocked in a rectangular fish tank (Scott, 1996; Detwiler, 2019).

Using **Fish Stock Calculator**: U.S. Units

Tank Length (across front) X Tank Width (front to back)

Here, surface area is used to determine fish stock levels. For slender fish, the value of 12 square inches of surface per length-inch of fish (or 30 sq.cm/1 cm of fish). For full-bodied fish the value of 20 square inches per inch (50 sq.cm/1 cm) of fish. For cold water fish 30 inches of surface area per inch of fish (76 sq.cm/1 cm) and marine fish 48 inches of surface area per inch of fish (122 sq.cm/1 cm).

Example:

Tank length=48 inches; Tank width = 25 inches

Fish stock level results:

Calculated surface area= 1200 inches

Maximum inches of slender fish a tropical freshwater tank can support= 100

Number of fishes= 100/5 (average length of fish) =20

Maximum inches of full-bodied fish a tropical freshwater tank can support= 60

Number of fishes=60/5 =12

Maximum inches of fish a cold freshwater tank can support= 40

Number of fishes= 40/5= 8

Maximum inches of fish a saltwater tank can support= 25

Number of fishes= 25/5= 5

- Suggested combinations of aquarium fishes:

  Compatible fishes can be kept together in a tank. Combinations are suggested based on their behaviour, food and feeding habits.

  1. Guppy, molly, swordtail, platy, tiger barb, danios, widow tetra, rosy barb, angels (small), loaches.
  2. Gold fishes all varieties except comet, with black or albino white shark.
  3. Koi carps, tiger shark, angels(big), tiger barbs, tinfoil, albino tinfoil barbs.
  4. Tetras, tiger barbs, rosy barbs, tinfoil barbs.
  5. Discus, angels, rainbows, pearl gourami.
  6. Gourami, small cichlids.
  7. Giant gourami, koi carps, cichlids, loaches, albino and black shark.
  8. Cichlids of more or less same sizes.

- How to select a healthy fish

  There are some tips which are to be followed for selecting fishes for aquarium. These are:

  - Fishes should have normal swimming movement.
  - Fishes must have normal appetite.
  - They should have rounded belly.
  - There should not be any increased respiration, scratching signs.
  - Fins of the fishes should remain erect and there should not be any cut in the fins.
  - There should not be any marks of external injury on the fish's body.
  - Fishes should be without any external parasites and must not have any signs of disease occurrence.

4.3 List of Accessories and Equipment Used

4.3.1 In small ornamental fish farms and home aquaria:

For small fish farms and for home aquaria, limited number of equipment and accessories are needed.

- **Glass tank**:

  A tank made of sheet glass, float glass or transparent fibre-glass, acrylic sheet may be kept on a strong stand made of wood, steel etc. Tanks are with hood, either made of glass, wood or tin plate.

- **Light**:

  Lights fixed inside the hood of the aquarium, may be fluorescent tubes, mercury vapour lamps or metal halide lamps. A lamp of 10 watt, and lighting for 10 to 15 hours a day is sufficient for an aquarium of 5 litre capacity.

- **Heaters**

  Immersion heaters with adjustable thermostats are used in cool places or during winter months to keep optimum temperature.

- **Air pumps** and accessories:

  Low to medium cost air pumps are available in the markets. Air tubes, controlling devices and air stones are also used for regulating air flow in the water and for efficient oxygenation.

- **Filters**:

  Biofilters and mechanical filters are commonly used in an aquarium. Under gravel filter, box filter, poly U filter, power filter etc., are the filters. Biofilters coupled with a sump and overhead tank is in use. A small electric pump is needed here to lift water from sump to overhead tank/ filtering units.

- **Hand nets**, **Feeding ring** and **Scraper**:

  Hand nets (Figure 4.3) used are generally made of mosquito nets, velon screen or bolting silk. Other accessories which are used for small aquatic units are different types of feeding rings and scraper for removing algae from the surface of the fish tank.

Figure 4.3 Hand nets

4.3.2 In large scale ornamental fish farms:

In large scale ornamental fish farms, a lot of equipment and accessories are required for maintenance and smooth running of the facilities.

- Aeration devices:

  Good aeration systems are necessary for large scale ornamental fish production and rearing units, because the facilities may be from a number of tanks to large pond. Aeration devices used are generally surface aerators (paddle wheel aerator), diffuser aerators and turbine aerators. The water inlet system is to be set up in such a way to facilitate gravity aeration (splash board, race way, transversely corrugated inclined plane etc. Aspirator aerators (air injectors) can be used if the depth of the culture facility is more than 2 m.

- Pumps and water pipe lines:

  Pumps and connecting pipelines are an integral part as good quality water is necessary to carry out the culture program. To fetch underground water, for water exchange, for periodic flushing and emergency irrigation, quality pumps are to be installed.

- Arrangements for ornamental fish seed production:

  Breeding units, hatcheries and rearing units are the needed facilities.

- Food or feed production units:

  Live food production units in the form of cement tanks, FRP tanks are to be set up. For the preparation of pellets etc., essential equipment should be installed.

4.4 References

Detwiler S. *How many Fish? Aquarium Stock Level Calculator*. 2019. www.howmanyfish.com Aquarium Estimator Tools.

Scott P.W. *The Complete Aquarium*. Published by Dorling Kindersley. 1996.

Swain S.K., Sarangi N. and Ayyappan S. *Ornamental fish farming*. Published by Directorate of Information and Publications of Agriculture. ICAR. 2010, pp. 1-145.

5

Equipment, Accessories and Decors

For setting up of an aquarium, one has to acquire equipment, accessories and decorative pieces for the tank. To start with, the most needed ones are chosen first. Some items which are used in different types of aquariums are conferred here.

5.1 Algae Scrubber

Algae scrubber is a water filtering device which removes undesirable chemicals from water by growing algae, using light (Figure 5.1). Algae scrubbers are useful in freshwater and saltwater aquariums and ponds.

In the ocean, both iron fertilization and ocean nourishment promote the primary production of algae in the ocean water. The algae so produced consumes massive amounts of nutrients and carbon dioxide. In an aquarium or pond, algae perform the same function.

In an algae scrubber, water is filtered by moving water fast over a rough, highly illuminated surface. This causes algae to start growing in large amounts. As the algae grow, they consume nutrients like nitrate, phosphate, nitrite, ammonia, ammonium and copper from the water. This removes the excess nutrients (scrubs the water) and diminish the nuisance algae thriving on the nutrients. The helpful algae in the algae scrubber is present inside the filter and can be removed later or fed back to the livestock. The nuisance algae that are removed are *Cyano* or slime, bubble, hair, *Chaetomorpha*, *Caulerpa*, film algae, dinoflagellates and *Aiptasia*.

5.1.1 Recent forms

Some of the modern algae scrubbers are:

- A simple PVC plumbing pipe is used to flow water down a piece of rough plastic knitting screen ("plastic canvas") on which algae get attached.
- **Green hair algae** or light-green hair algae scrubbers are very useful, as these algae allow light and water to penetrate all the way down to the screen. This allows the algae to grow faster and absorb more nutrients. Green hair algae grow naturally in a properly constructed algae scrubber and does not die easily and loose attachment to the screen.

In Up-flowing air bubble algae scrubber, the algae scrubber is placed underwater in the aquarium, sump or pond instead of above it as in case of waterfall model. There are three types of this model. These are: (i) Attached to and shine through the aquarium or sump glass; (ii) Float on top of the aquarium, sump or pond water surface; (iii) Go completely underwater like a submarine.

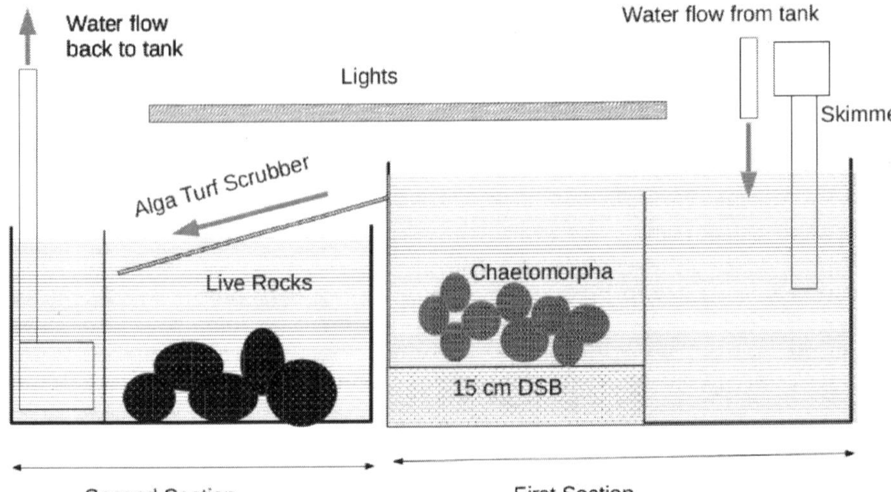

Figure 5.1 Algae scrubber.

5.1.2 Cleaning and Harvesting

Algae scrubbers, except for specific continuous-filtering or continuous-cultivating ones, need the algae to be removed (harvested) periodically from the scrubber. As the algae use the undesired nutrients to grow, the removal of algae has the effect of removing undesired nutrients. Removal takes place, every 7 to 21 days, or when it is black, or when it fills up the scrubber, or when it starts letting go, or when nutrients start to rise in the water.

5.2 Berlin Method

Berlin method (Figure 5.2) is a process of biological filtration by maintaining a clean and stable environment in a saltwater aquarium. Developed by Peter Wilkens and the Berlin Aquarium Club, this system use live rocks which have live marine organisms and bacteria on or in it. Aerobic bacteria covering the surface of the porous live rock and sand convert harmful ammonia from fish and invertebrate wastes into nitrites, then nitrates. The nitrates move deep into the rock where they are converted by anaerobic bacteria to free nitrogen gas. Regular partial water changes are done to remove left over nitrates. For the removal of dissolved organic compounds before they break down into ammonia, protein skimmer is used.

The main material live rock is used in about 0.45 to 0.9 kg per gallons (US) or 5 litres of aquarium water, depending on the density of the rock. Live rock acts as a biological filter, adding beneficial bacteria, introduces large number of marine organisms into the tank, it provides a natural reef appearance and also helps in balancing and stabilizing pH in the tank. Regular water changes, use of a calcium reactor, kalkwasser (calcium water-calcium hydroxide, mixed with pure water) or a balanced two-part solution, may be used to replace calcium, alkalinity and other trace elements.

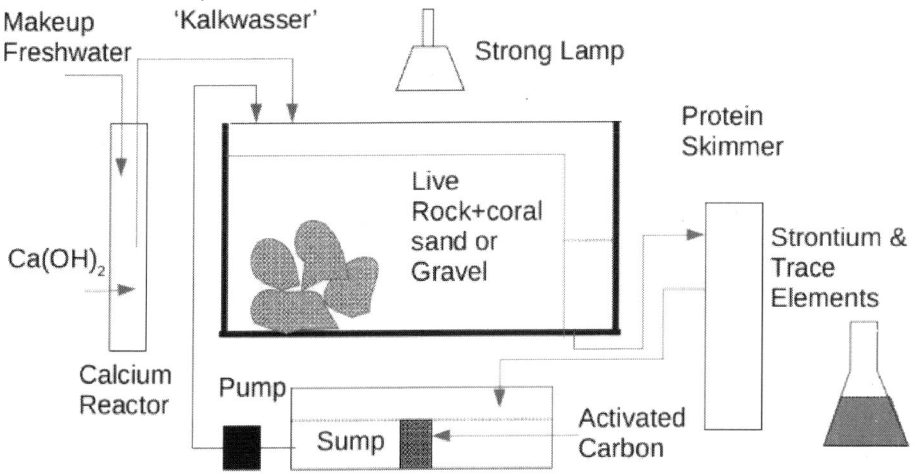

Figure 5.2 Berlin method.

5.3 Deep Sand Bed

Deep sand bed (Figure 5.3) is a filtration system used in some salt water aquariums. This consists of a bed of fine sand with a minimum depth of four to six inches ensuring that a portion of the sand at the bottom will not be exposed to significant circulation of water. A deep sand bed cultivates anaerobic bacteria in the bottom layers of the sand, converting nitrate to nitrogen gas. Besides bacteria and algae, other marine organisms burrow and overturn the top two to three inches of sand in search of food and this causes water circulation deeper in the sand.

Fine or super fine sand is used to make deep sand beds, with a grain size between 1mm and 0.05 mm. A larger particle size enhances circulation and a greater depth of it is required to establish anaerobic areas. Larger particles may have detritus accumulation and so periodic siphon cleaning is necessary.

5.4 Live Sand

Live sand (Figure 5.4) in salt water aquarium, is the natural reef coral sand which is inhabited by millions of beneficial bacteria and organisms. These help in dissolving organic wastes like ammonia, nitrites, and nitrates in the aquariums. **Live sand**s are available in aquarium stores. This may also be prepared at home by seeding dead sand with live sand from other aquaria.

5.5 Live Rock

Live rock (Figure 5.5) is the rock from the ocean that is made from the aragonite skeletons of long dead corals or other calcareous organisms which are part of the coral reef.

When brought from the ocean, it is usually encrusted with coralline algae and with large number of micro and macroscopic marine organisms that live on and inside the rock. In the aquarium (reef) trade, live rock is collected in the wild near reefs, from which this has been detached by storms. Live rocks are also seeded from small coralline rocks, in warm ocean water, or by adding base rock to an active reef aquarium.

5.5.1 Types of live rock:

Depending on the area from which it is originated, live rocks are typed and named. Live rocks are found in the Southern Pacific region, in areas such as Fiji, Tonga and the Marshall Islands and the Caribbean. Live rock from the Fiji region is often porous and large, and rock from the Tonga region is often dense and elongated.

5.5.2 Use of live rock

- Live rock a highly valued material in aquarium trade, serves as major decorative pieces, especially when encrusted with multiple coloured coralline algae.
- Live rocks are often modelled as caves, arches, overhangs or other structures for the purpose of aquascaping.
- In the marine aquarium, a large number of bacteria (including aerobic and anaerobic nitrifying bacteria), algae and invertebrates are introduced by the live rock and this acts as a superior biological filter. Thus, it becomes the main biological nitrification base or biological filter of a salt water aquarium.
- Live rocks provide a stabilizing effect on the water chemistry, specially maintaining constant pH by release of calcium carbonate.
- Live rocks also provide shelter for the aquarium inhabitants.

5.5.3 Base rock

Base rock is aragonite rock that has no organisms inside or on the rock. In the aquarium, the base rock or dry rock is often used as filter rock as it costs much less than live rock. Now a days, base rock is mined from inland ancient reefs. They are either grown in home aquarium, or maricultured and sold as live rock. Base rock is also made from aragocrete i.e., by combining crushed aragonite, sand and Portland cement. This is dried, acid washed and then allowed to soak in clean water for one or more months.

5.6 Substrate

Substrate of an aquarium refers to the materials present on the aquarium tank base. This, besides being an integral part of the tank's aesthetic appeal, affects the water chemistry, filtration and provides comfort to the inhabitants of the tank.

A substrate is essential, as fishes do not live over bare glass in nature but prefer a natural base to their world and also as the aquarium looks unfinished without it (Bailey and Sandford, 2006). Substrate also provides an anchorage to the plants and a home for essential nitrogen cycle bacteria. In freshwater aquaria the substrate is usually gravel (Figure 5.6 A), large river rocks (Figure 5.6 B), marbles (Figure 5.6 C) and sand (Figures 5.6 D) whereas it is coral sand or coral gravel in salt water aquarium (Figure 5.6 E). There are some specialized freshwater aquaria where the fishes require hard alkaline water, such as Rift Like Tanks (Sandford, 2007).

5.6.1 Types of substrate:

5.6.1.1 In freshwater aquaria

- **Gravel** (Figure 5.6 A)—is the most common substrate. The gravel should not be sharp and size may be from 1-2 mm to pea-sized. It may be naturally coloured or dyed and should have a polymer seal to ensure that it does not affect water chemistry. Gravel is composed of quartz or other lime-free minerals and is to be chemically inert. For planted aquaria, the Cation Exchange Capacity (CEC) is an important matter to be considered as CEC absorbs positively charged nutrient ions. This means the substrate will hold nutrients and make them available for the plant roots. Gravel comes in a variety of sizes, colours and even shape. Any gravel for use in aquaria is treated.
- **River rock** (Figure 5.6 B) --- is attractive and makes for a natural setting. It is used by itself or with gravel beneath it.
- **Marble** (Figure 5.6 C) -- used as breeding substrate when egg scattering species of fishes are kept in the aquarium. Marbles are also preferred in glass bowls, for their easy maintenance.
- **Laterite** or **vermiculite** are beneficial for aquarium live plants. They are usually used as a lower layer of the substrate and covered with a layer of gravel.
- **Sand** (Figure 5.6 D) --- is also good as substrate. Clay can be mixed with sand to provide extra nutrients for live plants. River stingrays prefer sand substrate as they bury themselves in the fine substrate.
- **Peat** or decomposed plant matter--- is used in aquaria representing soft water or black water river ecosystems. These are soft in texture and are suitable for bottom-dwelling species such as *Corydorus* catfish. This substrate softens water by acting as an ion exchanger also. It prevents algae growth and kill microorganisms.

Different substrates are also used at the same time in different parts of the tank, serving various purposes.

5.6.1.2 In marine aquaria and rift-lake aquaria

- Shell grit, crushed limestone, crushed marble, crushed coral skeletons, coral sand and oolitic aragonite substrates are composed of calcium carbonate and these increases water hardness and pH. These are used as substrates for African rift lake cichlids or for saltwater aquarium fish (Figure 5.6 E) and salt water aquarium invertebrates.

5.6.2 Substrate quantity

The substrate is generally filled to a depth of approximately one and half to two inches. For sand, the depth may be a bit less.

5.6.3 Substrate colour

Substrate colour should mimic natural habitats.

5.6.4 Advantages of substrate:

Substrates are useful in various ways. These are--

- The substrate plays an important role in the aquarium nitrogen cycle by acting as a medium, which beneficial bacteria colonize and grow.
- Substrate gives a natural look to the aquarium. Presence of substrate provides a natural habitat for the fish. Aquarium fishes found substrate as their shelter, hiding place, feeding and breeding ground.
- Some substrates are helpful in altering water chemistry. Crushed coral and coral sand both contain calcium carbonate which raise the carbonate hardness and buffer the pH. Peat as substrate in some aquaria represent some soft water habitats.
- Substrate is also used as part of a biological filtration system. Undergravel filter and the deep sand bed as such substrates acting as biological filtration system.
- Substrate provide support and nutrients to the live plants in aquaria. Plant roots penetrate into the substrate and for this the substrate should be at least 5 cm deep and better remain loose for accommodating the plant root system.

Figure 5.3 Deep sand bed.

Figure 5.4 Live sand.

Figure 5.5 Live rock.

Figure 5.6 A Gravel as substrate.

Figure 5.6 B River rocks as substrate.

Figure 5.6 C Marble substrate.

Figure 5.6 D Sand substrate.

Figure 5.6 E Substrate in salt water aquarium.

5.7 Bog-Wood

Bog-wood (Figure 5.7) is also called abonos and morta, is used in aquaria for ornamental purpose. Moreover, this provides a hiding place for fish and provides growing surface for plants such as Java fern. The leaching of organic compounds such as tannins from it, into the water causes a brown colouration.

Bog-wood is formed from the trunks of trees that have been buried in peat bogs and preserved from decay by the acidic and anaerobic bog conditions, for centuries even millennia. Deprived of oxygen, the wood undergoes the process of fossilization, though bog-wood is at the early stages of fossilization of wood and morta formation. The wood is usually stained brown by tannins dissolved in the acidic water. Bog-wood may come from trees naturally growing near or in bogs, like oak, pine, yew, swamp cypress and kauri. This rare form of timber is considered as one of the most expensive wood.

5.8 Drift Wood

Drift wood is a popular decorative piece in the ornamental fish tanks. It is the remains of trees, in whole or part, that have been washed into the ocean, due to flooding, high winds, or as the result of logging. Drift lumber is considered as subset of driftwood, includes the remains of wooden objects, remains of shipwrecked wooden shits and boats (flotsam) washed into the sea during storms.

5.9 Fish Cam

Fish cam is the broadcast programme which uses a video camera focussed to an ornamental fish tank. Famous Fish cams are broadcast programmes in channels in Australia, Canada, Hongkong, Turkey Netscape etc.

5.10 Sump

Sump is an accessory aquarium tank in which equipment are kept. It is generally used along a marine or reef aquarium. For a clutter-free fish tank, a remote sump is utilized.

A sump is housed below or behind the main tank for holding equipment like protein skimmer, calcium reactors and heaters and is also used as filter. It also helps to increase the water volume in the system so that there are less fluctuations of pH and salinity. A sump also helps in improving aeration of the water in the aquarium so that gas exchange takes place between water and air, which further helps in increasing dissolved oxygen and also in preventing cyanobacteria population increase.

5.11 Air Stone

Air stone (Figure 5.8) or aquarium bubbler is generally a piece of limewood or porous stone, which is used to diffuse air into the tank avoiding noise and large bubbles of conventional air filtration systems. Air stones are also made from bonded glass beads and synthetic products like fiberglass. Air stones are available in different shapes, sizes and levels of coarseness. The coarseness of air stones may be very fine producing minuscule bubbles or may be very rough producing larger bubbles.

Wooden air stones are used for protein skimmers to produce large quantities of micro bubbles. Wooden air stones are made up of limewood or basswood. The wooden blocks are drilled, tapped, fitted with an air fitting and connected by air tubing to one or more air pumps, delivering minimum 1 cfm (cubic feet per minute). At the bottom of a tall column of water, the air stone is placed. The tank water is pumped into the column, it is then allowed to pass by the rising bubbles and back into the tank.

5.12 Aquarium Heater

Aquarium heater (Figure 5.9 A) is an appliance used to warm the temperature of water in aquariums. The water temperature is kept at the range of 22-30°C (71-88°F) in most tropical freshwater and marine aquariums.

Types of aquarium heaters

- **Glass immersion heater**s are the most commonly used heaters in aquariums. In its basic form, used for smaller tanks, it consists of a submerged unit combining a heater and a thermostat. The heater consists of a glass tube containing a heating element wound around a ceramic or glass insert. Most of these are fully submersible (Figure 5.9 A). Some of these heaters also contain sand. The glass tube of the heater contains an adjustable thermostat which turns the heating element on. A bimetallic strip is often used as thermostat, as the strip contains two metals, the metals will expand at different rates when the temperature rises, and the strip bends. The strip carries the current and breaks the current at the correct temperature. Microchips are used in more advanced thermostats.

One heater is used in small aquarium, but in larger aquarium, two or more than two are used.

For larger set-ups separate heater(s) and thermostat are used, with connections made outside the tank via a terminal block. Larger the tank, the greater the capacity the heater(s) must have. For a 90 litres tank, a 100 W unit is usually sufficient, 200 W for 200 L tank. It is to be remembered that, the higher the capacity or wattage of the heater, the greater must be the capacity of the thermostat which controls it (Ward, 2004).

- **Undergravel heater**s are flexible heating cable buried underneath the aquarium substrate. This is preferred in heavily planted aquariums.
- **Undertank heater**s are the heating mats placed beneath aquaria. These are less efficient as glass is poor conductor of heat.

5.13 Thermometer

Thermometer is a necessary item to check the conditions within the tank. A traditional liquid-filled glass type can be used which is to be attached to the inside of the tank with a plastic sucker (Figure 5.9 B), or a dial-type stuck on the outside. Now a days, a plastic strip containing liquid crystals is used. It is stuck on the side of the tank and indicates temperature by changing colours.

5.14 Aquarium Lighting

Aquarium lighting is any kind of artificial lighting that is used to illustrate an aquarium (Figure 5.10). Artificial lightings are generally chosen on the basis of aquatic set-ups to which it is to fitted.

Freshwater aquarium lighting

Screw-in incandescent bulbs, fluorescent tubes and LED lamps, high -intensity compact fluorescent bulbs, high output fluorescent bulbs or metal halide lamps are frequently used to provide much stronger light source and also encourage better plant and algae growth. These days T2, T5, SHO and LED are efficient lighting that use less electricity and produce less heat.

Brackish water aquarium lighting

In brackish water aquariums, lighting used is similar to freshwater aquariums. In addition, Infrared or UVB bulbs (or both types) are also used.

Marine aquarium lighting

- In fish-only marine aquariums, lighting is generally used for illumination i.e., for viewing fishes. This lighting is usually of much lower intensity.
- In reef aquarium, lighting used is of highest intensity of all other aquarium types. Fluorescent, LED, metal halide and plasma lamps are used. Single lamp, multiple lamps of a single type and several lamp types and also lamps of different colours are utilized. Lamps of different colours are used not only for visuals but also to sustain life of reef organisms.

T5 High Output fluorescent bulbs, VHO (Very High Output) fluorescent tubes and compact fluorescent lighting, metal halide lighting with spectra from 5000 K on up to 20,000 K are used.

Some light fixtures are with light for both day light hours and night time viewing. These all in one fixture are with large fans to cool the bulbs and to achieve maximum light efficiency.

Reef aquariums and planted aquariums require specialized high intensity lighting to support the process of photosynthesis in the tank.

5.15 Aquarium Fish Feeder

Fish feeders are electric or electronic appliances used in aquariums, to feed fishes in the tank at regular intervals. Fish feeder consists of a hopper which is filled with a variety of dry food, a timer which rotates the hopper at regular intervals and dispensing the food. The interval between feeding and amount of food dispensed is set. Fish feeders are clamped to the wall of the aquarium just over the water. The electronic aquarium fish feeder should be kept clean and the electronic part of the feeder should not get wet.

Advantages

- Aquarium fish feeder is helpful in case when the aquarist is not at home.
- These electronic fish feeders, by feeding small portions of food at regular intervals, are useful in maintaining fish health.
- These feeders are useful in avoiding overeating by releasing the right quantity of food at scheduled times.

5.16 Power Head (Pump)

Powerhead is a water pump which circulates water in the aquarium. This water pump or powerhead remains completely submerged in the water of the tank and creates water flow throughout the tank. Water circulation is necessary for providing freshwater free-swimming fishes enough exercise. In most salt water aquariums water circulation is essential for proper biological filtration.

In freshwater aquarium, a single powerhead is used to simulate water flow in the river. Multiple powerheads can be fitted throughout the tank to create more turbulent movement. In reef aquarium a voltage system or switching or wavemaker is used to simulate ocean water flow.

5.17 Aeration

5.17.1 Principle

Aeration is the process of increasing or maintaining the oxygen saturation of water in both natural and artificial environment.

5.17.2 Aerator

Aerator (Figure 5.11) is an electrically operated air-pumping device. It consists of a vibrating air-pump, a tiny rubber diaphragm. When current passes a shaft bearing a magnet vibrates under a magnetic field effect. As the shaft is attached to the diaphragm, the latter is allowed to vibrate up and down like a piston producing a forceful stream of air. A plastic air tubing or thin pipe is used to flow air directed into the bed of the aquarium tank. A diffuser or air stone may be used at the opening to produce tiny bubbles of air. Such aerator is very useful for small home aquarium (Figure 5.8).

5.17.2.1 Power air pump

Power air pump, powerful electric motor driven piston-pumps or rotary power air pumps are used to produce aeration in large and deep aquarium or a battery of aquariums (public aquaria). These pumps are very efficient due to high output and noiseless operation, though these are expensive and needs care and maintenance regularly.

5.17.2.2 Spray bar

Spray bar is fitted to the outlet of a power filter (Figure 5.11). This spray bar is a perforated tube producing a rain of small droplets of water falling on the surface of tank water. There is better oxygen diffusion at air-water interface as the agitation of water is produced at the surface by falling water drops.

Figure 5.7 Bog-wood in a freshwater aquarium.

Figure 5.8 Air stone in home aquarium.

Figure 5.9 A Aquarium heater.

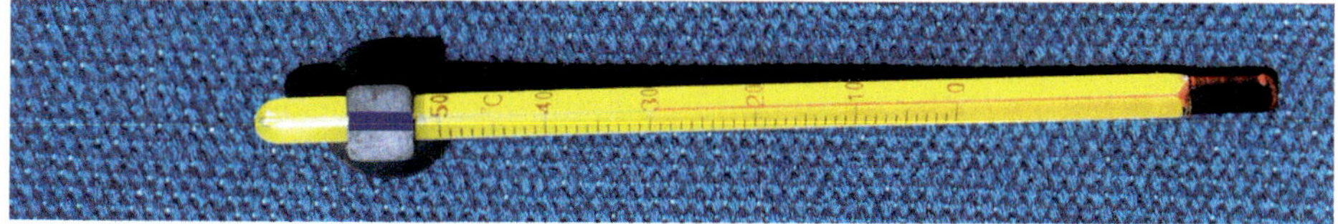

Figure 5.9 B Thermometer used in aquarium.

Figure 5.10 Aquarium lighting.

Figure 5.11 Aerator.

5.18 Aquarium Filters

Aquarium filters are high priority items in both freshwater and marine aquaria. Filters in the tank not only remove physical and soluble chemical waste products from aquaria but also helps in supporting aquatic life in the enclosed environment (Bailey and Sandford, 2006; Sandford, 2007; "Aquarium filter" n.d.).

Fish waste from excrement and respiration, uneaten food, decaying plant material and living and dead microorganisms form a thick sediment known as mulm (Ward, 2004). This settles to the bottom of the tank and accumulates in corners and behind plants. Mulm is unsightly and can also be a source of health problems.

5.18.1 Types of filtration:

- **Mechanical filtration** (Figure 5.12 A)—During the process of mechanical filtration, solids are trapped by the filter. Removal of solid wastes involve filtering water through some form of mesh. Removal of solid wastes may range from using hand nets or involving highly complex equipment. The solid wastes are first collected and then must be physically removed from the aquarium systems. If the solid wastes are not removed from the filter, mechanical filtration, is ultimately ineffective. The commonest media used are filter floss, plastic foam and gravel or sand.
- **Chemical filtration** (Figure 5.12 B)—Chemical filtration is used for the removal of dissolved wastes. The chemical composition of water is altered here. The alteration of pH for example, is done by using peat or lime rich material, reduction of hardness by ion-exchange resins, and removal of metabolic products by activated carbon or zeolite (which remove carbon

dioxide and ammonia respectively). Chemical media also trap solids mechanically. The use of activated carbon and foam fractionation are most popular.

- **Biological filtration** (Figure 5.12 C)—In this filtration method, water is circulated through a series of compartments each filled with different filtration media. It overflows from the aquarium

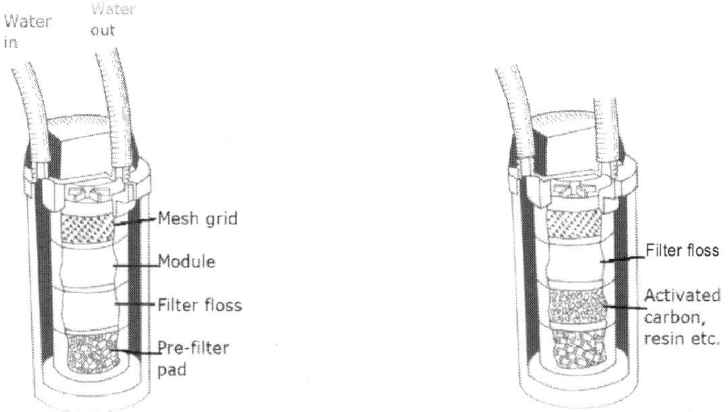

Figure 5.12 A Mechanical filtration. Figure 5.12 B Chemical filtration.

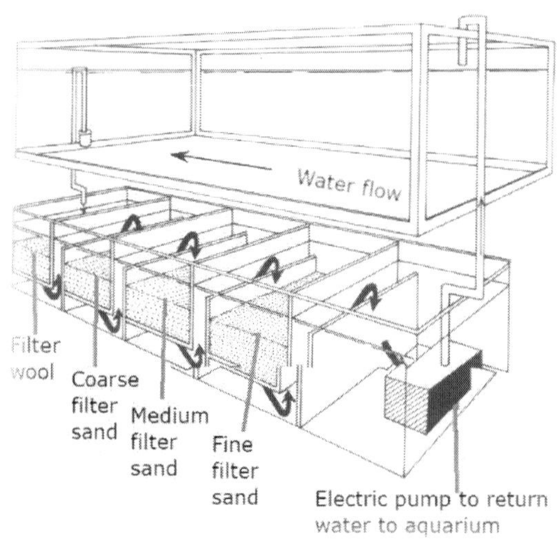

Figure 5.12 C Biological Filtration

into the filtration tank and a submersible pump returns it to the aquarium (Figure 5.12 C). Biological filtration is the technique of enhancing natural populations of the bacteria operating the nitrogen cycle (Bailey and Sandford, 2006). Bacteria are provided, the surfaces to colonize, supplies of wastes to process and for those which convert nitrites to nitrates, a constant supply of oxygen.

This filtration confirms a constant supply of wastes and oxygen. Effective biological filtration is the key to maintaining water quality on a day-to-day basis. Filter wool, filter sand, plastic, ceramic and glass media, all maximize the surface area available for colonization.

The nitrogen cycle must be considered before establishing a filtration system. It takes about two weeks for the nitrogen cycle to become properly functional in a new aquarium and filters. Fishes should be introduced only after this maturation period because of ammonia/nitrate toxicity. Even if the established nitrogen cycle is interrupted, there will again be risk of toxicity. Therefore, only part i.e., maximum 50% of the filter contents should be cleaned or replaced, so this will leave enough residual bacterial population. Feeding should be reduced for a few days to lessen the load on the biological system till it regains its full capacity.

5.18.2 Materials used as aquarium filtration media

All filter media perform mechanically in straining solids from the water. If allowed to mature all media will be colonized by beneficial bacteria. Different materials are used as filtration media. These are—

- Synthetic wools (filter wool) made of polyethylene terephthalate (PET) or nylon.
- Synthetic sponges or foams, various ceramic and sintered glass and silicon products along with igneous gravels in mechanical filters.
- Plastic bio-balls, filter wool, and filter sand are used in biological filters.
- Activated carbon absorbs smaller impurities. The adsorption of activated carbon can be restored by thermal regeneration, electrochemical regeneration and ultrasound.
- Hexagonal ceramic tubes are designed to optimize bacterial activity.
- Carbon and ammonia remover (zeolite) remove certain toxins.
- Peat, limestone chips, coral sand are used to adjust hardness and/or pH.

5.18.3 Filter media

Filter media are grouped on the basis of their mode of working and items removed (Table 5.1).

Table.5.1. Various filter media commonly used

Filter media	Mode of working	Controlling items
Nylon floss, filter wool, plastic foam, sand, gravel, diatomite, sponge, diatom skeleton	Mechanical	Solid suspended particles
Activated charcoal	Chemical	CO_2, pH, & hardness
Limestone chips, coral sand, peat	Chemical	pH & hardness
Resin	Chemical	Hardness & NH_3
Filter wool, filter sand	Biological	Ammonia, nitrites

5.18.4 Types of aquarium filters

There are various types of aquarium filters available for use. The most popular ones are—

Power filters:

Power filters or **Hang on Back (HOB)** filters are the most common, effective and easy to maintain. This filter is powered by an impeller, remove water from the tank and then push or pull through a series of different filter media and then returned to the tank. As this filter is placed outside the tank it is easy to clean without disturbing the inhabitants of the tank. A large number of filters can be used for this type of filter though capacity of this filter is small.

- **Diatom filters**

These filters are used in occasional cleaning of tanks and are not operated continuously in the tanks. Diatom filters utilize diatomaceous earth to create an extremely fine filter up to 1μm which removes particulate matter from the water column.

- **Canister filters**

Canister filters (Figure 5.13 A) are power filters which use a greater amount of filter materials and can utilize a wide range of filter material. A canister is filled with filter material and water enters at the bottom of the cannister through an intake pipe, passes through the material and is fed back to the aquarium through the return pipe. There is a pump, installed at the top of the canister and this forces water to circulate through the filter. In this filter, the inlet and outlet form two siphons counterbalancing each other. The filter pump therefore does not have to spend any effort to lift the water back to the aquarium. This filter provides a high volume of filter material without reducing the internal space in the aquarium. These can also be disconnected from the tank for cleaning/ maintenance and replaced without disturbing the interior of the tank and the inhabitants. Drawbacks of these filters are that the increased cost and complexity relative to internal filters and problems in cleaning the tubes which transfer water from the aquarium. Like other outside-tank filters, there is also a chance of leak.

- **Trickle filters**

  Trickle filters (Figure 5.13 B) are stacked above the tank and used in conjunction with an external canister filter. These filters consist of one or more trays with perforated bottoms, part-filled with filter media. The return from canister filter is sprayed on to the top tray and trickles back into the tank via the perforations. The water is exposed to air, its oxygen uptake is therefore optimized and so the bacterial activity. Its capacity i.e., the volume of media can be enhanced by the addition of trays without the need to increase water flow rate. Trickle filter may also be placed beneath the aquarium, with the water being siphoned into the filter and pumped back (Bailey and Sandford, 2006).

- **Baffle filters**

  These filters are usually fitted below the aquarium and resemble the wet and dry trickle filters. Baffle filters consist of a series of baffles that the water must pass through in order to reach the pump which in returning water to the aquarium. These baffles function like a number of canister filters and different filter media may be used here for various uses.

- **Algae filters**

  Algae filters are natural filtering technique which uses algae for removing chemical from the water. If needed, algae and other disease-causing organisms are removed using ultraviolet radiation.

- **Fluidized bed filter**

  In Fluidized Bed Filter (FBF) water is passed through sand (or other media) bed from below so the sand becomes fluidized i.e., below like a fluid. The total surface of all sand particles in the filter is very large and so a large surface is offered to the aerobic denitrification bacteria. The filter can be external or internal. In its simplest DIY (Do It Yourself) internal version, an FBF is very easy to build. A container, sand, pump and some plumbing are required. Shape and size of the container, quantity of sand or equivalent, particle sizes, pump's capacity and plumbing etc., may vary accordingly.

- **Internal filters**

  Internal filters are placed inside the aquarium attached with the help of suction cups. These include sponge filters, corner filters, foam cartridge filter and the undergravel filter. An electric pump may be present with it.

- **Sponge filters**

  Sponge filters (Figure 5.13 C) and corner filters (or box filters) work as internal filters. These work by airlift, using bubbles from an air pump rising in a tube to create flow. This filter consists of a perforated plastic tube fitted with a cylindrical sponge. Air is passed through the tube, drawing water through the sponge. Sponge filters act both mechanically and biologically. These filters are used in aquaria with low filtration needs.

- **Box filters**

  Box filters are run by the airlift principle and may be **internal box filter** (Figure 5.13 D) or **external box filter** (Figure 5.13 E). The internal box filters are still useful for small tanks with less loading.

- **Undergravel (UG) filters**

  In Undergravel filter (Figure 5.13 F), a plastic plate is placed in between the aquarium bottom and the substrate, with one or more uplift pipes. Water is drawn down through the substrate and returned via the uplift(s). These filters are generally powered by (i) the airlift principle i.e., rising bubbles of air draw water with them, an airline being inserted in the uplift tube. (ii) powerheads, electric pumps on the uplift tubes (iii) an external canister filter with its inlet tube inserted in the uplift. Greater flow of water through the gravel can be attained by the use of water pump rather than air displacement. The substrate (5-7.5 cm) deep acts as the filter medium and acts as mechanical, biological even chemical filters. Advantages of this filter is that the entire aquarium bottom acts as the "inlet". Wastes are processed maximum and large quantity of filter medium is available for bacterial colonization. Disadvantage of this filter is that cleaning poses a major problem. A reverse-flow UG system using an external canister filter is employed to over the problems of UG system. Though it creates other problems.

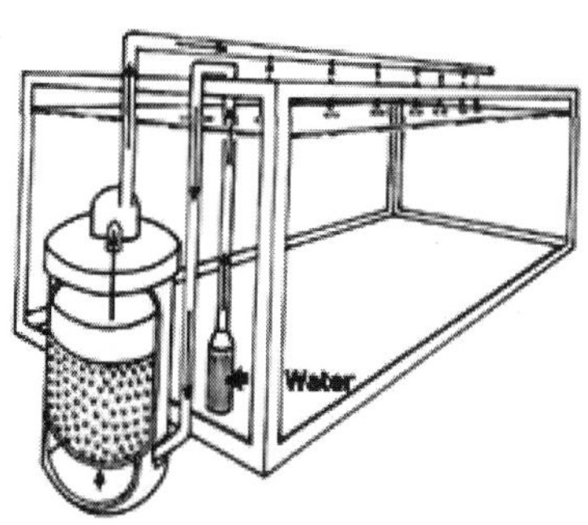

Figure 5.13 A Canister filter.

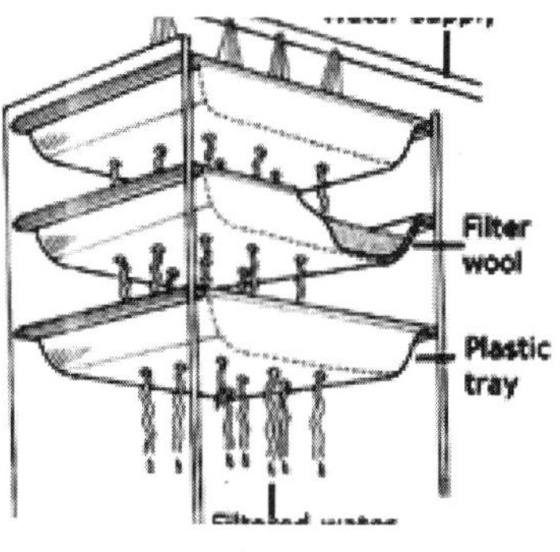

Figure 5.13 B Trickle filter.

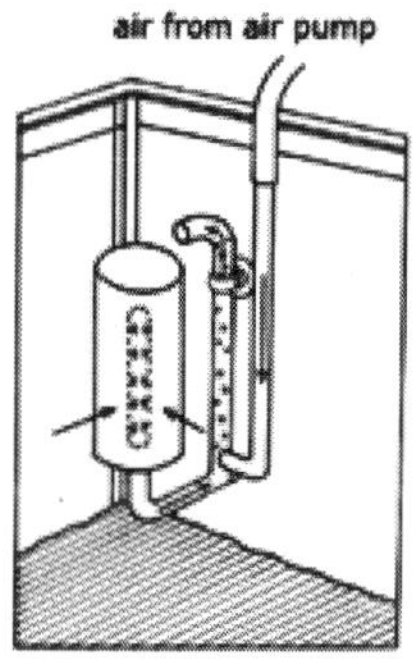

Figure 5.13 C Sponge filter.

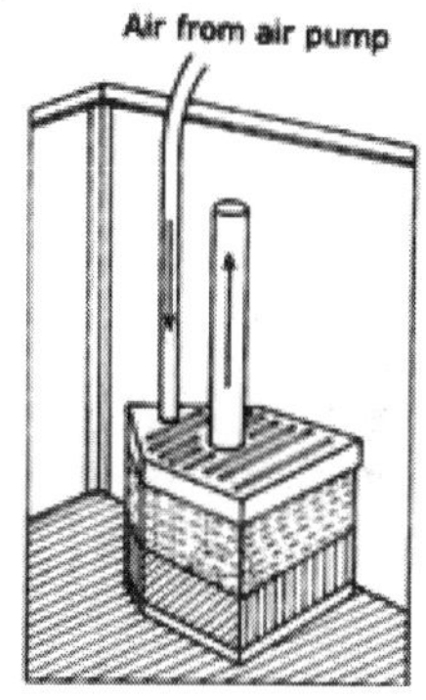

Figure 5.13 D Internal box filter.

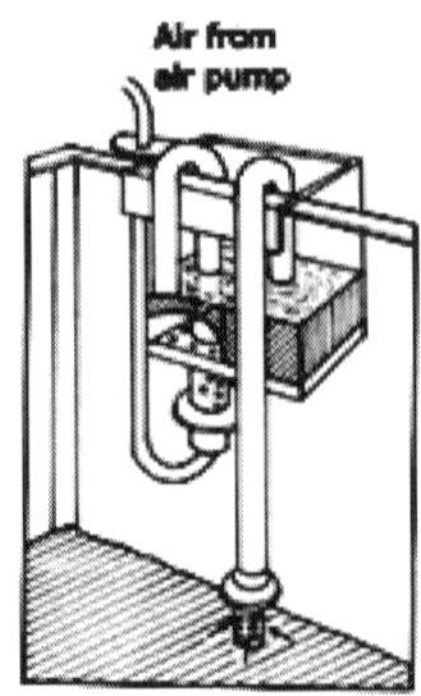

Figure 5.13 E External box filter.

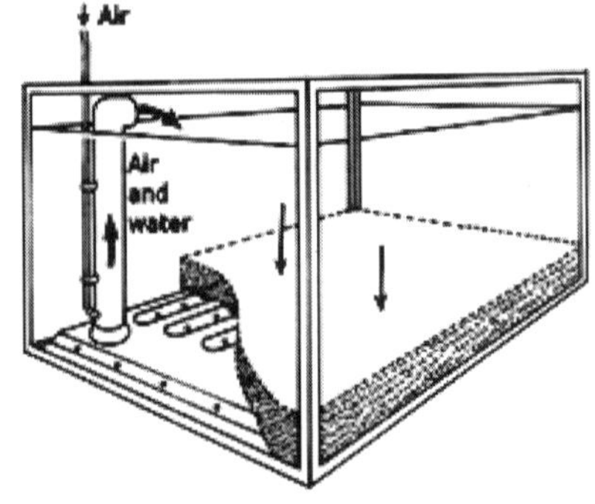

Figure 5.13 F Undergravel filter.

5.19 Systems in Marine Aquariums

Protein skimmer is the system followed in marine aquariums. The other two systems are Berlin method and deep sand bed.

Protein skimmer

Protein skimmer (Figure 5.14) or foam fractionator is a device used in marine aquarium. In this system, water flows through a chamber and is brought into contact with a column of fine bubbles. These bubbles collect proteins and other substances and carry them to the top of the device, where the foam is collected in a cup. The foam condenses in a liquid and this can be easily removed from the system. In a skimmer, the following functions may take place:

- Presence of a large air/water interface.
- Organic molecule should be collected at the air/water interface.
- The bubbles must form a foam.
- The water in the foam must be partially drained without the bubbles.
- The drained foam should be separated from the water and discarded.

Protein skimmers are classified into two depending on their mode of operation, i.e., either by co-current flow or counter current flow. In co-current flow system, air is introduced at the bottom of the chamber and comes in contact with water as it rises upwards towards the collection chamber.

5.19.1 Types

- Air stone protein skimmers may be constructed at low cost.
- Venturi pump or aspirator is used to introduce the bubbles.
- In counter-current system, air is forced into the system under pressure and moves against the flow of the water for a while before it rises up towards the collection cup.
- Aspirating skimmer is the most popular type of skimmer used with residential reef aquariums today.

- Downdraft
- Beckett skimmer
- Spray induction
- Recirculating skimmer designs. In this, the skimmer is fed dirty water from the aquarium as a means to recirculate water within the skimmer multiple times before it is returned to the sump or the aquarium.

5.19.2 Functions

Protein skimmer removes certain organic compounds including proteins and amino acids present in food particles, by using the polarity of the protein itself. Due to their intrinsic charge, water borne proteins are either repelled or attracted by the air/water interface. Protein skimmer work by generating a large air/water interface by injecting large numbers of bubbles into the water column. Organic molecules collect on the bubble surface and bubbles are removed from the water.

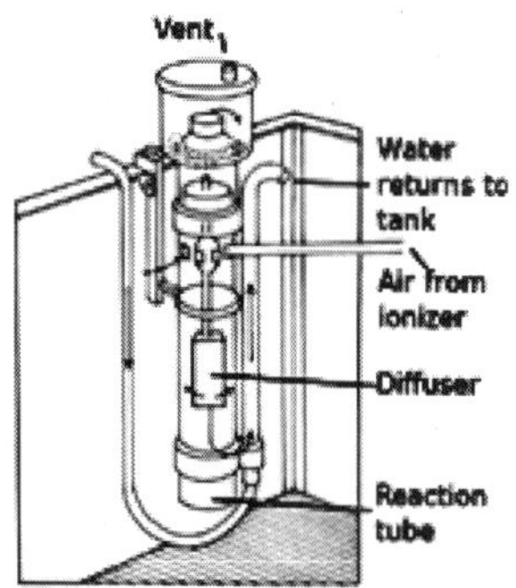

Figure 5.14 Protein skimmer

5.20 Calcium Reactor

Calcium reactor is used to produce a balance of alkalinity in marine and reef aquaria. The reaction dissolves the calcium containing media to provide bicarbonates HCO^- (alkalinity) and calcium (Ca++) ions. As an acidic pH is required, carbon dioxide (CO_2) is used. The reaction formula is: $CaCO_3$+H_2O+ CO_2=Ca+ $H_2CO_3^-$

A calcium rich media, mainly $CaCO_3$ comes in contact with water injected with carbon dioxide (CO_2) in order to create carbonic acid (H_2CO_3). This increases the solubility of the calcium carbonate. Calcium and carbonate are produced, and the aquarium gets Ca^{2+} and CO_3^- enough for maintaining alkalinity and calcium levels.

5.21 Refugium

Refugium (Figure 5.15) is an accessory connected to the main fish tank, to maintain ornamental fishes and other animal that would not survive in the main system. This is a separate sump linked to a marine, brackish or freshwater fish tank. Strong lighting is used in the refugium for up taking of nutrients by microalgae. Filter wool, Bio balls, sand and gravel are placed inside this sump.

Refugium is used for one or several uses such as denitrification, nutrient export, plankton production, circulation, surface agitation to improve oxygen exchange with the atmosphere or even aesthetic purposes. In some refugium, water flow is limited in order to protect organisms that require slow flow. The volume of refugium may be 1:10 to the main tank. The light cycle of the refugium may be operated opposite to the main tank, to keep total system pH more stable.

In home aquarium, the refugium is generally located in the table/cabinet underneath the main fish tank. The plants and creatures are kept in the refugium, the biological filtration takes place in the refugium and the main fish tank remains clean and delightful.

Tropical freshwater refugium generally contains large amounts of aquarium plants. These often contain scavengers such as several species of prolific snails and prolific shrimps. *Daphnia, Moina* and *Gammarus* are raised in some refugium to supplement the diets of some fish in the main tank.

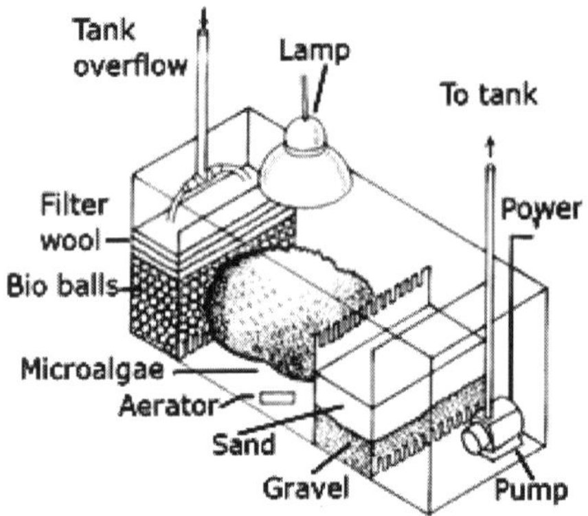

Figure 5.15 Refugium.

Tropical marine refugium is an additional tank that are generally kept under the display tanks which are mostly the show tanks and reef tanks. It serves to house filtrations, protein skimmers, reaction chambers, hoses, heaters and other such critical equipment. The sump, also serves to add volume to the total system reducing the effects of nutrient build up or unintentional introduction of foreign substances.

5.22 References

Aquarium filter. Wikipedia, the free encyclopedia. 24 June 2020. en.wikipedia.org/wiki/ Aquarium_ filter. Accessed 19 August 2020.

Bailey M. and Sandford G. *The ultimate encyclopedia of aquarium of fish & fish care*. Hermes House. London. 2006, pp. 1-256.

Sandford G. *An illustrated encyclopedia of Aquarium Fish*. Quantum Publishing. London. 2007, pp. 1-256.

Ward B. *Aquarium Fish, Survival manual*. Quantum Publishing. London. 2004, pp. 1-175.

6

Water Quality Management

Water is the most important part of an aquatic set-up, in which the fishes live and maintain all their metabolic activities. Potable water may be obtained from any source, pre-treated to make it suitable for the fish. Water quality of any water body plays a significant role in fish health, breeding, sustenance and production. For ornamental fish farming, water is generally collected from rain, river, artisan well, canal and reservoir. Aquariums should be filled with clear potable water depending on the type of fishes to be kept. Tap water is the safest source for most of the tropical fishes and plants. To remove chlorine which may be present in the water, water is allowed to mature (or aging) for a few days or aerated it overnight before addition. De-chlorination may also be done.

For the aquarist it is a challenge to try to re-create the natural environments in which the fishes thrive in the confined aquatic set-up.

6.1 Water Chemistry

Water chemistry is the term generally used to describe the physico-chemical parameters such as pH, chlorine, total hardness, free carbon dioxide, dissolved oxygen, temperature, sound and vibration, and salinity. A fish can survive well if its metabolism is in tune with these aspects of its environment (Bailey and Sandford, 2006; Adhikari, 2011). These physico-chemical characteristics of water play a pivotal role in the survival, breeding and production of many ornamental fish species.

pH

pH is an indication of acid or alkaline character and a vital index of productivity of water body. Below 7 is acidic while above 7 is basic. Optimum range for pH is 7.0-8.5. The optimal pH for the growth and breeding of majority of the ornamental fishes should be neutral or slightly alkaline i.e., 7-8. Water in the tanks should never fall below 5 or rise above 8.5. Rosy barb, tiger barb, tetra, angel and danio prefer slightly acidic pH (6.2-7). Oscar and discus will spawn only in slightly acidic pH and soft water. Various water softeners and inorganic acids are used for this. If pH of the water falls below the desired value, it may be increased by the addition of required doses of calcium hydroxide (slaked lime). Cichlids, goldfish, Koi and gourami prefer alkaline water. Small variations can have dramatic effects on fishes. To measure acidity/ alkalinity, pH papers (Figure 6.1), special test kits and electronic pH meters are available.

Chlorine

Tap water is used as aquarium water for majority of the tropical fishes and plants. But the tap water contains chlorine, which even in quite low concentration(0.1ppm) is toxic to fishes. To remove chlorine naturally, it is better to allow maturing of water for a few days (aging) or aerating it over night before addition. In urgent cases, dechlorination is done with commercially available chemicals.

Total Hardness

Hardness is the total soluble Ca and Mg salts (in some cases Fe salts). It includes sulphates and chlorides along with carbonate, bicarbonate and hydroxide salts. Designation of water as soft to hard is based on the concentration of salts present in the water (Table 6.1). Water with less than 12 ppm/L require liming. Hard water also influences feed intake and growth of cichlids such as angel, black zebra, blue morph, auratus and fire-mouth.

Table 6.1. Water hardness range, concentration of salts in water and designation of water depending on salt concentration

Hardness range	Water hardness (as $CaCO_3$) per Litre	Designation of water
0-4dH	0-70 ppm	Very soft
4-8dH	70-140 ppm	Soft
8-12dH	140-210ppm	Medium hard
12-18dH	210-320 ppm	Fairly hard
18-30dH	320-530 ppm	Hard
	50-200 ppm	Desirable hardness
	100-300 ppm	Optimal for growth of majority of ornamental fishes.

Free Carbon dioxide

Carbon dioxide in water regulates photosynthesis and is essential. Free CO_2 should be < 10 mg/L. A concentration of more than 15 ppm is detrimental to ornamental fishes.

Dissolved Oxygen (O_2)

Oxygen in any aquatic body is derived from atmosphere and through photosynthesis of aquatic plants. For the survival and growth of fishes, dissolved oxygen plays a vital role. Ideal range for dissolved O_2 is 6-10 mg/L. The solubility of oxygen in water depends on its temperature and also on the rate at which it is kept in contact with water. The oxygen level of tanks can be increased by constant aeration, circulation of water, sprinkling of water, surface agitation etc.

Temperature

Water temperature is one of the most important factors which influences the ornamental fish keeping. Temperature controls the health, rearing, breeding, production and transport of aquarium fishes.

Metabolic activity is doubled with every 10°C rise in temperature. Majority of the aquarium fishes tolerate water temperature between 21 and 30°C, though 28°C have been found to be most suitable for the breeding purpose.

Ammonia

Ionized ammonia, NH_3 (Free ammonia) and ionized ammonia, NH_4^+ comprises the total ammonia concentration. Together, these two forms of ammonia are called TAN which means total ammonia nitrogen. Unionized ammonia fraction is more toxic to fish and the amount of total ammonia in this form depends on the pH and temperature of the water.

- Range of ammonia and effects on fish:
  - 0.02-0.05 mg/litre—safe concentration for many tropical fish species.
  - 0.05-0.4 mg/litre—sub-lethal effects depending on the species.
  - 0.4-2.5 mg/litre—lethal to many fish species.
- Detection of ammonia:

  Various test kits (Figure 6.2) are available in the market for easy and quick detection of concentration of ammonia in the water of aquarium fish culture tanks.
- Measures to maintain safe ammonia concentration:
  - Aeration can reduce toxicity by increasing dissolved oxygen concentration and by decreasing pH.
  - Healthy phytoplankton population remove ammonia from water.
  - Biological filters may be used for converting ammonia to nitrite, and then to harmless nitrate through nitrification process.

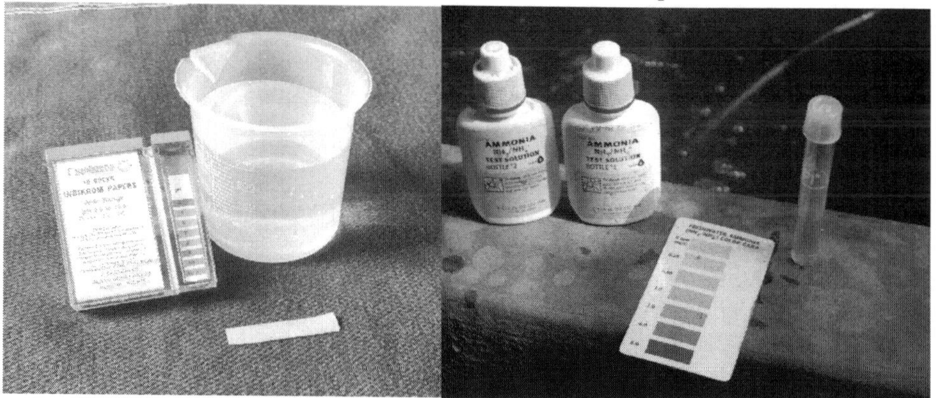

Figure 6.1 Measuring pH with pH papers.

Figure 6.2 Test kit for detecting ammonia.

Sound and vibration

Many fishes are actually sensitive to sound and other mechanical disturbances of the water (Datta, 2012).

Salinity

Salinity is the relative concentration of dissolved salts, usually sodium chloride, in given water. It is expressed as ppt or g/L and can be calculated from the value of chlorinity. Water can be classified on the basis of salinity as freshwater 0-0.5 ppt; brackish water 0.5-30 ppt; marine water > 30 ppt (Datta, 2012).

Salinity is normally related to salt water and brackish water. Hydrometer is used to measure it. Due to coastal pollution, these days, water for the marine and brackish water aquarium is normally created, rather than collected, using tap water and special marine salts, containing not only sodium chloride but also desirable trace elements. Though instructions are given, the salinity of the mix should be checked. Domestic salts should not be used as these may contain additives which are toxic to fishes (Bailey and Sandford, 2006).

6.1.1 Water quality parameters and their ranges

Water quality parameters and their optimum ranges (Table 6.2) are to be maintained in ornamental fish keeping facilities for sustainable farming.

Table 6.2. Water quality parameters, their optimum ranges for ornamental fish keeping

Physico-chemical parameters of aquarium water	Optimum ranges suitable for ornamental fishes
Temperature	17-38°C
pH	7.0-8.5
CO_2	<5 ppm
Alkalinity	75-120 ppm as CaCO3
Hardness	60-100 ppm as CaCO3
Dissolved Oxygen	6.0-8.0 ppm
Free ammonia	< 0.05 ppm
Ionised ammonia	< 0.1-0.4 ppm

Nitrogen Cycle

In nature as well as in any confined aquatic set-up, detritus (excreta from fish, plant leaves, food particles etc.) is broken down by bacteria during a series of processes known as the **Nitrogen cycle**. This biogeochemical cycle in general consists of steps such as nitrogen fixation, nitrogen assimilation, ammonification, nitrification and denitrification. In Nitrogen fixation nitrogen gas is converted into inorganic nitrogen compounds. In Nitrogen assimilation, these are then assimilated into specific tissue compounds of algae and higher plants. Organisms in aquarium excrete nitrogenous waste and decomposition of organic matter takes place, and their nitrogen is converted to ammonium ions (NH_4^+) by the action of bacteria. This process is known as **ammonification**. Next stage is **nitrification**, in which ammonia (NH_3) and ammonium (NH_4^+) are oxidized to nitrites by **Nitrifier** bacteria of the genus *Nitrosomonas* and others. In the second stage of **nitrification**, nitrite is oxidized to nitrate by bacteria *Nitrospira* and *Nitrobacteria*. Nitrates are relatively harmless, though sudden or long -term exposure to high concentrations can be harmful. In Denitrification nitrates are metabolized by denitrifying bacteria and form free atmospheric nitrogen, completing nitrogen cycle. Aquatic plants also eliminate nitrogen waste by metabolizing ammonia and nitrate. **Nitrogen cycle** normally operates in any mature aquarium (Figure 6.3). But due to high fish population density, the system may get unbalanced. Action is therefore necessary to redress the balance and avoid such problems.

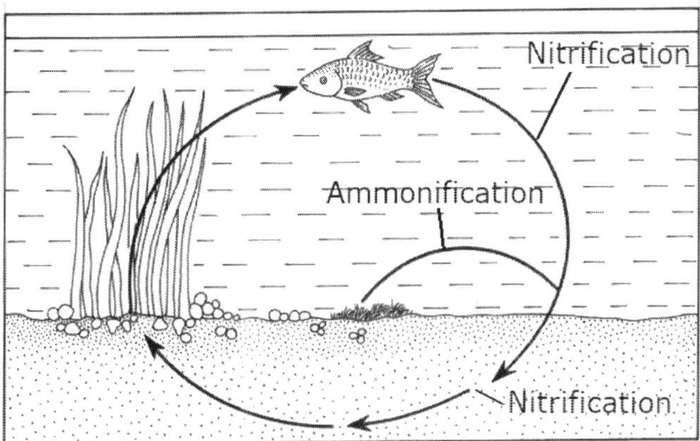

Figure 6.3 Nitrogen cycle in an aquarium.

6.2 Aquarium Care

General maintenance

General maintenance of the aquarium is largely a programme of continuous application of the rules learned, while setting up the aquarium. So, ensuring continued optimal living conditions for the tank inhabitants. The aquarist should know what is needed, how often is needed. The following routine may be adopted for the average general community (Bailey and Sandford, 2006).

- Daily—Temperature, health of fishes and proper working of equipment to be checked.
- Weekly—A 25% water change, siphoning any accumulated debris from substrate surface. Removal of dead vegetation, checking of nitrite level.
- Monthly—Cleaning of filter(s); If necessary "hover" gravel. Checking of nitrate level. The chemistry (physico-chemical parameters) of "new" water should match the tank conditions. Feeding should be reduced for a couple of days before and after monthly maintenance routine.
- Diagnosis of disease of the aquarium inmates and their treatment.
- Euthanasia—A fish is to be removed from the tank due to old age or illness and destroyed.
- Equipment Failure—Tank equipment such as heater, air-pump, filter may breakdown or tank breakage may happen at any odd hour, so spares of these items should be kept

- Power cuts—The electric company may be contacted if necessary. A battery- operated air pump may be used.
- During absence—Properly maintained set-ups need no special preparation. Most fishes survive a fortnight without food. If feeding is necessary, a container of feed should not be left, instead individually wrapped daily rations should be provided.

6.3 References

Adhikari S. *Soil and Water Quality Management in Aquaculture.* In Handbook of Fisheries and Aquaculture. ICAR. New Delhi. 2011, pp. 591-622.

Bailey M. and Sandford G. *The ultimate encyclopedia of aquarium fish & fish care.* Hermes House. London. 2006, pp. 1-256.

Datta S. *Management of Water Quality in Intensive Aquaculture.* 2012.In www.researchgate.net/publication/259175404_ Management_ of_ Water_ Quality_ in_ Intensive Aquaculture. Accessed on 26 August 2020.

7

Aquarium Plants

7.1 Introduction

Plants are kept in the aquarium to give it a natural look and to enhance beautification (Biju Kumar and Alappat, 1995). Plants of aquarium are adapted to live in aquatic environments (freshwater, brackish or marine water). These **hydrophyte**s or **macrophyte**s as they are often called, are common components of wetlands (Das, 2012). They can only grow in water or in moist soil that is permanently saturated with water (Naskar, 1990). **Aquarium plant**s are generally classified on the basis of their position in the aquatic systems. They are also categorized as **indigenous/ native** or **exotic** depending on their origin and distribution. Most of the **aquatic plant**s which are suitable for the aquariums are sold in the market.

The principal factor controlling the distribution of aquatic plants is the depth and duration of flooding. Nutrients, disturbance from waves, grazing and salinity are other factors which control their distribution, abundance and growth form.

7.2 Classification and Types of Aquatic Plants

Aquatic plants are classified according to ecological position. These plants are also categorised into types depending on their position in water.

7.2.1 Classification according to ecological position

- **Amphiphyte**s: Plants that are adapted to live either submerged or on land.
- **Helophyte**s: Plants rooted in the bottom, but with leaves above the water line.
- **Elodeid**s: Stem plants that complete their entire lifecycle submerged, or with only their flowers above water-line.
- **Isoetid**s: Rosette plants that complete their entire lifecycle submerged.
- **Nymphaeid**s: Plants rooted in the bottom but with leaves floating on the water surface.
- **Pleuston**: Vascular plants that float freely in the water.

7.2.2 Types of aquatic plants

- **Floating plant**s: Plants have their foliage (leaves) above the surface of water with roots hanging free. Examples, *Eichornia, Pistia, Spirodela, Lemna, Azolla, Wolffia, Salvinia, Riccia* sp.
- **Emergent plant**s: Plants having their leaves and flowers above the

water surface but rooted in the bottom. Examples, *Nymphaea, Nelumbo, Nymphoides.*

- **Submerged plants**: Leaves are submerged in the water, may or may not be rooted. Examples, *Hydrilla, Ceratophyllum, Myriophylum, Ottelia, Bacopa, Ludwigia, Saggitaria, Aponogeton, Cabomba, Limnophila, Ceratopteris.*

  Rooted plants: *Hydrilla, Najas, Ottelia, Vallisneria, Potomogeton, Lagarosiphon, Chara, Nitella.*

  Devoid of roots: *Ceratophyllum, Utricularia*

- **Marginal plants**: They are growing in the shallow areas of the water body (near shore). Examples, *Typha, Cyperus, Ipomea, Ileocharis, Cryptocoryne, Echinodorus.*

7.3 Importance of Aquarium Plants

Aquarium plants are an essential part of the aquarium fish keeping (Alderton, 2003; Ostrow,1995; Ward, 2004). Some of the benefits that they provide are mentioned here:

- Aquarium plants give natural look to aquarium.
- They provide food for fishes. Herbivorous fishes directly eat the plants. Other fishes eat the organisms attached to the leaves.
- They provide shelter, shade and hiding places for smaller fishes.
- These plants provide spawning sites for certain fishes.
- Plants and fishes have symbiotic relationship. The fish waste is used as an excellent fertilizer for plants. Plants in turn remove nitrate and CO_2 from water.
- Plants are best filters and thus help to maintain water quality.
- Some of the plants are believed to reduce pH of water which is suitable for breeding of certain fishes like tetra, angel etc.
- Aquarium plants are one of the most important items used in decorating all types aquatic set-ups.

7.4 List of Freshwater Aquarium Plant Species

7.4.1 Freshwater aquarium plants:

Most of the **freshwater plant** species are utilised by aquarists all over the world. Here, the genera of those aquarium plant species are listed (Table 7.1).

Table. 7.1: List of freshwater aquarium plant species

Acorus spp.	*Fittonia*	*Najas* spp.
Alisma spp.	*Fontinalis* sp.	*Nesaea* sp.
Alternanthera spp.	*Glossadelphus* sp.	*Nitella* spp.
Ammania spp.	*Glossostigma* spp.	*Nomaphila* spp.
Anubias spp.	*Gymnocoronis* sp.	*Nuphar* spp.

Acorus spp.
Aponogeton spp.
Armoracia sp.
Azolla spp.
Bacopa spp.
Baldellia sp.
Barclaya spp.
Blyxa spp.
Bolbitis spp.
Bucephalandra spp.
Cabomba spp.
Caldesia sp.
Calla sp.
Callitriche spp.
Cardamine spp.
Ceratophyllum spp.
Ceratopteris spp.
Cladophora sp.
Crassula spp.
Crinum spp.
Cryptocoryne spp.
Cyperus spp.
Damasonium sp.
Didiplis sp.
Echinodorus spp.
Egeria spp.
Eichhornia spp.
Elatine spp.
Eleocharis spp.
Elodea spp.
Equisetum spp.
Eriocaulon spp.
Eusteralis sp.
Fittonia
Hemianthus spp.
Heteranthera spp.
Hippuris sp.
Hottonia spp.
Hydrilla sp.
Hydrocharis sp.
Hydrocleis sp.
Hydrocotyle spp.
Hydrothrix sp.
Hydrotriche sp.
Hygrophila spp.
Hygroryza sp.
Iris spp.
Isoetes spp.
Isolepis sp.
Juncus sp.
Lagarosiphon spp.
Lagenandra spp.
Lemna spp.
Lilaeopsis spp.
Limnobium
Limnophila spp.
Lindernia spp.
Littorella sp.
Lobelia spp.
Lomariopsis sp.
Ludwigia spp.
Luronium sp.
Lycopodiella sp.
Lysimachia sp.
Marsilea spp.
Mayaca spp.
Micranthemum sp.
Microsorum sp.
Monosolenium sp.
Myriophyllum spp.
Najas spp.
Nymphaea spp.
Nymphoides spp.
Orontium sp.
Ottelia spp.
Phyllanthus sp.
Pilularia spp.
Pistia sp.
Pogostemon spp.
Pontederia sp.
Potamogeton spp.
Proserpinaca sp.
Ranunculus spp.
Regnellidium sp.
Riccia sp.
Ricciocarpos sp.
Rorippa sp.
Rotala spp.
Ruellia sp.
Ruppia sp.
Sagittaria spp.
Salvinia spp.
Samolus sp.
Saururus sp.
Selaginella sp.
Shinnersia sp.
Spiranthes sp.
Spirodela sp.
Staurogyne sp.
Stratiotes sp.
Subularia sp.
Synnema sp.
Taxiphyllum sp.
Tonina sp.
Trapa sp.
Triglochin spp.
Typha spp.

Acorus spp.	*Fittonia*	*Najas* spp.
		Utricularia spp.
		Vallisneria spp.
		Vesicularia sp.
		Wolffia spp.
		Wolffiella sp.
		Zannichellia sp.

7.4.2 Pseudo-aquarium plants

Several species of beautiful terrestrial plants are sold in the market as "aquarium plants". These plants survive for some time under water, and eventually die. They should be removed immediately as their decayed parts may contaminate the aquarium water. Some of these false or **pseudo-aquarium plant**s are follows:

Aglaonema modestum; Aglaonema simplex; Chlorophytum bichetii; Dracaena sauderiana; Hemigraphis colorata; Ophiopogon japonicus; Pilea cadierei; Sciadopitys verticillata; Spathiphyllum tasson; Syngonium podophyllum; Trichomanes javanicum.

7.5 Aspects of some Popular Freshwater Plants chosen as Ornamentals

Some freshwater aquatic plants are popular with aquarists for a long time. These are as follows:

7.5.1 *Acorus* L.

- Common name(s): calamus, sweet flag.
- Family: Acoraceae.
- Origin/ Distribution: The genus is native to North America and northern and eastern Asia, and naturalised in southern Asia and Europe from ancient cultivation.
- Habitat: Wetlands, marshes, shorelines and floodplains.
- Description: *Acorus* is a genus of monocot flowering plants and perennial herbs, 30 to 100 cm tall. Tufts of basal leaves are present arising from a spreading rhizome. The leaves are linear with entire margin, erect, yellowish-green, with pink sheathing at their bases. These are sword shaped, flat and narrow, tapering into a long, acute point, having parallel veins. A semi-erect spadix, a thickened fleshy axis, 4-10 cm long, emerges from one side of the flower stem. The spadix is densely crowded with tiny greenish yellow flowers. The bract can be ten times longer than the spadix.
- Propagation (natural): They spread by means of thick rhizomes.
- Species: Aquarium species are--

  Acorus calamus L., 1753 (sweet flag) Figure 7.1

  —Native to India, Central Asia, southern Russia, Siberia and eastern Europe. Widely cultivated and naturalised elsewhere. Varieties are—

Acorus calamus var. *americanus; Acorus calamus* var. *angustatus; Acorus calamus* var. *calamus*.

A. canaliculatum—Japan, Korea, the Ryukyu Islands, Taiwan, the Kuril Islands and China. *A. gramineus*.

A. lanceolatum—Europe, North Africa and Temperate Asia.

A. nanum—Xinjiang in western China.

A. orientale—Asia.

A. plantago-aquatica—Most of Europe and Asia.

A. subcordatum—Most of eastern and central United States and Canada.

A. triviale—Canada, the United States and Northern Mexico.

A. wahlenbergii—the northern Baltic Sea, Finland, Sweden, Estonia, Lativia, Lithuania, Belarus and north-western Russia.

7.5.2 *Alisma* L.

- Common name: Water-plantains
- Family: Alismataceae/ Alismaceae
- Origin/Distribution: Worldwide
- Habitat: The sides of ponds, beels, river marshy lands, water logged soils, shallow or deep water, flooded farmland.
- Description: It is a perennial herb. Normal scented leaf, with air-pocket, roots in soil, entire body may remain within water. These aquatic plants are with leaves either floating or submerged, found in a variety of still water habitats, around the world. The flowers are hermaphrodite and have single whorl. These emergent plants have broad leaves that can be either tapered or rounded at the base. When submerged, the plant produces ribbon-like leaves. They produce whorls of perfect flowers either white or pinkish.
- Propagation: By dividing root stock in the months of May, June.
- Species:

  Alisma gramineum—Europe, Asia, North Africa, minor naturalization in North America.

  A. calaliculantum—eastern Asia.

  A. juzepezukii—Russia.

  A. lanceolatum—Europe, Asia, North Africa.

  A. nanum—Xinjiang.

  Alisma plantago-aquatica L., 1753 Figure 7.2

  Europe, Asia, Africa.

  A. rhinocarpum—Western Europe.

  A. subcordatum—eastern North America.

A. triviale—North America.

A. wahlenbergii—Baltic Sea region.

7.5.3 *Alternanthera* Forssk.

- Common names: Joyweeds, Joseph's coat.
- Family: Amaranthaceae.
- Origin/Distribution: Tropical and sub-tropical open forests in Central and South America, Asia, Africa and Australia.
- Habitat: Found both in aquatic and terrestrial habitats.
- Description: These are annual or perennial herbs or sub-shrubs with attractive colourful leaves. They are found in different forms, from prostrate to erect to floating. The brightly coloured leaves are arranged oppositely. Flowers are insignificant. The inflorescence is a spike or a rounded head occurring in the leaf axils or the ends of branches.
- Species: About 200 known species. *Alternanthera* species kept in aquarium are—*Alternanthera bettzichiana, A.lilacina, A.philoxeroides, A.reineckii, A. rosaefolia, A.sessilis.*
- Propagation: In late winter, propagation is done indoors from seeds. Then they are transplanted outdoors. In late summer, step tip cuttings may be done.
- Remarks: Though some species of *Alternanthera* are popular aquarium plants, some are eaten as vegetables and others are treated as noxious weeds.

7.5.4 *Ammania* L. Figure 7.3

- Common name: Red stems.
- Family: Lythraceae.
- Origin/Distribution: America, Africa, Asia, Australia and Europe.
- Habitat: Moist areas, damp open soil, or sometimes in submerged conditions.
- Description: It is an annual, graceful stem plant. This small shrub has glabrous, colourful stems. Leaves are opposite, sessile, generally with basal ear-like lobes. Inflorescence is a cluster, axillary, flowers 3-10. Flowers are radial, hypanthium bell-to urn-shaped. Seeds are produced in large numbers. Many species are found to grow above the water surface.
- Species: About 30 known species. Aquarium species are, *Ammania gracilis, A.latifolia, A.senegalensis*.
- Propagation: New growths or side shoots are cut from main stem, then these are planted into substrate.
- Remarks: It is a popular aquarium plant.

7.5.5 *Anubias* Schott. Figure 7.4

- Common name: anubias.

- Family: Araceae
- Origin/Distribution: Native to tropical, central and western Africa.
- Habitat: They grow in rivers and streams, but are also found in the marshes.
- Description: This is the genus of slow-growing aquatic and semi-aquatic flowering plants. These plants have broad, thick, leather-like dark leaves which are in many different forms. Species under the genus are distinguished on the basis of the characteristics of the inflorescence present. These plants are hardy with thick rhizome and strong root structure.
- Propagation: Propagation in aquariums are accomplished by the division of stolon or from side shoots. Propagation is also possible by seeds. These plants can be grown above water and so are used in paludariums. Propagation in Anubias is carried out by cutting the rhizome also. After taking out the plant from water, the rhizome is to be cut into two or more pieces, making sure that at least three or four healthy leaves come out on each new section of rhizome. The roots should not be damaged in the process. The new pieces are then to be planted back in the substrate keeping in mind not to bury the rhizome.
- Species: Aquarium species are—

  Anubias afzelii—West Africa, Senegal, Guinea, Sierra Leone and Mali.

  A.barteri (5 varieties)—South-eastern Nigeria, Cameroon, and on Bioko.

  A. gigantean—West Africa.

  A. gilletti—

  A. gracilis—West Africa.

  A. hastifolia—Ghana, Nigeria, Cameroon, Gabon and Democratic Republic of the Congo.

  A. heterophylla—Cameroon, Equatorial Guinea, Gabon, Republic of the Congo, Democratic Republic of the Congo and Angola.

  A. pynaertii—Gabon, Republic of the Congo and Zaire.
- Remarks: In the market, varieties of *Anubias bartari* are available, and they are most commonly used in the aquarium. These are generally attached to rocks or bogwood. *Anubias* spp. prefer subdued lighting and can also produce flowers under water. In aquariums, these plants are placed in the shaded places to avoid growth of algae on the leaves. As the light and nutrient requirements of the *Anubias* spp. are very low, they are considered as the easiest plant to be grown in aquariums. These plants are not edible to herbivorous fishes also.

7.5.6 *Aponogeton* L. f. Figure 7.5

- Common name: Cape Pondweed.
- Family: Aponogetonaceae

- Origin/Distribution: South Africa, tropical to warm temperate regions of Africa, Asia and Australia.
- Habitat: Most of the species grow in temporarily still or flowing waters and live through the dry period as a dormant tuber.
- Description: Aquatic and flowering plants. An *Aponogeton* from Asia will have a single bloom stalk, while those from Africa, including Madagascar, will have multiple blooming stems on the same flower stalk.
- Species: Fifty- six known species. Among them popular aquarium species are, *Aponogeton crispus, A. fenestralis , A. ulvaceus, A. elongatus ,A. rigidifolius, A. robinsonii, A. undulate.*

7.5.6.1 *Aponogeton crispus* Thunberg (1781)

- Common name: ruffled/crinkled or wavy-edged Aponogeton.
- Origin/Distribution: Native to southern India and Sri Lanka
- Habitat: Seasonal ponds, in still and running waters. Become dormant during the dry season.
- Description: This plant is a seasonally submerged aquatic one. The leaves are light green to olive green-brown in colour, 8-14 inches long and 2.5 inches broad with a wavy margin. Flowers are produced on an erect stem. This plant does not have leaves that float on the surface of water.
- Propagation: Propagation is done either through seed or through splitting the rhizome. The seeds are with two prolongations. The horizontal one gets curved and stuck into marshy ground, forming the initial roots. Seeds may be obtained by pollinating flowers with soft brush. The seeds are then sown in a propagator at normal room temperature. They take several weeks to germinate. After leaves and roots come out, they can be potted in a peat-based compost and covered with water.

7.5.6.2 *Aponogeton madagascariensis* (Mirb.) H. Bruggen

- Common name: Madagascar lace leaf, lattice leaf or lace plant.
- Origin/Distribution: Madagascar.
- Habitat: Cool, permanent rivers in Madagascar.
- Description: It is an aquatic perennial plant with tuberous rhizomes. The leaves are oblong and spread horizontally beneath the surface of water. There are little more than a lattice-like skeletal network of veins with no tissue between.
- Propagation: Bulbs grow well in the first year but often fail in the second year. Seeds may be produced but often fail to germinate or thrive.

7.5.6.3 *Aponogeton ulvaceus* Baker

- Origin/Distribution: Throughout central and northern Madagascar.
- Habitat: It is a permanently submerged plant and grows both in still and moving waters. It grows in sunny and partially shaded environments.

- Description: This plant is a submerged aquatic one, with slightly hairy rhizome. The leaf blades have a base that tapers gradually. Leaf blades are pale green in colour, with a wavy margin on petioles of equal length. No floating leaves are found.
- Propagation: Seeds are collected. Sands or very fine gravel compost can be used for the germination of seeds. Seeds are germinated in about one or two weeks.

7.5.7 *Azolla* Lam.

- Common names: mosquito fern, duckweed fern, fairy moss, water fern.
- Family: Salviniaceae.
- Origin/Distribution: It is native to much of Africa, Asia (Brunei, Darussalam, China, India, Japan, Korea and the Philippines) and parts of Australia.
- Habitat: Grows on water surface of deep or shallow water.
- Description: It is an annual, aquatic, small free-floating fern. The plant remains above water and small weak modified roots stay in water. Leaves of this plant overlap giving a quilted look to the surface and this hide the weak stem. A single root protrudes from each stem. *Azolla* vary in colour from green to red. The triangular fond of 2.5 cm floats on the surface. The fond is made up of many rounded or angular overlapping leaves each 1 or 2 mm long. The colour is green, blue-green or dark red. These are covered with tiny hair giving them a velvety appearance. The hair-like roots extend out into the water. A dense layer of these plants covers a water body.
- Propagation: This plant reproduces in vegetative manner when branches break off the main axis, or sexually when sporocarps on the leaves release spores. *Azolla* can be grown in ditches, ponds and wetlands. It prefers shade and requires light (30 to 50% light). Sufficient water is required for its cultivation. Optimum temperature for *Azolla* growth is 20°-35° C. For cultivation, clean, fertile soil is mixed with cow dung and water and spread across the pond. To cover the 6 feet x 4 feet pond, 1kg of fresh *Azolla* culture is required. The culture is applied uniformly in the pond. The water depth of at least 5 to 6 inches should be maintained in the pond. There should not be high content of salt in the pond water where *Azolla* cultivation is done. *Azolla* will be ready for harvesting in 2-3 weeks after stalling the culture in the pond.
- Species: Fourteen known species.

  Aquarium species are: *Azolla caroliniana*—native to the eastern and southern United States.

  Azolla filiculoides Lam. Figure 7.6

  —native to the warm temperate and tropical regions of Americas, Asia and Australia.

Azolla pinnata—native to much of Africa, Asia (Brunei, Darussalam, China, India, Japan, Korea and the Philippines and parts of Australia.

- Remarks: These plants form a symbiotic relationship with the cyanobacterium, *Anabaena azollae,* which fixes atmospheric nitrogen, giving the plant access to the essential nutrients. The limiting factor in its growth is phosphorus. This super-plant grows at a great speed and readily colonize areas of freshwater. *Azolla* spp. are used to control mosquito larvae and this plant removes heavy metals from effluent.

7.5.8 *Bacopa* Aubl.

- Common names: Water hyssop
- Family: Plantaginaceae.
- Origin/Distribution: In tropical and subtropical regions of the world, particularly the Americas.
- Habitat: Found in moist amphibious conditions.
- Description: These are annual or perennial plants, with decumbent or erect stems. The leaves are opposite or whorled and sessile. The leaf blade is regular, round to linear. Its stems are hairy or smooth. The flowers are present either as solitary or in pairs. Colour of the flowers are white, blue or purple.
- Propagation: By seeds and stem fragments.

  Propagation by seeds—Some varieties of *Bacopa* are grown from seed.

  Method—Fill a container with moist compost. Sprinkle the seeds onto the compost, do not cover them with soil. Mist the seeds with a water spray. Put the container into a clear plastic bag and place it in a bright area out of direct sunlight. The seeds germinate quickly, and new growth can be seen within ten days. Remove the soft cover after the seeds germinate. Once it is big enough to handle, you can transfer the seedlings to individual plots.

  Propagation by cutting

  Method—In late summer, cut about three inches of fresh healthy stem from a *Bacopa* plant using a clean, sharp blade. Dip the cut end in a rooting powder or hormone. Remove the lower leaves on the stem, then insert an inch of cut stem into a pot containing cutting mix. Water your cutting, then enclose it in a clear plastic bag. Keep the cutting out of direct sunlight but in a well-lit area. The cuttings root after about five weeks. You can then transfer the cuttings to individual pots.
- Species: About 70-100 species are known.

  Popular aquarium species are—*Bacopa australis, B.myriophylloides,*

  Bacopa amplexicaulus—native to Southeast United States.

  Bacopa monnieri- originates from tropical Asia.

  Bacopa carolona—native to marshy areas in the Southeast United States and in South Korea.

Bacopa crenata—indigenous to tropical Africa and Madagascar.

Bacopa rotundifolia—native to water bodies of the central United States.

- Remarks: Some of the species of the genus are ornamentals. *B. monnieri* has medicinal value.

7.5.8.1 *Bacopa monnieri* Pennell Figure 7.7

- Common names: water hyssop, brahmi, herb of grace, Indian pennywort.
- Origin/Distribution: Southern and eastern India, Australia, Europe, Africa, Asia and North and South America.
- Habitat: Marshy areas, wetlands. It can tolerate slightly brackish conditions.
- Description: It is a perennial, creeping non-aromatic herb. The leaves of this plant are oblong and succulent. Leaves are of 4-6 mm thickness and are oblanceolate, arranged oppositely on the stem. The flowers are small and white with four to five petals.
- Propagation: Cuttings are often used for propagation.
- Remarks: Besides being used as an aquarium plant, this plant is utilised in Ayurvedic medicines for treatment of epilepsy, asthma and many other diseases.

7.5.9 *Cabomba* Aubl.

- Common name: Fanwort.
- Family: Cabombaceae.
- Origin/Distribution: North and South America.
- Habitat: Water bodies.
- Description: This submerged plant has stems up to 150-200 cm long. The leaves are finely divided and arranged in pale green fans. Leaves grow opposite each other at regular intervals.
- Propagation: This plant is propagated through cuttings, which should be at least 2 inches in length. It needs neutral water, moderate lighting and rich substrate. For propagation, trim the branches from the main *Cabomba* plant, and relocate them in another part of the aquarium.
- Species: About 5 species are noted, among them 3 are popular aquarium plants.

  Popular aquarium species are—

  Cabomba aquatica

  Cabomba caroliniana—native to North and South America.

  Cabomba furcate—native to South America and Cuba and Florida.
- Remarks: It is very popular with aquarists as an ornamental and oxygenating plant for aquariums.

7.5.9.1 *Cabomba aquatica* Aubl. Figure 7.8

- Common names: fanwort, giant Cabomba.
- Origin/Distribution: South America (Brazil, Mexico).
- Habitat: Fresh standing water or in lakes and rivers with slight currents.
- Description: It is a perennial aquatic plant, with stems having horizontal basal portions as well as ascendant portions. These parts reach the surface during flower production. The stems generally grow up to 50 cm. The floating leaves of this plant are at the apical part of the stem. The first floating leaf of this plant always appear at the same node as the submerged apical leaf. The orbicular leaves are present and the plants have fine and usually red longitudinal striations.
- Propagation: By detachment of stem portions from the mother plant and have adventitious roots that are produced in the nodes.
- Remarks: It is a popular aquarium plant. This plant helps in maintaining aquatic ecosystem and also acts as an efficient accumulator of heavy metals in water bodies.

7.5.10 *Ceratophyllum* L. Figure 7.9

- Common names: coontails, hornwort
- Family: Ceratophyllaceae.
- Origin/Distribution: Cosmopolitan distribution. Native to all continents except Antarctica.
- Habitat: Commonly found in ponds, marshes and quiet streams in tropical and temperate regions.
- Description: It is a perennial, submerged free floating plant. This rootless plant floats on the surface and has long (1-3 m) stems bearing side branches. Along nodes of the stem, at intervals rings of bright green leaves are produced, which are narrow and much branched. Stems bear whorls of at most 10 leaves growing at a slightly upward angle. Each leaf with 2-4 leaflets, forked and spiky reminiscent of a pair of horns. The flowers are small and inconspicuous.
- Propagation: Propagation is very easy. Stem cuttings are taken from anywhere on the plants and these will develop a basic root structure. Cuttings can be placed on the substrate. Remove some of the lower leaves to reveal a small portion of bare stem. Carefully place this into the substrate. In early autumn, the ripe seeds may be sown in a greenhouse with the pot immersed in water. Grown seedlings may be placed in individual pots and grown in a tray of water in the greenhouse.
- Species: About 30 reported species. Two species are favourite aquarium plants.

  Ceratophyllum demersum (hornwort)—native to all continents except Antarctica.

Ceratophyllum submersum (soft hornwort)—Europe.

- Remarks: Appearance of these plants and their high oxygen production make them favourite aquarium plants. They also offer excellent protection to fish-spawn.

7.5.11 *Ceratopteris* Brongniart

- Common names: Water Sprite, Indian Water Fern.
- Family: Pteridaceae
- Origin/Distribution: A widespread genus inhabiting the humid tropics.
- Habitat: Exclusively aquatic.
- Description: This plant is erect, aquatic fern of moderate size. Rhizome short, fleshy, horizontal and ascending to erect, loosely rooted. Sterile fronds are pale green, thin, flaccid and spreading, fertile fronds are pale green to brown, firm and held erect.
- Propagation: Either through adventitious roots or splitting off plantlets at the leaf edge. Rarely through spores. For maximum production, young plants with three-plus fronds may be broken by hand from mature leaves.
- Species: About seven species are reported under the genus. Aquarium plant species are

  Ceratopteris cornuta

  Ceratopteris pteridoides

  Ceratopteris thalictroides—native to Pan-tropical.
- Remarks: Widely used as an aquarium plant. Fronds of this plant are used as vegetable.

7.5.12 *Cryptocoryne* Fisch. Ex Wydler

- Common names: crypts, water trumpet.
- Family: Araceae.
- Origin/Distribution: Tropical regions of Asia and New Guinea.
- Habitats: Streams and rivers with not too rapidly flowing water, in the lowland forest. In seasonally inundated forest pools or on river banks submerged only at high water.
- Description: Bog or submerged plants with elongated, tough leaves. These are mid-green, deep olive coloured and red or purple-tinged undersides of the leaves. Depending on the species, they may be 7cm to 60 cm. Inflorescence is a spadix enclosed by a spathe, resembling a trumpet.
- Propagation: Remove a small portion of the root with some stem attached. This can then be planted in the substrate and will slowly develop into a new plant. Another method is to remove mineral wool from roots. Separate the full mature plant into several planting portions. Plant them into the aquarium substrate.

- Species: There are 64 known species under the genus. Popular aquarium species are listed in the Table. 7.2.

Table. 7.2: Popular ornamental species under the genus *Cryptocoryne*

Plant species	Native to	Plant species	Native to
C. affinis	S.W Malay Peninsula	*C. lutea*	
C. alba		*C. minima*	Malaysia, Sumatra
C. albida	Lao People's Democratic Republic, Myanmar, and Thailand.	*C. moehlmannii*	Sumatra
C. aponogetifolia	Philippines, on the islands Negros and Panay and the south-east of Luzon.	*C. nevillii*	Sri Lanka
C. axelrodii		*C. nurii*	Peninsular Malaysia
C. balansae		*C. pallidinervia*	Borneo
Cryptocoryne beckettii Thwaites ex Trimen (1885) Figure 7.10 A	Sri Lanka	*C. parva*	Sri Lanka
C. blassii		*C. petchii*	
C.bogneri		*C. potenderiifolia*	Sumatra
C. bullosa		*C. purpurea*	Peninsular Malaysia, and Borneo
Cryptocoryne ciliata (Roxb.) Schott Figure 7.10 B	India, Bangladesh, Indo China, Malaysia, Indonesia, and New Guinea.	*C. retrospiralis*	India, Bangladesh and Myanmar
Cryptocoryne cordata Griff Figure 7.10 C	Thailand, Malaysia, Borneo and Java.	*C. scurrilis*	Sumatra
C. crispatula	Southern China, northeastern India, and Bangladesh.	*C. siamensis*	
C. cruddasiana	Myanmar	*C. spiralis*	India, Bangladesh
C. dewitii	Papua New Guinea	*C. striolata*	Borneo
C.didrici		*C. thwaitesii*	Sri Lanka
C. elliptica	Malaysia	*C. tonkinensis*	
C. ferruginea	Sarawak	*C. undulata*	Sri Lanka
C.fusca	Borneo	*C. usteriana*	Philippines

Plant species	Native to	Plant species	Native to
C. gaserii		*C. venemae*	
C. grabowskii		*C. versteegii*	New Guinea
C. gracilis		*C. zewaldiae*	
C. griffithii	Kalimantan, and Peninsular Malaysia	*C. zonata*	
C. hudoroi	Kalimantan	*C. zukalii*	Peninsular Malaysia
C. keei	Sarawak		
C. legroi			
C. lingua			
C. longicauda	Borneo, Malaysia, and Sumatra		
C. lucens			

- Remarks: A large number of species are commercially cultivated aquarium plants.

7.5.12.1 *Cryptocoryne ciliata* (Roxb.) Schott Figure 7.10 B

- Common name: Water Trumpet.
- Origin/Distribution: From India to New Guinea.
- Habitat: Streams and rivers, in lowland forest. In seasonally inundated forest pools or on river banks submerged only at high water.
- Description: It is a strong bog plant which grows up to a height of up to 90 cm. Leaves are lanceolate, medium, green, fleshy yet stiff, long-stalked. This plant has a spathe (spathaceous bract) with long fringes or cilia along the margin. Two varieties are known to be cultivated. These are-*C. ciliata* var. *ciliata* and *C. ciliata* var. *latifolia*.
- Remarks: It is cultivated and used in the aquarium trade. Sometimes used in traditional medicine.

7.5.13 *Echinodorus* Richard ex Engelmann Figure 7.11

- Common names: bur head, Amazon sword plant.
- Family: Alismataceae.
- Origin/Distribution: North and South America.
- Habitat: In water bodies.
- Description: Most species under this genus are annual or perennial and are robust and large. These plants may grow immersed, with floating leaves or seasonally submersed leaves glabrous to stellate-puberscent. Stolons, corms, tubers absent. Rhizomes present or absent. Leaves sessile or petiolate, petioles triangular. Inflorescence racemes or panicles. Flowers bisexual.
- Propagation: This is by division or by adventitious new plants developing

on submerged flowering stems. Prying of adventitious plants are done from the rosette of leaves and re-planting them in the substrate. The seeds can also be grown in damp sand in warm, damp conditions.

- Species:

Popular aquarium species are listed in the Table. 7.3.

Table.7.3: Some of the popular species of the genus *Echinodorus*

Plant species	Native to	Plant species	Native to
Echinodorus africanus	South America	*E. parviflotus*	Cuba, Central America and South America
E. amazonicus (Amazon sword)	Cuba, Central America and South America	*E. pelliscidus*	
E. berteroi	Southern and central United States, South America and West Indies	*E. quadricostatua*	Southern Mexico, Central America, the West Indies and South America
E. bolivianum	Southern Mexico, Central America, the West Indies and South America	*E. radicans*	
E. brevipedicellatus		*E. rigidifolius*	
E. cordifolius	Mexico, the West Indies, Central America, South America and the southeastern United States	*E. 'Rubin'*	
E. fluitans		*E. rubra*	
E. grandiflorus	Brazil, Paraguay, Uruguay, Argentina, Venezuela and Florida	*E. schlueteri*	
E. grisebachii		*E. subalatus*	
E. horemanii	South America	*E. tenellus*	Eastern United States, southern Mexico, the West Indies, Central America, South America
E. horizontalis	Northern South America	*E. tunicatus*	Panama, Brazil, Costa Rica, Colombia, Ecuador, and Peru
E. humilis		*E. uruguayensis*	South America

Plant species	Native to	Plant species	Native to
E. latifolius	South America		
E. macrophyllus	Brazil and Bolivia		
E. martii	Cuba, Mexico, Central America, Guayna, Venezuela, Bolivia, Brazil and Paraguay		
E. major			
E. opacus	South America		
E. Osiris	South America		
E. 'Ozelot'			
E. palifolius	Brazil, Peru, Uruguay and Mexico		
E. paniculatus			

- Remarks: Due to their attractive forms and general hardiness, Amazon sword plants are one of the most popular aquarium plants.

7.5.13.1 *Echinodorus grisebachii* Britton & al.

- Common name: Amazon sword plant.
- Origin/Distribution: Cuba, Central America and South America.
- Habitat: Freshwaters
- Description: This aquatic plant has submergent leaves which are 40-60 cm long attached by petioles, flatly triangular. Leaf blades are lanceolate or narrowly oval. It develops large, conspicuous rosette with large number of leaves.
- Remarks: This plant is used in aquarium and in ponds. Fishes like angelfish, use the sword plant leaves as spawning ground.

7.5.14 *Eichhornia* L.

Eichhornia is considered now as subgenus to ***Pontederia*** L.

- Common names: water hyacinth.
- Family: Pontederiaceae.
- Origin/Distribution: Tropical and sub-tropical South America.
- Habitat: Ponds, rivers, marshy lands, shallow or deep water.
- Description: These perennial aquatic plants have erect stalk, each of which supports a single spike of up to 30 attractive flowers.
- Propagation: Seed setting plant. Propagate by runners or stolon.
- Species: Two aquarium species-

  Eichhornia crassipes

  E. diversifolia

- Remarks: Used in aquariums and ponds. Also used as fodder, organic manure, fuel and mulch. It is a significant invasive plant.

  An aquarium species of special mention-

7.5.14.1 *Eichhornia crassipes* (Mart.) Solms Figure 7.12

- Common name: common water hyacinth.
- Origin/Distribution: Amazon basin.
- Habitat: Ponds, rivers, marshy lands, water-logged soil, shallow or deep waters.
- Description: This plant is a free- floating perennial aquatic plant. Broad, thick, glossy, ovate leaves, 10-20 cm across on a stem. These are floating by means of buoyant bulb like nodules at its base above the water surface. Long, spongy and bulbous stalks are present and feathery, purple-black roots are freely hanging. A single spike has 8-15 attractive flowers lavender to pink in colour with six petals is present supported by an erect stalk.
- Propagation: It is by way of runners or stolon, which later form daughter plants. Each plant can also produce thousands of seeds each year.
- Remarks: Besides being used as an ornamental flower, it is also utilised as a source biogas, fertilizer, in phytoremediation, waste water treatment. It is also considered as an invasive species in some countries.

7.5.15 *Eleocharis* R. Br. Figure 7.13

- Common names: spikerushes, spikesedges.
- Family: Cyperaceae.
- Origin/Distribution: Amazon Rainforest, South American Andes, northern Australia, eastern North America, California, Southern Africa and subtropical Asia.
- Habitat: Aquatic or mesic.
- Description: These flowering plants are with photosynthetic stems but without green leaves. The leaves are reduced to sheaths surrounding the base of the stems. The flowers are borne on unbranched terminal spikelet at the apices of stem.
- Species: About 250 or more known species.
- Remarks: Some species are traded as aquarium species.

7.5.15.1 *Eleocharis acicularis* (L.) Roem& Schult

- Common names: needle spikerush, least spikerush.
- Origin/Distribution: Europe, central and south eastern Asia, North America and north eastern South America as far as Ecuador.
- Habitat: Marshes, vernal pools and bogs.
- Description: This is an annual or perennial plant with long, grass-like stems from a creeping rhizome. Stems are about 15 cm or shorter in height.

It forms short spikes of tiny flowers amongst flat overlapping bracts. The flowers are borne at the tip of each stem in single, sharply pointed, lanceoloid spikelets.

- Remarks: It is a favourite aquascaping plant used in artificial aquatic set-ups. It needs plenty of light and high concentration of carbon dioxide. There are two varieties available of this species. These are-*E. acicularis* var. *acicularis* and *E. acicularis* var. *porcala*.

7.5.16 *Hygrophila* R.Br.

- Common name: swamp weed.
- Family: Acanthaceae.
- Description: Origin/Distribution: Across the tropical and subtropical world.
- Habitat: Freshwater bodies.
- Description: These plants generally have long, pale green pointed leaves and tough stems. The leaves are either all of one form (homomorphic) or with different forms on one plant (heteromorphic). The flowers have white to purple corollas.
- Propagation: By cutting—take a cutting from the mother plant and remove the bottom leaves. This can then be placed into the substrate taking care not to crush the stem. In a short time, the cutting will produce roots and a new plant is formed. Ensure that the cuttings receive plenty of nutrients in the water to promote quicker root growth.
- Species: There are about 80-100 species of which many are aquatic plants. Some ornamental *Hygrophila* are listed (Table 7.4).

Table. 7.4: Some ornamental species of *Hygrophila*

Plant species	Plant species
Hygrophila angustifolia	*H.guianensis*
H.corymbosa crispa	*H.lacustris*
H.corymbosa glabra	*H.lancea*
H.corymbosa gracilis	*H.natalis*
H.corymbosa siamensis	*H.polysperma*
H.corymbosa strigose	*H.salicifolia*
H.difformis	*H.stricta*

- Remarks: These are easy to grow and so many species of this genus are favourite aquarium plants.

7.5.17 Ipomoea L. 1753 Figure 7.14

- Common name: water spinach
- Family: Convolvulaceae.
- Origin/Distribution: Southeast Asia.

- Habitat: Sides of ponds, bills, river, marshy land, waterlogged soil, shallow or deep water.
- Description: This perennial herb has weak stem with hollow pith. Roots are in soil, but the plant floats on water, leaves smaller to medium in size.
- Propagation: Seed setting plant, with normal germination.
- Species: *Ipomoea aquatica*
- Remarks: Used in ponds and aquarium. Also used as vegetable, medicine and fodder.

7.5.18 *Lagenandra* Dalzell

- Common name: Lagenandra
- Family: Araceae.
- Origin/Distribution: It is endemic to the Indian subcontinent (India, Bangladesh and Sri Lanka).
- Habitat: Freshwaters.
- Description: Leaf blade is simple. It is ovate to almost linear. Colour of the leaf is generally orange-brown or pink. Thick creeping rhizome is present.
- Species: About 15 species are reported under this genus. Aquarium species are—

  Lagenandra dewitii—Sri Lanka

  L. insignis

  L.koenigii—Sri Lanka

  L. lancifolia—Sri Lanka

  L.meeboldii—southwestern India

  L. nairii—southwestern India

  ***Lagenandra ovata* (L.) Thwaites** Figure 7.15

  —southwestern India, Sri Lanka

  L. thwaitasii—Sri Lanka

7.5.19 *Lemna* L.

- Common name: duck weed.
- Family: Lemnaceae.
- Origin/Distribution:
- Habitat: In ponds, ditch, canals, jheels.
- Description: It is an annual small herb, having simple free-floating thalli on or just beneath the water surface. Thalli have a single root. Roots are modified and modified leaves are with air pockets. Both roots and stems are weak.
- Propagation: By vegetative reproduction. Two daughter plants off from the

adult plant. This allows very rapid colonisation of the plant. Propagation is by seeds also.

- Species: About 14 species are known.

  Aquarium species are-

  Lemna gibba—Widely distributed.

  L. minor (duckweed)—Africa, Asia, Europe and North America.

  L. paucicostata

  L. perpusilla

  L. trisulca—Sub-cosmopolitan.

- Remarks: Besides being used as an aquarium plants also used as food for some aquatic animals.

7.5.19.1 *Lemna minor* L. Figure 7.16

- Common name: duck weed.
- Origin/Distribution: Sub-cosmopolitan distribution. Native to most of Africa, Europe and North America.
- Habitat: Ponds and slow-moving freshwater streams.
- Description: It is a floating freshwater aquatic plant, with one, two or three leaves each having a single root hanging in the water. The root is 1-2 cm long. Leaves are oval, 1-8 mm long and 0.6-5 mm broad, light green, with three (rarely five) veins and small air spaces to assist floatation. Flowers are rarely produced and measure about 1mm in diameter, with a cup-shaped membranous scale.
- Propagation: Vegetative by division.
- Remarks: Used as bioremediatory (removes heavy metals), as biofuel, as waste water nutrient and also as livestock feed.

7.5.20 *Limnophila* R.Br.

- Common name: marsh weeds.
- Family: Plantaginaceae.
- Origin/Distribution: Tropical and subtropical regions of Africa, Asia, Australia, and the Pacific Islands.
- Habitat: Grow in wet habitats, such as marshes and some waterbodies.
- Description: Annual or perennial herbs. Plants are erect to prostrate and with branching or unbranched stems. Submerged leaves are whorled, aerial leaves are whorled or oppositely arranged, some species have solitary leaves.
- Propagation: By cuttings.
- Species: About 40 known species. Aquarium species are—

  Limnophila aquatica—India, Sri Lanka.

L. aromatica—Southeast Asia.

L. glabra

L. heterophylla

L. indica—India.

L. sessiliflora—South East Asia, India, China, Japan, and the Philippines.

7.5.20.1 *Limnophila sessiliflora* (Vahl) Blume

- Common names: dwarf ambulia, ambulism, Asian marsh weed.
- Origin/Distribution: India, China, Japan, Philippines and South East Asia.
- Habitat: Mountain streams, lakes, rivers and damp soils.
- Description: Bright green coloured leaves are pinnate and present in proper whorls, about 1.2 inches (3cm) in diameter. The submerse leaves are dark green in colour and more or less lance-shaped. This plant can reach over 16 inches (40 cm) in height. The fruits are capsules.
- Propagation: By cutting.
- Remarks: It is used in the aquarium as a bunch plant. This plant prefers a nutrient-rich substrate, nutrient-rich water column and medium light.

7.5.21 *Ludwigia* L. Figure 7.17

- Common name: primrose-willow, water-purslane, water primrose.
- Family: Onagraceae.
- Origin/Distribution: Cosmopolitan though mainly tropical areas.
- Habitat: Marshy lands and waterlogged soil.
- Description: These plants are annual herbs with strong stem and small leaves. Roots are strong with air-pockets. Roots remain in water and entire plant body above water.
- Propagation: Seed setting plants. Vegetative propagation also takes place.
- Species: About 82 known species.
- Remarks: A common aquarium plant.

7.5.21.1 *Ludwigia palustris* (L.) Elliott

- Common names: marsh seed box, Hampshire-purslane, water purslane.
- Origin/Distribution: Some parts of the America.
- Habitat: Moist, wet or flooded areas.
- Description: It is an aquatic, semiaquatic perennial herb with a stem about half a meter long and which spreads to form mats. It has roots at nodes in contact with the substrate or floats ascending in water. The leaves are arranged oppositely and are green, red or purple in colour. Leaf axils are solitary flowers. There is no petals or tiny green sepals are present. Capsular fruits present which are with many tiny seeds.
- Remarks: It is cultivated as an aquarium plant.

7.5.22 *Marsilea* L.

- Common names: water clover, four-leaf clover.
- Family: Marsileaceae.
- Origin/Distribution: Worldwide in warm-temperate and tropical regions.
- Habitat: Grows in dense clumps in mud along the shores of ponds or streams or they may grow submerged in shallow water with some of the leaves extending to float on the water surface.
- Description: These plants have long stalked leaves with four clover-like lobes and are either held above water or submerged.
- Propagation: Takes place by division of root stock.
- Species: About 65 species are known. Aquarium species are—

  Marsilea hirsute

  M. pubescens

  M. quadrifolia (water clover)—Central and southern Europe, Caucasia, western Siberia, Afghanistan, southwest India, China, Japan and Vietnam.

  M. crenata—Southeast Asia.

  M. drummondii--Australia.

7.5.22.1 *Marsilea quadrifolia* L. Figure 7.18

- Common names: four leaf clover, European water clover, sushni (India).
- Origin/Distribution: South west India, China, Japan, Vietnam, central and southern Europe, Caucasia, western Siberia and Afghanistan.
- Habitat: By the sides of ponds, jheels, canals, lakes and marshy lands.
- Description: This aquatic perennial herb has 4 parted leaf resembling '4-leaf clover'. Leaves either float in deep water or remain erect in shallow water or on land. It grows on soils of ponds with roots in the soil and runs over water surface. Leaflets are obdeltoid, about3/4 inches long, glaucous. Petioles are about 8 inches long. Sporocarp are dark brown in colour, ellipsoid, attached to base of petioles by stalks of ¾ inches long.
- Propagations: By division of the root stock.
- Remarks: In the aquarium, this plant is grown fully submerged. It prefers sandy and loamy soils. It requires moist, wet soil or water to grow. It has been used as vegetable and as medicine.

7.5.23 *Myriophyllum* L.

- Common names: watermilfoil.
- Family: Haloragaceae
- Origin/Distribution: Cosmopolitan distribution. Found in most temperate and tropical regions.
- Description: Elongate stems are present with air canals and whorled very

fine leaves that are finely, pinnately divided. Leaves above the water are stiffer and smaller than the submerged leaves.

- Species: Aquarium species are—

Table. 7.5: Some species of the genus *Myriophyllum*

Plant species	Native to	Plant species	Native to
Myriophyllum alterniflorum	Europe, Asia and North America	*M. scabratum*	
M. aquaticum	Amazon River in South America	*M. spicatum*	Europe, Asia and North Africa
M. elatinoides		*M. tuberculatum*	
M. heterophyllum		*M. ussuriense*	
M. hippuroides	West coast of North America	*M. verticillatum*	North America, North Africa and Eurasia
M. mattogrossense	Wetlands of South America	*M. oguraense*	
M. proserpinacoides			

- Remarks: Very popular aquarium plants. They also act as shelter for the egg-scatterer aquarium fishes.

7.5.24 *Najas* L. (1753)

- Common names: water nymphs, naiads.
- Family: Hydrocharitaceae/Najadaceae.
- Origin/Distribution: Cosmopolitan.
- Habitat: Aquatic.
- Description: *Najas* grows in dense clusters and has highly branched stems. The small flowers are located in clusters along the leaf axis. The leaves are opposite, unbranched, strap-shaped and are around 4.5 centimetres in length. The leaves have serrations which are visible in the naked eye.
- Propagation: The stems fragment easily and this plant is capable of propagation from stem fragments or from small seeds which grow along its stem.
- Species: About 40 species are recognized. Aquarium species are—

  Najas graminea—Africa, Middle East, Central Asia, East Asia, Southeast Asia, New Guinea, New Caledonia and northern Australia.

  N. guadelupensis

  N. indica—India, China, Japan, Southeast Asia and New Guinea.

  N. marina—Europe, Asia, Africa, Australia, the Americas and many Oceanic islands.

  N. minor—Europe, Asia, North Africa, Japan, Philippines, Netherlands,

China, Siberia, Central Asia, Iran, Turkey, Ukraine, Germany, Canada, USA, Thailand.

7.5.25 *Nymphaea* L. Figure 7.19

- Common name: water lilies.
- Family: Nymphaeaceae
- Origin/Distribution: Cosmopolitan.
- Habitat: aquatic.
- Description: Perennial herbs. These rhizomatous plants are sometimes with stolon. Leaves mostly float on the surface and may grow from the rhizome on long petioles. The flowers rise out of water or float on the surface. They open during the day or night. Flower has at least eight petals in shades of white, pink, blue or yellow.
- Propagation: Either through seeds or by vegetative reproduction.
- Species: About 36 species are known. Aquarium species are—

  Nymphaea alba—North Africa, temperate Asia, Europe and tropical Asia.

  N. lotus—East Africa, Southeast Asia.

  N. micrantha—Tropics of West Africa.

  N. pubescens—temperate and tropical Asia (India, Bangladesh, Sri Lanka, Yunnan, Taiwan, Philippines, Cambodia, Laos, Myanmar, Thailand, Vietnam, Indonesia and Malaysia), north eastern Australia and Papua New Guinea

  N. pygmea

  N. stellate—Southern and eastern parts of Asia.

  N. zenkari—East Africa.

7.5.25.1 *Nymphaea pubescens* Willd.

- Common name: hairy water lily.
- Origin/Distribution: India, Bangladesh, Cambodia, Indonesia, Lao People's Democratic Republic, Malaysia, Myanmar, Nepal, Pakistan, Philippines, Sri Lanka, Thailand, and Viet Nam.
- Habitat: Shallow pools and ponds.
- Description: This is an aquatic, perennial herb with erect perennial rhizomes or root stocks that anchor the plant to the bottom. Stolons are produced from the rhizomes. Leaf blades of this plant are round above and heart-shaped below. These are papery and "abaxially densely pubescent". Some of the leaves are held by their stem to keep them slightly above water whereas other leaves just float on the surface. Those floating with undy edges make a crenellate effect. Flowers are quite large.

7.5 26 *Nymphoides* Seg. 1754

- Common name: floating heart.

- Family: Menyanthaceae.
- Origin/Distribution: Native to United States, Europe, Eurasia, Asia.
- Habitat: Aquatic.
- Description: These aquatic plants are with submerged roots and floating leaves that hold the small flowers above the water surface. Flowers of this plant are sympetalous divided into five petals which are either yellow or white.
- Propagation: Propagates sexually by seeds. Asexually via vegetative splitting from separated leaves or cuttings.
- Species: About 50 species, 8 species in India. Aquarium species are—

  Nymphoides aquatica—South-eastern United States.

  N. humboldtiana

  N. indica—Tropical areas around the world.

  N. peltata—East Asia and the Mediterranean.

7.5.26.1 *Nymphoides indica* (L.) Kuntze

- Common names: water-snowflakes, robust marshwort, banana plant.
- Origin/Distribution: Tropical areas in the world.
- Habitat: Shallow water in permanent or seasonal pools and slow-flowing rivers. It may also occur in rice fields.
- Description: It is a perennial herb. Strong roots are located in the soil, while plants are present over the water. Stems are weak. Clusters of leaves with white flowers are present. Leaves are with air-pockets and these remain above water. Flowers have five or more petals.
- Remarks: Besides being used as an ornamental plant, it is also used as vegetable.

7.5.27 *Ottelia* Pers. 1805

- Common name: duck-lettuce
- Family: Hydrocharitaceae
- Origin/Distribution: Native to tropical and subtropical regions Africa, Asia, South America and Australia.
- Habitat: Waterbodies.
- Description: Species belonging to this genus, are submersed aquatic herbaceous plants. Flowers project above the surface of water. Small terminal flowers are held on a long peduncle.
- Species About 19 known species. Aquarium species are:

  Ottelia alismoides—native to Asia and northern Australia; *O. mesenterum*; and

  O. ulvifolia.

7.5.28 *Pistia* L. Figure 7.20

- Common names: Water lettuce, Tropical duck-weed, water cabbage, Nile cabbage.
- Family: Araceae
- Origin/Distribution: Nile near Lake Victoria in Africa.
- Habitat: Fresh waterbodies.
- Description: This plant is a perennial herb with thick green, soft leaves that form a rosette. It floats on the surface of the water, its roots hanging submerged beneath floating leaves. The leaves can be up to 14 cm long and have no stem. These are light green, with parallel veins, with wavy margins and are covered with short hairs. These hairs form basket-like structures which trap air bubbles. The plant's buoyancy is increased this way. The flowers are hidden in the middle of the plant amongst the leaves, leaf with air-pockets, above water.
- Propagation: It is through asexual and sexual propagation. Asexual propagation is common. Small daughter plants form and float connected by a short stolon to the mother plant. These young plants may or may not get detached, but grow into an adult plant. Sexual reproduction is very rare in the home aquarium. Each plant has either male or female flowers. Successful fertilization results in a small, many-seeded green berry.
- Species: Only one species under the genus.

  Pistia stratiotes (water lettuce)

7.5.28.1 *Pistia stratiotes* L.

- Common name: water lettuce.
- Origin/Distribution: Pantropical species, possible origin in Gondwana. Nile near Lake Victoria in Africa, Pantropical.
- Habitat: Lakes, ponds, other open, still freshwater bodies.
- Description: Free-floating plant that is capable of forming dense mats on the surface of water.
- Propagation: Vegetative.
- Remarks: Grown as an ornamental in lakes, ponds, aquaria and gardens. It has medicinal properties and can be used as fodder for cattle and pigs. It acts as a shelter for fry and small fishes. This plant helps in preventing massive algal blooms. This invasive species provides habitat for mosquitos, such as *Anopheles* and *Mansonia* and so should be carefully handled.

7.5.29 *Riccia* L., 1753

- Common names: liverworts. Crystalwort, riccia moss.
- Family: Ricciaceae.
- Origin/Distribution: Tropical areas.
- Habitat: Ponds.

- Description: This is a clumping floating plant, actually composed of hundreds of tiny plants interlinked. These form dense, bright, green mats on the surface of water. These plants ordinarily float but can also be attached to underwater objects.
- Propagation: By division of the thallus. Any single branch or antler bud can reproduce into a large colony of the plant if kept in proper condition.
- Species:

  Aquarium species—

  Riccia fluitans

7.5.29.1 *Riccia fluitans* L.

- Common name: floating crystalwort.
- Origin/Distribution: Tropical areas.
- Habitat: Ponds, present on or under the water surface.
- Description: It is a floating plant often forms thick mats on or under the water surface. It normally floats but can get attached to any under-water object.
- Propagation: Any single branch or antler bud can produce a large colony if the plant is kept in proper conditions.
- Remarks: It is a popular aquarium plant preferred as a shelter for young fry and is used in live-bearing tanks.

7.5.30 *Sagittaria* L.

- Common names: arrowhead, duck potato, Katniss.
- Family: Alismataceae.
- Origin/Distribution: Most are native North, Central and South America. Some species are from Europe, Africa and Asia.
- Habitat: Fresh waterbodies.
- Description: Leaves of this plant are either submersed or emersed, with long leaf stalks, bladeless or with unlobed or arrow-shaped blades. The plant stock or the perennial rhizome is a horizontal creeper.
- Propagation: Takes place by producing multiple runners, by offshoot or by seed also.
- Species: There are about 40 known species. Aquarium species are:

Table. 7.6: Species of *Sagittaria*

Plant species	Native to	Plant species	Native to
Sagittaria chapmani		*S. montevidensis*	North America and South America
S. eatonii		*S. natans*	Northern Europe and Asia
S. filiformis	Eastern United States	*S. papillosa*	South-central United States

Plant species	Native to	Plant species	Native to
S. graminea	Eastern North America	*S. platyphylla*	Eastern United States
S. guyanensis		*S. pusilla*	Republic of Colmbia, the District of Columbia, Venezuela
S. isoetiformis	Cuba and to the south eastern United States	*S. sagittifolia*	Europe, Russia, Ukraine, Siberia, Japan, Turkey, China, Australia, Vietnam, and the Caucasus
S. latifolia	Southern Canada, most of the United States, Mexico, Central America, Colombia, Venezuela, Ecuador, and Cuba	*S. subulate*	The Republic of Colombia, the District of Colombia, Venezuela
S. microfila			

7.5.31 *Salvinia* Sig. Figure 7.21

Common name: water moss.

- Family: Salviniaceae.
- Origin/Distribution: Tropical in North America, Mexico, West Indies, Central America, Eurasia and Africa including Madagascar, and South Borneo.
- Habitat: Fresh water bodies.
- Description: Floating fern and heterosporous. The upper side of the floating leaf which appears to face the stem axis is morphologically abaxial.
- Propagation: By spores.
- Species:

  Aquarium species are—

  Salvinia auriculate—Native to the Americas.

  S. cucullate

  S. minima—South America, Mesoamerica and the West Indies.

  S. natans—Africa, Asia, central Europe and South America, in New York State.

  S. oblongifolia

  S. rotundifolia
- Remarks: Popular ornamental plant.

7.5.32 *Synnema* sp.

- Common name: water wisteria
- Family: Acanthaceae.
- Origin/Distribution: South Asia, India, Thailand, and Malaysia.
- Habitat: Fresh waterbodies.
- Description: This plant has large leaves. The substrate needs nutrients.
- Propagation: Is by cuttings from the side shoots of the mother plant.
- Species: Aquarium species is---
- *Synnema triforum*—South Asia, India, Thailand and Malaysia.
- Remarks: Popular aquarium plant.

7.5.33 *Taxiphyllum* (Cardot & Copp.) Z.Iwats.

- Common name: Java moss.
- Family: Hypnaceae.
- Origin/Distribution: Native to Southeast Asia.
- Habitat: Freshwaters, attaches to rocks, roots and driftwood.
- Description: It has small irregularly branched stems with rows of tiny overlapping oval-shaped leaves. Without any true roots, it absorbs nutrients primarily through its leaves. It grows free-floating or send out rhizoids to attach itself.
- Propagation: By division. The main plant may be divided into pieces and placed in different locations.
- Species: Aquarium species

  Taxiphyllum barbieri (Cardot & Copp.) Z. Iwats. Southeast Asia.

  Remarks: It is suitable for both aquarium and vivarium. Popular in aquariums for raising fry by protecting them from cannibalistic adults. It also provides food for newly formed fry.

7.5.34 *Typha* L.

- Common name: bulrush, reedmace, cattail.
- Family: Typhaceae.
- Origin/Distribution: Northern Hemisphere.
- Habitat: Variety of wetlands.
- Description: These are aquatic, semi-aquatic, rhizomatous, herbaceous, and perennial plants. Leaves are linear, alternate and mostly basal on a simple, jointless stem that bears the flowering spikes.
- Propagation: By division and through seeds. The seeds germinate rapidly under hot, moist condition and are then ready for transplant in a few weeks.

- Species: Aquatic species are—
- *Typha angustifolia*—Asia, North America, Central America, Caribbean, South America and Europe.
- *T. latifolia*—North and South America, Europe, Eurasia and Africa.
- Remarks: These plants provide protective cover and nesting areas for aquatic animals. Many species are cultivated for ornamental purpose.

7.5.35 *Vallisneria* L.

- Common names: eelgrass, tape grass or vallis.
- Family: Hydrocharitaceae.
- Origin/Distribution: Tropical and subtropical regions of Asia, Africa, Europe and North America.
- Habitat: Shallow or deep water in ponds or lakes.
- Description: Annual, submerged herb. Leaves arise from the roots and form clusters. Leaves have rounded tips and definite raised veins. Strong roots present, leaves are like long, narrow belts. Female flowers are single, white with very long stalks and grow to the water surface. Male flowers are with short stalks after getting detached, float to the surface. The fruit has many tiny seeds.
- Propagation: By runners. These plants have arched stems which develop small planters at their nodes.
- Species: About 14 known species.

  Aquarium species are-

  Vallisneria americana—Eastern North America and many parts of Asia and America.

  V. asiatica

  V. gigantica—Tropical subtropical regions of Asia, Africa, Europe and North America.

  V. neotropicalis

  V. rubra

  V. spiralis—Tropical and subtropical regions worldwide.

  V. fortifolia

  V. tortissima

  Remarks: Species of *Vallisneria* are popular tropical and subtropical aquaria plants. These plants are tolerant and adaptable. These are among the few aquarium plants which can tolerate brackish water.

7.5.35.1 *Vallisneria spiralis* L. Figure 7.22

- Common name: straight Vallisneria, tape grass, eel grass.
- Origin/Distribution: Tropical and subtropical regions worldwide.

- Habitat: Ponds, lakes, rivers, marshy lands. Shallow or deep water.
- Description: It is a perennial herb which has narrow, linear leaves, whose colour varies from pale-green to reddish. Leaves are long and about 0.75 inches broad, flat resemble ribbons. Root remain in soil and leaves are within water. This plant is monoecious with flowers on long spiral stalks.
- Propagation: By runners and by seeds.
- Remarks: This plant is a decorative and favourite aquarium plant though it is considered as an invasive species in some countries. It requires good light and a nutrient rich substrate. It serves as an oxygenator in aquatic set-ups and used as a medicinal plant.

Figure 7.1 *Acorus calamus*

Figure 7.2 *Alisma plantago aquatica*

Figure 7.3 *Ammania* sp.

Figure 7.4 *Anubias* sp.

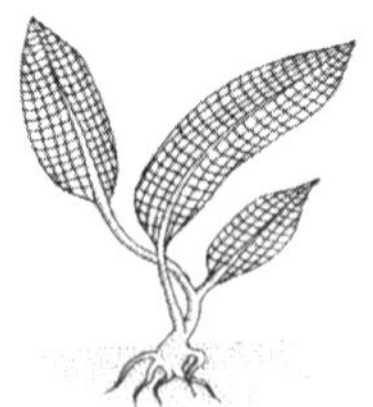

Figure 7.5 *Aponogeton* sp.

Figure 7.6 *Azolla filiculoides*

Figure 7.7 *Bacopa monnieri*

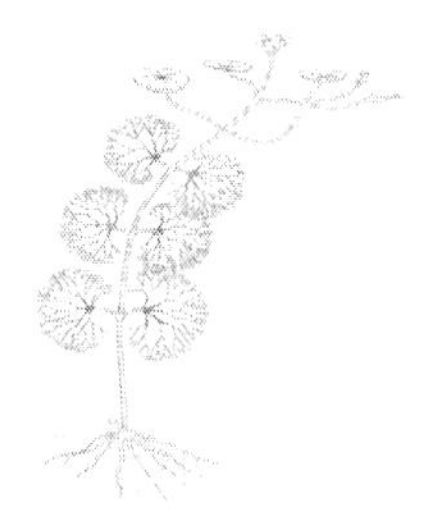

Figure 7.8 *Cabomba aquatica*

Figure 7.9 *Ceratophyllum* sp.

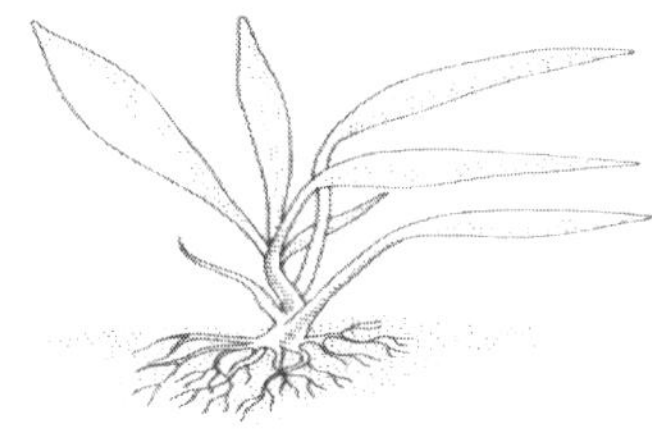

Figure 7.10 *A Cryptocoryne becketii*

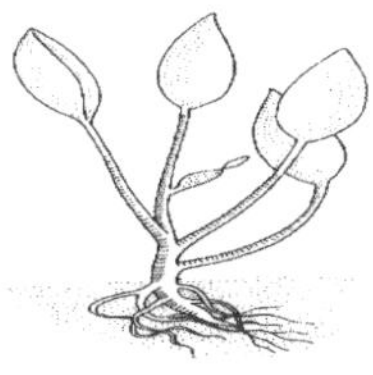

Figure 7.10 B *Cryptocoryne ciliata*

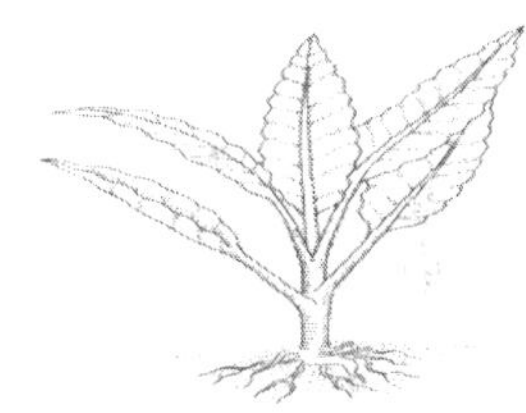

Figure 7.10 C *Cryptocoryne cordata*

Figure 7.11 *Echinodorus* sp.

Figure 7.12 *Eichornia crassipes*

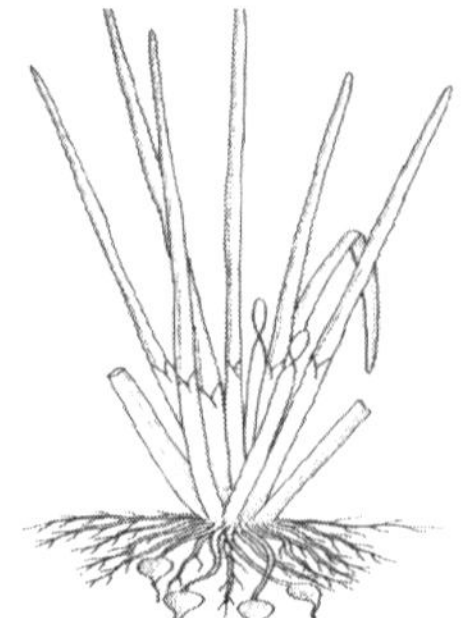

Figure 7.13 *Eleocharis* sp.

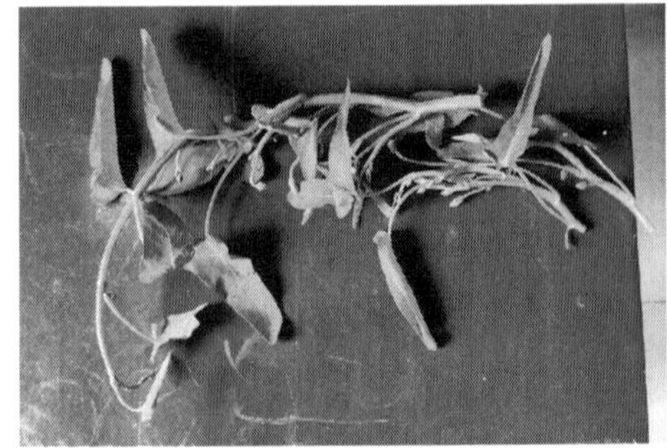

Figure 7.14 *Ipomoea* sp.

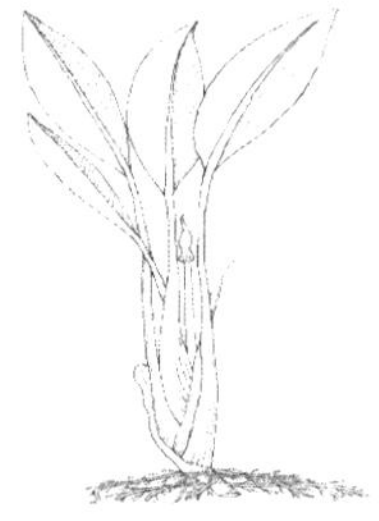

Figure 7.15 *Lagendra ovata*

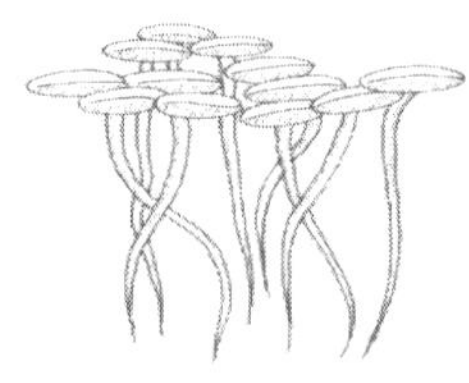

Figure 7.16 *Lemna minor*

Figure 7.17 *Ludwigia* sp.

Figure 7.18 *Marsilea quadrifolia*

Figure 7.19 *Nymphaea* sp.

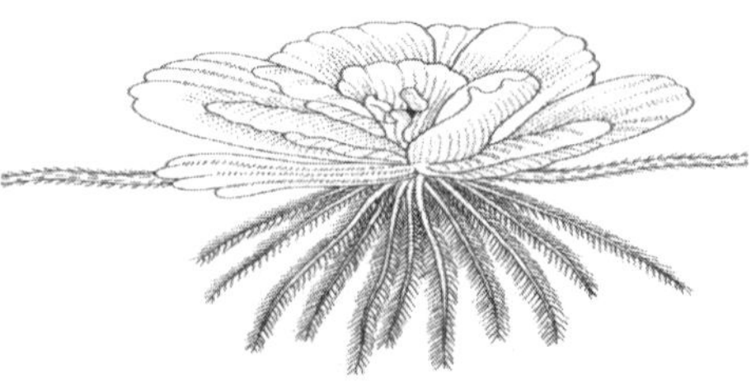

Figure 7.20 *Pistia* sp.

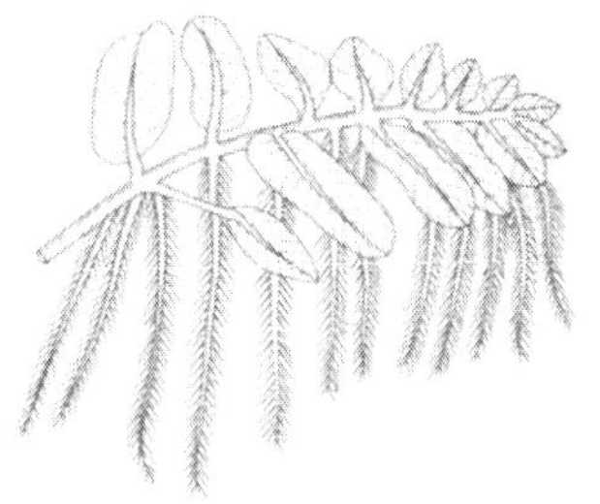

Figure 7.21 *Salvinia* sp.

Figure 7.22 *Vallisneria spiralis*

7.6 Brackish Water Ornamental Plant Species

Plants which grow naturally in brackish water, are used in the brackish water aquatic set-ups. Some of the preferred brackish water aquarium plants are listed below (Table.7.7).

Table 7.7: Brackish water aquarium plants

Plant species and their origin/distribution	
Bacopa monnieri—southern and eastern India, Australia, Europe, Africa, Asia and North and South America.	*Najas marina* -Europe, Asia, Africa, Australia, the Americas and many Oceanic Islands.
Crassula aquatica—North America and Eurasia.	*Sagittaria* spp.—South, Central and North America.
Cryptocoryne ciliata—India to New Guinea.	*Spartina alterniflora* -Atlantic coast of the Americas.
Cymodocea nodosa -shallow parts of Mediterranean Sea and certain adjoining areas of the Atlantic Ocean.	*Vallisneria* spp.—Asia, Africa, Europe, North America.
Helanthium tenellum -eastern United States, southern Mexico, West Indies, Central America, and South America.	*Zostera marina*—coastlines of northern sections of North America and Eurasia.
Juncus roemarianus -North America.	*Zostera noltii*—shallow coastal waters in north western Europe, the Mediterranean Sea, Black Sea, Capspian Sea and Aral Sea, and islands in the Atlantic off the coast of northwest Africa.

Mangrove plants and their origin/distribution
Black mangrove—*Avicennia germinans* -coastal lagoons and brackish water estuaries. Tropical and subtropical regions of the Americas, on both the Atlantic and Pacific coasts, and on the Atlantic coast of tropical Africa.
Red mangrove—*Rhizophora mangle* -estuarine ecosystems throughout the tropics.
White mangrove—both *Avicennia marina*—in the intertidal zones of estuarine areas. Africa's east coast, south-west, south and south-east Asia and Australia. *Laguncularia racemose*—coasts of western Africa.

7.7 Aspects of some Brackish Water Aquarium Plants

7.7.1 *Crassula aquatica* (L.) Schonl Figure 7.23

- Common names: water pigmy-weed, common pigmy weed, pigmy weed.
- Family: Crassulaceae.
- Origin/Distribution: North America and Eurasia.
- Habitat: Salt marshes, vernal pools, wetlands, fresh to brackish water bodies. Along muddy banks and in tidally active areas of estuaries.
- Description: Annual plant, small and forms thin mats or small patches on mud and sand. This plant is greenish-yellow to bright red or purplish in colour. Leaves are tiny, fleshy and pointed and of a few mm length. Leaves are oppositely arranged and in pair. A flower with a short stalk arises from the intersection of these leaves. Fruit is minute and contains several seeds.
- Propagation: Through cuttings from young stems.

7.7.2 *Cymodocea nodosa* (Ucria) Asc. (1870) Figure 7.24

- Common name: little Neptune grass.
- Family: Cymodoceaceae.
- Origin/Distribution: Atlantic-northeast, Atlantic-eastern central, Mediterranean and Black Sea.
- Habitat: It is restricted to growing underwater, shallow parts of marine ecosystem. It grows at depths of 10m down in sandy sediments in sheltered area of sea.
- Description: This species of seagrass has light green or greyish-green leaves. They are very narrow but very long as much as 40 cm. There are seven to nine veins running along the length of each leaf. Rhizomes are present and these have leaf scars at intervals. Grass -like flowers are produced at the end of long stems in the spring.
- Propagation: Through seeds.
- Remarks: They are important rearing ground for juvenile fish.

7.7.3 *Helanthium tenellum* (Mart. Ex Schult.f.) J.G. Sm. Figure 7.25

- Common name: not known.
- Family: Alismataceae.
- Origin/Distribution: Eastern United States, southern Mexico, West Indies, Central America, and South America.
- Description: The stem of this plant is thin, erect, often curved. The stem ends in inflorescence. The leaves of the plant are narrowly elliptical or lanceolate. The blade of the leaf resembles the petiole, sometimes.
- Species: Three species are known.

- Propagation: Spreads by runner chains, with plantlets forming every 2 to 5 cm. These plantlets can be detached and planted separately.

7.7.4 *Zostera noltii* Hornemann Figure 7.26

- Common name: dwarf eelgrass.
- Family: Zosteraceae.
- Origin/Distribution: North western Europe, the Mediterranean Sea, Black Sea, Caspian Sea and Aral Sea and on islands in the Atlantic off the coast of northwest Africa.
- Habitat: Shallow coastal waters, intertidal and shallow subtidal ecosystems of estuaries, bays and lagoons.
- Description: This seagrass has a creeping rhizome that runs along under the surface of the seabed. From the nodes on the rhizome, two to five ribbon-shaped leaves grow out. A tuft of short roots come out from each node that anchor the plant in the sediment. Leaves are up to 22cm and contain air spaces. Three irregular, longitudinal veins are present on leaves, which are with blunt, notched ends. Short, spear-shaped lateral stems are present bearing several separate male and female flowers.
- Propagation: Part of the rhizome from the parent plant can be detached and then implanted.
- Remarks: In aquaria, provide refuge for many developing juvenile fishes.

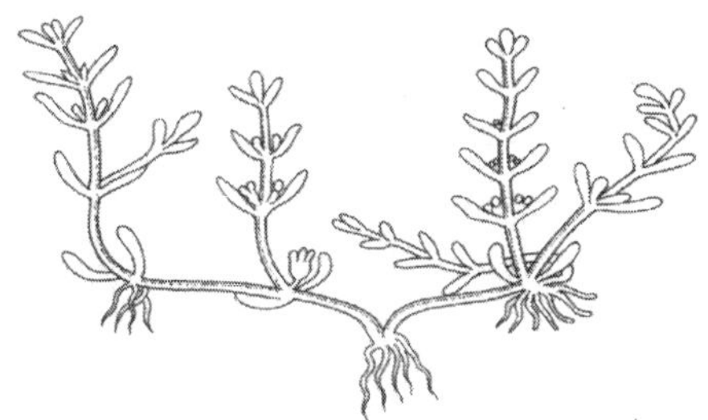

Figure 7.23 *Crassula aquatica*

Figure 7.24 *Cymodocea nodosa*

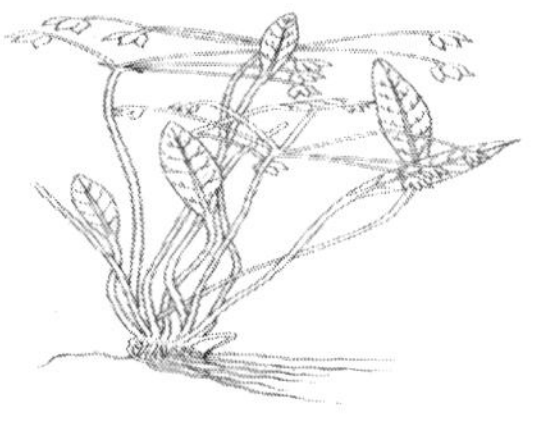

Figure 7.25 *Helanthium tenllum*

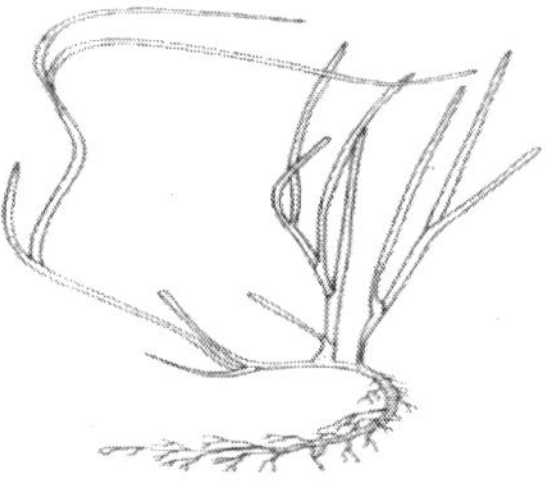

Figure 7.26 *Zostera noltii*

7.8 Marine Aquarium Plant Species

Thera are some marine water plants which are found suitable for keeping in marine water aquariums. Those plant species are listed below.

Table 7.8: Marine aquarium plant species

Common name	Scientific name
Brown macroalgae:	
False sargassum	*Cystoseira* sp.
Scroll algae, potato algae	*Padina* sp.
Green macroalgae:	
Chaeto	*Chaetomorpha* sp.
Blade caulerpa	*Caulerpa prolifera*
Grape Caulerpa	*Caulerpa racenosa*
Green calcareous algae	*Halimeda* sp.
Sawblade Caulerpa	*Caulerpa serrulate*
Sea lettuce	*Ulva latuca*
Red macroalgae:	
Ogo algae	*Gracilaria parvispora*
Coralline algae	
Mangroves:	
Black mangrove	*Avicennia germinans*
Red mangrove	*Rhizophora mangle*
White mangrove	*Laguncularia racemose*
Sea grasses:	
Turtle grass	*Thalassia hemprichii* & *T. testudinum*
Manatee grass	*Syringodium filiforme*

7.9 Aspects of some Marine Aquarium Plants

7.9.1 *Cystoseira* C. Agardh, 1820

- Common name: false sargassum
- Family: Fucaceae
- Origin/Distribution: Widely distributed genus. Found mostly in Temperate regions of the Northern Hemisphere, such as the Mediterranean, Indian and the Pacific Oceans.
- Habitat: Marine water bodies.
- Description: Species under this genus of brown algae have tall, stiff branches, light brown in colour. Basal and apical regions are highly differentiated. Lower parts are strongly flattened into foliar expansions or basal leaves. The aero cyst or air vesicles present on the outer branches keep the plant erect by causing it to float in strong currents.
- Propagation: By fragmentation and sporulation.
- Species: 51 species known.
- Remarks: Requires brightly lit environment in the aquarium.

7.9.2 *Chaetomorpha* Kutzing Figure 7.27

- Common names: spaghetti algae, green hair algae.
- Family: Cladophoraceae.
- Origin/Distribution: Tropical and subtropical waters of Atlantic, Indian Ocean Islands, Asia, south east Asia, Australia and New Zealand, and Pacific Ocean.
- Habitat: Salt waters.
- Description: The plant body is made up of macroscopic filaments of cylindrical cells. Each cell grows end to end creating long, stiff strands. It grows in filamentous clumps. This green macroalgae has unbranched filaments.
- Propagation: By fragmentation and sporulation.
- Species: There are about 50 known species.
- Remarks: These green macroalgae are very popular with aquarists. These plants are favourite with marine aquarium and refugium. They help to absorb nitrates and phosphates from the water and help in maintaining water quality.

7.9.3 *Gracilaria* Greville, 1830 Figure 7.28

- Common name: gracilaria seaweeds.
- Family: Gracilariaceae.
- Origin/Distribution: Widely distributed all over the world, in temperate and tropical sea water.
- Habitat: Intertidal to subtidal areas.
- Description: Long, cylindrical, alternating branches that are attached by a single holdfast.
- Propagation: By fragmentation or by sporulation.
- Species: The most popular species of Red *Gracilaria* is *Gracilaria parvispor.*
- Remarks: Popular with marine aquarists.

7.9.3.1 *Gracilaria parvispora* I.A. Abbott Figure 7.29

- Common name: ogo
- Origin/Distribution: Gulf of Mexico, Atlantic, Caribbean, South Pacific.
- Description: This red macroalgae has short, bushy fronds with bright red colour and burgundy in hue.
- Remarks: It is used in the refugium as it can uptake nutrients. This macroalgae is hardy and fast growing. Many marine herbivorous fishes like tangs and others find this plant highly palatable.

7.9.4 *Rhizophora* L.

- Common name: mangrove.

- Family: Rhizophoraceae
- Origin/ Distribution: Tropical and subtropical regions worldwide. In Indian **Mangroves** (Figure 7.30 A), members of the family Rhizophoraceae are dominant and represented by 11 species (Das, 2001).
- Habitat: Can grow in fresh, brackish and salt waters.
- Species: Popular aquarium species is *Rhizophora mangle*

7.9.4.1 *Rhizophora mangle* L. Figure 7.30 B

- Common name: red mangrove.
- Origin/Distribution: Estuarine ecosystems throughout the tropics.
- Habitat: On coastlines in brackish water and in swampy salt marshes.
- Description: It is a small aquatic tropical tree, with prop roots that extend down into the water from higher up on the trunk of the plant. Leaves are opposite, simple and bright green. It has viviparous seeds called propagules.
- Propagation: Viviparous seeds or propagules become fully mature and drop off the parent tree. These are dispersed by water and get embedded in shallows.
- Remarks: This red mangrove can be planted in the sand, or in between rocks in any open topped aquarium.

7.9.5 *Thalassia* Banks ex K.D. Konig

- Common name: turtle sea grass.
- Family: Hydrocharitaceae.
- Origin/Distribution: Through the whole tropical region of the Indian Ocean and the western Pacific, also further to the north in the Red Sea.
- Habitat: Shallow sandy or muddy locations in seawaters.
- Description: Rhizomes greenish to light brown. Leaf sheath persistent whitish, or translucent. Blade wide and margin is entire.
- Propagation: By vegetative and sexual means.
- Remarks: Most abundant and good plant for any sea water aquarium.
- Species: Aquarium species are—

  Thalassia hemprichii (Ehrenberg) Ascherson, 1871 Figure 7.31

  –Pacific turtlegrass

  --Shores of western and central Indo-Pacific.

  T. testudinum

  —Caribbean turtlegrass

  —Gulf of Mexico, Caribbean Sea and Bermuda.

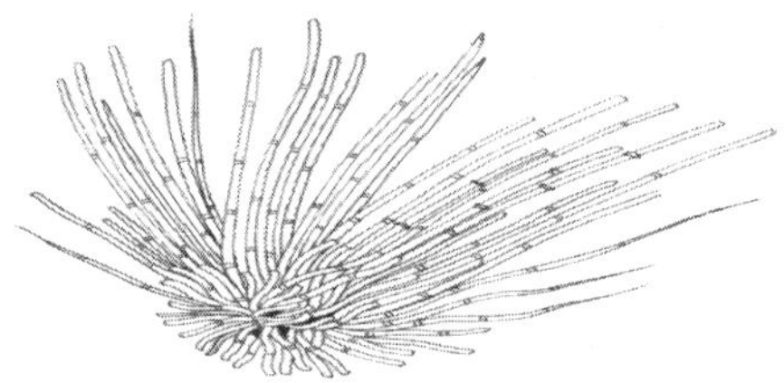

Figure 7.27 *Chaetomorpha* sp.

Figure 7.28 *Gracilaria* sp.

Figure 7.29 *Gracilaria parvispora*

Figure 7.30 A Mangrove in Sundarbans, India.

Figure 7.30B *Rhizophora mangle*

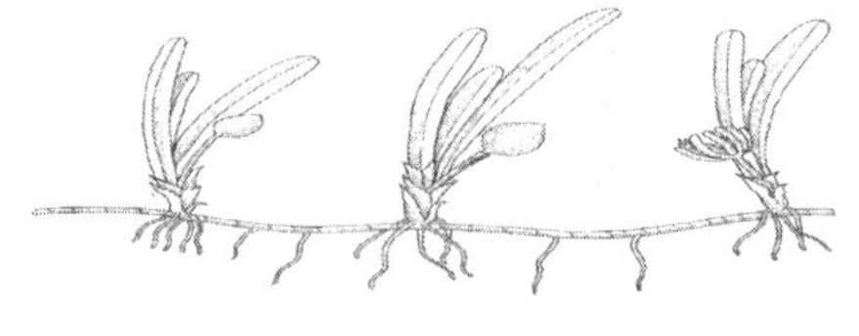

Figure 7.31 *Thalassia hemprichii*

7.10 Propagation of Aquarium Plants

Aquarium plant production is at present a powerful and growing industry. Adoration and significance of aquarium plants are gaining momentum. Many entrepreneurs, throughout the world are now involved in this activity. More manpower, money and time have now been engrossed for its further development.

7.10.1 Types of propagation:

There are two basic ways by which plants multiply naturally. Sexually and asexually. Sexual reproduction involves the production of spores and seeds that germinate to form new plants. Asexual reproduction involves a number of vegetative processes by which new plants are produced from various parts of the parent plant.

Sexual propagation

- In lower plants such as algae, mosses, liverworts and ferns, the zygotes (produced from the union of male and female sex cells) develop into spores. In the higher plants, such as herbs, shrubs and trees, the zygotes ripen into seeds. Through sexual propagation, genetic make-up of new plants is influenced by the characteristics of their respective parents, so changes can be introduced into the next generations. In case of aquarium plants though asexual reproduction play the important role than the sexual methods still there are few aquarium plants that can be propagated by sexual propagation. There are aquarium plants which readily produce flowers and those that do not. In the former category, deep marginal plants such as *Aponogeton* and the water lilies are included. Floating flowers are pollinated largely by insects, the flowers set seeds and these then germinate readily in the substrate. Plants of the second category produce flowers and seeds only when the water level falls low enough for them to grow emersed. *Cryptocoryne, Echinodorus, Hygrophila* and *Ludwigia* belong to this category. For commercial production of seeds for few aquarium plants like, *Echinodorus berteroi, Ludwigia* sp. and *Samolus parviflorus*, in the greenhouse, the level of water in which the plants are growing, is allowed to fall. The plants then produce stiff aerial leaves and then flowering shoots. After cross pollination, ripe seeds are collected and sprinkled on the surface of shallow parts of the sowing compost.

  Besides insect pollination of the flowers in aquarium plants, fine paint brush is often used to transfer pollen from the stamens to the stigmas. But these seeds should be sown earliest as these have no storage life.

Vegetative propagation

- The plant is propagated by using part of the plant itself either a stem cutting or bulb or other part of the plant. This is the easiest and cheapest method and is most widely used for reproducing aquarium plants. Vegetative propagation takes place either by natural means or by artificial means. Modified stems, leaves, and roots allowing natural vegetative propagation are, runners (stolon), bulbs, tubers, corms, suckers, plantlets, keikis (additional offshoots) and apomixis. Artificial vegetative propagation applied for better crop with desirable qualities, are cutting, grafting, layering, suckering, offset and tissue culture.

  - **Runners** (Figure 7.32 A)—Elongated outgrowths, produced by most aquarium plants are called runners. These shoots arise from leaf axils i.e., the junction of a leaf and the stem, on the side shoots but sometime from its apical shoots. In *Vallisneria* sp. and *Echinodorus tenellus*, these shoots either run along the surface of the substrate and in case of *Cryptocoryne*, these grow underground for a while before coming up to the surface. Slip, the baby plant, develops at the tip of each runner. When shoots are developed from them, and become established in the substrate, it also produces one or more runners. Each baby plant should be allowed to develop undisturbed for a time before detaching it from the parent plant. The connections between the parent and baby plants break down eventually, if left undisturbed. Floating plants like *Limnobium lavigatum* and *Pistia stratiotes*, produce several runners at the same time and spread quickly.

- **Stolon** (Figure 7.32 B)—These are runners or horizontal stem growing above the ground. Tiny plantlets form with roots which eventually can be detached for a new plant.
- **Offsets** (Figure 7.32 C)—These are short and stout outgrowths from which new plants are produced. These are similar to runners. To propagate, these outgrowths or clump are divided into smaller plants and then set in the substrate. Many species of *Echinodorus* gradually form large clumps by continually producing offsets.
- **Rhizomes**—These are underground stems with scale-like leaves and buds. For propagation of plants with rhizomes, the rhizome is simply cut or split into pieces and each piece will develop new shoots and roots. Examples are *Acorus gramineus, Nuphor* sp. And *Nymphae* sp.
- **Sucker** (Figure 7.32 D)—It is a shoot that develops from a lateral bud on the rhizome, emerging from the soil near the parent plant.
- **Adventitious shoots**-- These are plantlets that arise from any organ of mother plant. *Ceratopterus thalictroides* produces large numbers of tiny ferns at the leaf margins. In Java Fern, *Microsorium pteropus*, planters develop on older leaves. *Nymphaea daubenjana, Eleocharis vivipara* and certain species of *Echinodorus* propagate by the formation of adventitious plantlets.

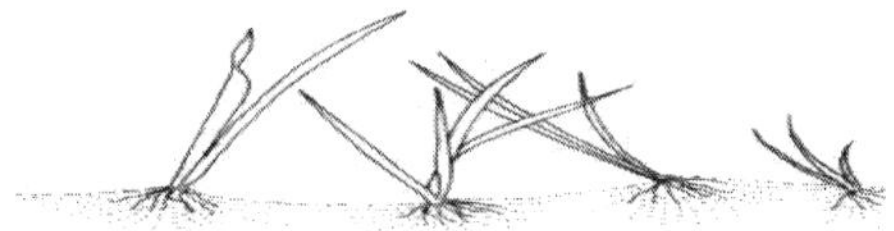

Figure 7.32 A Runner for vegetative propagation.

Figure 7.32 B Stolon in vegetative propagation.

Figure 7.32 C Offset in vegetative propagation.

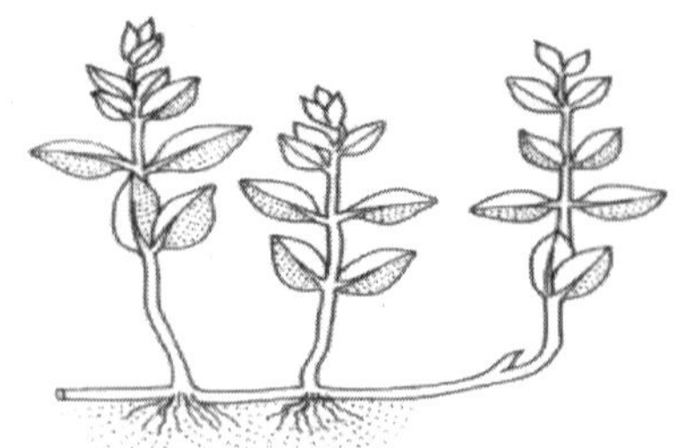

Figure 7.32 D Sucker in vegetative propagation.

- **Artificial propagation**—This mode of propagation is applied to many aquarium plants.
    - **Cuttings**—This is a familiar technique for a vast majority of aquarium plants. Cuttings may be top cutting, taken from the top, taken from the side shoots or from the middle portions of the plant. Each cutting should have one or more nodes. Lower leaves are to be removed and dip the base in a fungicide. The cutting may be dipped in hormone rooting powder. The lower third of each cutting should be inserted into moist growing medium and to be kept in humid conditions. *Myriophyllum* sp., *Ludwigia* sp., propagate through stem cuttings. *Synnema triflorum* propagate from leaf cuttings.
    - **Rhizome dividing**—Plants with rhizomes can be artificially propagated by dividing the rhizome into two. The plant after removing from the substrate, the rhizome can be cut carefully. Both pieces should have at least one good shoot growing from them. Each piece of plant can then be re-planted

    Both the sexual and vegetative modes of propagation are well known and most widely used in the aquarium plant trade. They are cheap and fairly successful with most plants. But the disadvantages are both require stock or mother plants to collect the propagation material from whether it is seed or cuttings or other parts of the plant. This takes up a lot of space which could be used for plants that will become saleable. Moreover, prevailing methods of propagation in open areas encounter threats from serious pests, diseases and infestation. Such cultivation systems also make it difficult to meet the balanced fertilizer program, and therefore shelf life is a major factor in the aquarium plant trade. The time factor can with some species like the slower growing *Anubias* can be a problem slowing down production. Tissue culture propagation of these plants is gaining popularity among aquarists, hobbyists and traders as these plants are available to the end user in plastic containers with its roots, well anchored in a balanced nutrient medium, with enhanced shelf life of the plants, and free of diseases (Unnikrishnan, 2019).

- **Micropropagation** or **Tissue culture** (Figure 7.33)

  This is the mode of rapidly multiplying stock plant material to produce a large number of progeny plants, using modern plant tissue culture method (Figure 7.33). It is used to multiply plants such as those that have been genetically modified or bred through conventional plant breeding methods. It is also used to provide a sufficient number of plantlets for planting from a stock plant which does not produce seeds or does not respond well to vegetative reproduction.

  In this method plants are propagated in a sterile environment using part of plant like the meristematic region or the undifferentiated prothallus of a fern. Plants are generally grown in clear plastic or glass containers under controlled lighting and temperature. Normally a sterile jelly like medium is used that has nutrients, antibiotics or hormones etc., to control the plants growth. Once the plants have reached a suitable size they are taken out of the container and hardened in greenhouse conditions. Most *Anubias* species are produced this way. Growth medium used are sucrose, plant hormones, agar etc. Different methods such as meristem culture, callus culture, suspension culture, embryo culture, protoplast culture, are used. This commercial venture is applied to rare and expensive plants.

  It allows a large turnover of plants in a very short time with very little space and ensures healthy disease- free plants that generally give a better-looking plant once it has reached saleable size. It may also help with preservation of the wild plants that are collected for propagation material. Certain *Cryptocoryne* species which increase only slowly by other methods, tissue culture technique is applied.

  The disadvantages are that it is a specialist type of production and a lot of experimentation is needed. Set-up costs are generally high too, making the end products or plants expensive. Still it is used by the large producers and promises to play a bigger role in the growing future of aquarium plants.

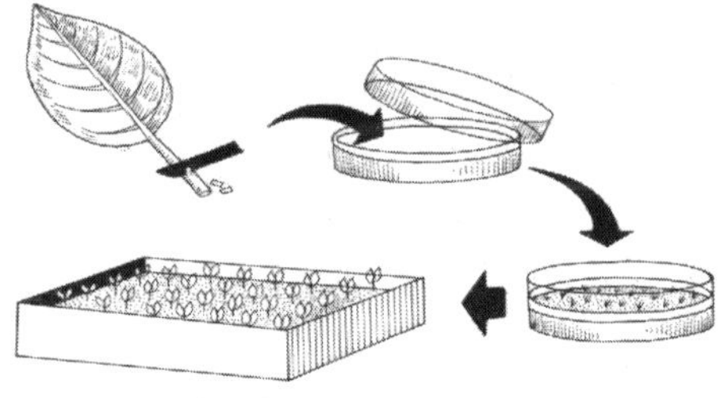

Figure 7.33 Micropropagation or Tissue culture.

7.10.2 Promising methods of producing plants without soil

Advanced methods in climatically controlled conditions are adopted nowadays by some farmers. These methods are as follows.

- **Controlled-environment agriculture (CEA) system**—It is the practice in which the modification of natural environment takes place in enclosed structures to increase crop yield or extend the growing season.
- **Hydroponics**—It is the technique of growing plants without soil. Here, the roots of plants are submerged in liquid solutions containing macronutrients, such as nitrogen, phosphates, sulphur, potassium, calcium and magnesium, as well as trace elements, including iron, chlorine, manganese, boron, zinc, copper and molybdenum. Additionally, inert (chemically inactive) medium such as gravel, sand and sawdust are used as soil substitutes to provide support for the roots. The advantages of hydroponics are the ability to increase yield per area and reduce water usage (Resh, 2016). The most common hydroponic method is using **Drip system** which delivers nutrient enriched water to plants grown in rockwool slabs (Figure 7.34).

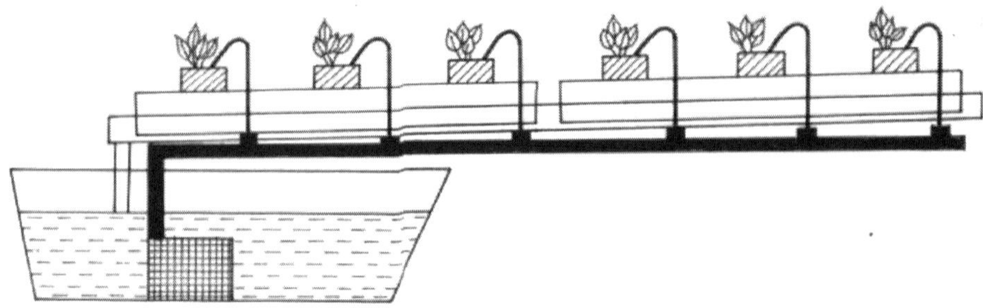

Figure 7.34 Drip system which delivers nutrient enriched water to plants grown in rockwool slabs.

- **Aquaponics**—In this method, aquaculture is combined with hydroponics i.e., the production of plants is integrated with the production of aquatic animals in a closed-loop system. Nutrient-rich waste water from the fish tank is filtered by a solid removal unit and then led to a biofilter, where toxic ammonia is converted to nitrate. The plants absorb nutrients and then purify the waste water, which is recycled back to fish tanks. The plants also consume carbon dioxide produced by the fish, and water in the fish tanks gets heat and helps the greenhouse to maintain temperature at night (Hai et al., 2018).
- **Aeroponics**—Motivated by the initiative of NASA (the National Aeronautical and Space Administration) to find an efficient way to grow plants in space in the 1990s. In this method liquid solution mixed with nutrients is misted in air chambers where the plants are suspended (Mytton-Mills, 2018).
- **Vertical farming**—It is the practice of growing crops in vertically stacked layers, often incorporating CEA, hydroponics, aquaponics and aeroponics. The advantages of this technology, is the increased crop yield that comes with a smaller unit area of land requirement (Benke and Tomkins, 2017), being eco-friendly with smaller carbon footprint as produce is grown locally.

7.10.3 Growing media used for aquarium plant production

The medium in the aquarium generally serves as an anchorage and as a source of nutrients for the plants.

There are several growing media available for aquarium plants, among these rockwool and cultiwool are most commonly used nowadays.

- **Cultiwool**—This fibre glass medium is inert in character so neither affect the water quality nor pollute the aquarium environment. It has high air fill porosity, wets easily and has a high nutrient exchange capacity. Moreover, this clean and excellent growing medium is easy to work with.
- **Rockwool**—This popular growing medium is made from basalt rock and chalk. It is sterile, porous and non-degradable medium. It is melted back and poured into a spinner, to make fibres which are formed into different shapes and sizes such as rigid slabs, blocks and cubes. It is also available as a highly refined and consistent hydrophilic or hydrophobic granulate which is basically water absorbent and water repellent. It retains water, holds air, comes in a variety of sizes and shapes, clean and convenient. But problems are it is not environmentally friendly, dust and fibres are a health risk.

Other media for the aquaria—

- Organic and animal material and manure
- Peat moss
- Mineral soils—made up of clays, silts and sands. Most useful.
- Red clay soils have a high soil nutrient capacity and have a high iron content.
- Commercial mixes—prepared commercially. These often named as commercial laterite is basically red clay.

Aquarium plants are often sold fresh without any medium. If any medium is present, for example rockwool, that should be carefully removed before planting in the aquarium. Plant roots are often trimmed and fast- growing plants are cut off before placing them in the tank.

7.10.4 Fertilizers used for aquarium plants—

Aquatic ornamental plants absorb nutrients also through their leaves and their roots. Since there is no danger of drying out, these plants either have a very thin or no cuticle on their leaves and stems. This, though helps in the absorption of the nutrients but exposes the plants to damage and diseases.

7.10.4.1 Nutrients—

- **Macronutrients**—nitrates, phosphates, calcium, chlorides, sodium, magnesium, potassium, sulphates.
- **Micronutrients**—iron, boron, copper, zinc, molybdenum, manganese.

  Aquarium plants show symptoms if there is deficiency in the required nutrients (Table. 7.9).

Table 7.9: Essential nutrients and their deficiency symptoms in aquarium plants

Nutrients	Deficiency symptoms
Calcium	New leaves are stunted
Carbon dioxide	Stunted growth of plants, white deposits on leaves.
Iron	Young leaves are yellow or white in colour.
Magnesium	Lower leaves of the plant turn yellow.
Manganese	Yellow spots and elongated holes between veins of the leaves.
Nitrogen	Leaves of the lower part of plant turn yellow.
Phosphorus	Loss of leaves and leaves become darker in shade than normal.
Potassium	Tender leaves turn yellow at the tips and edges. Dead or yellow patches or spots develop on leaves.

7.10.4.2 Aquarium plant fertilizer mixes:

A good aquarium plant **fertilizer mixes** usually contain very few or no macronutrients. This is simply to avoid algae generation. The macronutrients for the plants should come from water changes. Tap water normally contain these nutrients in small amounts and some amount comes from fish. As water changes are done at least every 2 weeks, about 15 % should be drained and replaced. This presumably supply sufficient macronutrients. Micronutrients cannot be supplied by tap water in good amounts. More over micronutrients are very unstable in their inorganic form and normally precipitate and become unstable. All micronutrients and iron should be in the chelated form.

Chelate—The chelate is an organic molecule which binds the metal-ions tightly and allows them to be dissolved in water with no precipitation. This makes the nutrient available to the plant.

Types of chelate--Two among six types of chelates are used for the production of aquarium plants. These are--

DTPA (Diethylenetriaminepentaacetic acid)—It is widely used as it is stable to pH of 7.5, which suits the aquarium conditions. This chelating agent, has the ability to extract exchangeable, carbonate and organically bound metal fractions i.e. potentially mobile and phytotoxic fractions.

EDTA (Ethylenediaminetetraacetic acid) –It is unstable at a higher pH than 6.0, but aquarium water pH remains between 6.0-8.0. This is in use being a cheaper variety. Forms soluble complexes with heavy metal ions (Aboulroos, 1981).

7.10.4.3 Importance of carbon dioxide and oxygen in aquarium:

When number of plants are more in an aquarium, the utilization of carbon dioxide for carrying out photosynthesis is more. As a consequence, deficiency of carbon dioxide may be realized in the tank. Carbon dioxide trapped as bicarbonates is used then. The pH rises above 7.5-8.0 in freshwater, and calcium may form a fine layer on plants and suffocate them. To overcome this, carbon dioxide diffusers are used in the aquaria. Specific plants are also chosen for aquariums with various water qualities (Table.7.10).

Table 7. 10: Plants suitable for aquariums with different water qualities

Aquaria with different water quality	Aquarium plant species
Aquarium with slightly hard and alkaline water	*Cabomba, Bacopa, Ludwigia, Myriophyllum, Sagittaria, Cryptocoryne*
Aquarium with soft and acidic water	*Cabomba, Echinodorus, Eichornia, Myriophyllum, Salvinia*
Aquarium with variable water quality	*Anubias, Nymphaea*
Aquarium with slightly soft and acidic water	*Hydrophila, Nymphaea, Rotala, Vallisneria, Acorus*
Aquarium with moderate hardness	*Aponogeton*
Aquarium with wide range of water quality	*Ceratophyllum, Elodea, Lemna, Ludwigia, Pistia, Riccia*
Aquarium with temperate water quality	*Ceratophyllum, Elodea, Lemna, Riccia*

7.10.3.4 Light requirement in aquarium:

To know how much light is needed, a reasonable rule of thumb is about "one watt per litre" of aquarium water. Though it may vary with different plant species (Table.7.11).

Table 7.11: Amount of light required by different ornamental plant species

Light required	Aquarium plant species
Low-moderate	*Anubias* sp., *Cryptocoryne* spp., *Echinodorus* sp., Amazon swords, Java fern, Java moss
Bright	*Ceratopteris* sp., *Hygrophila* spp., *Ludwigia* sp., *Myriophyllum* sp., *Nymphaea* sp., *Vallisneria* sp.
Very bright	*Bacopa, Cabomba, Rotala* sp.

7.11 References

Aboulroos S.A. *Reaction of EDTA, DTPA, and EDDHA Complexes of zinc, copper, and manganese with a calcareous soil*. Journal of Plant Nutrition and Soil Science. 144(2). 1981, pp. 164-173.

Alderton D. *The Aquarium Owner's handbook*. Cassell' Illustrated. London. 2003, pp. 1-144.

Benke K. and Tomkins B. *Future food production systems: vertical farming and controlled-environment agriculture*. Sustainability Science, Practice and Policy. 13(1). 2017, pp. 13-26.

Biju Kumar A. and Alappat H.J. *A Complete Guide to Aquarium Keeping*. D.K. Publishers Distributors (P) Ltd. I-xv. 1995, pp. 1-80.

Das A. K. Mangroves. *Ecosystems of India*, edited by Alfred, J.R.B., Das, A.K. and Sanyal, A.K. ENVIS-Zool. Surv. India, Kolkata. 2001, pp. 239-259.

Das N.R. *Introduction to aquatic and Semi-aquatic Plants of India*. Kalyani Publishers. India. I-xxix. 2012, pp. 1-335.

Hai F.I., Visvanathan C. and Boopathy R. (Editors). *Sustainable Aquaculture*. Springer International Publishing. 2018, pp. 1-322.

Mytton-Mills H. *Smart Futures, Challenges of Urbanisation and Social Sustainability*. Springer International Publishing. 2018, pp. 169-191.

Naskar K.R. *Aquatic and semi-aquatic plants of lower Ganga Delta*. Daya Publishing House, Delhi, India. 1990, pp. 1-408.

Ostrow M. *Goldfish*. Barron's Educational Series., Inc. 1995, pp. 1-94.

Resh, Howard M. *Hydroponic food production: a definitive guidebook for the advanced home gardener and the commercial hydroponic grower*. Seventh ed. Boca Raton, FL. 2016.

Unnikrishnan S.K. *Latest trends in Propagation of Aquarium Plants*. Biotopeone.com. 2019.

Ward B. *Aquarium Fish, Survival marual*. Quantum Publishing, London, 2004, pp.1-175.

8

Ornamental Invertebrates and Vertebrates

8.1 Introduction

There are many living creatures other than fish and plants, which provide grace to the aquariums. They also provide shelter, help in procuring food, maintain tank environments and serve as breeding ground. Aquarists favour these decorative animals and they are very much part of the aquarium trade.

8.2 Freshwater Aquarium Invertebrates and Vertebrates

There are a number of adorning animals that are kept in freshwater aquariums along with fishes and plants. They are classified accordingly (Young, 1981; Ruppert and Barnes, 1994; Kardong, 2007; Pechenik, 2010) and are listed here (Table. 8.1).

Table 8.1: List of Freshwater Aquarium Invertebrates and Vertebrates

Classification	Common Name (S)
Phylum: Arthropoda Subphylum: Mandibulata Class: Crustacea	Shrimps: European dwarf shrimp, green lace shrimp, bamboo shrimp, green shrimp, Malayan dwarf shrimp, striped shrimp, red-nose, white-nose, Amano bee, blue tiger, crystal black, crystal red, tiger shrimp, tangerine shrimp, Chinese zebra shrimp, Indian dwarf shrimp, Indian white-banded shrimp, Sulawesi, harlequin, Sulawesi cardinal, bumble bee, black midget, Ninja, Malawa shrimp, Sri Lankan dwarf shrimp, Red claw shrimp, Guinea swamp shrimp, Amazon zebra shrimp, Rusty longarm, Fuzzy claw, Sand shrimp, Riceland prawn, Caribbean dwarf filter, Wild type cherry, Cherry, Yellow, Taiwan Pale Blue, Marbled dwarf, Blue pearl shrimp, Snowball, Amazon shrimp, American ghost, purple zebra, Yellow nose. Crayfish: Blue crayfish, Red swamp, Marble crayfish, common yabby, Australian red claw, Acocil, Cajun dwarf, Mexican dwarf, Signal crayfish. Crabs: Thai Micro crabs, Panther crabs. Branchiopods: Golden, tadpole shrimp, longtail tadpole shrimp, Australian tadpole, European tadpole, Asian tadpole, desert tadpole, Thai fairy, alkali fairy, giant fairy, beavertail fairy, red tail, dry lake fairy, winter fairy, large water flea, common water flea, Japanese water flea, clam shrimp. Hermit crab: *Clibanarius fonticola.*

Phylum: Mollusca Subphylum: Conchifera	Gastropods: Zebra apple snail, Colombian ram's horn, spike-topped apple snail, channelled apple snail, bladder or tadpole snails, pond snails, great pond snail, Great ram's horn, Malaysian trumpet, spike-tail trumpet, Japanese trap door, horned nerite, zebra nerite, red onion, olive nerite, red lips nerite, freshwater limpet, dusky nerite, piano snail, giant lower cap, miniature red ram's horn, Tylo snails, black Sulawesi, tiger tuba, west African freshwater snail, bella snail, hairy tower lid snail. Bivalves: Asian clam.
Phylum: Chordata Subphylum: Vertebrata Class: Amphibia	Frogs: African dwarf frog, African clawed frog, Indonesian floating frog, Western clawed frog, Surinam toad, Poison dart frog. Newts: Paddle-tail newt, Fire belly newt, spotted paddle tail newt. Salamanders: Mole salamander, Axolotl, mud puppy, amphiuma, hellbender, giant salamander. Caecilians: Cayenne caecilian, rubber eel.
Phylum: Chordata Subphylum: Vertebrata Class: Reptilia	Turtles: Red-eared slider, yellow-bellied slider, Cumberland slider, pig-nosed turtle, spotted turtle, razor-backed musk turtle.
Phylum: Chordata Subphylum: Vertebrata Class: Mammalia	Cetaceans: Amazon river dolphin, Tucuxi, Yangtze river dolphin, Ganges river dolphin, Indus river dolphin, Irrawaddy dolphin, Finless porpoise. Sirenians: African manatee, Amazonian manatee.

8.3 Account of Freshwater Aquarium Species

8.3.1 Blue poison dart frog Figure 8.1

Dendrobates tinctorius (Cuvier, 1797)

Dendrobates tinctorius "azureus" (Hoogmoed, 1969)

Common Name(s):

English—Blue poison dart frog, blue poison arrow frog.

IUCN Red List Status: Least Concern (LC).

Systematic Position:

Kingdom: ANIMALIA

Phylum: CHORDATA

Class: AMPHIBIA

Order: ANURA

Clade: HYLOIDEA

Family: DENDROBATIDAE

Genus: *Dendrobates*

Species: *D. tinctorius*/ *D.tinctorius "azureus"*

Scientific Name: *Dendrobates tinctorius* (Cuvier, 1797)

Dendrobates tinctorius "azureus"(Hoogmoed, 1969). It was considered a valid species for a long time, until recently it is considered as a morph of *D. tinctorius*.

Description:

- Body is of bright blue colour, darker around its limb and stomach.
- The black spots on the head and back are unique to each frog.
- It has a distinct hunch -backed posture.
- Each foot has four toes, each with a flattened tip, having a suction cup pad which is used for gripping.
- The glands of poisonous alkaloids are present.
- Tadpoles, as in other frogs, differ greatly in appearance from others.
- Sexual Dimorphism: Males are smaller than females, but have larger toes. Tips of toes are heart-shaped in males, and round-shaped in females.

Species Details:

- Distribution:

  In Sipaliwini Savanna, located in southern Suriname and adjacent far northern Brazil.
- Habitat: It inhabits the forest floor of tropical rain forest.
- Food and feeding: Carnivorous. Feed on ants, flies, beetles, mites, spiders, termites. In captivity, feed on springtails, fruit flies and rice flour beetle larvae.
- Maximum length: 3.0-4.5 cm.
- Life span: 5-7 years in the wild. About 10 years in captivity.

Aquarium Aspects:

- Minimum tank size: 40 gallons.
- Substrate: Nonorganic substrate with plants, dried leaves, rocks and logs.
- Tank conditions: Paludarium with adequate water and land area with gentle sloping sides.
- Temperature: 72-80° F.

Behaviour:

Social—Though a mainland animal, it stays close to water sources. They are very territorial and aggressive both towards their own species and others. A series of calls, chases and wrestling are used by them to ward off intruders. They are active during day and can be found hiding among boulders and debris. Poison dart frogs are known for their skin toxins, used on the tips of darts or arrows by the natives.

Reproductive—Breed seasonally when the weather is rainy. Males produce quiet calls for females. Fertilization occurs externally. Eggs are laid out of water, and then the tadpoles are carried by the males to puddles (tree holes etc.), where they develop further. At each mating 5-10 offspring are produced. The eggs hatch after

14 to 18 days, the tadpoles attain maturity after 10 to 12 weeks. These frogs are bred in captivity since the 1970s.

Remarks: It is kept in semi-aquatic home or public paludariums, which simulate rain forests.

8.3.2 Red-eared slider Figure 8.2

Trachemys scripta (Thunberg, 1792)

Common Name(s):

English—Red-eared slider, red-eared terrapin, red-eared slider turtle, red-eared turtle, water slider turtle.

IUCN Red List Status: Near Threatened (NT).

Systematic Position:

Kingdom: ANIMALIA

Phylum: CHORDATA

Class: REPTILIA

Order: TESTUDINES

Suborder: CRYPTODIRA

Superfamily: TESTUDINOIDEA

Family: EMYDIDAE

Genus: *Trachemys*

Species: *T. scripta*

Scientific Name: *Trachemys scripta* (Thunberg, 1792)

Description:

- Medium-sized, semi-aquatic turtle.
- The carapace is oval in shape with a little rounding, a central crest with knobs.
- The carapace is olive green to orange brown or brown and the plastron is a dull yellow or darker.
- The head, legs and tail are green with fine, irregular, yellow lines.
- There is a red stripe on each side of the head. This stripe is located behind the eyes.
- Some have a small mark of this colour on the top of their heads.
- Sexual Dimorphism: Males are normally smaller (carapace diameter 10 cm) than the females (carapace diameter 15 cm). Males have longer claw on their front feet than the females.

The tail of male turtle is thicker and longer. The red eye patches in male are brighter and bigger. The male's plastron (lower ventral carapace) is slightly concave, while that of the female is completely flat.

Species Details:

- Distribution:

  South Eastern United States, the Mississippi River and the Gulf of Mexico. Southeast of Colorado to Virginia and Florida.
- Habitat: Found in swamps, in marshy areas with warm and calm water such as ponds, lakes, slow-moving rivers, creeks etc.
- Food and feeding: Omnivorous. But as they grow older, they become more herbivorous. In aquarium, both plant-based and animal-based items are to be given and good quality commercial pellets also.
- Maximum length: Carapace diameter: in male—10 cm; in female—15 cm.
- Life span: 20-40 years. Up to 80 years in the wild.

Aquarium Aspects:

- Minimum tank size: 10 gallons.
- Water conditions: Clean tap water with--Temperature: 77-80° F (25.5-26.5° C); pH: 7.0-7.5.
- Light: Enough light, 5% or more UVB light.

Behaviour:

Social—Harmless and very peaceful. They are almost completely aquatic, very good swimmers. Gregarious, and very much alert about the surroundings. They are territorial also. These turtles do not hibernate but brumate (remain inactive).

Reproductive: Females lay 2 to 30 eggs at a time. The eggs are laid inside a burrow. Gestation period is from days to weeks, and the incubation time is 59 to 112 days.

Remarks: They are very popular around the world as an aquarium pet or for outdoor ponds.

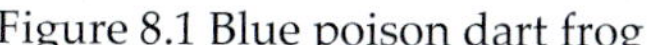

Figure 8.1 Blue poison dart frog

Figure 8.2 Red-eared slider

8.4 Brackish Water Aquarium Invertebrates and Vertebrates

Table 8.2: List of brackish water aquarium invertebrates and vertebrates

Classification	Common Name
Phylum: Arthropoda Subphylum: Mandibulata Class: Crustacea	Fiddler crab
	Red clawed crab
	Amano shrimp
	Grass shrimp
	Blue crab
Phylum: Mollusca Subphylum: Conchifera Class: gastropoda	Olive Nerite
Class: Cephalopoda	Dart squid
	Argus brief squid
	Atlantic brief squid
	Panama brief squid
Phylum: Chordata Subphylum: Vertebrata Class: Reptilia	Diamondback terrapin
Phylum: Chordata Subphylum: Vertebrata Class: Mammalia	West Indian manatee
	African manatee

8.5 Detail of Brackish Water Ornamental Vertebrate Species

8.5.1 Diamondback terrapin

Malaclemys terrapin (Schoepff, 1793)

Common Name(s):

English—Diamondback terrapin, terrapin.

IUCN Red List Status: Vulnerable (VU).

Systematic Position:

Kingdom: ANIMALIA

Phylum: CHORDATA

Class: REPTILIA

Order: TESTUDINES

Suborder: CRYPTODIRA

Superfamily: TESTUDINOIDEA

Family: EMYDIDAE

Subfamily: DEIROCHELYINAE

Genus: *Malaclemys*

Species: *M. terrapin*

Scientific Name: *Malaclemys terrapin* (Schoepff, 1793)

Description:

- The carapace is wider at the back than in the front.
- They have large webbed feet.
- Carapace has a diamond pattern, though overall pattern and colouration vary greatly.
- The shell colour may be brown to grey, and the body colour may be grey, brown, yellow or white.
- They have a unique pattern of wiggly, black markings or spots on their body and head.
- Sexual Dimorphism: Males are smaller than females. Males with carapace length of 10-14 cm and females with carapace length of 15-23 cm. Females have proportionately larger head.

Species Details:

- Distribution:

  In a narrow strip of coastal habitats on the Atlantic and Gulf coasts of the United States, from as far north as Cape Cod, Massachusetts, to the southern tip of Florida and around the Gulf Coast to Texas.
- Habitat: Live in Spartina marshes that are flooded at high tide and in mangrove swamps in Florida.
- Food and feeding: Omnivorous. Feeds on fish, marine snails, clams, mussels, some plants, other molluscs, hard-shelled molluscs, and crustaceans. In captivity, turtle pellets, dried shrimps and sea food may be provided.

Aquarium Aspects:

- Minimum tank size: 75 gallons.
- Water condition: Temperature: > 70° F; light: Heat light required.

Behaviour:

Social—They are quite docile in nature. In colder months they may hibernate in the mud of creeks and marshes.

Reproductive—Fertilization is internal. Nests are built in sand dunes or scrub vegetation near the ocean. Females lay 2-3 clutches of eggs annually, with clutch sizes ranging from 4-23 eggs. Eggs hatch in about 60-85 days.

Remarks: They are popular inmates of public aquariums.

8.6 Marine Ornamental Invertebrates and Vertebrates

A large number of marine invertebrates and vertebrates are exhibited in home and public aquariums. They are listed as follows (Table 8.3).

Table 8.3: List of marine ornamental invertebrates and vertebrates

Classification	Common Name	Scientific Name
Phylum: Porifera Subphylum: Cellularia Class: Demospongiae (sponges)	Branching vase	*Callyspongia vaginalis*
	Red ball	*Dragmacidon lunaecharta*
	Red tree	*Amphimedon compressa*
Class: Calcarea (sponges)	Pineapple	*Sycon* sp.
Phylum: Cnidaria Class: Scyphozoa (jellyfish)	Purple striped	*Chrysaora colorata*
	Sunbathing	
	Sea Nettles	
	Lion's mane	
	Blue Blubber	*Catostylus mosaicus*
	Moon jellyfish	*Aurelia aurita*
	Upside down jellyfish	*Cassiopea sp.*
Class: Anthozoa (sea anemones)	Snakelocks anemone	*Anemonia viridis*
	Green surf anemone	*Anthopleura xanthogrammica*
	Magnificient anemone	*Heteractis magnifica*
	Aiptasia	*Aiptasia sp.*
	Bubble-tip anemone	*Entacmaea quadricolour*
	Delicate sea anemone	*Heteractis malu*
	Long tentacled anemone	*Macrodactyla doreensis*
	Rock flower anemone	*Phymanthus crucifer*
	Tube anemone	*Carianthus sp.*
	Condy anemone	*Codylactis gigantea*
Class: Anthozoa (corals) (corallimorphs)	Jewel anemone	*Corynactis viridis*
	Knobby mushroom coral	*Ricordea florida*
	Knobby mushroom coral	*Ricordea Yuma*

Classification	Common Name	Scientific Name
	Disk anemone	*Discosoma sp.*
	Strawberry anemone	*Corynactis californica*
Class: Hydrozoa (hydrocorals)	Fire coral	*Millepora sp.*
	Lace coral	*Distichopra sp.*
Class: Anthozoa (large-polyp stony corals)	Black sun coral	*Tubastraea micrantha*
	Bubble coral	*Plerogyra sinuosa*
	Candy cane coral	*Caulastrea furcate*
	Elegance coral	*Catalaphyllia jardinei*
	Flower pot coral	*Goniopora sp.*
	Frogspawn coral	*Euphyllia divisa*
	Hammer coral	*Euphyllia ancora*
	Lobed brain coral	*Lobophyllia hemprichii*
	Open brain coral	*Trachyphyllia geoffroyi*
	Pineapple brain coral	*Favia sp.*
	Sun coral	*Tubastraea sp.*
	Torch coral	*Euphyllia glabrescens*
	Whisker coral	*Duncanopsammia axifuga*
Small-polyp stony coral	Pink bird's nest coral	*Seriatopora hystrix*
	Cauliflower coral	*Pocillopora sp.*
	Dimpled encrusting Montipora	*Montipora verrucose*
	Finger coral	*Montipora digitate*
	Millepora coral	*Acropora millepora*
	Plating montipora	*Montipora capricornis*
	Staghorn coral	*Acropora cervicornis*
Soft corals	Cabbage leather coral	*Sinularia brassica and S.dura*
	Clove polyps	*Clavularia* sp.
	Devil's hand leather coral	*Lobophytum sp.*
	Finger leather coral	*Sinularia sp.*
	Jasmine polyp	*Knopia*
	Pulse coral	*Xenia sp.*
	Red chili coral	*Nephthyigorgia sp.*
	Spaghetti leather coral	*Sinularia flexibilis*
	Star polyps	*Clavularia viridis*
Zoanthids	Stick polyps	*Acrozoanthus*

Classification	Common Name	Scientific Name
	Button polyps, zoas	*Zoanthus*
	Button polyps, palys	*Palythoa*
	Button polyps, protopalythoa	*Protopalythoa*
Phylum: Mollusca Class: Bivalvia	Atlantic thorny oyster	*Spondylus americanus*
	Bear paw clam	*Hippopus hippopus*
	Carditid	*Cardites floridanus*
	Blue clam	*Tridacna crocea*
	China clam	*Hippopus porcellanus*
	Electric flame scallop	*Ctenoides ales*
	Flame scallop	*Ctenoides scaber*
	Fluted giant clam	*Tridacna squamosa*
	Giant clam	*Tridacna gigas*
	Maxima clam	*Tridacna maxima*
	Southern giant clam	*Tridacna derasa*
	Thorny oyster	*Spondylus sp.*
	Pecten	
Class: Gastropoda	Abalone	*Haliotis sp.*
	Arabian cowrie	*Cypraea arabica*
	Astraea snail	*Astraea sp.*
	Bumble bee snail	*Engina mendicaria*
	Cerith snail	*Certhium sp.*
	Gold ring cowrie	*Cypraea annulus*
	Lettuce sea slug	*Elysia sp.*
	Nassarius snail	*Nassarius sp.*
	Queen conch	*Eustrombus gigas*
	Sand conch	*Strombidae*
	Sea Hare	*Aplysia sp.*
	Tiger cowrie	*Cypraea tigris*
	Turbo snail	*Turbo sp.*
Class: Cephalopoda	Common tropical octopus	*Octopus vulgaris*
	Dwarf cuttlefish	*Sepia bandensis*
	European common cuttlefish	*Sepia officinalis*
Phylum: Annelida Class: Polychaeta	Christmas tree worms	*Spirobranchus giganteus*

Classification	Common Name	Scientific Name
	Cluster duster	*Bispira brunnea*
	Feather duster worm	*Sabellastarte sp.*
Phylum: Arthropoda Subphylum: Chelicerata Class: Merostomata	Atlantic horseshoe crab	*Limulus polyphemus*
Class: Arachnida	Sea spider	*Pycnogonids*
Subphylum: Mandibulata Class: Crustacea	Anemone crab	*Neopetrolisthes maculatus*
	Arrow crab	*Stenorhynchus seticornis*
	Hermit crab	*Paguroides sp.*
	Emareld crab	*Mithraculus*
	Pom-pom crab	*Lybia tessellate*
	Sally lightfoot crab	*Percnon gibbesi*
	Spider decorator crab	*Camposica retusa*
	Spiny lobster	*Panulirus versicolor*
	Brine shrimp	*Artemia salina*
	Sexy shrimp	*Thor amboinensis*
	Snapping shrimp	*Alpheidae sp.*
	Peacock mantis shrimp	*Odontodactylus scyllarus*
	Coral banded shrimp	*Stenopus hispidus*
	Camel shrimp	*Rhynchocinetes durbanensis*
	Harlequin shrimp	*Hymenocera sp.*
	Peppermint shrimp	*Lysmata wurdemanni complex*
	Skunk cleaner shrimp	*Lysmata amboinensis*
Phylum: Echinodermata Subphylum: Echinozoa Class: Holothuroidea	Sea apple	*Pseudocolochirus axiologus*
	Yellow sea cucumber	*Colochirus robustus*
	Florida sea cucumber	*Holothuria floridana*
	Pink and black sea cucumber	*Holothuria edulis*
	Tiger tail sea cucumber	*Holothuria hilla*
Subphylum: Asterozoa Class: Stelleroidea	Common starfish	*Asterias rubens*
	Spiny starfish	*Marthasterias glacialis*
	Chocolate chip sea star	*Protoreaster nodosus*
	Mottled linckia	*Linckia multifora*

Classification	Common Name	Scientific Name
	Blue and pink sea star	*Astropecten sp.*
	Brittle star	*Ophiomastix sp.*
	Bun star	*Culcita novaeguinea*
	Blue linckia	*Linckia laevigata*
	Indian sea star	*Fromia indica*
	Little red star	*Fromia elegans*
	Purple linckia	*Linckia teres*
	Red sea star	*Fromia millepora*
	Red-knobbed starfish	*Protoreaster linckii*
	Sand sifting sea star	*Astropecten polyacanthus*
	Tiled sea star	*Fromia monilis*
Subphylum: Crinozoa Class: Crinoidea	Feather star	*Himerometra robustipinna*
Subphylum: Echinozoa Class: Echinoidea	Common sea urchin	*Echinus esculentus*
	Black longspine urchin	*Diadema setosum*
	Collector urchin	*Tripneustes gratilla*
	Globe urchin	*Mespilia globulus*
	Slate pencil urchin	*Eucidaris tribuloides*
	Purple short spine pincushion urchin	*Pseudoboletia maculate*
	Red slate pencil urchin	*Heterocentrotus mamillatus*
	Reef urchin	*Echinometra sp.*
	Variegated urchin	*Lytechinus variegatus*
Phylum: Chordata Class: Aves	Penguins	
	Gulls	
Phylum: Chordata Class: Mammalia (cetaceans)	whales, dolphins and porpoises	
(sirenians)	manatees and dugongs	

8.7. Detailed Account of some Marine Aquarium Invertebrates and Vertebrates

8.7.1 Branching vase sponge

Callyspongia vaginalis (Lamarck, 1814)

Common Name:

English—Branching vase sponge.

Systematic Position:

Phylum: PORIFERA

Class: DEMOSPONGIAE

Subclass: CERACTINOMORPHA

Order: HAPLOSCLERIDA

Family: CALLYSPONGIIDAE

Genus: *Callyspongia*

Species: *C. vaginalis*

Scientific Name: *Callyspongia vaginalis* (Lamarck, 1814)

Description:

- Long, erect tubes of this sponge have a wide vent up to 2.5 cm in diameter.
- The erect tubes taper slightly and has thin wall.
- Body is very elastic and surface rough with its irregular pits and nubs covering its surface.
- Colour ranges from brownish-grey, greenish grey, light tan to lavender.

Species Details:

- Distribution:

  Caribbean, Florida, Bermuda and the Bahamas.
- Habitat: Inhabit shallow and mid-range reefs, at temperature of 20-24° C, on walls and rocky areas. The species is frequently infested with sponge zoanthids *Parazoanthus parasiticus,* a colonial anemone, and *Ophiothrix suensonii*, a brittle star.
- Food and feeding: Feeds on plankton and detritus.
- Maximum length: 30.0 cm height.

Behaviour:

Social—This sponge and the brittle star *Ophiothrix lineata* show mutualistic symbiotic interaction.

Reproductive—It is hermaphroditic. The zygote develops into parenchymula larva before settling down on a substrate. Then it grows into a young sponge.

8.7.2 Purple-striped jellyfish

Figure 8.3

Chrysaora colorata (Russell, 1964)

Common Name(s):

English—Purple-striped jellyfish, purple jellyfish, purple-striped sea nettle.

IUCN Red List Status: Not Evaluated.

Systematic Position:

Phylum: CNIDARIA

Class: SCYPHOZOA

Order: SEMAEOSTOMEAE

Family: PELAGIIDAE

Genus: *Chrysaora*

Species: *C. colorata*

Scientific Name: *Chrysaora colorata* (Russell, 1964)

Description:

- The bell (body) has typically a radial pattern of stripes.
- There are eight marginal long dark arms and four central frilly oral arms.
- In adult stage, maroon colour appears as stripes on the bell.
- In very young stage, there is a pinkish colour and its tentacles are long and dark maroon.
- Four frilly arms in young adults become longer. The tentacles thicken and the purple stripes start to darken eventually and the tentacles start to look pale.
- Its oral arms tend to disappear.

Species Details:

- Distribution:

  Southwest Atlantic and Eastern Central Pacific: Brazil and USA.
- Habitat: Found in oceanic or slope waters.
- Food and feeding: They are carnivorous. Feed on copepods, larval fishes, ctenophores, salps, fish eggs, other medusa jellies.
- Maximum length: 80.0 cm WD male/ unsexed.

Behaviour:

Social—They are more active in lively current. Tentacles and oral arms are used in catching prey. This jellyfish is often used as living place for cancer crabs, which in turn eat the parasitic amphipods, causing damages to the jellyfish. Leatherback turtles are the main predator of purple-striped jellyfish.

Reproductive—Two-phase reproductive cycle is shown, asexual and sexual. They are gonochoric. Egg is laid by the mature medusa. The eggs develop into a free-living planula, then scphistoma, strobila and then to a free- living young medusa.

8.7.3 Sea nettle

Figure 8.4

Pacific sea nettle

Chrysaora fuscescens Brandt, 1835

Common Name(s):

English—Pacific sea nettle, West Coast sea nettle.

Systematic Position:

Phylum: CNIDARIA

Class: SCYPHOZOA

Order: SEMAEOSTOMEAE

Family: PELAGIIDAE

Genus: *Chrysaora*

Species: *C. fuscescens*

Scientific Name: *Chrysaora fuscescens* Brandt, 1835

Description

- The bell is big, can grow to be larger than 1m, in diameter, though generally less than 50 cm in diameter.
- The oral arms are white, long and spiralling.
- Twenty-four undulating maroon tentacles trail as far as 15 feet.
- The bell is with distinctive golden-brown colour with a reddish tint.

Distribution

- Eastern Pacific Ocean from Canada to Mexico.
- Habitat: Coasts.
- Food and feeding: Carnivorous. Feeds on variety of zooplankton, crustaceans, salps, pelagic snails, small fish, fish eggs, larvae and other jellyfish.
- Maximum length: Diameter of the bell, 50.0 cm.

Behaviour

Social-- They spread their tentacles like a large net, and are able to catch food as it passes by. Nematocysts are released as the prey touches the tentacles, launching barbed stingers releasing a paralysing toxin. The oral arms start digestion as they transport the prey into the mouth. Their population reach peak during late summer. Predators of sea nettle include sea turtles, several species of fishes and marine birds. They swim against current, although most of the time they just float.

Reproductive—They are capable of both sexual reproduction in the medusa stage and asexual reproduction in the polyp stage.

Remarks: It is very popular for display at public aquaria due to their bright colours and easy maintenance.

8.7.4 Snakelocks anemone

Figure 8.5

Anemonia viridis (Forsskal, 1775)

Common Name(s):

English—Snakelocks anemone

IUCN Red List Status: Not Evaluated (NE)

Systematic Position:

Phylum: CNIDARIA

Class: ANTHOZOA

Order: ACTINARIA

Family: ACTINIIDAE

Genus: *Anemonia*

Species: *A. viridis*

Scientific Name: *Anemonia viridis* (Forsskal, 1775)

Description

- Radially symmetrical body with long greenish tentacles. This anemone has some of the longest tentacles of all sea anemones.
- There are over 200 sticky tentacles and the tentacles are lined with venomous sting cells called cnidocysts.
- Under the tentacles, tubular polyp body is present.
- Tentacles are deep green colour with purple tips due to the presence of symbiotic algae within the tentacles.
- The tentacles may be grey in colour in deep water.

Species Details

- Distribution:

  Northeast Atlantic and the Mediterranean: France.
- Habitat: Inhabit infralittoral zones. Remain attached to the bottom in shallow salt water and intertidal pools. They are found as far down as 20 meters, but are rare below 10-12 meters.
- Food and feeding: Eats fish, non-insect arthropods, molluscs, zooplankton, phytoplankton and detritus.
- Maximum length: 20 cm diameter, 8 cm tall.

Behaviour

Social—They need high intensity sunlight near the surface for their symbiotic algae zooxanthellae. Tentacles are with cnidocytes. Mutualist species present in the anemone is zooxanthellae and commensal species are of majid spider crab and Buchichi's goby. Snakelocks anemone is known to have an alloimmune memory.

Reproductive—These anemones are either gonochoric or hermaphroditic. No parental investment is seen. The zygote develops into a planktonic planula larva. Metamorphosis begins with early morphogenesis of tentacles, septa and pharynx before larval settlement on the aboral end.

Remarks: It is a popular aquarium pet.

8.7.5 Green surf anemone

Figure 8.6

Anthopleura xanthogrammica (Brandt, 1835)

Common Name(s):

English—Green surf anemone, giant green sea anemone, green anemone.

IUCN Red List Status: Not Evaluated.

Systematic Position:

Class: ANTHOZOA

Order: ACTINARIA

Family: ACTINIIDAE

Genus: *Anthopleura*

Species: *A. xanthogrammica*

Scientific Name: *Anthopleura xanthogrammica* (Brandt, 1835)

Description

- It has a broad, flat oral disc surface, without any markings.
- It appears bright green when under water.
- The tentacles are short, conical and are arranged in six or more rows surrounding the oral disc, either pointed or blunt at tips.

Species Details:

- Distribution:

  Pacific Ocean, from Alaska to southern California and sometimes downwards to Panama.
- Habitat: Inhabits sandy or rocky shorelines, where water remains for most of the day. Found in tide pools up to 15 m deep, in deep channels of more exposed rocky shores and concrete pilings in bays and harbours.
- Food and feeding: Carnivorous. Detached mussels, sea urchins, crabs and small fishes.
- Maximum length: 30.0 cm H male/unsexed.

Aquarium aspects

- Substrate: Rocks should be present.
- Depth range: 0-30 m.

Behaviour

Social—Live a solitary life. Sometimes can be seen in groups.

Reproductive—Reproduce sexually via external fertilization of sperm and eggs in the late fall. Pelagic, planktotrophic larvae float in the water. Then dispense and settle in mussel beds.

Remarks: A popular display at marine public aquariums.

8.7.6 Magnificent sea anemone Figure 8.7

Heteractis magnifica (Quoy & Gaimard,1833)

Common Name(s):

English—Magnificent sea anemone, Ritteri anemone, bulb-tip anemone.

IUCN Red List Status: Not Evaluated.

Systematic Position:

Class: ANTHOZOA

Order: ACTINIARIA

Family: STICHODACTYLIDAE

Genus: *Heteractis*

Species: *H. magnifica*

Scientific Name: *Heteractis magnifica* (Ouoy & Gaimard,1833)

Description:

- It has a sticky foot at the base of the pedal column. This foot is much larger than the column in comparison to other anemones.
- Tentacles are present all over the surface of the oral disc. About 3″ long tentacles are finger-shaped and some may be branched. The disc and tentacles are brown in colour but can be green, yellow or white.
- These are topped with spherical or bulb shaped tips which are lighter and generally coloured in iridescent yellowish green.
- The column is deeply coloured, from avocado green, greenish brown and white to sky blue and magenta.
- When these feel threatened, they retract all its tentacles and form a large ball.
- The mouth is at the centre of the oral disc. It remains closed and tight, open only when hungry.

Species Details

- Distribution:

  Indo-Pacific.
- Habitat: Inhabit waters at depths of 3 to 33 feet (1to 10 m) in shallow reefs. They are found singly or in groups in dense colonies.
- They are accompanied by clownfish. This forms a symbiotic relationship for the protection and nourishment of both the animals. These anemones use their venomous cells or nematocysts found in the tentacles to sting and deflect any possible threats or attacks and also use them to catch prey.
- Food and feeding: Carnivorous. They derive nutrition from symbiotic algae, zooxanthellae, wastes from clownfish, food brought by clownfish, shrimp, crustaceans. In captivity, mix of chopped fresh shrimp, fish and mussels.
- Feeding frequency: Daily.
- Maximum length: 99.06 cm.
- Life span: 80-100 years.

Aquarium Aspects:

- Minimum tank size: 50 gallons.
- Suitable for Nano tank: No.
- Live rock requirement: Typical amount.

- Substrate type: Any.
- Lighting needs: High -strong light. Need strong light.
- Temperature: 72.0 to 82.0° F (22.2 to 27.8° C).
- Specific gravity: 1.023-1.025 SG.
- Water movement: Moderate.
- Water region: Top.
- Water changes: Bi-weekly.
- Iodine levels: Addition of iodine and trace elements.

Behaviour:

Social—They are semi-aggressive. In the tank, they are to be kept at least 2-3 feet away from each other. In nature, they host 12 clownfish species.

Reproductive—They do not breed in captivity. In general, they can multiply by sexual and asexual means.

Remarks: Popular public aquarium exhibit.

8.7.7 Fire coral

Figure 8.8

Millepora alcicornis Linnaeus, 1758

Common Name(s):

English—Fire coral, branching fire coral, stinging coral, bladed fire coral.

IUCN Red List Status: Least Concern.

Systematic Position:

Kingdom: ANIMALIA

Phylum: CNIDARIA

Class: HYDROZOA

Order: HYDROIDA

Suborder: ANTHOMEDUSAE

Family: MILLEPORIDAE

Genus: *Millepora*

Species: *M. alcicornis*

Scientific Name: *Millepora alcicornis* Linnaeus, 1758

Description:

- Fire corals have a sharp, calcified external skeleton.
- On these, there are very small nematocysts which contain tentacles that protrude from numerous surface pores.
- These have a bright yellow-green and brown skeleton covering.
- There exist several common growth forms—branching, plate and encrusting.

Species Details:

- The Caribbean Sea, the Gulf of Mexico, Florida, the Cape Verde Islands and along the coast of Central and South America as far as south as Brazil.
- Habitat: It is found at depths up to 40 metres.
- Food and feeding: Feed on zooplankton and phytoplankton zooxanthellae also provide the coral with food.
- Maximum length: 50 cm H.

Aquarium Aspect:

- Substrate: Rocks, wood, dead corals, live corals.
- Light: Bright.
- Water movement: high.

Behaviour:

Social—A number of species of shrimp and fish take shelter among the branches of the fire coral. Fire worm, certain nudibranch and filefish are its predators. It may be aggressive towards gorgonian sea fan, if comes in close proximity.

Reproductive—It reproduces either by asexual or sexual means.

Remarks: It is kept in reef aquariums.

8.7.8. Open brain coral

Figure 8.9

Trachyphyllia geoffroyi (Audouin,1826)

Common Name(s):

English—Open brain coral, green open brain coral.

IUCN Red List Status: Near Threatened (NT).

Systematic Position:

Phylum: CNIDARIA

Class: ANTHOZOA

Order: SCLERACTINIA

Family: TRACHYPHYLLIIDAE

Genus: *Trachyphyllia*

Species: *T. geoffroyi*

Scientific Name: *Trachyphyllia geoffroyi* (Audouin, 1826)

Description:

- These solitary or colonial corals are small, rarely reach more than 20 cm in diameter.
- They are free living and have distinct valley regions separated by walls.
- In colonial forms, the valley regions can have multiple individual polyps.
- They exhibit bilateral symmetry.
- Polyps and mantle are very fleshy.

Species details:

- Distribution:

  Indo-West Pacific.
- Habitat: Inhabits tropical, inter-reef habitats. Found on soft substrates, soft sandy or muddy substrates on sheltered reef slopes or in lagoons.
- Food and feeding: They receive nutrients from zooxanthellae and planktons. In captivity, they eat *Mysis*, fortified brine shrimps, rotifers, cyclops and other meaty foods.
- Maximum length: 20 cm in diameter.

Aquarium Aspects:

- Minimum tank size: 50 gallons.
- Marine lighting: Low to moderate.
- Temperature: 74-83° F (23-28° C).
- Salinity/specific gravity: 1.025.
- Water movement: Low to moderate, gentle.
- Water region: Bottom of the tank.
- Inorganic salts: Calcium: 400-430 ppm; alkalinity: 3.5 MEQ/L; phosphates: 0; magnesium: 1200-1350; strontium:8-10.
- Water changes: 20% a month, 10% biweekly, 5% weekly.

Behaviour:

Social—It is peaceful. Generally solitary, found along with other free-living corals. Sometimes form vast colonies where these large colonies are found in particular protected, shallow island embayments.

Reproductive—They are either gonochoric or hermaphroditic. Mature gametes are spawned through the mouth. The zygote develops into a planula larva.

8.7.9 Torch coral

Figure 8.10

Euphyllia glabrescens Chamisso & Eysenhardt, 1821

Common Name(s):

English—Torch coral.

IUCN Red List Status: Near Threatened.

Systematic Position:

Class: ANTHOZOA

Order: SCLERACTINIA

Family: EUPHYLLIIDAE

Genus: *Euphyllia*

Species: *E. glabrescens*

Scientific Name: *Euphyllia glabrescens* Chamisso & Eysenhardt, 1821

Description:

- It is a colonial coral with a phaceloid formation of corallites of 20-30 mm in diameter. These are spaced at 15-30 mm apart.
- Walls of these are thin and with sharp edges.
- Tentacles of the polyps are large tubular having knob- like tips.
- Tentacles and polyp-tips are generally of two contrasting colours.
- It has long sweeper tentacles tipped with potent cnidocytes.

Species Details:

- Distribution:

  Indo-Pacific. Widely distributed coral species, though rarely found in Red Sea and the Gulf of Aden.
- Habitat: Found in depths of 1-35 m, in a wide range of reef environments. Common in protected lagoons and on reef slopes. On fringing reef.
- Food and feeding: Feeds on variety of food. Planktonic organisms, dissolved organic matter are taken in nature. In captivity, mysis, krill, and brine shrimp can be given.
- Maximum length: 1 meter.

Aquarium Aspects:

- Minimum tank size: 50 gallons.
- Substrate: live sand and live rock.
- Lighting needs: Medium to high.
- Temperature: 74-83° F (23-28° C).
- Salinity: 1.023-1.025.
- Water movement: Moderate.
- Water region: All areas of the aquarium depending on light.

Behaviour:

Social—They are aggressive towards coral of other genera.

Reproductive—Either gonochoric or hermaphroditic. The zygote develops into a planktonic larva. Tentacles, septa and pharynx are formed before larval settlement on the aboral side.

Remarks: A common coral species in marine aquaria.

8.7.10 Whisker coral

Figure 8.11

Duncanopsammia axifuga Milne-Edwards & Haime, 1848

Common Name(s):

English—Whisker coral, duncanops coral, Duncan coral.

IUCN Red List Status: Near Threatened (NT).

Systematic Position:

Phylum: CNIDARIA

Class: ANTHOZOA

Order: SCLERACTINIA

Family: DENDROPHYLLIDAE

Genus: *Duncanopsammia*

Species: *D. axifuga*

Scientific Name: *Duncanopsammia axifuga* Milne-Edwards & Haime, 1848

Description:

- It has large polyps having a wide, flat, circular disc.
- The disc is light brown or green in colour, and is generally patterned by faint stripes or lighter or darker colour.
- The tentacles are thin and usually of brown or green colour.
- Each polyp resembles a miniature sea anemone. The diameter of each polyp can reach a diameter of 2-3 cm.
- The tentacles have rows of stinging cells, the nematocysts.

Species Details:

- Distribution:

  Western Central Pacific: Philippines
- Habitat: Sandy and rocky sea beds. Usually grows at depths up to 30 meters.
- Food and feeding: Phytoplankton and zooplankton.
- Maximum length: 3 cm each polyp height.

Aquarium Aspects:

- Water temperature: 75-80° F (24-27° C).
- Light: Low to moderate.
- Substrate: low-lying rocks.
- Water movement: Low to moderate.

Behaviour:

Social—Grow in colonies of branching or clustered individuals.

Reproductive—Either gonorchoric or hermaphroditic. Planktonic planula larva after formation of tentacles, septa and pharynx settles on the aboral end.

Remarks: They are popular in aquarium trade due to the relatively easy care and fast growth rate even in captive conditions.

8.7.11 Millepora coral

Figure 8.12

Acropora millepora (Ehrenberg, 1834)

Common Name(s):

English—Millepora coral.

IUCN Red List Status: Near Threatened (NT).

Systematic Position:

Phylum: CNIDARIA

Class: ANTHOZOA

Order: SCLERACTINIA

Family: ACROPORIDAE

Genus: *Acropora*

Species: *A. millepora*

Scientific Name: *Acropora millepora* (Ehrenberg,1834)

Description:

- Short branches are cylindrical.
- This small colonial coral grows in clumps.
- The radial corallites are all the same size with projecting lower rims.
- The colour may be green with orange tipped branches, pale pink, orange, plain green or blue.

Species Details:

- Distribution:

  Western and central Indo-Pacific. Ranges from the Red Sea, Keya and South Africa to India, Malaysia, Japan, Indonesia and Australia.
- Habitat: Inhabits shallow water, between 2-12 metres deep, on reef flats and also on upper reef slopes and in lagoons.
- Food and feeding: Phytoplankton, zooplankton, detritus and microbes.
- Maximum length: Each polyp 1.5 cm height, 2 mm diameter.

Aquarium Aspects:

- Water temperature: 22.5° C (72.5° F); with high oxygen concentration.

Behaviour:

Social—Like all other coral, the polyps of this coral are social and sessile animals. They secrete the minerals together to form the skeleton of their colony. Dino flagellates living in the gastrovascular cavity of the corals provide their photosynthetic products to the corals. Predator of this coral is the crown-of -thorns starfish.

Reproductive—Hermaphroditic. Reproduce sexually in "mass spawning". This occurs once a year, around 3 nights in early summer when the moon is nearly full. Large quantities of eggs and sperm are released simultaneously from huge numbers of coral colonies (Hatta, 1999). This coral spawns once per year, but can reproduce by fragmentation at any time. No parental involvement is present. Zygote develops into planktonic planula larva. Larvae settles on aboral end.

Remarks: It is a favourite in reef aquarium trade.

8.7.12 Plating Montipora

Figure 8.13

Montipora capricornis Veron, 1985

Common Name (s):

English—Plating montipora, leaf plate montipora, vase coral, cup coral, montipora plate coral.

IUCN Red List Status: Not Listed.

Systematic Position:

Phylum: CNIDARIA

Class: ANTHOZOA

Order: SCLERACTINIA

Family: ACROPORIDAE

Genus: *Montipora*

Species: *M. capricornis*

Scientific Name: *Montipora capricornis*

Description:

- These are flat, plating colonies of small polyp stony corals.
- Each polyp appears as small bumps on the surface of the skeleton.
- The polyps are generally transparent.
- Zooxanthellae, are symbiotic algae living in the coral tissue. These impart colour to the polyps.
- It grows in colonies of thin plates or scrolling shape, and attached to encrusting base on rock.
- The colours are usually purple, gold or green topped with purple or deep blue polyps. The edge may be a lighter pink or purple.

Species Details:

- Distribution:

  Indian and Pacific Oceans, as well as in reefs in the Red Sea.
- Habitat: It is found on the top half of the reef with adequate sunlight and photosynthesis occurs. It also inhabits in coral reefs and prefer warm sunny temperature.
- Food and feeding: Montipora corals receive nutrients from zooxanthellae. Also feed on planktonic organisms, microscopic food particles. In captivity, they eat zooplankton, tiny plankton, copepods, artemia nauplii, invert larvae and new strains of rotifers.
- Maximum length: 17.78 cm.

Aquarium Aspects:

- Minimum tank size: 10 gallons or larger.
- Lighting: All, high lighting levels.

- Temperature: 74-83° F (23-28° C).
- Salinity/ specific gravity: 1.023-1.025.
- Water movement: Moderate to high.
- Water region: Middle to bottom of the aquarium.
- Water changes: 5% water changes once a week.
- Inorganic salts needed in the tank: Calcium-400-450 ppm; alkalinity-3.2 to 4.5 MEQ/L (8 to 10 dKH); phosphate-0; magnesium-1350-1500; strontium-10.

Behaviour:

Social—They do not have strong defences and are not aggressive corals.

Reproductive—In the wild they reproduce sexually. These corals reproduce asexually also and propagation is rather simple.

Remarks: It is very attractive small polyp stony (SPS) coral, and so it is a favourite with reef aquarists.

8.7.13 Staghorn coral

Figure 8.14

Acropora cervicornis (Lamarck,1816)

Common Name(s):

English—Staghorn coral.

IUCN Red List Status: Critically Endangered (CR).

Systematic Position:

Phylum: CNIDARIA

Class: ANTHOZOA

Order: SCLERACTINIA

Family: ACROPORIDAE

Genus: *Acropora*

Species: *A. cervicornis*

Systematic Position: *Acropora cervicornis* (Lamarck,1816)

Description:

- Branching stony coral having cylindrical branching which may be a few centimeter to over two meters in length and height.
- There are golden tan or pale brown with white tips and they get their colour from the algae that live within their tissue. They have antler-like branches that stem out from a central trunk and angle forward.
- They grow in dense strands and form an interlocking framework known as thickets.
- Each colony is made up by many individual polyps that grow together.
- Each polyp is an exact copy of all the polyps on the same colony.

Species Details:

- Distribution:

  Florida Keys, the Bahamas and the Caribbean islands, western Gulf of Mexico. East coast of Florida, around Palm Beach, Florida. South East Asia: Coasts of Sabah, Malaysia.

- Habitat: It inhabits the clear, shallow water back reef and fore reef environments from 0 to 30 m depth. The upper limit is governed by wave.
- Food and feeding: They procure food from photosynthetic algae that live inside coral cells. They also feed planktons by capturing with the polyp's tentacles.
- Maximum length: 1.2192 meters in height; 1.8288 meters in diameter.

Behaviour:

Social—Form dense groups called "thickets" in very shallow waters. Colonies form strands several meters across. Arborescent corals, they subdivide infrequently. They provide important habitat for other reef animals, especially fishes.

Reproductive—They reproduce sexually once per year after the full moon, in late summer by broadcast spawning, eggs and sperm into the water column. They also form new colonies by fragmentation.

Remarks: Common coral in reef aquariums.

8.7.14 Devil's hand leather coral

Figure 8.15

Lobophytum **Marenzeller**, 1886

Common Name(s):

English—Devil's hand leather coral, finger leather coral, lobed leather coral.

IUCN Red List Status:

Systematic Position:

Phylum: CNIDARIA

Class: ANTHOZOA

Order: ALCYONACEA

Family: ALCYONIIDAE

Genus: *Lobophytum*

Species: *Lobophytum* spp.

Description:

- They have a thick and heavy outer surface with lobed projections forming low encrusting colonies.
- The clusters at the top are crowded together forming finger like structures with stalks, that are fat and short.
- The lobes vary in size and number depending on the species.
- The retractable polyps present are tufts of 8 tentacles.

- The colours range from dull grey, pink, cream, brownish, green and yellow.

Species Details:

- Distribution:

  Indo-Pacific and Red Sea.
- Habitat: Found near the shore in shallow waters on reef flats at depths of 6-18 metres. They also inhabit turbid lagoons.
- Food and feeding: They feed on planktons, microscopic food particles from the water column, dissolved organic matter. They also receive nutrients from the marine algae zooxanthellae, with which they have a symbiotic relationship. In captivity, they may be fed phytoplankton and micro-zooplankton.
- Maximum size: Up to 3 feet (1 metre) in diameter.

Aquarium Aspects:

- Minimum tank size: 50 gallons (190 Litres) or larger.
- Water conditions: Temperature: 76-82 F (24-28 C); salinity:/specific gravity: 1.023-1.025.
- Lighting: Moderate to strong.
- Water movement: Moderate to strong.
- Water region: Top in case of dim light and bottom in case of strong light.

Behaviour:

Social—Semi-aggressive. Not as toxic as some other leather corals.

Reproductive—It is very easy to propagate. They bud or pinch out to form daughter colonies and grow quickly under strong light.

Remarks: A common coral in reef aquarium.

8.7.15 Button polyps "zoas"

Figure 8.16

Zoanthus Lamarck, 1801

Common Name(s):

English—Button polyps, green button polyps, sea mat, zoas.

Systematic Position:

Phylum: CNIDARIA

Class: ANTHOZOA

Order: ZOANTHARIA

Family: ZOANTHIDAE

Genus: *Zoanthus*

Description:

- Soft bodied *Zoanthus* corals with a leathery covering.

- Polyps have short stalks with small flat oral disc.
- From the oral disc, delicate tentacles arise.
- The oral disc has a mouth, or siphonoglyph in the centre.
- The polyps are connected to each other and they grow as mats.
- These corals are found in different colours from brown, grey to bright orange, red , yellow, green, blue and many other colour morphs.
- Like other zoanthid corals, this coral may also have some amount of poison (palytoxin) in them.

Species Details:

- Distribution:

  Atlantic Ocean, and Pacific Ocean.
- Habitat: Shallow tidal areas, lagoons, back reefs, and many other rural areas. Some as the fluorescent orange varieties are found in deeper waters.
- Food and feeding: They feed on planktonic organisms, microscopic food particles and dissolved organic matter. These corals receive majority of their nutrients from their symbiotic partner zooxanthellae. In captivity, they feed on bacteria, algae, dissolved organic material, zooplankton and brine shrimp nauplii.
- Maximum length: Each polyp, 1.25 cm across the top.

Aquarium Aspects:

- Minimum tank size: 10 gallons (38 litres) or larger.
- Water conditions: Temperature: 72-82° F (22-27° C); salinity/specific gravity: 1.023-1.025.
- Lighting: High.
- Water movement: Moderate.
- Water region: All.

Behaviour:

Social—They are not aggressive. They can be kept with small coral reef fish and shrimps.

Reproductive—They are hermaphrodite or of separate sexes. After fertilization, a swimming larva called zoanthina is formed. Buds are also formed at the bases of the coral. Propagation takes place simply by separating the mat or chip from the body.

Remarks: These button polyps are quite favourite with the aquarists. And are the easiest coral to keep.

ARTHROPOD

8.7.16 Caribbean spiny lobster

Figure 8. 17

Panulirus argus (Latreille, 1804)

Common Name(s):

English—Caribbean spiny lobster.

IUCN Red List Status: Data Deficient (DD).

Systematic Position:

Phylum: ARTHROPODA

Subphylum: CRUSTACEA

Class: MALCOSTRACA

Order: DECAPODA

Infraorder: ACHELATA

Family: PALINURIDAE

Genus: *Panulirus*

Species: *P. argus*

Scientific Name: *Panulirus argus* (Latreille, 1804)

Description:

- Body long, cylindrical and covered with spines.
- There are two large spines above the eyestalks.
- They have no claws or pincers.
- The first pair of antennae are slender and biramous.
- The second pair of antennae longer than the body and are covered with forward pointing spines.
- The bases of these antennae are thick and are covered with rows of spines.
- The legs are striped and end in a single spine-like point.
- Colour red to brown and blue in colour.
- Adult lobsters have brown, white or yellow spots on their tails and orange-yellow and black stripes on their tail fans.
- The legs are striped with blue.
- Sexual Dimorphism: In males, the sternum is broader and lacks the striations that are found in females. The second pair of walking legs in males are more elongate than the other legs, and bears longer, curved dactyls. The raised genital openings of males are located ventrally, on the base of the fifth pair of walking legs.

  Females have small chelae on the dactyls of the fifth pair of walking legs. In females, the gonopores are set at the bases of the third pair of walking legs.

Species Details:

- Distribution: Atlantic—eastern, western, central and southwest. From Bermuda and east coast USA, south to Rio de Janeiro in Brazil, including the Caribbean Sea, the Gulf of Mexico and also from Cape Verde.

- Habitat: Inhabits reefs and mangrove swamps. Benthic, lives at depths up to 90 m.
- Food and feeding: Foragers. Feed on gastropods, bivalves and chitons, worms, sea urchins, crustaceans, macroalgae, detritus and carrion.
- Maximum length: 60 cm.
- Life span: 12-20 years.

Behaviour:

Social-- They are nocturnal. Groups of hundreds take migratory march across the floor of Florida, the Bahamas, ranging from North Carolina to Rio de Janeiro, Brazil, the Gulf of Mexico and throughout the Caribbean, during day hours of autumn months. They can produce a loud rasping sound. The sound is produced using a stick-slip method as in the *P. interruptus* (Patek and Baio, 2007).

Reproductive—The female carries eggs for about 4 weeks. Larvae are called "phyllosoma" and are microscopic and free swimming. Eggs hatch within 3 weeks of fertilization. Larvae undergo 11 distinct phyllosoma stages, and then metamorphose into pueruli. After 6 months and after molting and metamorphosis, they become juvenile lobsters. Throughout their lives, these lobsters molt as they get larger.

ECHINODERMS

8.7.17 Chocolate chip sea star

Figure 8.18

Protoreaster nodosus (Linnaeus, 1758).

Common Name(s):

English—Chocolate chip sea star, horned sea star.

IUCN Red List Status: Not Evaluated.

Systematic Position:

Phylum: ECHINODERMATA

Subphylum: ASTEROZOA

Class: STELLEROIDEA

Subclass: ASTEROIDEA

Order: VALVATIDA

Family: OREASTERIDAE

Genus: *Protoreaster*

Species: *P. nodosus*

Scientific Name: *Protoreaster nodosus* (Linnaeus, 1758)

Description:

- The appearance with the small horns on its dorsal side, gives this sea star a look like a bumpy cookie.
- The rows of spines or horns, black conical points are arranged in a single row, radially on the dorsal side.

- These are a roughly rigid five-pointed star-shape with tapering arms to the end.
- Tube feet are present on the ventral side.
- Colour may be red, brown or light tan.

Species Details:

- Distribution:

  Indo-West Pacific: from Seychelles to Australia and Japan.
- Habitat: Inhabit warm shallow waters seagrass beds and seaweed areas, on reef flats. Sheltered, sandy or slightly muddy bottoms.
- Food and feeding: It is an extra-oral feeder or microphagus feeder. Feeds on microbial, microalgal films, living and decomposing seagrass blades and filamentous green algae.
- Maximum length: 30.0 cm WD male/unsexed.

Aquarium Aspects:

- Water conditions: Temperature: 72-78° F; dKH: 8-12; pH: 8.1-8.4; sg: 1.023-1.025.

Behaviour:

Social—Peaceful sea star.

Reproductive—Breeding in aquarium is extremely difficult.

Remarks: Very popular ornamental sea star.

8.7.18 Mottled linckia

Figure 8.19

Linckia multifora (Lamarck, 1816)

Common Name(s):

English—Mottled linckia, spotted linckia, Dalmatian linckia, multicolour sea star, multipore sea star.

IUCN Red List Status: Not Evaluated.

Systematic Position:

Subclass: ASTEROIDEA

Order: VALVATIDA

Family: OPHIDIASTERIDAE

Genus: *Linckia*

Species: *L. multifora*

Scientific Name: *Linckia multifora* (Lamarck,1816)

Description:

- It has a small disc and five long, slim, cylindrical arms that taper slightly towards the tips.
- The surface is covered with granulations and the surface has a rough texture.

- The colour may be brown, pink, red or grey with small red spots.

Species Details:

- Distribution:

  Indo-Pacific.
- Habitat: Inhabit atoll, lagoons. Often found on reef flats and in seaward reef, on subtidal rocks and coral reefs.
- Food and feeding: Cryptic. Feeds on algae and microbes.
- Maximum length: 13 cm in diameter.

Behaviour:

Social—Exhibits autotomy i.e., self-amputation, if necessary. Parasitic snails are sometimes found on the body of this sea star.

Reproductive—Exhibits both asexual (regeneration and clonal) and sexual (gonochoric) means of reproduction. Embryos hatch into larvae and these planktonic creatures metamorphose into pentamerous juveniles which develop into young sea stars with stubby arms.

Remarks: This sea star is more tolerant to aquarium conditions than other sea star.

8.7.19 Common sea star

Figure 8.20

Asterias rubens Linnaeus, 1758

Common Name(s):

English—Common starfish, common sea star, sugar star fish.

IUCN Red List Status: Not Evaluated.

Systematic Position:

Phylum: ECHINODERMATA

Class: ASTEROIDEA

Order: FORCIPULATIDA

Family: ASTERIIDAE

Genus: *Asterias*

Species: *A. rubens*

Scientific Name: *Asterias rubens* Linnaeus, 1758

Description:

- It has five arms, which are broad at their base, and gradually tapering to a point at the tips. These are generally turned up slightly.
- Body surface is either oral (bearing the mouth) or aboral (not bearing the mouth).
- From the mouth a wide furrow (ambulacral groove) extends into each arm. Each ambulacral groove contains two or four rows of small, tubular projections, called tube feet or podia.
- The margins of the ambulacral grooves are guarded by movable spines.

- The aboral surface bears both the anus and large, button like structure, the madreporite.
- It is usually orange or brick red on the aboral surface and paler on the oral surface but can also be purple or pale brown.

Species Details:

- Distribution:

  North Atlantic Ocean.
- Habitat: Epibenthic. Found in intertidal to subtidal zones. Inhabit exposed areas with sandy substrate.
- Food and feeding: Predatory. Feeds on the blue mussel, *Mytilus edulis*. Feeds on a variety of benthic organisms. These are bivalve, molluscs, polychaete worms, barnacles, gastropods, other echinoderms and carrion.
- Maximum length: 52.0 cm WD male/unsexed.
- Life span: 7-8 years.

Behaviour:

Social—A common free- living species.

Reproductive—Show both asexual (regeneration and clonal) and sexual (gonochoric) means of reproduction. A mature female produces 2.5 million eggs. Embryos hatch into planktonic larvae which metamorphose into pentamerous juveniles, later to become young sea stars.

Remarks: A common exhibit of marine aquariums.

8.7.20 Spiny starfish

Figure 8.21

Marthasterias glacialis (Linnaeus, 1758)

Common Name(s): Spiny starfish.

IUCN Red List Status: Not Evaluated

Systematic Position:

Phylum: ECHINODERMATA

Class: ASTEROIDEA

Order: FORCIPULATIDA

Family: ASTERIIDAE

Genus: *Marthasterias*

Species: *M. glacialis*

Scientific Name: *Marthasterias glacialis* (Linnaeus, 1758)

Description:

- It is a large starfish with a small central disc and five slender, tapering arms.
- Each arm has three longitudinal rows of conical, whitish spines, usually with purple tips.

- Each of these are surrounded by pedicellariae.
- The body colour may be brownish or greenish-grey, tinged with yellow or red and sometimes with purple at the tips of the arms.
- Sexual Dimorphism: The male starfish are smaller and the female starfish are larger in diameter.

Species Details:

- Distribution:

  Eastern Atlantic Ocean.
- It inhabits the subtidal zone down to about 200m. Found on both sheltered muddy substrates and rocks.
- Foods and feeding: It is a predator. Feeds mostly on bivalve molluscs and other invertebrates.
- Maximum length: 70 cm.

Behaviour:

Social—It is a predator and feeds mostly on bivalve molluscs and other invertebrates.

Reproductive—These starfish gather in very shallow water in July and August. After a few days, on a warm afternoon, they release spawn into the sea.

8.7.21 Common sea urchin

Figure 8.22

Echinus esculentus Linnaeus, 1758

Common Name(s):

English—Common sea urchin, European edible sea urchin.

IUCN Red List Status: Near Threatened (NT).

Systematic Position:

Phylum: ECHINODERMATA

Class: ECHINOIDEA

Order: CAMARODONTA

Family: ECHINIDAE

Genus: *Echinus*

Species: *E. esculentus*

Scientific Name: *Echinus esculentus* Linnaeus, 1758

Description:

- Body spherical and slightly flattened at both poles.

Species Details:

- Distribution:

  Coastal areas of western Europe.
- Habitat: Found down to a depth of 1,200 m.

- Food and feeding: Feeds on algae, encrusting invertebrates, worms, barnacles, hydroids, tunicates, bryozoans, algae such as *Laminaria* spp., sludge and detritus.
- Maximum diameter: 15-16 cm.

Behaviour:

Social—Adult sea urchins are well protected by their strong spines. Their predators are sea otters, starfish, wolf eels, triggerfish and humans.

Reproductive—Spawning occurs in the spring. A female may release upto 20 million eggs into the water column. Planktonic larvae take about 45 to 60 days for development in captivity. In autumn and winter settlement take place and large number of juveniles are found at a greater depth than the kelp zone.

Remarks: A common display in public aquariums.

AVES

8.7. 22 Gentoo penguins

Figure 8.23

Pygoscelis papua (Forster, 1781)

Common Name(s):

English—Gentoo penguin, Johnny penguin, rump-tailed penguin.

IUCN Red List Status: Least Concern (LC).

Systematic Position:

Kingdom: Animalia

Phylum: CHORDATA

Class: AVES

Order: SPHENISCIFORMES

Family: SPHENISCIDAE

Genus: *Pygoscelis*

Species: *P. papua*

Scientific Name: *Pygoscelis papua* (Forster, 1781)

Description:

- There is a wide white stripe across the top of the head.
- This aquatic bird has bright orange bill.
- The webbed feet is pale whitish-pink and a fairly long tail.
- Throat is black and tail is like a brush.
- Sexual Dimorphism: Males are heavier than females.

Species Details:

- Distribution:

  Antarctica; Australia; Falkland Islands; French Southern Territories; Heard Island and McDonald Islands; South Africa; South Georgia and the South Sandwich Islands; New Zealand; Argentina and Chile.

- Habitat: Antarctic Peninsula and numerous islands around the frozen continent. Found in ice-free areas, coastal plains, sheltered valleys and cliffs.
- Food and feeding: Carnivorous. Feed mainly on crustaceans such as krill. They are opportunistic feeders and feeds on fish, squat lobsters and squid.
- Maximum length: 75 cm.
- Life span: 15-20 years.

Behaviour:

Social—They form sizable colonies of a few hundred adults. They can call in different ways, but often as a loud trumpeting sound. They are congregators and dispersive, but not migratory in nature. They are most fast underground swimmers of all penguins. They are well adapted to extremely cold and adverse climates.

Reproductive—Egg-laying months are from June to November. For nesting, they prefer shallow coastal areas, and among tufts of grass. Gentoos breed monogamously and it is strictly followed in the colony. Nests are made with roughly circular stones and are quite big. Two eggs are laid. Both the parents take part in incubation. The eggs hatch after 34 to 36 days. The chicks remain in the nests for about 30 days and then join other chicks in the colony to form creches. The chicks go out to sea in about 80 to 100 days.

Remarks: Gentoo penguins are often seen in marine public aquariums.

MAMMAL

8.7.23 Beluga whale

Figure 8.24

Delphinapterus leucas Lacepede, 1804

Common Name(s):

English—Beluga whale, white whale, sea canary.

IUCN Red List Status: Least Concern (LC).

Systematic Position:

Kingdom: ANIMALIA

Phylum: CHORDATA

Class: MAMMALIA

Order: ARTIODACTYLA

Infraorder: CETACEA

Family: MONODONTIDAE

Genus: *Delphinapterus*

Species: *D. leucas*

Scientific Name: *Delphinapterus leucas* Lacepede, 1804

Description:

- Body is almost round in shape, stocky and tapers smoothly to the head than the tail.
- The sudden tapering to the base of its neck gives it the appearance of shoulders.
- The flippers are broad and short.
- The tailfin grows and becomes increasingly and ornately curved.
- Dorsal fin is absent.
- There is a distinctive protuberance at the front part of the head. This part has an echolocation organ, having fatty tissue and is known as melon. In this species, it is large and deformable.
- The melon is extremely bulbous, lobed and visible as a large frontal prominence.
- The whole body is white in colour.
- Sexual Dimorphism: The male (3.5 to 5.5 m) are longer than the females (3 to 4.1 m). Males are sturdier than females and weighs in males 1,100 and 1,600 kg; females 700 to 1,200 kg.

Species Details:

- Description:

  Arctic Ocean, and the seas and coasts around North America, Russia and Greenland.
- Habitat: The majority of belugas spend the winter around the Arctic ice cap. And in summer, they move to warmer river estuaries and coastal areas.
- Food and feeding: They are opportunistic feeders, and their feeding habits depend on the season and on their locations. They may eat, Arctic cod, rose fish, Greenland halibut, northern shrimp, Pacific salmon, capelin, smelt, sole, flounder, herring, sculpin, shrimp, squid, crabs, clams, octopus, sea snails, bristle worms.
- Maximum length: Males: 3.5 to 5.5 m; Females: 3 to 4.1 m.
- Life span: 30 years.

Behaviour:

Social—They are high sociable and form small groups or pods, having two and 25 individuals. They are cooperative and often hunt in coordinated groups. They constantly play, vocalise and swim around each other. They show a great deal of curiosity towards humans. They are slow swimmers. Belugas use sounds and echolocation for movement and communication. Belugas show seasonal migratory pattern.

Reproductive—Female belugas give birth to one calf every three years. Gestation lasts from 12 to 14.5 months. Births usually take place in bays or estuaries where the water is warm. After birth, new-borns swim alongside their mothers and nurse underwater. The calves generally continue nursing until they are 20 months old. Alloparenting (care by females other than the mother) has also been observed.

Remarks: Belugas are one of the most common cetaceans housed in marine aquariums. They are very popular with the public because of their appearance and behaviour.

Figure 8.3 Purple striped jellyfish

Figure 8.4 Pacific sea nettle

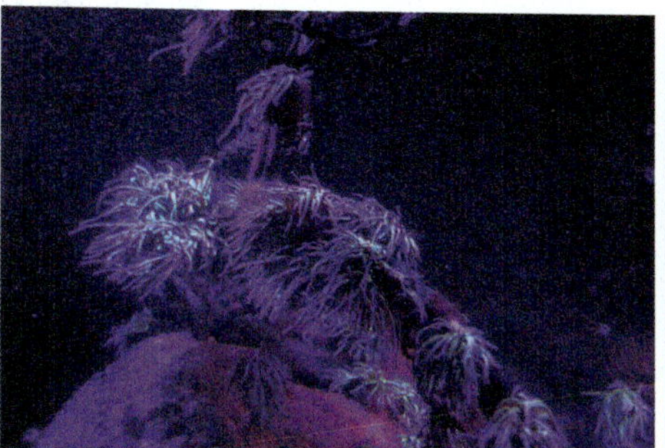

Figure 8.5 Snakelocks anemone

Figure 8.6 Green surf anemone

Figure 8.7 Magnificent anemone

Figure 8.8 Fire coral

Figure 8.9 Open brain coral

Figure 8.10 Torch coral

Figure 8.11 Whisker coral

Figure 8.12 Millepora coral

Figure 8.13 Plating montipora

Figure 8.14 Staghorn coral

Figure 8.15 Devil's hand leather coral

Figure 8.16 Button polyp zoas

Figure 8.17 Caribbean spiny lobster

Figure 8.18 Chocolate chip seastar

Figure 8.19 Mottled linckia

Figure 8.20 Common star fish,

Figure 8.21 Spiny star fish

Figure 8.22 Common sea urchin

Figure 8.23 Gentoo penguin

Figure 8.24 Beluga whale

8.8 References

Hatta M. *Reproductive and genetic evidence for a reticulate evolutionary history of mass spawning corals*. Molecular Biology and Evolution. 1999, 16(11) pp. 1607-1613.

Kardong K.V. *Vertebrates: Comparative anatomy, function, evolution*. 4th edition. Tata McGraw-Hill Publishing Company Limited. 2007, pp. 1-782.

Patek S.N. and Baio J.E. *The acoustic mechanics of stick-slip friction in the California spiny lobster (Panulirus interruptus).* Journal of Experimental Biology. 2007, 210(20) pp. 3538-3546.

Pechenik J.A. *Biology of Invertebrates*. McGraw-Hill International Edition. Sixth Edition. 2010, pp. 1-606.

Ruppert E.E. and Barnes R.D. *Invertebrate Zoology*. Sixth Edition. Harcourt Asia Pvt Ltd. 1994, pp. 1-1056.

Young J.Z. *The life of Vertebrates*. Oxford University Press. 3rd Revised edition. 1981, pp. 1-660.

9

Nutriment and Feeding of Ornamental Fishes

For successful ornamental fish farming, the type of food, species-specific nutritional needs, feeding behaviour of the fishes to be cultured and reared, should be known well in advance and taken care of. In a closed environment of an aquarium, season -specific large variety of food are not available, so aquarists are to provide live food in addition to pelleted supplementary feed to meet the nutritional needs of the fishes for maintenance and also for proper breeding function. Live food is substantial for the proper culture and better breeding of aquarium fishes and are therefore to be produced uninterrupted in the culture facility. Though a wide range of prepared feeds are available in the market, for obvious reasons, they are not cost-friendly. These feeds imported from Singapore, Hong Kong, Korea, Thailand and other countries are costly and not suitable for commercial production of ornamental fishes.

9.1 Nutritional Requirements

Food should ideally provide the fish, proteins, lipids, carbohydrates, vitamins and minerals.

9.1.1 Proteins and Amino acids:

Proteins are the major organic compound in fish tissue. About 65-75% of the total dry weight is protein, which helps in replacement of worn-out tissues and other proteinaceous matters, in the body of the fish. Protein, the nitrogenous, macromolecule is composed of many amino acids and is the most expensive part of fish feed.

This structural component and source of energy is affected by nutritional qualities of dietary protein and the level of energy from non-protein sources. Size of the fish, water temperature, dissolved oxygen, pH and feeding rate etc., affect protein requirement of fish. For herbivorous and omnivorous ornamental fishes, 25-30% protein is optimum. Out of 20 amino acids, 10 are essential and since not biosynthesized, are required by the fish. It is essential to determine the protein requirements for each species and size of cultured fish.

9.1.2 Lipids:

Lipids supply about twice the energy as that by proteins and carbohydrates. Lipids exist as simple lipids, compound lipids and derived lipids. Lipids comprise about 6-15% of fish diets, supply essential fatty acids (EFA) and serve as transporters of fat-soluble vitamins. A recent trend in fish feeds is to use higher levels of lipids in the diet. Fish require omega 3 and omega 6 fatty acids. About 7-9% dietary lipids are generally considered optimum for practical diets of carps. In nature, the primary food source of marine fishes, the marine algae and planktons and others, are rich in EPA and DHA. Whereas, freshwater algae, contain higher levels of 18:3 n-3 fatty acids than EPA and DHA (Tasbozan and Gokce, 2017). Freshwater fish requires either dietary 18:2 n-6 (linoleic) or 18:3 n-3 (linolenic) acids or both. Marine fish have dietary requirement of eicosapentaenoic acid (EPA, 20:5 n-3), docosahexaenoic acid (DHA, 22:6 n-3). These two fatty acids are termed as n-3 HUFA (Highly Unsaturated Fatty Acid) or LC (Long Chain) n-3 PUFA. Dietary phospholipids have beneficial effects on growth and survival of fish (Mohanty et al., 2011).

9.1.3 Carbohydrates:

Carbohydrates are either simple sugars or compound carbohydrates. In fish, carbohydrates are stored as glycogen. Carbohydrates are included in the diets of fishes as they are inexpensive source of energy and hence utilized as pellet binder. Though carbohydrates are not utilized efficiently by fish, they are included in the fish diet as they are the most economical and inexpensive sources of energy. Carbohydrates act as precursors for formation of various metabolic intermediates needed for growth. In the absence of adequate dietary carbohydrates, fish utilize protein for energy at the cost of growth (Mohanty et al., 2011).

Ornamental fishes at different stages of their life cycle are provided the nutrients in pre-determined amounts (Table 9.1).

Table 9.1: Nutrient requirements (percentage) at different stages of the life cycle of ornamental fish

Nutrients required	Amount in percentage in young fish	Amount in percentage in brood fish	Amount in percentage in all developmental stages of fish
Proteins	40-45	30-35	30-45
Lipids	4-6	6-8	4-9
Carbohydrates	40-45	40-45	30-50

9.1.4 Energy:

It is the property of nutrients, which is released during metabolic oxidation of proteins, lipids and carbohydrates. It refers to the muscle activity, energy for chemical reactions in body, to enable movement of molecules against a concentration gradient and for other biological and physiological body functions. As the maintenance of body temperature requires no energy expenditure and its neutral bouncy, fish requires very little energy. Proteins, lipids and carbohydrates contain 5.6, 9.4 and 4.1 Kcal of gross energy (GE/g) respectively. Physical activities of fish affect energy requirements. Excess proteins in the diet is wasteful, uneconomical and pollutes the water and causes stress to fish. If there is excess energy, lipid accumulation happens which results in fatty fish. In fish diet, so protein and energy should be balanced properly.

9.1.5 Vitamins and Minerals:

Vitamins, required in trace amounts helps in growth, fight against diseases, and in metabolism. Eleven water-soluble vitamins and four fat-soluble vitamins are required in the fish diet.

There are at least six minerals along with ten trace elements which are required by the fish in the diet.

9.1.6 Additives:

Cholesterol, mould inhibitor, antioxidants, pellet binder, monocalcium phosphate, minerals, vitamins, choline chloride, vitamin C (coated), inositol, immunostimulants, pigments are some of the additives which are used in small amounts in the fish -diet to better its quality.

9.1.7 Practical diets:

If commercial feeds could not be procured, most farmers use farm-made feeds of rice bran and ground nut oilcake (1:1) or mustard oilcake or improved type of feed mixture. These feeds are prepared purely on the basis of ingredients availability and feeding experiences.

9.1.8 Colour Enhancers:

One of the greatest challenges of the ornamental fish industry is to replicate the accurate natural colour of the fish in captive conditions. Carotenoids are the primary source of colour in the skin of fish. Since carotenoids are only synthesized by plants and get modified in animal tissues, fishes must obtain them from their diet. A number of ornamental fish species when fed on natural sources had better colouration than when fed on synthetic zeaxanthin and canthaxanthin. The fish utilises more efficiently blue green algae spirulina and algae from ponds. Effect of different foods on colour of various ornamental fishes are studied since long time.

9.2 Feeding Practices Followed for Ornamental Fishes

- No feed input is based on natural production.
- Fertilizers used in ponds are organic fertilizers and green manure.
- Supplementary diets consist of low- cost agricultural by-products and animal by-products.

9.3 Feeding Rate and Frequency

- Feeding rates and frequencies are related to fish size.
- Small larval fish and fry are fed a high protein diet frequently and in excess.
- Feeding rates and frequencies should be lowered and protein content reduced as fish grow.
- Growth and feed conversion increase with feeding frequency.
- Some factors like time of day, season, water temperature, dissolved oxygen levels, and other water quality variables affect the feeding rates of fish.

- Feed acceptability, palatability and digestibility vary with the ingredients and feed quality.

9.4. Feeds and Feed Management

9.4.1 Nutritional requirement of ornamental fish species:

Nutritional requirements of ornamental fish species of different developmental stages, various feeding habits etc., have been adjusted through long term experiments and studies. Feeds, amount in percentage and the type of fish concerned are presented here (Table. 9.2).

Table 9.2: Nutritional requirement of ornamental fish species

Types of fish	Amount of feed (percentage)	Feeds provided
Herbivorous and omnivorous Juveniles/grower and brood stock	30-40	Soybean meal, mustard meal, groundnut meal, wheat/maize gluten
Carnivorous	Above 45	Fish meal, squid meal, shrimp meal, clam meat
Larval stage (most of the species)	Above 50	Natural/live food
Most of the species—maintenance feeds	25-30	Mixture of plant and animal-based feed ingredients

9.4.2 Fish feed according to the feeding habit of the concerned fish:

- Surface feeders: Dry mash/meal
- Column feeders: Mixture of dry/moist feed
- Bottom feeders: Moist/wet/paste feed

9.4.3 Fish feed ingredients:

- Rice bran, corn flour, wheat flour as energy source and binding agent.
- Animal protein source: Fish meal, squid meal, shrimp meal etc.
- Plant protein sources: Soybean meal, mustard meal, ground nut meal.

9.4.4 Commercial formulated feeds available in the market:

- Good quality and cost friendly readymade feed can be purchased from the market, if not prepared in the farm.
- Before purchasing any fish feed, it is necessary that the nutrient composition along with additives like carotenoids and probiotics be checked.
- Fish feed should be selected having proper size and water stability.
- Instead of live feeds, freeze-dried/refrigerated form of *Chironomus* larvae (red worm), Tubifex (sludge worm), decapsulated *Artemia* cysts can be used.

9.4.5 Formulation of nutritionally balanced and cost- friendly feed:

To make the ornamental fish business more profitable it is always better to prepare feed at the farm, using locally available ingredients (plant/animal origin) and which are cost effective.

- Selection of the feed ingredients:

  Feed ingredients should be chosen on the basis of locality (preferably local), easy availability, nutritional status, cost friendliness and fulfilling the nutritional requirement of the ornamental fish.

- Types of formulated feeds:

  The formulated feeds are basically two types—dry feeds and non-dry (moist/wet) feeds.

  - **Dry feeds**: Moisture content remains between 6-10% and these are made from dry ingredients. Dry feeds are available in two forms:
    - **Mash meal**—used for small sized fish larvae (fry). It is simple mixture of dry ingredients.
    - **Pellets**—dry feeds compacted into definite shape are pellets (Figure 9.1) In small farms, hand operated or electric pellet making machine are used, in large farms feed mills are used.
    - **Flakes**—flat in shape, floats in water first, and then sinks slowly.
    - **Tablet form**—this can stick at different water levels.
    - **Granular feed**—small granules (Figure 9.2) which can be fed to small fishes and larvae.
    - **Freeze dried feed**—cuboid in shape which stick to the glass tank. After it get dissolved, fishes nibble at it.
  - **Non-dry feeds**: These are available in two forms:
    - **Moist feeds**--The moisture content remains between 18-40%. These are mixture of either both wet and dry ingredients or only dry ingredients with added moisture.
    - **Wet feeds** or **Paste feeds**—These are used to feed young fishes, carnivorous species and brood fishes. These are made from wet ingredients such as trash fish, shrimps, beef heart etc. or live food with 45-70% moisture.

Figure 9.1 Pellets as dry fish feed.

Figure 9.2 Small granules as dry feed.

9.4.6 Selection of feeds according to the stage and size of fish:

- Larvae/fry—mash or meal, wet/paste feed/live food.
- Fingerlings /grown -ups/brooder—pellet feed/live food.

Selection of feeds according to feeding habits of fish:

- Herbivorous & omnivorous—dry/moist (plant- based ingredients/live food).
- Carnivorous—dry feed/moist/live food.

9.4.7 Dietary ingredients for preparation of fish feed:

For preparation of fish feed, feed ingredients are selected on the basis of their availability, physical properties and composition (Table. 9.3). For least-cost diet formulation, screening of food stuffs in relation to raw material quality processing, and storage conditions, and for formulating diets to minimize aquatic pollution, the assessment of ingredient digestibility is essential.

Table 9.3: Dietary ingredients in per cent dry matter (DM) and their proximate composition (in percentage) (Source: Swain et al., 2010)

Source	Crude protein	Crude lipid	Crude fibre	Nitrogenous faecal excretion
Plant source Soybean meal (solvent extract)	45-55	1-2	4-6	30-35
Soybean cake	36-40	8-10	4-6	30-35
Spirulina	50.5	1.0	2.1	26.7
Groundnut cake (expeller)	40-45	4-8	6-7	30-35
Sesame cake	29-43	7-13	6-18	19-22
Gingelly oil cake	30-35	6-8	6-7	30-40
Rice-bran	8-11	12-16	10-15	40-45
Rice-polish	11-14	10-15	7-9	40-45
Maize-bran	9-11	4-6	2-3	70-80
Wheat-bran	9-15	2-8	4-12	30-50
Wheat flour	13-14	3-4	2-3	50-60
Tapioca flour	2-3	1-2	2.0	78-86
Animal source: Fish-meal	40-50	6-10	2-3	5-8
Prawn head-meal	28-30	8-10	1-5	2-4
Silkworm pupae meal	60-65	18-20	3-5	3-5
Liver meal	60-65	3-4	1-2	20-21
Earthworm meal	50-52	3-4	12-13	14-15

9.4.8 Energy metabolism in fish:

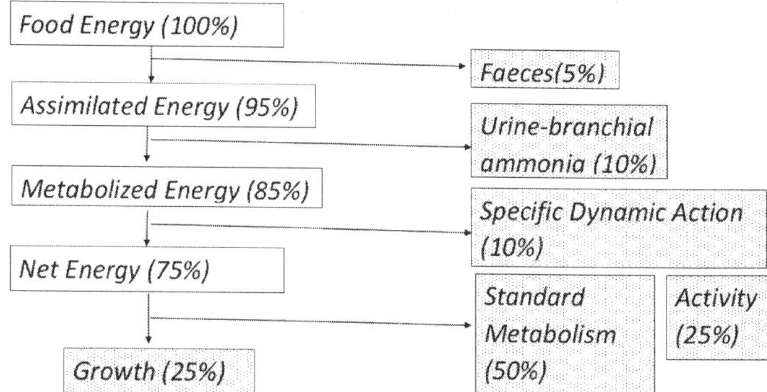

Figure 9.3 Energy metabolism in fish.

The food energy (100%), in fish is the basic energy and is converted into assimilated energy (95%) after a loss of 5% through the excreta. The assimilated energy gradually turns into metabolizable energy (85%) after a loss of 10% through urine and branchial ammonia. After a loss of 10% through specific dynamic action (SDA) or heat increment, the metabolizable energy becomes net energy (75%). About 50% of the net energy is used by the fish for maintenance of standard metabolism and rest is utilised towards activity (about 25% of net energy) and growth or meat formation (about 25% of net energy) (Figure 9. 3).

9.4.9 Feed formulations:

The most expensive portion of the fish diet is protein, and so, it is usually computed first in diet formulation. First of all, the crude protein and energy levels are balanced. If the formulation is low in amino acids, necessary alterations have to be made by the addition of ingredients with high level of the required amino acids. A final check is necessary to ensure that the balance of protein and energy levels is not altered.

9.4.9.1 Square method:

The most commonly used methods for balancing crude protein levels are the Pearson square method and algebraic equations.

A diet contains 30% protein, using fish meal (60% protein), soybean meal (51% protein), rice bran (8 % protein) and corn meal (10% protein), in the proportions 2 parts soybean meal: 1part fish meal and 1part rice bran to 1part corn meal. Using the square method (Figure 9.4) as in the above example, the desired protein level (30%) is placed in the middle of the square. The ingredients are separated into two groups and the protein level of each group is calculated accordingly:

Protein source: fish meal=1x60=60%

Soybean meal=2x51=102%

Average= 162/3=54%

Energy source: rice bran=1x8=8%

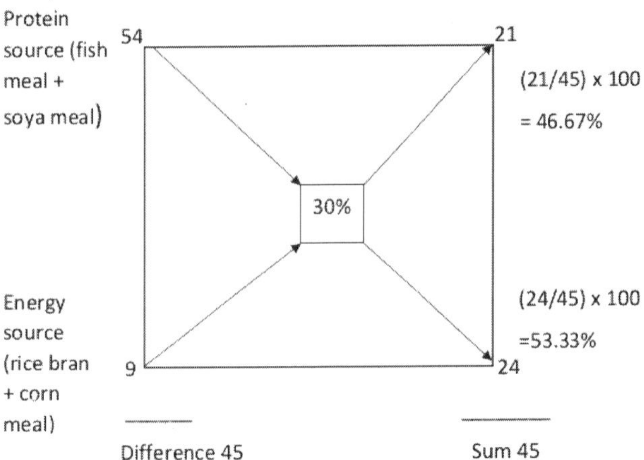

Figure 9.4 Square method.

Corn meal= 1x10= 10%

Average= 18/2=9%

The protein and energy sources, along with their calculated protein contents are placed on each corner at the left-hand side of the square. After diagonal subtraction, the calculations are as follows. The final figure for the protein source is divided into 2/3 soybean meal and 1/3 fish meal.

Protein source= 46.67%

Fish meal= 46.67x1/3= 15.56%

Soybean meal= 46.67x2/3=31.11%

And that for the energy source into ½ rice-bran and ½ corn meal:

Energy source= 53.33%

Rice bran= 53.33/2= 26.665%

Corn meal= 53.33/2= 26.665%

9.4.9.2 Efficiency of fish feeds:

While formulating a feed, it is essential to determine its efficiency in terms of food conversion ratio (FCR), protein efficiency ratio (PER) etc.

- Food Conversion Ratio /Absolute conversion ratio (FCR):

  The FCR (in kilos) denotes the amount of dry feed necessary to produce 1kg of fish.

  FCR= Total dry-weight of feed/ Total wet-weight gain (growth of fish)

- Relative Food Conversion Ratio (RFCR) = The quantity of total feed (natural feed plus supplementary feed) consumed by the fish/ Total production.

  FCR depends not only on the feed but also on the stocking density, weight of individual fish, age of fish, temperature of water and methods of feeding, quantity and frequency of food distribution etc.

- Protein Efficiency Ratio (PER)

  By this, the efficiency of protein in the conversion as meat could be determined.

  PER= Total wet-weight gain (growth of fish)/ Total dry-weight protein fed

9.4.9.3 Farm-made feed formulation:

Farm-made feed formulation (Table.9.4) may be followed for feeding ornamental fishes in the farm as it is locally procured and cost -friendly.

Table 9.4: Farm-made feed formulation (Source: Mohanty et al. **2011)**

Ingredients	Crude protein %	Crude lipid %	Ratio
Groundnut oilcake (OC): rice bran (RB)	25	8.5	50:50
Mustard OC: RB: linseed OC	25	7.0	40:40:20
Sesame OC: RB	25	6.4	60:40
MOC: LOC: RB: silkworm pupae	24	10	45:15:35:5
Sunflower OC: RB: silkworm pupae	25	9	65:30:5

9.4.9.4 Preparation of pelleted diets:

Pelleted diets are prepared as such-

- Raw materials are procured, and ingredients are mixed according to the nutritive requirement of the specific species.
- Commonly available, good quality ingredients are to be selected to make a good and palatable feed.
- Next, crude protein and energy levels are balanced using square method.
- Some additives like vitamin-mineral mixture, probiotic supplements and enzymes at required levels may be added. For essential fatty acids, 1% each of vegetable oil and fish oil/cod liver oil may be added.
- Ingredients are milled and mixed together in quantities as required and kept in trays.
- A portable mini grinder cum pulveriser is used for grinding and mixing.
- Wheat flour about 2-3% is used as binder.
- To remove the anti-nutrient factors, the mixture is steamed with some water at 400 ml for every 1 kg of feed to make dough.
- It is cooked for 10-15 minutes in a pressure cooker or autoclave for sterilization that improves digestibility.
- After cooling down, feed additives are added by spraying over them.
- The dough is passed through pelletizer to make nodules.
- Different sizes of sieves are used for different age-groups of fishes. For big-size fishes, sieve of 2-3 mm diameter is usually used.
- Moist pellets are then dried at 40-45° C to avoid oxidation and protein denaturation at high temperature. Pellets are dried either by using solar drier or in a fully ventilated place.

- Pellets are broken through mesh/sieve into different forms for different sizes of fishes.
- Pellets are stored in air-tight container, and are kept in cool, dark place.
- Required amount of fresh feed for fishes can be prepared every two months.
- Preservatives like sodium benzoate (C_6H_5COONa) at 0.1% are generally added in commercial productions.

9.4.10 Storage of fish feeds:

Fish feeds should be stored properly under hygienic conditions.

- Storage of dry feeds:

  Due to low moisture content, the dry feeds are less vulnerable to spoilage. These can be stored in a moisture free place, at room temperature for a longer period of time.

- Storage of moist feeds:

  The moist feeds are highly vulnerable to fungal, bacterial and parasitic infestation. The wet feeds are either to be consumed fresh/ within shortest possible time or should be stored under refrigeration only for a short period.

  Different fish feeds used in the ornamental fish farm are with various characteristics like nutritional value, storing method, hygiene, effects of use (Table 9. 5).

Table. 9.5. Characteristic features of different fish feeds

Characteristics	Pellets / Dry mash/meal	Moist feed / Wet feed
Nutritional value	Highly /Moderately Nutritious	Nutritional degradation may happen
Effects	Less pollution	Wastage and pollution
Storage	Can be stored for 2-3 months in moisture free ventilated facility	Can be stored at low temperature (-20°C) for 1 week. Better to use in fresh condition.
Hygiene	Chance of less microbial growth	Chances of microbial growth.

9.4.11 Feed dispensing methods:

- Hand feeding: Fish feed can be dispensed manually by hand, in a required amount, at a fixed time and at a fixed place. Preferably by the same person. To avoid wastage and pollution, fishes should not be fed too frequently and in too much quantity.
- Tray feeding: Dry/dough/ non-dry feed can be dispensed in meshed plastic trays at different places in the culture tanks or ponds.
- Automatic demand feeders: To dispense calculated amount of feed on the basis of fish stock in the pond, automatic demand feeders are used.

9.4.12 Amount of feed, feeding rate, time of feeding and feeding schedule:

- Amount of fish feed is an important factor, and is calculated as such:

  Total amount of feed= Average fish size (body weight) x feeding rate (%) x total number of fish in the aquatic set up/ 100.

- Feed and feeding schedule at different developmental stages of the egg-layers:

  After hatching stored yolk provides nourishment to larvae for 2-3 days. After that a feeding schedule is generally followed (Table 9.6).

Table 9.6: Approximate Feeding schedule of different developmental stages of the aquarium fish

Developmental stage of fish	Foods	Feeding frequency/ per day	Quantity of feed(g)/ per day
1st week	Infusoria or chicken egg yolk milk	3-4 times	6.0-9.0
2nd week	Infusoria and rotifers	3-4 times	6.0-9.0
3rd week	Rotifers and Artemia nauplii. Slowly introducing powdered feed	3-4 times	6.0-9.0
4th week	Powdered feed	3-4 times	6.0-10.0
Up to 3 months	Conventional feed	3 times	30.0-50.0
Up to 4 months	Conventional feed	2 times	30.0-50.0
After 4 months	Conventional feed	1-2 times	200.0-300.0

- Feeding during pre-breeding and post-breeding periods:

  In pre-breeding and post breeding periods of ornamental fishes highly nutritious diet containing 40-45% protein is required. During pre-breeding stage, fishes can be provided with live-foods like brine shrimp, earthworms, tubifex, bloodworm, mosquito larvae, *Moina, Daphnia* along with protein-rich formulated feed. About 100 mg vitamin E/kg of diet of goldfish is better for early gonadal maturation (Swain et al., 2010).

- Maintenance feed:

  A maintenance feed with 20-25% crude proteins and 50-55% carbohydrates and of course palatable to the fish, is given to the aquarium fish. High carbohydrate ingredients like rice-bran, rice-polish, wheat-bran, maize-bran, tapioca, etc., along with kitchen wastes like vegetable peels, edible human waste food, meat and bone waste, prawn shells, cooked chicken, raw and cooked fish, are all processed together carefully to make aquarium fish feed for maintenance.

- Underfeeding:

  Underfeeding can lead to poor health and ultimately death of the fish. So, regular checking of the feed formulation, feed quality, water quality, health of the fish and its environment is needed to avoid underfeeding of the fish.

- Overfeeding:

  A fish should consume the feed in 20 minutes time. The unconsumed food if remains spoils the water quality eventually, and which ultimately affects the health of the fish.

9.5 Marine Ornamental Fish Diets

- The Central Marine Fisheries Research Institute (CMFRI), Kochi, has developed and launched "Cadalmin TM varna" which are dry formulated feeds for marine ornamentals. Formulated feed for marine ornamental fishes is usually imported at the price of R 4,000/kg. These are imported from South East Asian countries.
- Varna feeds contain 40% protein, 9% fat, 39% carbohydrates, 7% ash and less than 2% fibre, marine protein, soya protein, wheat flour, oil, colour imparting ingredients like carotenoids, nutraceuticals (immune promoters), microbial products (probionts) and antioxidants.
- These are available in three particle sizes, 0.25 mm,0.75 mm and 1 mm. These feeds are produced through twin-screw extrusion technology. These are available through the Agriculture Technology Information Centre (ATIC) of CMFRI.
- These feeds are also used for feeding freshwater ornamental fish in farms of Kerala.
- Thumb rule feed is @ 2-3 % of the fish body weight once in a day.
- Indicators (Table 9.7) are used for calculating feeding quantity (Mohanty et al., 2011).

Table 9.7: Length -weight indicator and feeding rate in clown fish and damsel fish

Length (mm)	Weight (mg)	Feeding rate
<10	<200	6 mg
>10-20	300-800	15-24 mg
>20-30	900-1,500	18-45 mg
1 pinch of 0.25-0.50 mm feed= 100 to 150 mg (sufficient for feeding 20-30 fish of <10mm)		
1 pinch of 0.75 mm feed= 150-200 mg (sufficient to feed 10-20 fish of >10-20mm		
1 pinch of 1.0 mm feed= 200-250mg (Can be fed to 10-12 fish of > 20 mm)		

- SILO Feed:

  SILO Feed was developed jointly by the CIFT and CMFRI under NAIP.

This is prepared from the waste generated during processing of tuna.

9.5.1 Guidelines for use of appropriate fish feed in ornamental fish farms:

- Locally available non-conventional agro- by-products can be used as fish feed ingredients.
- Digestibility and nutrient availability feed ingredients may be improved.
- Farm made feed preparation should be popularized.
- More specialized fish feed plants are to be established.
- More cottage feed/mini feed industries should be established.
- For quality assurance existing regulations should be imposed on fish feed industries.

9.6 Live Fish Food Organisms

Live fish food may be considered as a "living nutritious capsule" as it contains all essential nutrients that enhance feeding efficiency and excellent growth of fish, besides providing vibrant colours (Swain et al., 2010).

Though a number of live food organisms are used for aquarium fishes, still feeding only live food excluding all other kinds of food is not advisable, since it may lead to nutritional or other internal disorders for the fish. Live food may bring with them aquarium pests or disease-causing organisms, so they may either be collected from fish free water or disinfected before use.

Live fish food organisms are extensively used for rearing larval fish in aquaculture. The nutritional quality of these food organisms very much depends upon the conditions of their culture. Very often high mortality of fish larvae is encountered by feeding live food organisms deficient in essential fatty acids (EFA) (Izquierdo and Hernandez-Palcios, 1997). Larvae of fish cannot be fed artificial supplemented feed. Small-size live foods are required for their nutrition. Live foods can be purchased from the market. But it is a costly affair and may not be always available. So, culture of live foods may be easy and economical.

Providing appropriate live food at proper time play a major role in achieving maximum growth and survival. To achieve maximum production and profitability, the nutritional components of natural foods must be identified and quantified. (Mandal et al., 2009; Das et al., 2012)

Live food organisms generally used in ornamental fish farming are infusoria (*Paramecium* sp.), White worm (*Enchytraeus* sp.), brine shrimp (*Artemia salina*), West African nightcrawler (*Eudrilus eugeniae*) and branding worm (*Eisenia fetida*), sludge worm (*Tubifex* sp.), water flea (*Moina* sp.), bloodworm (*Chironomus* sp.), mosquito larvae etc.

In aquaculture practices, live food organisms constitute one of the most important resources. In nature, most of the fish and shell fish larvae feed on small phytoplanktonic and zooplanktonic organisms.

Feeding habits of fishes may be different but all the fishes require protein rich live food for their better growth, efficient breeding and survival (Mandal et al., 2009). Moreover, the success in the hatchery production of fish fingerlings for stocking in the grow-out production system is largely dependent on the availability of suitable live food for feeding fish larvae, fry and fingerlings (Lim et al., 2003).

9.6.1 Feeding of young ones (fry stage):

Fish fry consumes smallest size of food. Fry of live bearers are bigger in size than the fry of egg layers and eat food items immediately after release from the parents. Fry in case of egg layers depend on yolk after hatching. As they become free swimming after 3 to 5 days, they search for food.

Feeds of fry—"Oxbow theory" on feeds of fry:

After becoming free swimming, the fry can be fed immediately with any of the following food items or all the items alternatively.

- Green water—Green water is collected from pools or ponds or tanks. To produce green water scrapings of green algae are kept in small containers, and are inoculated. Urea and super phosphate are added as fertilizers. The green water is ready after 5 days and can be fed to fry.
- Infusoria—Infusoria are either collected or cultured and fed to fry.
- Baker's yeast suspension—Yeast is dissolved in water and this milky liquid is fed to fry.
- Egg yolk suspension—Yolk of boiled egg is dissolved in the fry tank through bolting silk cloth. In this case, water exchange should be done every day to avoid water pollution.

9.6.2 Feeding frequency:

The feeds for fry are continued for 7 to 10 days, 5 or 6 times daily. Cichlid fry and live bearers may be given feed according to their size. *Daphnia* and *Artemia* nauplii which are just hatched, can be given. *Daphnia,* blood worms, earthworms, tubifex, mosquito larvae, *Artemia* and formulated feeds can be given after 25 to 30 days of birth.

9.7. Some Live/Natural Fish Food Organisms

Some of the natural food organisms include phytoplankton, zooplankton, annelids, worms, insects, molluscs etc.

9.7.1 Plankton:

Plankton, the floating organisms whose movements are more or less dependent on currents, is most essential for many fishes as food. Application of organic manure and chemical fertilizers increase plankton in a water body.

9.7.2 Infusoria:

Kingdom: Protozoa; Phylum: Ciliophora and Phylum: Euglenozoa

Microscopic, single celled aquatic creatures like *Paramecium* (Figure 9.5), *Euglena* that exist in freshwater ponds, are collectively called infusoria. Besides being small in size, they are soft bodied and nutritionally very rich and therefore serve ideally as starter diet for early stages of ornamental fishes (larval stage). In the early developmental stages of fish larvae, infusoria or small live organisms are indispensable. *Paramecium* and *Stylonychia* are the most common forms of freshwater infusoria while *Fabrea* and *Euplotes* are of marine ones. Infusoria are found in ponds, tanks, and ditches. They are either collected with 0.12 mm fine mesh cloth, from foul smelling water bodies or cultured for constant supply.

Culture of infusoria:

- Banana peelings, cabbage, potato, hay, lettuce leaves may be kept in a container filled with water.
- The container is covered in such a way that air is allowed but not mosquitoes or any other insects.
- The container is to be kept in a cool place.
- The water turns milky in two days and smells foul. This is due to the multiplication of bacteria which decay the material.
- Then a film of slime forms on the water surface.
- In next 4 to 5 days, the water turns clear with light yellow colour. This is due to the floating spores of infusoria which have settled in the water and are feeding upon the bacteria and multiplying.
- The film of slime on the water surface gradually break up and disintegrate.
- The infusoria culture is now ready for feeding the early stage of fish larvae.
- If a few drops of milk are added, the culture will continue to flourish.

9.7.3 Rotifers:

Kingdom: Animalia; Phylum: Rotifera; Class: Monogononta

These are popularly known as wheel animalcules and are about 0.2 and 0.6 mm in length. About 1850 species are known from freshwater, brackish water and sea water.

- *Brachionus* (Figure 9.6) is the most known rotifer, which is used as a starter diet for early larval stages of many fish.
- *Brachionus plicatilis* is the most commonly used species used as food for fish larvae in hatcheries. This small sized euryhaline species is of good nutritional value.

Culture of rotifers:

- Continuous culture of rotifers is used in general.
- The culture tank is fertilized with ammonium sulphate (100 gm/1000 l), single super phosphate (10 g/1000 l) and urea (10 g/1000 l), the tank is inoculated with *Chlorella*.
- After *Chlorella* reaching its peak density, tank is re-fertilized with the previous medium.
- After *Chlorella* cells are consumed, Baker's yeast is introduced a the rate of 1.0 gm/ million *Brachionus plicatilis* /day.

9.7.4 *Tubifex*:

Kingdom: Animalia; Phylum: Annelida; Class: Clitellata.

Tubifex (Figure 9.7) can grow up to 20 cm in length and are commonly found in sediments of lakes, rivers and in sewage drains. The red worms (*Tubifex tubifex*) are one of the most ideally suited natural food organisms.

- These annelid worms are used as diets for ornamental fish brooders.

Culture of tubifex:

- For mass scale culture of *Tubifex*, a container is taken with 50 to 75 mm thick pond mud at the bottom, blended with decaying vegetable matter and masses of bran and breed.
- In the container, a continuous mild water flow is to be maintained.
- The container is inoculated with *Tubifex* worms collected from muddy canals or sewage canals.
- Clusters of *Tubifex* worms are developed within 15 days.
- Worms are collected and washed with water.
- After proper cleaning the worms are used as food.

Figure 9.5 *Paramecium* sp.

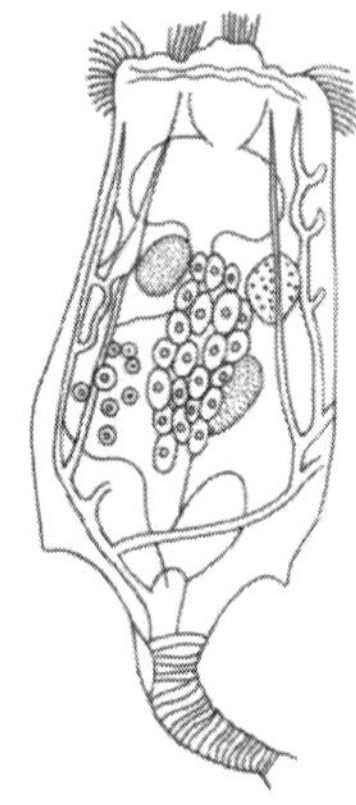

Figure 9.6 *Brachionus* sp.

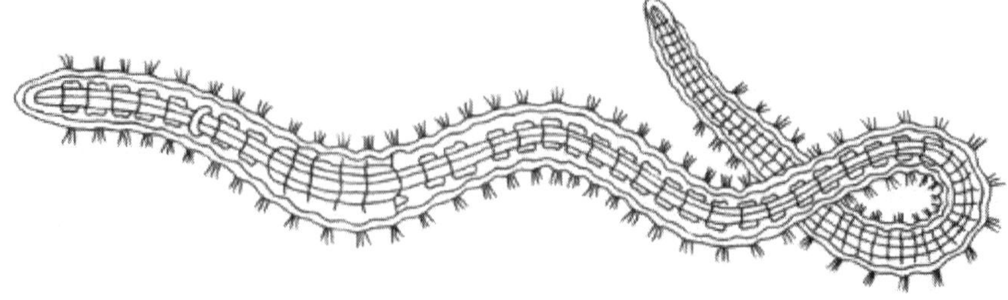

Figure 9.7 *Tubifex* sp.

9.7.5 White worms:

Phylum: Annelida; Class: Clitellata

About 40 Species of annelid worms are included in the genus *Enchytraeus*. *Enchytraeus albidus* is the white worm (Figure 9.8), though all the species in general are referred to as white worms.

- These live foods are 0.5-1 cm in size.
- They may either chopped into small pieces or fed as whole according to the size of the fish.
- It is very difficult to get a pure culture of white worms.
- Cool, dark place, food and a moist substrate is needed by white worms to thrive and reproduce.
- White worm culture media should contain a high portion of organic matter and soil- based media.
- If available, then growing and harvesting white worm is easy.

9.7.6 Earthworms:

Phylum: Annelida; Class: Clitellata

Earthworm species *Eudrilus eugeniae, Eisenia fetida, Perionyx excavatus* and common earthworm are excellent food for all most all kinds of ornamental fishes, including goldfish.

- Earthworms can be collected from garden or waste ground and then should be kept for a few days in a sealed container. Container should have small air holes. These worms clean themselves of solid and wastes and will then be more palatable for the fish.
- Common earthworms can be cultured by making a bed of organic composting method. Other earthworms are to be collected through vermicompost farms.
- These are then inoculated at the rate of 1,000 worms per square metre in compost pits of 10mx 4 m x 9" with semi-decomposed cow dung manure and soft -dried leaves. Earthworms can be harvested according to needs.
- The worms can be given as whole or chopped, depending on their size and size of the fish.

9.7.7 Mosquito larvae:

- Mosquito larvae (Figure 9.9) are of small size (about 5 mm-12 mm) and fish can shallow them easily.
- Mosquito larvae as well as pupae can be fed to fish.
- These can be collected from stagnant water bodies containing decaying organic matter or stored clear water. Cow dung is often used for culture of mosquito larvae.
- They are collected with a scoop /plankton net.

- Larvae and pupae should be deep-treated with $KMnO_4$ and then washed in freshwater before feeding.

9.7.8 *Chironomus* larvae:

Kingdom: Animalia; Phylum: Arthropoda; Class: Insecta.

- *Chironomus* (non-biting midges) larvae (Figure 9.10) are aquatic arthropod, which are commonly called as blood worms due to presence of haemoglobin in their body fluid. Fully grown larvae are dark red in colour and about 1 cm in length.
- These are excellent source of protein, lipid, vitamins and mineral and are very popular live foods for aquarium fishes.

Culture of *Chironomus* larvae:

- Trays with soil, compost manure or organic matter or decaying vegetable matters are filled with water and kept for *Chironomus* flies to deposit eggs.
- The larvae, live in soft tubes made up of organic matter at the bottom of the track.
- They come out of the tubes after 2 to 3 days and swim in water vertically.
- The larvae are collected with scoop net and washed thoroughly.

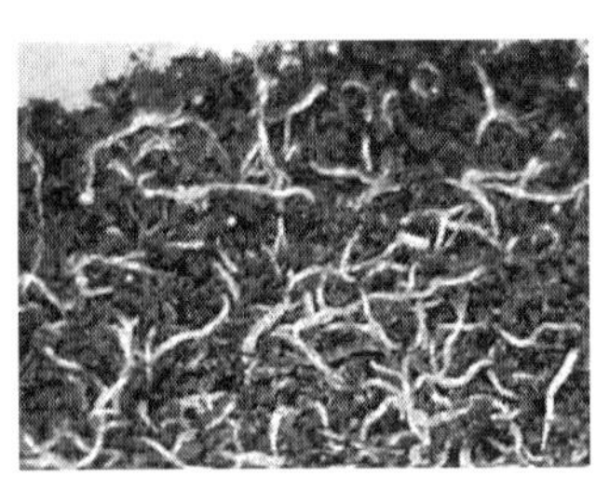

Figure 9.8 White worms

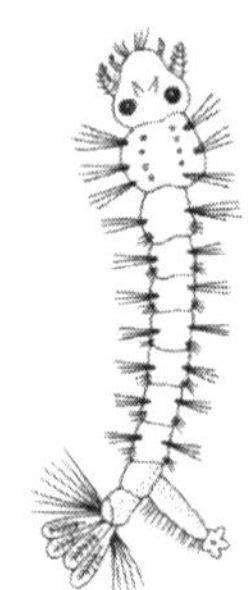

Figure 9.9 Mosquito larva

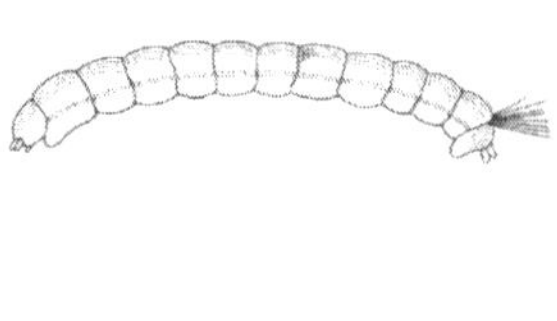

Figure 9.10 *Chironomus* larva

9.7.9 *Artemia*

Phylum: Arthropoda; Subphylum: Crustacea; Class: Branchiopoda

Artemia (Figure 9.11) lives in saline lakes, lagoons and man-made solar salterns. It takes 15 days for the new born larva to grow to adult size. The adults release cysts every 5 days in the form of floating spherical brown particles. The cysts can be kept viable up to 20 years. Permanent availability of cysts, simple hatching techniques, small size, slow swimming behaviour, softness of body tissue, presence of essential fatty acids, amino acids and high protein content makes the brine shrimp very popular among the aquarium fish breeders.

- *Artemia* or brine shrimp are zooplankton which are used as live food in the aquarium trade. More than 50 geographical strains of *Artemia* are available.

- *Artemia* is the most widely used food organism and can be produced from dry and storable powder i.e., dormant *Artemia* cysts which if put in seawater, regain their metabolic activity and within 24 hours, release free swimming larvae (nauplii) of about 0.1 mm length.
- Cysts (after decapsulation), nauplii juveniles, sub-adults, all the stages of *Artemia* are used as feed. This with high nutritive value and high conversion efficiency are widely used by aquarists. *Artemia* nauplii, instar I and instar II are widely used, as they are easy for operation, rich in nutrients and small.
- *Artemia* trading is a growing business in several parts of the world.

Culture of *Artemia*:

- Glass jars can be set up in warm room with water of salinity 30-35 parts per thousand.
- Room temperature of 26-30° C is perfect for hatching cysts within 24 hours. Cysts are kept in water.
- A light source can be used to attract freshly hatched nauplii. These are collected in a very fine-meshed net. These are then fed to fish larvae.
- Batch culture method and flow-through culture methods are also used for culture of *Artemia*. Batch culture method is carried out in a rectangular tank with a central partitioning. In flow-through culture method, the medium is continuously renewed as against the former system where the same water is used.

9.7.10 *Daphnia*:

Kingdom: Animalia; Phylum: Arthropoda; Class: Crustacea; Subclass: Branchiopoda; Order: Cladocera; Family: Daphniidae.

These small, planktonic crustaceans (0.2-5mm in length) are commonly called water fleas. *Daphnia* (Figure 9.12) are found in acidic swamps to freshwater lakes, ponds, streams and rivers. The two most commonly available species of *Daphnia* are *D. pulex* (small and most common) and *D.magna* (large).

- *Daphnia* contains a broad spectrum of digestive enzymes, even cellulase which serve as coenzymes in the gut of fish. It is larger in size than *Moina*, so serves as live food for advanced stages of fishes.

Natural collection of *Daphnia*:

- Ponds, ditches and any stagnant water bodies are the places where *Daphnia* can be collected.
- As they swim on the water surface before sunrise, they should be collected during early morning hours.
- *Daphnia* are collected with the help of a scoop net having 100-200μ mesh.

Stock and pure culture:

- For stock and pure culture preparation, *Daphnia* are to be collected and this sample is to be diluted and taken in a glass beaker.

- Each *Daphnia* is picked with the help of a dropper and placed one in each tube containing 10 ml of filtered freshwater.
- Yeast or ground nut oil cake at 200 ppm daily is provided as food to the *Daphnia*.
- These tube cultures are then transferred to 1 litre jar and feeding is continued.
- The jar culture, after 5 to 6 days, is inoculated in mass culture tanks.

Mass culture:

- If required, *Daphnia* is also cultured in 500 to 20,000 litre capacity cement or plastic tanks.
- The culture tanks are properly cleaned and filled with filtered freshwater. Medium is to be prepared before a mass culture.

  Preparation of medium:

  - To prepare slurry, about 10 kg of chicken droppings, 5 kg of groundnut oil cake and 2.5 kg of single super phosphate is added in 250 litres of freshwater.
  - To remove obnoxious gases, fermentation and for the release of nutrients,
  - Continuous aeration is given for 3 days.
  - After 3 days, the slurry is used as fertilization solution in the mass culture tanks. In the mass culture tanks, the medium is added at 3 to 4 ml per litre of water regularly for 3 to 4 days.
  - *Daphnia* is inoculated at 50 individuals per litre, on 4th day.
  - *Daphnia* multiplies within 7 days and density reaches from 10,000 to 25,000 individuals per litre.
  - *Daphnia* is collected in the early morning or late evening when they are on the water surface, with the help of 100-200 μ mesh scoop net. *Daphnia* is then washed thoroughly and used as fish fry feed.

Figure 9.11 *Artemia* sp.

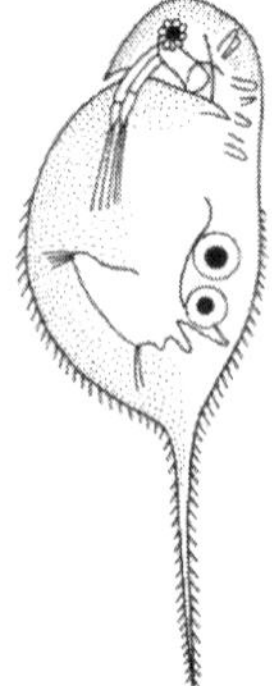

Figure 9.12 *Daphnia* sp.

9.7.11 *Moina*:

Kingdom: Animalia; Phylum: Arthropoda; Class: Crustacea; Subclass: Branchiopoda; Order: Cladocera; Family: Moinidae.

Moina (Figure 9.13) and *Moinadaphnia* (Figure 9.14) are primarily inhabitants of temporary ponds or ditches. *Moina* is smaller in size (0.5 to 2 mm) than *Daphnia* though containing 70% more protein. *Moina* and *Moinadaphnia* are extensively used as live food for aquarium fishes.

- High reproduction rate, wide temperature tolerance and thriving ability on phytoplankton and organic wastes make the cladocerans popular live food. They are mainly freshwater zooplankters except *Diaphanosoma celebensis*.

Culture of *Moina*:

- For pure culture of *Moina,* stock culture is developed first. For that *Moina* is collected from freshwater ponds and tanks with the help of scoop net with 250 to 500μ mesh. The content of the net is then placed in a plastic bucket.
- After bringing it in the laboratory, the sample is diluted by adding clear freshwater. It is then examined under microscope.
- Moinas are picked up with the help of fine dropper. Each *Moina* is then inoculated in 20 ml glass tube containing 10 ml of filtered water.
- Yeast @ 200 ppm or chlorella are provided as food at a cell density of 10x106 cells/ ml. Each gravid *Moina* produces 8 to 10 offspring in about 24 hours.
- This test tube culture is diluted daily through several 100 ml beakers. The volume is increased to 1 to 2 litre beakers or jars. Feeding is done in the same way as in the test tube culture. These cultured *Moina* after 4 to 5 days, are used as inoculums in mass culture tanks.
- In mass culture system of *Moina,* groundnut oil cake (75 ppm), single super phosphate (20 ppm) and urea (8 ppm) are put in the culture tanks.
- After fertilization takes place, the tank is inoculated with *Chlorella* or mixed phytoplankton.
- Within 3 to 4 days, when algal blooms are developed, *Moina* is inoculated @ 40 to 50 numbers/litre depending on the availability of stock culture.
- *Moina* multiplies rapidly, attaining a peak density of 20,000 to 25,000 nos./ litre in 5 to 7 days after inoculation. It is then regularly harvested to feed the larval stage of fishes.

9.7.12 *Cyclops*:

Kingdom: Animalia; Phylum Arthropoda; Subclass: Copepoda; Order: Cyclopoida.

Copepods are the most important planktonic constituent and form an essential link in the food chain of both freshwater and marine ecosystems (Khan, 2003). Out of 6 orders of Copepoda, the free- living forms belong to the orders Calanoida and Cyclopoida. Four genera are commonly found in the freshwater wetlands of India. These genera are, *Microcyclops, Mesocyclops, Paracyclops* (Figure 9.15) and *Tropocyclops*. Among these, *Mesocyclops leuckarti* is the most abundant zooplankter species (Khan, 2003).

Collection from the wetlands:

- These zooplankters can be collected from the wetlands, if they are plentiful in the water.

Culture of cyclops:

- Pour approximately 250 ml of non-chlorinated water in one glass dish.
- Add four autoclaved wheat seeds.
- Use a pipette to capture a single egg-bearing female cyclops and place her in a dish with few drops of clean, non-chlorinated water.
- Fill the dish about halfway.
- Recapture that female and place her into dish containing wheat seed.
- Temperature should be between 20 to 30° C.
- Nauplii hatch from the eggs within a few days and they grow to mature adults within two weeks.

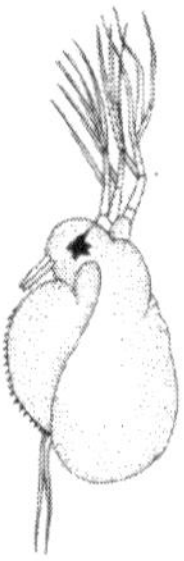

Figure 9.13 *Moina* sp.

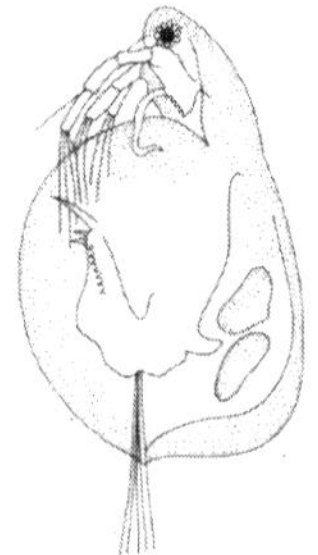

Figure 9.14 *Moinadaphnia* sp.

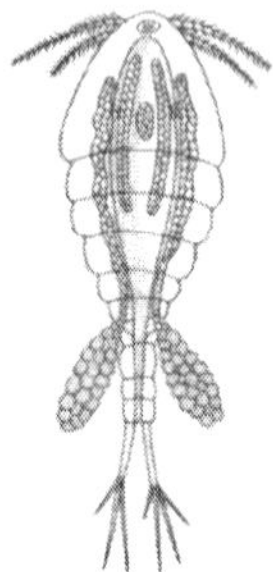

Figure 9.15 *Paracyclops* sp.

9.8 Colour Enhancers in Ornamental Fish Feed

An essential prerequisite for the quality of ornamental fish is the wide spectrum of colours, and so colour is the major factor which determine the price of aquarium fish in the world market.

Fishes since cannot synthesize their own colouring pigments de novo, the colouring agents synthesized by some plants, algae and microorganisms better be incorporated in their diet. There are varieties of colouring agent used in ornamental fish industry to impart colour to the muscle and skin of fishes.

One of the greatest challenges in the ornamental fish market is to recreate the accurate natural colour of the fish in captivity. Though several products have been launched to abate this problem, only carotenoid pigment has been found to be most effective. Different types of carotenoid pigments are used in fish diet for colour enhancement (Gupta et al., 2007). Astaxanthin, is the most promising carotenoids in improving colour in most species of brightly coloured ornamental fishes like Tetra, Cichlid, Gourami, Goldfish, Koi, Danios etc.

9.8.1 Carotenoids:

Carotenoids are a class of 800 natural fat-soluble pigments found principally in plants, algae, photosynthetic bacteria, yeast moulds. The majority of carotenoids are derived from a 40-carbonpolyene chain. This chain may be terminated by a cyclic end-groups and may have oxygen containing functional groups. The hydrocarbon carotenoids are known as carotenes example, beta-carotene, in carrots while oxygen derivatives of these hydrocarbons are xanthophylls such as, lutein, the major yellow pigment of marigold petals, a common xanthophyll.

9.8.1.1 Carotenoids in fishes:

Carotenoids commonly occurring in fishes with their colours are—tunaxanthin (yellow), Lutein (greenish-yellow), beta carotene (orange), alpha, beta dordexanthins (yellow), zeaxanthin (yellow-orange), canthaxanthin (orange-red), astaxanthin (red), echineuone (red) and taraxanthin(yellow). Among these astaxanthin is the dominant carotenoid common in red fishes.

9.8.1.2 Different carotenoids:

Different carotenoids both naturally occurring and synthetic are either procured or developed in ornamental fish farms. Carotenoids from natural sources contain mixture of several carotenoids like α-carotene, β-carotene, zeaxanthin, lutein, cryptoxanthin etc., and through synthetic process only β-carotene is available (Kaur et al., 2016). Natural carotenoids are classified as plant source and animal source carotenoids (Table 9.8). Contents of these natural carotenoids and synthetic carotenoids are shown in the Table. 9.9. Carotenoid source and corresponding colour is also given (Table 9.10) (Sharma et al., 2019).

Table 9.8: Carotenoids from natural sources and the amount (mg/Kg)

Plant source	Amount	Animal source	Amount
Marigold	7000	Crab meal	75-1300
Chlorella	4000	Crayfish meal	30-800
Yeast	1000	Shrimp meal	100-130
Sea weed	390-900	Shrimp oil	25-125
Corn gluten	290		
Alfa alfa	280		

Table. 9.9. Natural carotenoids, sources and synthetic carotenoids

Natural carotenoids:
Animal origin—Zooplankton, *Tubifex, Chironomus* larvae, *Artemia,* Crayfish meal, Shrimp meal, crab meal, yeast.
Plant origin—Algae, several types of flowers and vegetables.
Synthetic carotenoids (commercially available carotenoids): Astaxanthin, β-carotene, lutein, zeaxanthin etc.

Table 9.10: Carotenoid, source and colour of the particular source

Carotenoid	Source	Corresponding colour
Lutein	Maize, kiwi fruit, grapes, spinach, orange juice, broccoli, green peas, red pepper	Greenish-yellow
Zeaxanthin	-do-	Yellow-orange
B-carotene	Carrot, spinach, broccoli, green peas, tomato, Spirulina	Orange
Canthaxanthin	Mushrooms	Orange-red
Astaxanthin	Crustacean meal, Yeast (Phaffia sp.)	
Taraxanthin	-do-	Yellow
Tunaxanthin	White tuna	Yellow
Eichineuone	Spirulina	Red

9.8.1.3 Amounts of carotenoids in the fish feed:

Carotenoids are incorporated in the formulated feeds in particular amounts (Table. 9.11). Carotenoids from plant sources are grated, dried under shade, grounded and mixed with other ingredients before palletization (Kaur et al., 2016), and are supplemented with dry feed for 2-3 months. Carotenoids are again added at the end of rearing period for 1-2 months.

Table 9.11: Amounts of carotenoids in the fish feed (source: Kaur et al., 2016)

Carotenoid source	Amount in percentage (%)
Beet root	3-4
Carrot	4-5
Goose berry (amla)	1-2
Green peas	1-2
Lettuce	1
Marigold petals (powdered)	3
Rose petals (powdered)	3-4

Natural carotenoids in fish feed is a success to enhance the colouration in fishes without hampering the growth rate and survival (Sinha and Das, 2004; Sinha and Asimi 2007; Ramamoorthy et al., 2010).

9.9 Recommendations

Some recommendations regarding the food and feeding programme (Mohanty, et al., 2011), better be followed in ornamental fish farming are:

- Locally available, non-conventional agro-by products should be used as fish feed ingredients.
- Farm made feed preparation, feeding methods and feed storage methods are to be popularized among ornamental fish farmers.
- For sustainable and eco-friendly farming, use of feed pellets are suggested.
- Conventional and non-conventional fish feed ingredients should be

improved in terms of digestibility and nutrient availability.

- More specialized fish feed plants, mini feed and cottage feed plants should be established.
- There should be quality assurance for fish feed and this is to be monitored.
- The aquaculture feed industry should also be granted a waiver on tax concession for the fish feed ingredients.

9.10 References

Das P., Mandal S.C., Bhagabati S.K., Akhtar M.S. and Singh S.K. *Important live food organisms and their role in Aquaculture. Frontiers in Aquaculture*. Edited by M. Sukham. Narendra Publishing House. 2012, pp. 69-86.

Gupta S.K., Jha A.K., Pal A.K. and Venkateshwarlu G. *Use of natural carotenoids for pigmentation in fishes*. Natural Product Radiance. Vol.6(1). 2007, pp. 46-49.

Izquierdo M. and Hernandez- Palacios H. *Nutritional requirements of marine fish larvae and brood stock*. In Tacon A.G.J. (ed.), Basurco B. (ed.). Feeding tomorrow's fish. 1997, pp. 243-264.

Kaur S., Kaur V.I., Holeyappa S.A., Khairnar S.O. *Effect of Dietary Supplementation of Synthetic and Natural β-carotene on survival, Growth and Pigmentation in Freshwater Ornamental Koi (Cyprinus carpio L.) carp*. Indian J. Anim. Nutr. 33 (4). 2016, pp. 448-455.

Khan R.A. *Faunal Diversity of Zooplankton in Freshwater Wetlands of South eastern W.B.* Rec. zool. Surv. India, Occ. Paper No. 204. 2003, pp. 1-107.

Lim C.L., Dhert P., Soregloos P. *Recent developments in the application of live feeds in the freshwater ornamental fish culture*. Aquaculture, 227. 2003, pp. 319-331.

Mandal S.C., Das P., Singh S.K. and Bhagabati S.K. *Feeding of aquarium fishes with natural and artificial foods: available options and further needs*. Aqua International, 3. 2009, pp. 20-33.

Mohanty S.N., Giri S.S., Rath S.C., Mohanta K.N. *Nutrition of Finfishes and Shellfishes*. In Handbook of Fisheries and Aquaculture. ICAR, New Delhi. 2011, pp. 734-761.

Ramamoorthy K., Bhubaneswari S., Sankar G. and Sakkaravarthi K. *Proximate composition and carotenoid content of natural carotenoid sources and its colour enhancement on marine ornamental fish Amphiprion ocellaris (Cuvier, 1880)*. World Journal of Fish and Marine Sciences 2(6). 2010, pp. 545-550.

Sharma A. and Dhiman A.K., Mittal S., Attri S. and Dubey N. *Carotenoid Pigment: Significance as a Natural Food colourant and Factors Affecting its Isolation*. Int. J. Curr. Microbiol. App.Sci.8(2). 2019, pp. 44-51.

Sinha A. and Asimi O.A. *China rose (Hibiscus rosasinensis) petals: A potent natural carotenoid source for goldfish (Carassius auratus L.)*. Aquaculture Research. 38 (11). 2007, pp. 1123-1128.

Sinha A., and Das R.C. *Ornamental Fishes*. CIFE. 2004, pp. 1-54.

Swain S.K., Sarangi N. and Ayyappan S. *Ornamental Fish Farming*. ICAR. New Delhi. 2010, pp. 1-145.

Tasbozan O. and Gokce M.A. *Fatty Acids in Fish*. Fatty Acids, Angel Catala, IntechOpen, DOI: 10.5772/68048. 2017. https://www.intechopen.com/books/fatty-acids/fatty-acids-in-fish. Access Date 28 September 2020.

10

Reproductive Strategies in Ornamental Fishes

10.1 Introduction

Fishes exhibit great variety in their mode of reproduction. Most of the fishes which reproduce more than once in their life-time, are known as **iteroparous**, and a few which reproduce only once and die soon after spawning, are denoted as **semelparous** (Pal et al., 2011). Some fishes exhibit **ovuliparity**, in which the female lays unfertilized eggs which are to be fertilized externally. Majority of the fishes are **oviparous**, producing yolk-containing eggs, and fertilization occurs internally. Some are **ovoviviparous** (eggs develop inside the body of the female, but receive little or no nourishment directly from the mother) and rest are **viviparous** (the eggs receive nourishment either from other tissue or directly from the parent).

Many fishes are gonochoristic, that is, male and female gonads are in separate individuals and genders are distinct throughout the life. Some fishes are **hermaphrodite**, **protogynous** (develops first as female and after one or more spawning seasons may change to male) and **protandrous** fish that develops first as male and change later to female. In some species, simultaneous hermaphroditism is displayed with gonads showing simultaneously testicular and ovarian components.

Aquarium fishes are broadly grouped into two categories on the basis of breeding behaviour, **egg-layers** (oviparous) and **live-bearers** (ovoviviparous). Another grouping of these fishes, are as egg-layers with no care (non-guarders), and egg-layers with care (guarders). Egg-buriers, egg-carriers, mouth incubators and nest builders are the categories of the guarder fishes based on the of the type of parental care shown by them (Swain et al., 2011).

10. 2 Breeding of Egg Layers

The majority of aquarium fish are egg-layers with external fertilization. Egg-layers can be divided into five groups depending on their breeding habits. These are: 1. Egg scatterers; 2. Egg buriers; 3. Egg depositors; 4. Mouth brooders and 5. Nest builders.

10.2.1 Egg scatterers:

These fish species simply scatter their adhesive or non-adhesive eggs, and these eggs fall to the substrate, into plants or float on the surface. These species do not look after their brood and even eat their own eggs. These fish spawn in groups or in pairs. Often there is a large number of the small eggs laid. The fry hatch quickly.

Examples of the egg scatterer fishes laying non-adhesive eggs are Danios (giant danio, pearl danio, zebra danio), and Brachydanio. Egg scatterers laying adhesive eggs—Goldfish (all varieties), barbs, tetras and rainbowfishes.

10.2.2 Egg buriers:

These species usually inhabit waters that dry up at some time of the year. The majority of egg buriers are annual. Among egg buriers, killifishes, *Aplocheilus panchax*, *A. lineatus* and *A. blochii* are notables. They are shade-sensitive fishes.

10.2.3 Egg depositors:

These species deposit their eggs on any of the substrate such as a tank glass, wood, rocks, plants etc. Egg depositors usually lay less eggs than egg scatterers, although the eggs are larger. Examples of egg depositors are—Rasboras (slender rasbora, harlequin, red-striped rasbora and scissor tail), Angelfish, Discus, Central American cichlid fire-mouth.

10.2.4 Mouth brooders:

These fishes are usually polygamous and usually only the female carries the fertilized eggs and newly hatched eggs in the mouth. Mouth brooding or oral incubation is common in several groups of African cichlids and in few species in Central and South America (Nelson et al., 2016). In some labyrinth fishes, mouth brooding with either male or female parental care is seen.

10.2.5 Nest builders:

Nest builder fishes build some sort of nest for their eggs. In labyrinth fishes (such as *Osphronemus*) and catfishes (such as *Hoplosternum littorale* and other catfishes) the nest is usually in the form of bubble-nest formed from the plant debris and saliva-coated bubbles, with male parental care (Andrade, 1997). In species of most genera of labyrinth fishes, bubble nesting with eggs deposited in the bubbles and either male parental or biparental care have been observed. excavated pit in the substrate. Nest builders practice brood care. Fish that build and guard bubble nests are known as aphrophils. **Aphrophils** include gouramies, Siamese fighting fishes, paradisefishes, synbranchid eel *Monopterus alba* in Asia, *Microctenopoma, Polycentropis, Hepsetus odoe* in Africa, callichthyines and the electric eel in South America. The most recognised bubble nest maker fishes are, gouramies (dwarf gourami, banded gourami, pearl gourami, Siamese fighting fishes, and paradise fish.

10.3 Captive Breeding Protocols

Breeding of top-quality fishes depends mainly on three factors, selection of parent stock, avoiding indiscriminate inbreeding and rigorous culling of inferior offspring. Fishes intended for breeding should be vigorous, of good size and colour, with no inherited flaws. Breeding aquarium fishes require a good amount of care and patience.

10.3.1 Selection of spawners:

Fishes that are healthy and disease-free should be selected as spawners. These fishes should have required / desired qualities and are to be without visible anatomical or genetic deformities. To get best result, selective breeding may be followed. Spawners are to be selected from different sources to minimise inbreeding in case of livebearers. To avoid inbreeding, the males are separated from females, once the sex could be distinguished.

10.3.2 Conditioning of parent fish:

Before placing the parent fish together for spawning, they should be conditioned. Conditioning is feeding the fish with a variety of healthy foods to get them in top condition for spawning. Many species can be conditioned using a well-balanced flake food, though other should be conditioned on live foods such as brine shrimp, insect larvae, and flying insects. The parent fish can be separated while conditioning. This way when the fish are re-introduced, they are eager to spawn.

10.3.3 Spawning tank/ Breeding tank:

Though some species readily spawn in the aquarium, the eggs or fry often do not survive because of the predatory parents or other fish. Often fry die because of unfavourable water conditions. Many species that practice brood care will harm other tank mates in attempting to guard the eggs. Because of all these problems, most aquarists who breed fish use a separate spawning tank. The spawning tank should be like the hospital tank with protected heater so the fish are not burned, a slow-moving filter (sponge filter), so the eggs or fry are not sucked up, and good aeration. Depending on the spawning method, the spawning tank can be set up in a number of different ways.

10.3.3.1 Spawning tank/ Breeding tank set up for egg scatterers:

As the egg-scatterers often eat their own eggs, the spawning tank or breeding tank has to be set up in a way, that the parents cannot eat the eggs. A 25-50 litre tank is sufficient for spawning of most of egg-scatterers. For egg-scatterers/ adhesive eggs, the spawning tank/ breeding tank can be furnished with a substrate consisting of two layers of the tank floor. As the eggs are laid, they fall through marbles which act as fish-proof egg trap (Figure 10.1 A) or the finely meshed netting placed (Figure 10.1 B) so either eggs are out of the reach of the parents or the parents are removed. For egg scatterers that lay adhesive eggs like tetras, eggs get attached with plants (Figure 10.1C). Rainbowfishes are bred by the use of artificial mops (Figure 10.1 D). The tank should either be planted with fine-leafed plants or artificial mops are to be placed. The eggs are laid amongst the plants or on the mops. The parents should be removed after spawning. Another useful device used is breeding trap which is placed in the main tank (Figure 10.1 E).

10.3.3.2 Spawning tank set up for egg depositors:

Depending on the type of egg-depositor, the tank should be furnished differently. For those egg-depositors that remain in the tank after spawning, tanks should be with leafed plants or flat stones at the spawning sites. For some species such as Discus and Angelfish plastic plants, coconut shells, flowerpots turned on their sides, and rocky caves are arranged to give the fish a sense of security, and to provide suitable spawning sites.

10.3.3.3 Spawning tanks set up for egg buriers:

A peat moss substrate is one of the best substrates for egg-burying species. The peat moss can be removed after spawning and placed in a plastic bag to be stored for weeks to months (depending on the species). A new peat moss substrate can be placed in the tank for further spawning. In order to initiate hatching, the stored peat can be immersed in soft water. Spawning mops or artificial mops (Figure 10.1 D) can be used for breeding killifishes.

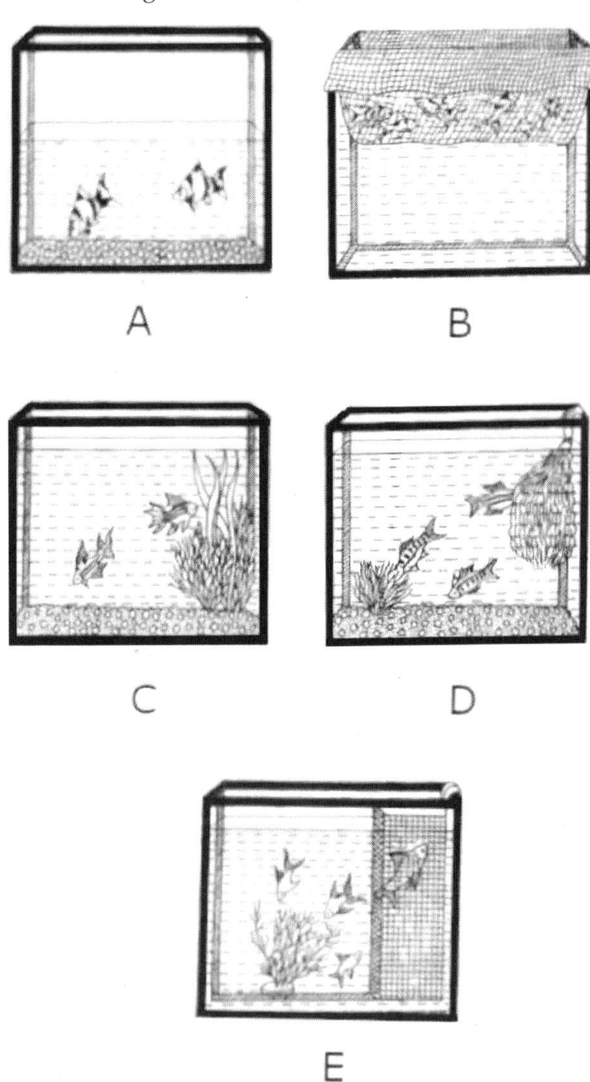

Figure 10.1 A Marble substrate set in the breeding tank for protecting eggs. B Finely meshed netting placed in the breeding tank. C In the breeding tank plants are placed for the eggs to get attached. D Artificial mops placed in the tank for rainbow fishes. E Breeding trap placed inside the main tank.

10.3.3.4 Spawning tank set up for mouth brooders:

Ovophile mouth-brooders can be bred in the main aquarium because the eggs are protected in the mouth cavity. However, it is better to separate mouth-brooders with eggs because of their potential aggressive behaviour. There are no special breeding tank requirements other than the usual tank set-up for the species. **Larvophile** mouth-brooders should be placed in a breeding tank because the eggs are not protected in the mouth, but laid on the surface.

10.3.3.5 Spawning tank set up for nest builders:

Nest-builders should be provided with material with which the nests are build. For bubble-nest builders, fine leafed and floating plants should be provided. Fish species that build nests in the substrate should be given fine gravel or sand.

10.4 Stimulating Factors for Spawning

One of the best ways to induce fish to spawn, especially difficult-to-spawn species, is to simulate natural conditions. Among factors that encourage fish to spawn are the environment, the food and the rainy season.

10.4.1 Water conditions:

The right water conditions are among the most basic requirements in spawning fish. Thus, the water conditions should be similar to those in the natural environment of the species. Another important environmental condition is the right tank set-up including hiding places, spawning sites, lighting, water current, and social conditions (schooling).

10.4.2 Food:

The right food is important to encourage spawning. Without proper food, natural conditions cannot be entirely recreated. Live food often can make a difference in spawning success.

10.4.3 Rearing tank:

A rearing tank is not required for species that take care of their young, though these are still recommended. For species that do not take care of their young, the rearing tank can be the same tank as the spawning tank as long as the parents are removed. The rearing tank should have a protected heater, a sponge filter and plastic or live plants. In addition, three of the tank sides should be covered with black paper, because a light can encourage fungal infections and cause discomfort for the fry. The water in the rearing tank should be similar to the water used for spawning.

10.4.4 Raising the fry:

As the egg hatch, the emerged larva has a large yolk sac to feed upon. Once the yolk sac disappears, the fry begins to look for food. Infusoria, green water or egg yolk can be first fed to the fry of small fish. These fry later on can be fed larger food like white worms, *Daphnia*, *Artemia* nauplii and ground flakes. These foods are good for slightly larger fry such as those of cichlids. Once the fish grow larger, larger foods like brine shrimp, larger *Daphnia*, flakes, insect larvae and chopped

Tubifex worms are accepted. In the rearing tank 25-50% of the water should be changed daily. The water added should be of same quality as that of the water taken out because fry is more sensitive to sudden changes in water. The fry should be fed several times a day. Periodic sorting on the basis of size may be needed for some fish species so that larger fish do not cannibalize smaller fish. With favourable water conditions, regular water changes and generous feeding, the fry grow quickly. Unhealthy and deformed fishes should always be removed.

10.5 Breeding of some Egg Laying Ornamental Fishes in Captivity

10.5.1 Breeding of egg scatterers laying non-adhesive eggs:

Captive breeding of *Danio rerio* (zebra danio)

- Zebra fish have large broods, breed throughout the year, are easily maintained and have transparent embryos.
- The female is larger in size than the male which is slim and of more vibrant colour.
- Medium size, all glass aquarium tanks (60 cmx30 cmx 30 cm) provided with a perforated tank divider, may be used for conditioning of the zebra danio.
- The males are introduced on one side and the females on the other side of the tank divider in the aquarium.
- High quality live foods are to be given to the danios. Live foods for danios include *Daphnia*, bloodworms, and mosquito larvae. Frozen food can be used as an alternative.
- After 1 to 2 weeks of conditioning, adult danios are ready to breed. Breeding tank should be set up during the first week of conditioning. Breeding tank can be between 5 and 10 gallons and should contain a filter with an air stone. A heater is to be placed in the tank to keep the temperature between 72° F and 82° F.
- The marble method, the breeder net method, or the spawning mop method can be chosen for the configuration of the breeding tank.
- After 1 to 2 weeks, danios may be moved to the breeding tank. The ratio of males and females are to be 2:1 or 3:1. Females may be introduced in the breeding tank one day earlier than males.
- Normally danios will breed within 24 hours of being placed in the breeding tank. The female danios have laid eggs and the male danios have fertilized the eggs.
- Once they have spawned, the adult danios are removed. Once the danio eggs are seen in the breeding tank, the adult danios are removed and are put back in their regular tanks. If a breeder net is used removal is very easy. If the marble or mop methods are used a fish net is used to capture the danios in order to move them.

- To protect the eggs in the breeding tank, marbles are used to create a safe place for the eggs (Figure 10.1 A). In this method that helps in protecting danio eggs, a 2-inch layer of glass marbles are placed at the bottom of a 5 to 10 gallons aquarium. Then the aquarium is filled with water until it reaches about 1 inch above the marbles. The eggs will then fall in between the marbles, where they will be safe from the adults.
- A breeder net is to be placed inside the breeder tank (Figure 10.1 B). A breeder net is basically a basket that can be attached to the rim of the tank. The mesh of the basket allows the eggs to fall through, into the main part of the tank, but the mesh keeps the adults from being able to get the eggs. The bottom of the breeder net should be in water.
- Use of spawning mop: For a 10-gallon tank, 10 to 20 spawning mops are needed. The spawning mops (Figure 10.1 D) are placed at the bottom of the tank and fill the tank with water that goes no further than a couple of inches above the top of the mops. The yarn will partially float in the water creating a protection area for the eggs.
- Hatching of the eggs normally takes 1.5 to 2 days for the eggs to hatch.
- The free -swimming fry are extremely small and transparent fry may be swimming around or attached to the side of the tank.
- Feeding the fry: A specialized powdered egg-layer fish food or infusoria, several times a day.
- A sponge filter and few snails may be added to keep the breeding tank clean. About 10% to 25% of the water should be changed daily.
- Feeding the fry with brine shrimps: After 10 days of hatching, fry are given brine shrimp once a day along with powdered egg-layer fish food twice a day.
- To move baby danios to a larger tank: When baby danios are about half an inch long, it is time to move them to a bigger tank.

10.5.2 Breeding of egg-scatterers laying adhesive eggs

Captive breeding of *Carassius auratus*, goldfish

- Male fish is slimmer than the female fish, and the rear of the anal fin is concave.
- Mature male and female goldfishes are selected and kept in circular tanks or ferrocement tanks after disinfecting containers with 1 ppm solution of potassium permanganate ($KMnO_4$). Water temperature should be 68-72° F (20-22°C) and pH should be neutral. The water should be 50% groundwater and 50% filtered pond water. The container better get sunlight in the early morning only for some time.
- Goldfish eggs are sticky in nature, so they require some surface for adherence of eggs. Some artificial nests of different types are provided. Natural, submerged aquatic plants like Hydrilla, or artificial plants are used. Artificial nests are made with split nylon ropes with blunt one end.

Food-grade polythene strips are also used. The nest should float close to water surface. The water temperature should be maintained between 20-28° C.

- Spawners and milters are released in the breeding tank in the ratio 1:2, in the late evening hours. The male fertilizes eggs and egg laying generally takes place within 6-12 hours of releasing males and females.
- Female lays about 2,000-3,000 eggs. Healthy eggs are golden, transparent in the beginning.
- The moment spawning is over either nest is transferred to a different container or parent fishes are transferred to another tank.
- Within three days, if ideal conditions remain, eggs hatch out with a hatching rate of 80-90%. After hatching tiny larvae (2-3 mm) remain clinging to nest.

10.5.3 Breeding of egg depositors:

Captive breeding of *Pterophyllum scalare* (Angel fish):

- During spawning time, the genital papillae appears in both male and female. These nipple-like projections are called ovipositors or egg-placers. The ovipositor of female is large and blunt whereas, the ovipositor of male is slender and pointed. These are used respectively for depositing eggs and fertilizing them.
- For conditioning of the fishes, clean water and a varied diet are necessary. For successful breeding the water should be soft with an alkalinity of 50-100 $CaCO_3$ mg/litre and pH 6.8. Live foods such as adult brine shrimp, mosquito larvae and finely chopped tender earthworms are to be provided. In mature fish breeding can be stimulated by a partial change in water and a rise in temperature between 26° and 28° C (Swain et al., 2010).
- Breeding behaviour: Angel fish select their own partner and the pair remains together for years, lay eggs two to three times in a year. The usual breeding season falls between March and May and in September and sometimes in early October (Swain et al., 2010).
- Angelfishes generally pair up and show spawning behaviour between the ages of 8 to 12 months or more in colour forms. Male and female angels spend a lot of time swimming side by side and defend their territory.
- Tank set up: The breeding tank should be of capacity of 20-30 gallons and it should be at least 18 inches in height and without substrate. A mature sponge filter, heater and some broad leaved live or plastic plants complete the set up. Since Angels usually choose vertical surfaces to lay their eggs, some breeders place pieces of slate in the breeding tank as spawning sites. Temperature of water should be 80-85° F (27-29° C) and water should be soft and acidic.
- Spawning: The pair will choose a spawning site and begin cleaning it with their teeth. After two or three days of cleaning the female lays rows of

eggs on the spawning site and the male follows behind her, fertilizing them. There may be 700-1200 eggs. The eggs are delicate in the first 24 hours.

- Hatching: There are two options—(a) Either leave the eggs with the parents and let them guard (b) Or remove the eggs and hatch them artificially.
- If the eggs are left with parents, water changes should be carried out as usual.
- In the second option some large jars/ and or a hatching tank is needed with clean, dechlorinated tap water and temperature at around 80° F (27° C).
- If Angels have chosen a plant leaf or piece of slate on which to spawn, the eggs are to be transferred quickly and easily to the jar. Angel eggs are light sensitive and methylene blue solution is used to darken the water in the jar. A water stone should be placed underneath the eggs to create a strong water flow.
- Raising the fry: Angel fish become free swimming on day 6 to day 7 when feeding should begin. Live baby brine shrimp is the best food for newly hatched Angel fry. Micro-worms, frozen brine shrimp, dried foods are also used. Exceptionally clean water is of great importance for Angel fish. After about three weeks or so, the fry can be weaned onto frozen adult brine shrimps and crushed flake food. By this time, they look like mini Angel fish and now they are separated according to size.
- In community tanks, large number of Angel fish are kept in community tanks. They lay their eggs on the filter tube or tank wall. After a few days, slates full of wigglers are found. They slowly turn into tiny fishes.

Captive breeding of Rasbora:

- Rasboras are small shoaling fishes, ideal for well-planted community aquarium.
- Rasboras including harlequin are in general fairly difficult to breed.
- Soft, slightly acidic (pH 6.5) water with temperature of 28° C is required.
- For conditioning adults for breeding, variety of good quality dried foods and different types of live foods are fed.
- After conditioning, male and females are placed in aquarium with flat-leaved plants.
- These fishes especially harlequin, prefer calm and peaceful environment with low-level lighting.
- Male and female fishes are placed together in the tank for a week for the purpose of breeding.
- If there is no response, brooders are to be separated and reintroduced after a fortnight.
- If spawning occurs, both parents are removed from the tank.

- Smaller rasboras lay up to 100 eggs, whereas, larger ones lay about 250 eggs. The eggs are laid on the underside of the flat leaves. Eggs hatch after 24-30 hours. And the resultant hatchlings become free swimming after 3-5 days.
- The tiny hatchlings are to be fed infusoria and newly hatched brine shrimps.
- As these hatchlings grow bigger, they are to be fed zooplanktons like *Moina* and *Daphnia*.

10.5.4 Breeding of egg buriers:

Captive breeding of killifishes:

- A tank of 10 to 20 litres is to be taken with fresh dechlorinated water and a sponge filter. The fishes are kept in the tank with spawning mops to receive eggs.
- After 10 days adult fishes are removed and eggs are left to hatch. If adult fishes are well fed, then they may not eat eggs.
- Fry can be raised in the tank until they attain stockable size.
- They accept brine shrimp nauplii and infusoria with addition of finely crumbled protein-rich supplementary diets, when they get little bigger.
- Eggs are laid in leafy environment of a well-planted aquarium. They should be kept in covered aquarium as they are good jumpers.
- Eggs remain viable even after dried conditions, and hatching may be possible even after some weeks or even months, when placed in water.
- They rarely grow over 3-4 cm in length and their life span remains between 2 and 3 years.

10.5.5 Breeding of mouth-brooders:

Captive breeding of Tilapia:

- The egg depositors take the fertilized eggs into their mouth and incubate them.
- The aquarium bottom should be made of sand or fine gravel as these mouth brooders usually prepare breeding pits on the bottom sand on which spawning takes place.
- In Tilapias, the female carries the eggs in her mouth.
- The eggs hatch inside the mouth and the parent carry the fry also for some time.
- To prevent swallowing of the eggs, the females are trapped in a fine net without disturbing the fish much.
- The fish is to be held in the hand with the head immersed in a small bowl.
- The mouth is opened slowly by applying pressure on the lower jaw with the index finger. The fish is to be shaken gently in water.

- Soon the eggs/fry are released into the bowl. These are then to be transferred to an aquarium with very little water column and should be given profuse aeration.

10.5.6 Breeding of nest builders:

In the anabantids or labyrinth fishes parental care has evolved and there is an exceptional wide diversity of behaviours. In 16 genera there may be substrate spawning with male parental care, nest building on submerged plants with male parental care, bubble nesting with eggs deposited in the bubbles and either male parental or biparental care and mouth brooding with either male or female parental care (Nelson et al., 2016). The species which are bubble nesters and incubate their eggs in floating nests specially made by male fish are—dwarf gourami (*Colisa lalia*), banded gourami (*C. fasciata*), pearl gourami (*Trichogaster leeri*), Siamese fighter fish (*Betta splendens*), and Chinese paradise fish (*Macropodus opercularis*).

Captive breeding of *Colisa lalia* and *C. fasciata.*

- The female is less brightly coloured than male and has more rounded dorsal and anal fins than the male.
- For breeding purposes, males and females are kept separately in different tanks for a few weeks with all kinds of prepared feed and also live-foods.
- When the abdomen of the female becomes bulged out with eggs, it is transferred to a small breeding tank with water level of 12.7-13.2 cm at a temperature of 28-30° C.
- Lots of fine-leaved plants such as *Cabomba* or *Myriophyllum* and some floating weeds like *Eichhornia* or *Pistia* may be kept in the tank. The ideal hardness is 100-200 ppm and pH of water should be 7.0-7.5 respectively.
- A good male is introduced in the breeding tank, after one or two days.
- A transparent perforated plastic sheet or a glass is placed over the tank to keep humidity and temperature at high level which helps maintain bubble-nest in a good condition.
- The male takes a large gulp of air at the water surface and converting it into many smaller bubbles that are passed into gill chamber and coated with an anti-burst agent before release.
- During and after making the nest, the male displays this to female, and then both male and female fish embracing near nest resulting deposition of a large number of eggs in the nest.
- Female is removed after breeding. Eggs are guarded by the male fish and eggs remain attached to floating bubble nest. Hatching takes place within 24-30 hours.
- As soon as, fry begin leaving the nest, male is removed from the tank.
- After 36 hours, when young ones remain in free-swimming stage, they are provided with infusoria as starter food.
- After a week, fry, are given newly hatched artemia and small filtered

Cladocera. Fry should be given huge feeding of live food. When fry grow bigger, they can be stocked in bigger cement tanks for further growth. They now accept all kinds of prepared feed.

Captive breeding of *Betta splendens* (Siamese fighter fish):

- The dorsal, anal and ventral fins of male fish are longer and may become extremely veil-shaped. In female fish, egg tube is visible.
- When sexually mature, males become very aggressive.
- These fishes attain sexual maturity at approximately 3-4 months of age. But breeding is generally best attempted with fishes of 9-12 months.
- For every male fish, two to three female fishes are kept and which are of the same sizes as male.
- Male fish is kept individually in small container/aquarium with a capacity to hold 2.5 litres of water. Aeration is not needed but regular change of water (50-75%) each week should be done. Less aggressive females are also kept together in one aquarium containing 25-50 litres of water.
- Breeding tank containing 50 litres of water (depth 15 cm) is set with plants like *Myriophyllum* or *Cabomba*. A fine meshed net is placed in the middle of the tank to keep mature female in one portion and mature male in another. Best breeding temperature is 26-28° C, pH of water 6.5-7.5 is suitable.
- A bubble-nest building is started by the male fish and the partition net is removed. Spawning occurs in early morning and 15 eggs may be laid each time. Total number of eggs may be 200-300 even up to 600.
- As the eggs are shed and fertilized, they sink to the tank floor. The eggs are collected by the male fish in his mouth and subsequently spitted into the bubble nest.
- The female fish is removed at the end of spawning. As the male fish guards the nest, he is left in the tank.
- After three days of occurrence of spawning, male fish is removed.
- After 36-48 hours, eggs hatch. After 5 to 6 days the fry become free swimming and are fed with infusoria or egg-yolk milk several times a day. Fry are given fine dry food after three or four days. When they are about 3-4 weeks old, development of labyrinth organ takes place.
- A glass or plastic sheet is placed over the tank to make the ambience warm and humid.
- Males and females should be kept separately before they are mature and start fighting.

10.5.7 Large scale commercial breeding:

- Well-balanced diet with more of animal proteins and live foods are provided as feed to the fighter brood stocks.
- Each breeding tank is provided with a single mature male and a piece of banana leaf for making bubble nest.

- After making of the bubble nest, a mature female may be kept, and the tank is to be covered with a card-board or any other material to make the tank atmosphere a bit humid.
- The eggs are laid by females and placed in the bubbles.
- The male guards eggs for 3-5 days. The male is removed and free-swimming larvae are fed on infusoria or rotifers.
- The larvae of 10-15 mm then may be stocked for 3-4 months to reach marketable size.

10.5.8 Captive breeding and breeding months of ornamental fishes:

Breeding of ornamental fishes in captive conditions are generally arranged, following their respective breeding months (Table. 10.1).

Table 10.1: Breeding months of some popular egg-laying ornamental fishes

Fish	B	R	E	E	D	I	N	G	M	O	N	T	H
	Jan	Feb	Mar	April	May	June	July	Aug	Sept	Oct	Nov	Dec	
Danio													
Goldfish													
Angel													
Gourami													
Siamese fighter													
Rosy barb													
Tiger barb													
Neon tetra													
Firemouth cichlid													

10.6. Breeding of Livebearers

10.6.1 Livebearers:

Livebearers are fish that bear live young ones. There are two types of livebearers—viviparous, where no eggs are formed, and the young are nourished through the umbilical-like cord or from secretion by the female (Table 10.2), in ovoviviparous the eggs form and hatch within the female before birth (Table 10.3). Livebearers are generally prolific and are easily bred.

Table 10.2: Viviparous livebearer fishes and their distribution

Order	Family	Common Name Scientific Name	Distribution
Beloniformes	Zenarchopteridae	*Dermogenys pusillus*	Southeast Asia.
		Long-snouted halfbeak *Hemirhumphodon pogonognathus*	Southeast Asia.

Order	Family	Common Name Scientific Name	Distribution
		Celebes halfbeak *Nomorhamphus liemi*	Sulawesi.
Cyprinidontiformes	Anablepidae	Striped four-eyed fish *Anableps anableps*	Central and northern South America.
		Jenynsia lineata	
	Goodeidae	Butterfly goodeid *Ameca splendens*	Mexico.
		Rainbow goodeid *Characodon lateralis*	Mexico.
		White's Ilyodon *Ilyodon whitei*	Mexico.
		Red-tailed goodeid *Xenotoca eiseni*	Mexico.
		Jewelled goodeid *Xenotoca variata*	Mexico.

Table 10.3: Ovoviviparous livebearers and their distribution

Order	Family	Common Name Scientific Name	Distribution
Cyprinodontiformes	Poeciliidae	Guppy *Poecilia reticulata*	Central America to Brazil.
		Sailfin Molly *Poecilia latipinna*	Southern USA (southern Virginia, Carolina, Florida, Texas).
		Common Molly, Mexican Molly *Poecilia sphenops*	Mexico to Colombia.
		Yucatan Molly *Poecilia velifera*	Mexico, Yucatan.
		Caucon Molly *Poecilia caucana*	Panama, Colombia, and Venezuela.
		One spot Molly *Poecilia vivipara*	West Venezuela to Argentine and some off shore islands.
		The caudo *Phalloceros caudimaculatus*	Central Brazil, Paraguay, Uruguay.
		Porthole livebearer *Poeciliopsis gracilis*	Central America.
		Mayon Blue-eye *Priapella compressa*	Mexico.

Order	Family	Common Name Scientific Name	Distribution
		Oxacan Blue-eye *Priapella intermedia*	Coatzacoalcos river (Mexico).
		Black-barred livebearer *Quintana atrizona*	Cuba.
		Green Swordtail *Xiphophorus helleri*	Central America.
		Southern Platy *Xiphophorus maculatus*	Eastern coast of Mexico and Guatemala and into northern Honduras.
		Catemaco Platy *Xiphophorus milleri*	Lake Catemaco, Mexico.
		Slender pygmy swordtail *Xiphphorus pygmaeus*	Axtla river (Mexico).
		Variable Platy *Xiphophorus variatus*	Southern Mexico.
		Spike tail Platy *Xiphophorus xiphidium*	Mexico.

10.6.2 Characteristic features of some livebearer families related to internal fertilization:

- Fishes of family Zenarchopteridae: These fishes have a modified anal fin (**Andropodium**) for internal fertilization. In some species there is a unique method of sperm packaging.
- Fishes of the family Goodeidae subfamily Goodeinae: Anterior rays of anal fin in males are crowded, shortened and slightly separated by a notch from the rest of the fin. This gonopodium like structure is termed as **Pseudophallus** or andropodium.
- Fishes belonging to the family Anablepidae, subfamily Anablepinae: **Gonopodium** is tubular in shape and is formed from anal fin rays and associated with sperm duct. In some males the gonopodium can move only to the left in others to the right. In females the genital aperture is open to the right or to the left. Dextral and sinistral mating types are known in both sexes of all species. Fertilization is internal, bear young alive.
- In family Poeciliidae, subfamily Poecilinae, male fish has elongated anterior anal fin-rays (gonopodium, primarily formed from the third, fourth and fifth rays) with internal fertilization. Eggs are with large yolks, have live birth (ovoviviparous) except in *Tomeurus*, which is egg-laying.

10.6.3 Breeding of livebearer fish species:

Young ones produced by a live-bearer are normally 50-60 only, though some larger swordtails may produce as many as 100. If the livebearers are fed properly with natural food, supplemented with better artificial feed, the mother produces more than 100 young ones. Soon after larvae come out from the mother, they are to be separated to check predation/ cannibalism. To keep their colour strain pure, one must isolate them from any possible contamination (crossbreeding). For this a number of tanks are required, one for male fish and one for female fish, as well as a few breeding tanks for each selected pair and some tanks for young ones. Livebearers are popular with aquarists due to their prolific breeding habit and acceptance of all kinds of food.

10.6.3.1 Protocols followed in the breeding of livebearers:

- Maturity determination: Usually livebearers mature between 4 and 6 months. However, guppy and platy may mature even within two months.
- Sex identification:
    - Male fish—Smaller; brightly coloured and attractive; fins especially dorsal and caudal fins are longer; belly region is flat; anal fin is modified into gonopodium which is a rod or tube- like structure.
    - Female fish—Larger; dull coloured, fins comparatively smaller; belly region is bulged; anal fin is normal in shape.
- Conditioning of parent fish:
    - Before placing the parent fishes together for spawning, they should first be conditioned. Conditioning is feeding the fish with a variety of healthy foods to make them attain suitable conditioning for spawning. Many species may be conditioned using a well-balanced flake food, though others may be conditioned with live foods such as brine shrimp, insect larvae, and earthworms.
    - The parent fishes should be separated while conditioning. Such fishes when reintroduced, will be ready to spawn.
    - Feeding for breeding: Livebearers are omnivorous in general. They normally accept both animal and plant-based foods with some exceptions. The feed composition should be balanced, providing proteins, carbohydrates, lipids, vitamins and mineral salts (trace elements) is needed and a bit of roughage. During growth or recuperation, higher protein diet of 35-40% is found to be suitable. Otherwise, lower levels of protein (around 25-30%) can be given in granular form.
    - Feed containing less or zero fish meal content showed better growth and early maturation. This probably shows that fish has tendency to consume and utilize plant proteins better than animal proteins.
- Breeding of the livebearers:
    - In livebearers, the eggs are situated in the egg duct where they are fertilized. Upon hatching, fry, are not immediately delivered, but they remain in the safety of the mother's body until they reach a stage of

development equivalent to the young of egg layers that have absorbed the yolk sac and become free swimming. The ideal temperature must be 27° C.

- As the male mature the anal fin becomes more pointed and straightened into a rigid tube-like projection, which is called gonopodium. The gonopodium is carried normally close to the body end pointing rearwards. However, it is a mobile organ and can be angled in almost any direction. The males court the females with their fins erect and they chase the female until the opportunity presents itself for a lightening thrust of the gonopodium.
- The female is fertilized by the nearest touch of the gonopodium on her vent, and one fertilization will last for several broods. Unlike the male, the female has normal anal fin. Females can have up to eight broods from one fertilization, therefore it is necessary to re-mate after the first brood. The period of gestation is being constant for every species.
- Females about to give birth are said to be ripe. This condition can be determined by the appearance of the dark, crescent shaped area in the female body close to the vent known as the gravid spot, which is accompanied by a general fattening of the belly when viewed from above.
- The eggs develop and actually hatch out inside the mother and leave her body as fully developed fish. The number of young in a brood is largely dependent upon the size of the female.
- Irrespective of the number in the brood, fry are approximately all the same size at birth.
- Livebearers are notorious cannibals. They will devour their young ones as soon as they are born. They can be prevented by having an abundance of cover for the young ones in the form of bunched fine leaved plants or to use breeding trap (Figure 10.1 E), which restricts the female to a small part of the aquarium, but allows the fry to escape into the wider reach of the breeding tank.

- Breeding trap used for livebearer species:
  - When females of livebearer species are ready to drop young ones, they should be taken to breeding tank where individual or group of fishes can be kept in perforated nylon-bag cages, from which young ones escape from their mothers through the mesh to outside to avoid parental predation (cannibalism).
  - Various types of net-cages, perforated dust-bins, plastic perforated baskets (Figure 10.2) or fabricated perforated containers (Figure 10.3) of required size can be used for this. Depending upon the size of the net-cages or traps, the number of females may vary.
  - A box-type or cylindrical but not a cage, can be fitted or hooked on to the wall of the aquarium for one or more mothers whose new-borns

drop through mesh-opening into lower space of the aquarium. When it is observed that females have stopped dropping young ones, they should be removed and removed separately with balanced nutritious diets.

- Number of young ones:
    - The number of young ones i.e., brood varies from fish to fish and also it depends on the size, feed given and tank environment. In an average 20-40 young ones can be expected. Guppy—100; Platy—50; Molly—100; Swordtail—200.

Figure 10.2 Perforated plastic baskets. Figure 10.3 Container as breeding trap.

    - After 2-3 days the female again becomes pregnant even without the contact of male. The sperm transferred during first mating is stored in the female body and when eggs are formed the sperm will join with eggs and form young ones. So, by single mating young ones can be released 8-10 times. In a year, it is possible to get young ones 10 times from one female.
- **Feeding of fry:**
    - Baby fish i.e., fry can swim and eat from the moment they are born. It accepts infusoria, finely crushed dried food, brine shrimp, nauplii etc.
- **Culture of livebearers:**
    - Young ones are to be stocked at 500 numbers per square metre in cemented tanks.
    - Water depth may be 30.4-60.9 cm under shade net or in a cool place for one month till male matures sexually.
    - Males are then reared separately to avoid premature breeding.
    - After rearing for one month, they are stocked in rearing tanks at 150 numbers per square metre till they reach the marketable size after 4 months.

10.6.3.2 Propagation of some livebearers:

- **Breeding of Guppy:**
    - Maturity size: Male—2.5-3.5 cm; Female—5-6 cm.

- Sex identification: Male is smaller with more flowing fins and pointed anal fin or gonopodium. No two guppies are alike in colour.
- Selection of the fish which one wants to breed. One male and two or three female guppies are to be selected for breeding. This makes breeding a less stressful process for the females.
- Conditioning: Guppies thrive in a large well-planted tank with a steady temperature, within the range of 20-25° C. During summer, a special cooling care has to be maintained to sustain them.
- Water condition in spawning tank: Temperature:20-28° C; water hardness: 50-100 mg $CaCO_3$/ litre (moderately soft water).
- Spawning: The female is fertilized by the nearest touch of the gonopodium on her vent, and one fertilization will last for several broods.
- The male fish is placed back in the regular tank when the female(s) are pregnant. All female fish develop a dark mark called a gravid spot on her abdomen, when pregnant. It becomes noticeably darker when the eggs have been fertilized.
- Breeding tanks: From community tank gravid females are removed, and those starting to swell with developing young ones are placed in the breeding tanks (30x20x20 cm) or 10 to20 gallon tanks, individually or in pairs. *Cabomba* plants, heater and a gentle filter may be placed in the aquarium. Aeration is not required now, but at the time when individual female gives birth to 20-200 young ones, aeration is needed.
- The gestation period is 26-31 days. When the female guppy is ready to give birth, her stomach will be very large and her gravid spot will be deep coloured. After she gives birth live young ones, she should be removed.
- Guppy parents can turn cannibalistic, so the fry should be provided with hiding places after they are born. Guppy fry tend to sink, so low-floating plants are used to cover them. Java moss or spawning moss provides a nice hiding spot for guppy fry. Some high plant cover is also required as healthy fry will swim upwards.
- A bare bottom tank without substrate is good for fry as it can be cleaned easily and fry can be observed properly.
- For mass breeding of guppies, a tank of 100cm x 60 cm is ideal where a perforated bucket dust-bin can be provided at one side of the tank, rapped with fibrous plastic flowing filaments where females can drop young ones.
- Soon after the birth, young ones escape from their mothers and enter into the perforated basket, and latter young ones are collected from the basket and are placed in a separate tank for further rearing.
- A temperature of 78° F (25.5° C) in the tank is needed for the fry till

they are full grown. The tank should also be cleaned regularly. About 40% of water should be changed every few days.

- Guppy fry eat brine shrimp, micro-worms or powdered flakes. Guppies enjoy both meat and vegetables. They should be fed twice a day. Excess food should not be given. New born fry should be fed recently hatched brine shrimp and also a small amount of boiled spinach.
- Fry are to be placed in the normal tank when they get big enough. When fry, are of good size, about a month and half to two months old, they will be ready to fend for themselves outside the breeding tank. They can now be kept in the normal tank with non-aggressive fishes.

Breeding of Platy fish:

- Male is smaller and slimmer than the female, and has gonopodium. Male to female ratio is 1:3. Breeding season is March to November.
- Once a female platy has become impregnated, her belly will quickly swell to massive proportions and a dark gravid spot will appear near her anal fin.
- These fish can withstand fluctuations much better than any other fish. Minimum tank size should be 10 gallons, temperature 20-26° C (68-79° F), pH 6-8, moderately soft water, water hardness should be 50-100 mg $CaCO_3$/ litre.
- Gestation period is 28-42 days. Young ones per female are 10-100.
- After the female has given birth, either the fry or the adult fish should be removed to another fish tank.
- If kept together, in order to save the fry, it is important to have a heavily planted tank. Some of the best plants to use are Java moss, Water sprite or any other dense plant that the fry can hide in.
- The new born fry can be immediately fed powdered flake food, and live food such as baby brine shrimp, micro-worms and banana-worms.

Breeding of swordtail:

- Male is smaller and slimmer with gonopodium and pronounced sword-like projection on the caudal fin. Dark gravid spot near anal fin in female. Male—8 to 9 cm; female—10 to 11 cm. Male and female ratio is 1:3 or even 1:4. Breeding season is May to October.
- Minimum tank size, 29 gallons. Water conditions in the tank should be, temperature 21-28 °C (72-82°F); water hardness 50-100 mg/litre (moderately soft water), pH 7-8.
- Sex reversal: Each immature fish could develop into male or a female depending on whether male or female reproductive organs develop first. If ovaries develop first, these will develop into a female. The case may be same for males. During their lifetime, a female may

change to male though male changing to female is very rare. External environmental conditions may play a role in sex reversal. Low pH of 5-6 helps in developing more males in broods of swordtail and pH value of more than 7.0 develop more females.

- Once a female swordtail has been impregnated, she will begin to swell with the quickly growing fry. A pregnant female can be recognized by its massive belly size and a dark gravid spot near her anal fin.
- Gestation period is 28 to 42 days. Young ones per female 20-100.
- A heavily planted tank is a must if the parents are kept with the fry. The best plants that can be kept should be dense or bushy, and Java moss provides excellent refuge for new born fry. Otherwise all the adult fish should be removed from the tank after the fry has been born.
- Starter diet for the fry are zooplanktons and fine powdered dried feed.

Breeding of black molly (Poecilia sp.):

- Sex identification: Male fish with gonopodium. Dorsal fin is flowing and bigger. Males are slimmer and smaller than females. Male is 8-9 cm and female is 10 cm. The pregnant molly can be identified by a dark gravid spot near their anal fin and a considerably plump stomach.
- Male and female ratio should be 1:2. Breeding season is April to October.
- Minimum tank size: 29 gallons. Optimum water conditions: temperature 21-26° C (70-79° F), pH 7.0-8.5 and hard water, aquarium salt at 0.5-1.0 g/litre.
- Gestation period is 40-70 days. Young ones per female 30-70.
- After the female has given birth, she should be removed from the fry tank.

Starter diet for the fry are zooplanktons and fine-powdered dried food.

10.6.3.3 Success in captive breeding of livebearers:

Livebearers can be successfully bred in captivity, if knowledge regarding certain conditions/inputs (Table. 10.4) and information about their breeding months (Table. 10.5) are acquired beforehand.

Table 10.4: Inputs regarding breeding of some popular livebearers (Source: Swain et. al., 2010)

Livebearer fish	Water conditions required	Gestation period (days)	Young ones per female fish	Starter diet
Guppy	Temperature: 20-28° C; water hardness: 50-100 mg $CaCO_3$/ litre (moderately soft water)	21-35	20-100	Zooplanktons and fine-powdered dried feed.

Livebearer fish	Water conditions required	Gestation period (days)	Young ones per female fish	Starter diet
Platy	Temperature: 23-28° C; water hardness 50-100 mg $CaCO_3$/ litre (moderately soft water)	28-42	10-100	Zooplanktons and fine-powdered dried feed
Swordtail	Temperature: 23-28° C; water hardness 50-100 mg $CaCO_3$/litre (moderately soft water)	28-42	20-100	Zooplanktons and fine-powdered dried feed
Black molly	Temperature: 23-28° C; aquarium salt at 0.5-1.0 g/litre	40-70	30-70	Zooplanktons and fine-powdered dried feed.

Table 10.5: Breeding period (months) of some popular livebearers

Fish	Jan	Feb	March	April	May	June	July	Aug	Sept	Oct	Nov	Dec
Guppy												
Platy												
Swordtail												
Molly												
	B	E	E	D	I	N	G	M	O	N	T	H

10.7 Success in Captive Breeding of Indigenous Ornamental Fish

Success has been achieved in breeding and larval rearing of some indigenous ornamental fishes such as Indian rosy barb, Black line rasbora, Indian flying barb, Dwarf gourami, and Giant danio by scientists at the Central Institute of Freshwater Aquaculture (CIFA). Technology had been developed by the College of Fisheries, Kochi in collaboration with National Bureau of Fish Genetic Resources (NBFGR), for captive breeding, larval rearing and domestication of 12 ornamental fishes from Western Ghats. These fishes are *Puntius fasciatus, Puntius filamentosus, Puntius exclamatio, Puntius melanostigma, Puntius sarana subnasutus, Danio malabaricus, Puntius pookodensis, Pristolepis marginata, P. fasciatus, Garra gotyla stenrhynchus* and *Nemacheilus triangularis* and *N. monilis*. Other achievements are, domestication of *Horabagrus brachysoma* and *Gonoproktopterus curmuca*, standardization of the breeding and larval rearing of *Puntius denisonii*, broodstock development and breeding of *Puntius chalakkudiensis, P. jerdoni and Osteobrama bakeri* (Swain et al., 2011).

10.8 Marine Ornamental Fish Culture and Breeding

Our oceans accommodate more than 25,000 fish species, of which 4,000 fish species are reef associated. About 800 reef fishes and 290 invertebrates are in aquarium trade (Swain et al., 2011). More than 90% of the total marine organisms are collected from wild and only 100 are bred in captivity. More than 50 marine ornamentals out of 200 varieties are brilliantly coloured and have export potential. As most of these traded species are collected from wild, there is an urgent need to develop technology for captive breeding of these species, so there is decreased dependence on wild-collected specimens and development of alternative source of specimens for ornamental fish trade.

Hatchery technology for 16 species of marine ornamental fishes—has been developed by the Central Marine Fisheries Research Institute (CMFRI), for the first time in India.

10.8.1 Captive breeding of marine ornamental fishes in India:

The export of marine ornamental fishes in India, could not attain the expected goal till now, so, the Central Marine Fisheries Research Institute (CMFRI) has taken initiatives in captive production and development of hatchery technology for 16 species of marine ornamental fishes. These fishes are clown fishes—*Amphiprion percula, A. ocellaris, A. sandaracinos, A. sebae, A. frenatas, Premnas biaculeatus,* Damsels—*Dascyllus trimaculatus, D. aruanus, Pomacentrus caeruleus, P. pavo, Neopomacentrus nemurus, N. filamentosus, Chromis viridis, Chrysiptera unimaculata, C. cyanae* and *Pseudochromis dilectus* (Swain, et al., 2011).

Breeding of clown fish:

- Pair formation: Five fishes of each sex of different size groups are to be stocked together along with a single host sea-anemone in a 500-litre FRP tank fitted with a biological filter.
- Sex change and pairing: All clown fishes are protandrous (male first) sequential hermaphrodites, and a female is dominant in the peck order. The male and the female form a monogamous pair bond.
- Brood stock rearing: For brood stock rearing suitable environmental conditions, proper feeding and best management practices should be followed.
- Tank set-up: A suitable tank of size 3 ft x 2 ft x 2 ft is to be selected with coral sand, few live rocks, a healthy anemone, bright lighting and an efficient protein skimmer. There should be suitable substrate for egg deposition.
- Breeding: Brood stalks under suitable rearing conditions and fed with nutritious food, start breeding within 4 to 6 months of rearing under captivity.
- Egg hatching and larval rearing: Clown fish parent fishes look after eggs, so parents can be allowed to remain in the tank till hatching takes place. Eggs take 6 to 15 days to hatch depending on the water

temperature. Larval rearing can be carried out in three ways—(a) using same tank or parental tank (b) transferring the eggs to hatching tank and then rearing the larvae (c) transferring the larvae to the larval rearing tank. Second method has been found to be most suitable. The larval rearing is generally carried out under green water system. Feeding is to be done with super small (SS) rotifer and newly hatched artemia nauplii.

- Larval feeding and rearing conditions: Nutritionally, sufficiently enriched feed in high density is one of the important factors. Maintenance of suitable water quality is essential for rearing clown fish larvae.
- Juvenile rearing and feeding: The larvae become juvenile after 19-20 days of post hatching. The size of the rearing tank, stocking density, quality and quantity of food provided and water temperature are important factors for proper growth of juveniles. Wet feed such as mussel meat, prawn muscle, fish eggs and minced fresh of trash fish at 15 to 20 % of body weight are to be provided. *Artemia* nauplii (10-15 nos./ml) and *Brachionus plicatilis* (50-55 nos./ml) are to be given twice a day to retain colour of fishes. Adult artemia (2-4 nos./ml) should also be given as a supplementary diet.

10.9 References

Andrade D.V. *Foam nest production in the armoured catfish.* Journal of Fish Biology. 50(3). 1997, pp. 665-667.

Nelson J.S., Grande T.C., Wilson M.V.H. *Fishes of the World.* John Wiley & Sons, Inc. 2016, pp. 1-707.

Pal A.K., Dasgupta S. and Sahu N.P. *Fish Physiology.* Published by Directorate of Knowledge Management in Agriculture. ICAR, New Delhi. 2011, pp. 658-704.

Swain S.K., Madhu K., Madhu R. and Gopakumar G. *Ornamental Fish Breeding and Culture.* Published by Directorate of Knowledge Management in Agriculture. ICAR, New Delhi. 2011, pp. 500-532.

Swain S.K., Sarangi N. and Ayyappan S. *Ornamental fish farming.* Published by Directorate of Information and Publication of Agriculture. ICAR. 2010, pp. 1-145.

11

Transport of Ornamental Fish

11.1 Introduction

In ornamental fish farming, it has become obligatory to appraise the most suitable and ideal condition for transportation. This is of course, to obviate the problem of high mortality of fish species during transportation. There is fish mortality during collection and transportation. Making proper planning, application of suitable methods and appropriate technology, overall using best management practices, can help in reducing the high rate of fish mortality during packaging and transportation.

Nowadays, post-shipment mortality (death-on-arrival) from exporter to importer has been greatly reduced to less than 5% (Belema et al., 2017). Still, the transportation of live fish is not successful until the fish arrives at the stocking or receiving site in healthy conditions.

Fishes either caught from wild sources or reared at farms are transported to different places. These may be transported locally, interstate or overseas.

There are factors which should be considered in advance for successful transport. These are proper conditioning of fish prior to packaging, duration of transport, number of fish to be packed, species to be packed, size of packaging box, location of destinations and means of transport (Mathew et.al., 2016).

Ornamental fish packaging systems are now branded by very high fish loading densities and high metabolic wastes in the transport water after transport. Moreover, post-transport mortality could be reduced through enhancement of the stress resistance of the fish. So, quality of transport water, preparation of the fish for transport and recovery of the fish after shipment are important factors which should be considered (Lim et. al., 2003). There should be well- coordinated system for packing in order to ensure that arrival of fish safely. Success of transport is usually measured by Dead on Arrival (DoA) and high DoA results in loss of customers (Willis, 2011; Metar et al. 2018).

11.2 Basic Principles for Transport of Ornamental Fish

- There should be a proper planning regarding conditions, packaging and transportation systems. This helps to enhance the number of fishes to be transported and safeguard bio-security (Mandal et al., 2010).
- Fishes should be healthy and free of disease-causing organisms prior to transport.

- Fishes are to be starved (purged) before transport to prevent excrement from contaminating the water.
- There should be adequate dissolved oxygen levels and so aeration and/or oxygenation of transport water to be undertaken.
- To maintain relatively constant temperatures, temperature control measures should be carried out.
- Mucous and metabolic excretory products should be minimised.
- Fishes should be protected from mechanical damage.

11.3 Pre-Conditioning Preparations—Step 1

- All the fish whether wild caught or farm reared should be handled carefully and kept in the tank.
- Before harvesting, the fish should be fed with vitamin C supplement diets (8-10%) for 2 weeks before harvesting. This reduces stress and provide resistance to diseases.
- There must be proper consideration and preparation before and during harvest. The number of tanks, number of fishes, number of workers available, time, weather should be taken into consideration.
- Harvesting method followed should be according to the species of the fish.
- The netting material should be soft. Mesh size should be 1/20″ for smaller fish (up to 2″) monofilament nets and ¼″ mesh of knotless nylon for larger fish (4-5″).
- Water in the transport container are either taken directly from the pond or aerated well water or a half-pond/half aerated well water mixture. Fishes must be handled properly, least touched by hand and should remain in water while shifting to the tanks.

11.4 Conditioning of Fish before Packing—Step 2

Conditioning is done by acclimatizing the fish to the tank environment and clearing the fish of all the gut contents.

- As the fishes are shifted to tank from pond, dead fishes should be removed and disposed. Damaged or injured fishes are also to be removed or shifted to quarantine tanks.
- Fishes are to be graded according to the size. Grading is done either manually or by auto grading systems that uses screens of different mesh size.
- The water of the conditioning tanks should be treated with common salt @ 3 gm/litre. About 70% water is exchanged daily. Fishes are kept in this tank for 1-2 days. These tank fishes should be examined regularly.
- Fish are normally shifted to glass aquaria, to make it easy to check for signs of disease, quality etc. Glass tanks are to be provided with aeration and water re-circulation facility. The glass tank water is treated with

methylene blue @ 0.0035 ppm and common salt @ 0.1 ppt. Fishes are to be kept here for 1-2 days.

- Purging of fish can be done in conjunction with prophylactic treatment before packing.
- Purging (starving) is important for several reasons such as
  - Purging voids digestive tract. Fish may vomit or defecate in the bags when not purged.
  - Purging slows the metabolic rate of the fish. It reduces oxygen requirement and reduces ammonia and carbon dioxide output.
  - For most fish, one day purging is sufficient, but larger fish and herbivorous fish may need more.

11.5 Packing—Step 3

- Selection of packing material:
  - Thickness and shape of packing bags: The LDPE (Low Density Polyethylene) bags (Figure 11.1) of 250 μ thickness may be used. These packing bags are used in domestic as well as in international markets for transporting (Figure 11.2). Use of bags with straight corners are not good for small fishes, as they may aggregate at these corners. So, the corners should be tied with a rubber band.
  - Size of packaging bags: The number of fish that could be packed in a bag is directly proportional to size. In domestic trade, the bag sizes are 5″x12″, 12″x20″ and 18″x24″. In International trade size of the bags are 3″x9″ to 13″x26".
  - Size and material of packaging box: The box in which polybags are placed should not be of large sizes as polybags are arranged in a layer and not stacked one upon the other. Large boxes are difficult to handle also. In domestic trade corrugated boxes of 3-5 ply are used while in international trade, Polystyrene boxes (Figure 11.3) are used (Figure 11.4). Exporters of Southeast Asian countries use polystyrene boxes of 60 (L) cmx42 (W) cmx30 (H) cm and 49(L) cmx 38 (W) cmx 38 (H) cm.

Figure 11.1 LDPE bags.

Figure 11.2 Jewel cichlid in packing bag.

Advantage of polystyrene boxes: The polystyrene boxes are useful in providing insulation against temperature and these also reduce the risk of water leakage from box. The wall thickness of boxes is 2.5 mm, but may be more if the temperature of the importing country is very low. International Air Transport Association (IATA) guidelines are followed to fix the maximum weight that can be shipped in a box of specific size.

Figure 11.3 Polystyrene box.

Figure 11.4 Stacked polystyrene boxes in a facility.

- Determination of the packing density

  About 200 g (25 fishes of 3″ length) or 30 g (100 fishes of 1.25″ length) could be packed in one litre of water under standard conditions.

  Factors affecting packing density of fish:

  - Species tolerance to stress.
  - Size of fish which is inversely proportional to the packing density (Table 11.1).
  - Transit time—Longer transit time means lower the packing density.
  - Temperature—Lower temperature reduce metabolic activities and increase transit time.
  - Health and stress status of fish prior to packing—Before transit, health status of fish is to be determined.
  - Chemicals and other additives—Salts, anaesthetics, ammonia-binding agents are added in the water.
- Water quality:
  - Dissolved oxygen—It is important to have a 75% volume of oxygen in the bag to ensure adequate diffusion of oxygen at the surface of the water.
  - pH—The pH of the transport water will also affect the toxicity of ammonia.
  - Carbon dioxide—Carbon dioxide, a by-product of respiration will react with water to form a weak acid. This weak acid will in turn decrease the pH of water.

- Ammonia—It accumulates in the packing bags due to excretion and bacterial action on the excreta. Ammonia is highly toxic to fish in its un-iodised form. About 0.05 ppm could be harmful for the fish. By adding granules/rings of zeolite @ 15-20 g/ litre of water, it may be controlled.
- Additives—The most common additives used are Eugenol (5 mg/l), Quinaldine (5mg/l) and MS-222 (20mg/l). The dosages should be proper and the sedatives should be within the permitted limit of the importing country.

Table 11.1: Size of fish and the packing density

Fish length (in inches)	Number of fish (transport time 48 hrs)	Number of fish (transport time 72 hrs)
1.5	200	150
2.0	150	100
2.5	100	75
3.0	75	50
4.0	50	35
5.0	20	12
6.0	15	10

- Packing conditions:
  - Volume of water and oxygen is in 1:3 ratio.
  - Packing water temperature between 22-24° C.
  - Tranquilizer/water conditioner.
  - Heat packs/icepacks.
  - Purging 2-3 days prior to transport.
- Process of packaging:
  - Preparation of water for packaging—The required quantity of clean water is stored in clean tanks and is treated with common salt (3 gm/l) and methylene blue (2mg/l) or acriflavine (7 mg/l). The common salt helps in osmoregulation and methylene blue or acriflavine are anti-microbial agent.
  - Pre-packaging activities
    - Once counted, fish are packed in bags.
    - Bags are sealed and oxygenated bags are often kept on trolleys.
    - Pure oxygen should be used.
    - Bags should be marked according to species, size, consignment, especially when dealing with multiple transportation.
    - Pre-packed bags must be kept in a cool dark area to acclimate.

- Ideally cool down in an air-conditioned room to desired temperature.
- Bags of fish normally put into cool room and chilled to 22-23° C for tropical species and 15-18° C for cold water species to acclimate them to packing conditions for minimum of 4 to 6 hours.
- Salts may be added at between 1 to 3 ppt depending on species.

- Final packaging:
  - Once the fish have finished pre-packing process, they are ready for final packing.
  - Fish that show signs of sickness or stress should not be transported.
  - Pre-treated water is to be filled up in the polybags. These bags are then placed in another bag of same dimensions inserting newspaper in between the two bags to provide additional safety against water leakage and also to reduce stress to fish due to excess light. Fish are transferred to the new bag with the help of a hand net made up of very soft material. The air inside the bag is replaced with oxygen. These bags are sealed by twisting the top of the bag and folded over with rubber bands or metal clips used to fasten the top of the bag.
  - In domestic trade, bags are tied with rubber bands and in international trade, metal clips are used.
  - The species details, number of fish and average size are again mentioned on the bags with the help of marker pens.
  - All aggressive fish, fish with delicate finnage, or costly fish are to be packed individually.
  - Packing of polybags in boxes:
    - Polybags stocked with fish should be kept in boxes properly.
    - Bags are to be kept straight and not to be stacked one upon the other.
    - There should not be any sharp object inside the box.
    - After the polybags are kept inside the box these are to be covered with newspaper before closing the box.
    - For long distance transportation, ice packs may be kept inside the box.

11.6 Transporting and Transferring—Step 4

- Getting fish to the customer as quickly as possible is an essential step to safeguard health and quality.
- The air transport of ornamental fish is governed worldwide by the International Air Transport Association (IATA).
- Labelling requirements:

- Labelling must meet IATA and airline requirements, and the following must be included on each box:
    - Consignment note must be filled out legibly.
    - Boxes must be marked with—This end up, Live Fish, Handle with Care stickers.
    - Customer contact details.
    - Number of boxes in consignment
    - Proper labelling is required for all types of transport.
- Paperwork and documents:
    - Accurate and required paperwork should be done after consultation with the receiver, otherwise the material may be destroyed or sent back.
    - Airway bill, Invoice, packing lists and fish health certificate are among the required documents.

11.7 Conclusion

- Transportation is a critical component in the success of ornamental fish production. Live, healthy and good quality fish are only to be transported.
- Correct conditioning (prophylactic treatment, purging, prepacking) is needed to get fish ready for transportation.
- Correct preparation of packing water is essential for final pack, correct ratio of water and oxygen needed.
- Best management practices should be followed to improve the competence of the ornamental fish transport.
- Communication and correct paperwork are required for the accomplishment of assignment.
- Biosecurity is vital for future sustainability.
- Supply should strictly follow the demand, i.e., the order placed by the customer.

11.8 References

Belema M., Idowu K.O., Aghogho K.D., Ndubuisi A., Oluwakemi A. and Stella U. *Handling and packaging of ornamental fishes for successful transportation*. International Journal of Fisheries and Aquatic Studies. 5(5). 2017, pp. 263-265.

Lim L.C., Sorgeloos P. and Philippe D. *Recent developments and improvements in ornamental fish packaging systems for air transport*. Aquaculture Research 34(11). 2003, pp. 923-935.

Mandal S.C., Kakati B.K., Das P., Barman D. *Packaging and transportation of ornamental fishes*. Aqua International. 2010, pp. 28-31.

Mathew H.K., Mercy T.V.A. and Mahapatra B.K. *Conditioning and packaging of ornamental fish for transportation. In: Best management practices for freshwater ornamental fish production*. (Eds: Jain AK, Saini VP and Kaur VI). National Fisheries Development Board (Ministry of Agriculture and Farmer's Welfare, GOI, Hyderabad, India. 2016, pp.142-152.

Metar S., Shinde K., Chogale N. and Satam S.B. *Transportation of Live Marine Ornamental Fish*. www.researchgate.net/publication/325169781.2018. Access Date 28 September 2019.

Willis S. *Live Transport of Ornamental Fish*. National Aquaculture Training Institute. www.aquaculturetraining.comau/pdf/live-transport-of. 2011. Access Date 28 September 2019.

12

Ornamental Fish Diseases and their Control

12.1 Introduction

Disease is one of the most important hassles encountered in the ornamental fish keeping. This critical limiting factor not only hampers the growth of the ornamental fishes but also causes massive mortalities.

Since ornamental fish trade is a flourishing business, diseases affecting fishes have a serious impact on the trade ultimately. The culture practices are getting intensive and there are frequent transboundary introductions of many exotic species, and chances of new diseases and pathogens affecting this industry have therefore been enhanced (Sanil and Vijayan, 2008). Huge losses in the export consignments are the diseases affecting the fishes. Moreover, chemotherapy treatments also reduce the economic viability of the ventures.

Realising the potentiality of the development of ornamental fish industry, a thorough knowledge of the fish diseases, and the method of their treatments along with the prophylactic measures are therefore indispensable. It is also better and safe to opt for a proactive approach through better health management approaches for prevention and control of diseases.

Diseases are generally categorized as non-communicable diseases (caused by environmental, unhygienic handling, nutritional and genetic factors), communicable diseases (caused by microbes and parasites), and neoplastic diseases (tumours and cancers) (Pelczar et al. 1998; Cheng, 1999).

12.2 How to Prevent Fish Diseases

Some precautions may be taken to reduce the possibility of the fish getting a disease.

- Good quality, compatible fish should be bought.
- To follow the practice of quarantining new fish before adding them to the aquarium (A hospital tank may be used for this).
- Rough handling, sudden changes in conditions or bully tankmates should be avoided as these causes stress to the fishes.
- Overfeeding the fishes should be avoided.
- Sick fishes are to be kept in hospital tank.
- Water from quarantine tank should not be transferred to the main tank.

- Metal should be avoided to come in contact with the aquarium water.
- Stock of fish should be at the optimum level. Overcrowding should be totally avoided.
- Disinfection of fish is to be by bath treatment with chemicals such as 2 to 3 % sodium chloride or 20 mg/litre potassium permanganate.
- Good quality and optimum quantity of food should be provided.
- Surveillance and monitoring should find essential place in the fish-health management programme.
- To maintain optimum water quality the incoming water should be first channelled into a storage-cum-filter tank and treatment may be done for setting organic load and killing of pathogens. Mohua oil cake @ 250 mg/ litre and lime are used as disinfectants in ponds. It kills predators and fertilizes the water and soil as well.

12.3 Stress Related Diseases in Ornamental Fishes

Changes in the physical, chemical and biological components in the aquatic environment along with the culture technologies cause stress to the morphological and physiological system of fishes. Stressors are the agents causing harm to the fishes. These can be categorized as such:

Physical- Temperatures, supersaturated gas.

Chemical- Water quality, pollutants and metabolic wastes.

Procedural- Handling, transportation, stocking or disease treatment.

Biological- Population density, confinement, diet composition and availability.

12.3.1 Stress-mediated diseases

Some of the stress-related diseases are, impaired growth, exophthalmia, gas embolism, epithelial ulcerations, epithelial tumours, skeletal anomalies, reddish blotches, parasite infestations, gill necrosis, furunculosis, eye disease and gill disease. Stressed fishes are not able to resist the pathogens that they normally can ward off or coexist with. Factors which act as the most effective stressors are as follows:

- Temperature: Rapid temperature changes, though sub-lethal may produce a stress of considerable magnitude on the defence mechanism of the fish as well as the susceptibility to disease.
- pH: Any change in the pH range of 6.5 to 9.0 may pose threat to fish growth and health.
- Ammonia: Ammonia (NH_3) is toxic and even relatively low levels in confined water pose a threat to fish health.
- Light: Sudden changes in light intensity in confined tank condition may produce stress in fishes.
- Dissolved Oxygen: Depletion of dissolved oxygen in water below 2-4 ppm will result in asphyxia and death in fishes. The minimum

amount of dissolved oxygen required for good growth in fish is 5 mg/litre. Supersaturation of both oxygen and nitrogen also cause gas embolism. High CO_2 concentrations are directly proportional to low DO concentration.

- Cultural Procedures: Cultural procedures such as handling, crowding, transportation or a combination of all the three may pose stress to the ornamental fishes.

12.4. Non-Infectious Diseases

- **Gas-bubble disease-** When gases are present in the form of bubbles in the water, young fishes gulp them and that accumulate in the intestine. This can be avoided by not giving excess food and adding freshwater.
- **Gill disease**—Irritants like suspended matters in the water, chemical irritants cause heavy respiration and opercular movement in fish. Gills swell ultimately due to hyperplasia and fishes cannot fully close the opercula. Water quality should be properly managed to avoid this.
- **Algal toxicosis**—Blue green algae Microcystis and Anabaena present in the waterbody may clog the gills, causing respiratory distress and mortality in some cases. This can be avoided by applying copper sulphate (@ 20.5 mg/litre), sprinkling cow dung @ 200 kg/ha over surface of water or covering with water hyacinth which block sunlight.
- **Nutritional diseases**
    - **Starvation**

      Lack of food, insufficient supply of feed and non-acceptability of feed cause starvation in fishes. Symptoms are very slender body, large head, retarded growth, darker body colour, emaciated bodies, soft flesh and pale gills. Large number of parasites may also be present on the body. This condition can be considered first by understanding the nutritional components, the feed requirements of the candidate species and then by providing nutritionally balanced diet.
    - **Scoliosis/ Lordosis**

      Spinal deformities occur due to irregular development of skeleton and cartilage. Horizontal or vertical deformities may have effect on the vertebral column.

12.5 Infectious Diseases

Infectious diseases are caused by living agents or parasites/pathogens. A parasite or pathogen first gets attached, evades host's defence mechanism and multiplies before onset of the disease. According to their nature or aetiology, these diseases are categorized.

12.6 Protozoan Diseases

Protozoan diseases that are caused by **external protozoan parasites**.

12.6.1 Ichthyophthiriasis, Ich or White spot disease

This is the most common disease affecting all varieties of aquarium fishes.

- **Aetiology:**

  Ichthyophthirius multifiliis (Figure 12.1) causing the disease is a histozoic parasite of fish.

- **Description:**

  This ovoid parasite measures 100-1000 μm long and is ciliated uniformly. The distribution of *I. multifiliis* is cosmopolitan, found in freshwater ponds and hatcheries. It is the most common parasite of fishes in aquaria and hatcheries.

- **Symptoms:**

  It lives in the skin, gills, fins and eyes of freshwater fish and causes a skin disease-ichthyophthiriasis (commonly known as ich). This is characterized by the formation of lesions, minute sugar grain like white nodular spots, which often become infected with microbes, resulting in the death of the host.

- **Life cycle of the parasite:**

  The trophozoite forms of the parasite are found in thin-walled cysts underneath the host's epidermis. They increase in size and divide. After reaching a certain size they escape from their hosts and drop to the bottom of the aquarium or pond where they secrete gelatinous cysts around themselves. The cysts are generally attached to the substratum. Within the cyst wall, minute ciliated cells known as tomites are formed. They eventually rupture out of the cyst wall and are known as the thoronts. They are free-swimming and find out new fish hosts. They can attack during the first 96 hours after encystation, but are most effective the initial 48 hours.

- **Control measures and Treatment:**

  Control of this parasite is attempted on the life stages on the dermal tissue of the host.

  - A daily bath-treatment for 1 hr in 2-3% NaCl solution or 1:5000 formalin for 30 min daily for 7 days.
  - To add 1 ml of 1% methylene blue solution in 5 litres of aquarium water. This is to be repeated for 1-2 days alternately till recovery.
  - In ponds, 0.5 ppm $CuSO_4$ is added every week or 0.15 ppm malachite green.
  - By interrupting the life cycle of the parasite, by removing the fish from the infected water for 3 days.

12.6.2 Cryptocaryoniosis, White spot disease

- **Aetiology:**

  Cryptocaryon irritans is the causative agent. This parasite is closely related to Ichthyophthirius, and is the histozoic parasite of marine fish. The disease caused by this parasite is commonly referred to as the white spot disease.

- **Symptoms:**
  This ciliate causes slimy skin on the host as well as lesions on the body and gills. Cryptocaryoniosis may also lead to blindness.
- **Control measures and Treatment:**
  Same as Ich.

12.6.3 Trichodiniasis:

This is one of the most common disease in the ornamental fish farms of Asian countries affecting a wide variety of ornamental fishes.

- **Aetiology**:
  The causative agents are ciliate species of genus ***Trichodina, Tripartiella*** and ***Trichodinella***.
- **Description:**
  In ***Trichodina*** (Figure 12.2) there is a basal disc present containing a crown of hard pointed teeth like structures that help these parasites in attaching to their hosts. Average size is about 50 μm.
- **Symptoms:**
  Trichodina spp., cause some damage to the gills. Affected fishes have pale gills with a cream-like coating due to excessive mucus secretion and mild hyperplasia. *Trichodina* spp., infecting fishes multiply rapidly resulting in mass mortalities of young fishes.
- **Control measures and Treatment:**
  - Improvement in the water quality.
  - Controlling excess stocking density of fish.
  - Bath treatment of fishes with 2-3% NaCl or 100ppm formalin.
  - Pond treatment with 4-5 mg /litre $KMnO_4$ or 25 mg/litre formalin.

12.6.4 Ichthyobodosis (Costiasis)

One of the most common and widespread disease, caused by *Ichthyobodo* spp. Affecting the most freshwater and marine species of aquariums.

- **Aetiology:**
  There are two species of ***Ichthyobodo*** (Figure 12.3) responsible for Ichthyobodosis. These are *Ichthyobodo necator* and *Ichthyobodo pyriformis*. The parasite is 5-10 μm, small, bean shaped and flagellated. This obligate parasite has an adhesive disc on the posterior end by which the parasite attach itself to the hosts epidermis or gill. There are present two flagella, nucleus and contractile vacuoles. They are obligate parasite with one host cycle and the trophozont cannot survive long without the host. Hence transmission occurs from fish to fish.

- **Predisposing factors:**

  Overcrowding of the fish, mishandling, poor water quality, traumatic injuries and physiological damage.

- **Symptoms**:

  Listlessness, anorexia and flashing behaviour. The gills are usually swollen and covered with mucous which lead to respiratory distress. Thickening of mucous on the body and fins and dull-spots appear on the infested part. High mortality is associated with heavy infection.

- **Control measures and Treatment:**
  - Formalin 250 mg/litre for 1 hour by bath method.
  - Bath treatment with 2-3% NaCl or 50 mg/ litre $KMnO_4$ or 100mg/litre of formalin with aeration.
  - Pond treatment with 25 mg/litre formalin or 5 mg/litre $KMnO_4$.
  - Spawning fish must be given prophylactic bath before stocking.
  - Use of malachite green, Acriflavine, strong salt baths of 3%.

12.6.5 Tetrahymena infection:

Guppy killer disease affects fry of fishes like guppy, northern pike etc.

- **Aetiology:**

  Tetrahymena corlissi and ***Tetrahymena pyriformis*** (Figure 12.4), free living oval ciliated, 50-70 μ long protozoa.

- **Symptoms:**

  Necrosis and haemorrhages of the skin. In extreme cases, rupture of the body walls of the fish occurs and the fish eviscerate. The ventral abdominal wall is severely affected.

- **Control measures and Treatment:**

  Prevention of overcrowding and poor water quality.

12.6.6 Chilodonellosis

This disease causes high morbidity and mortality among freshwater tropical fishes.

- **Aetiology**:

  Causative agent is ciliated protozoa of the genera *Chilodonella*.

- **Symptoms:**

  Infects the skin and gill tissues of fishes. Dull colouration due to excessive slime, fraying of fins, weakness and damage to gills. Heavy infections under stressful conditions leads to mortality.

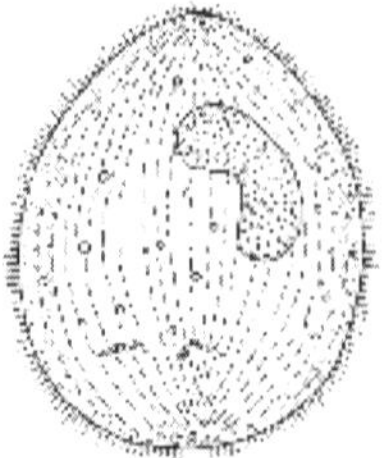

Figure 12.1 *Ichthyophthirius multifiliis*

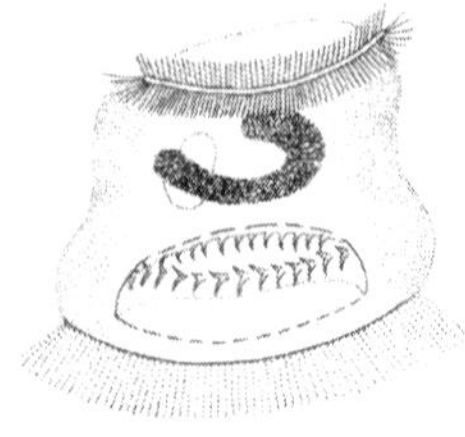

Figure 12.2 *Trichodina* sp.

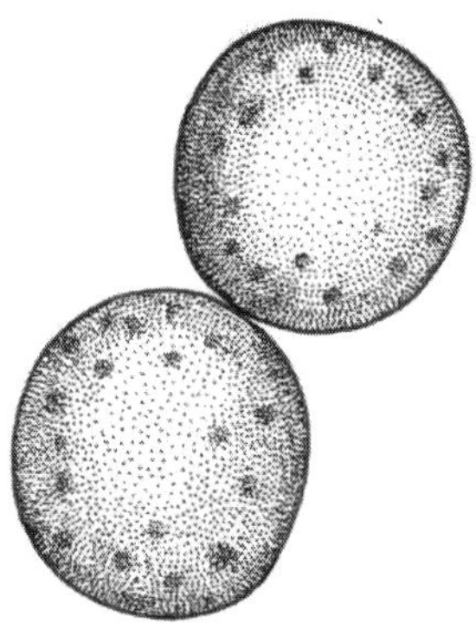

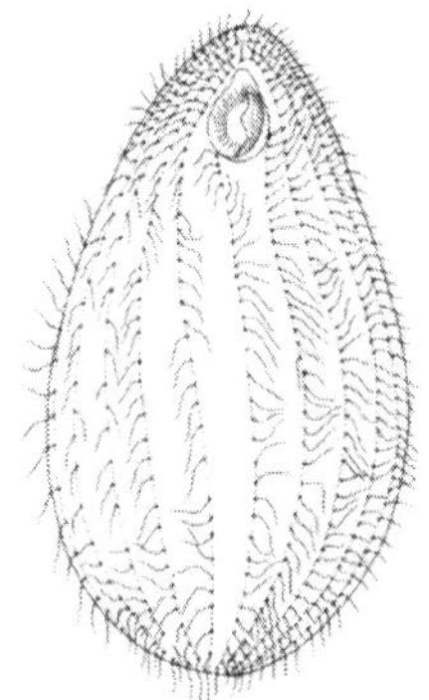

Figure 12.3 *Ichthyobodo* sp.

Figure 12.4 *Tetrahymena pyriformis*

- **Control measures and Treatment:**

  Acriflavine, methylene blue, malachite green, formaldehyde, sodium chloride, copper sulphate, potassium permanganate and acetic acid have been reported to be successful in treating *Chilodonella* infection.

12.6.7 *Epistylis* (Heteropolaria) infection

- **Aetiology:**

  A stalked ciliate *Epistylis* sp., which is commonly found in freshwater containing a high organic load.

- **Symptoms:**

  Lesions appear pale and white in colour and resemble a fungal disease.

- **Control measures and Treatment:**

  The classic treatment for this infection is un-iodized salt (sodium chloride).

12.6.8 Dinoflagellate infection

12.6.8.1 Oodiniasis (Velvet disease)

- **Aetiology**:

  Causative agents are ***Oodinium*** sp., causing velvet disease and *Amyloodinum* sp., causing coral fish disease. ***Oodinium ocellatum*** (Figure 12.5) is responsible for the velvet disease. This flagellate is an ectoparasite on the gill filaments of marine fish, especially coral fish. It causes the velvet disease which may reach epizootic proportions in aquariums. It is a small, ovoid organism encased within a shell or test that completely surrounds the body cytoplasm except for a terminal aperture through which a broad flagellum and cytoplasmic processes protrude. It is by these processes that the parasite gets attached to its host. These organisms remain attached to the fish while growing and finally drop off. The unattached parasite withdraws its flagellum and processes and gets enclosed by test. Within the test, the organism undergoes division and result is the formation of flagellated swarm cells, which rupture out of the test and actively seek the gill of a fish host to which they become attached.

- **Manifestation of the disease:**
  - Fishes have a dusty appearance on the skin.
  - It infests the skin and induces excess mucous production.
  - Young fishes may succumb to infection within two weeks.
  - It may lead to death resulting from suffocation as the parasites attached in large numbers to the surface of the gills.
- **Control measures and Treatment:**
  - Short dips in common salt solution for 1to 2 minutes at 35,000 ppm; 2-3 ppm Methylene blue treatment for 3 to 5 days and 1-2 ppm Copper sulphate bath for some time.
  - Tank should be disinfected before filling with water.
  - Newly purchased fishes should be quarantined for two weeks.

12.7 Internal Protozoan Parasites Causing Diseases

Small flagellate parasites that frequently found in the lumen of the intestinal tract of freshwater and marine tropical fishes are:

- *Spironucleus* sp.
- *Hexamita* sp.

Microsporidean parasite producing spore-filled cysts is:

- *Pleistophora hyphessobryconis*—causing a syndrome called "neon tetra disease" in zebra danios, cichlids and cyprinids.

Myxosporideans common parasites of many species of fish causing infections are:

- *Eimeria* sp. and *Goussia* sp.
- Flagellated hemoprotozoans.
- *Trypanosoma* sp.
- *Cryptobia inbilans*

12.7.1 Myxosporean Disease or Nodular Disease

- **Aetiology:**

  ***Myxobolus* spp.**, such as *Myxobolus bengalensis, M.catlae, M.sphericum* and *Thelohanellus catlae* affect the fishes.

- **Infection:**

  Gills, scales and body surfaces are infested with the cysts of parasites. The infective stage of the myxosporeans is the mature spore which is ingested by the fish from the water. After entering the fish, the infective sporoplasm of the spore comes out as a small amoebula and penetrate the gut wall. The amoeba then reaches the tissue through blood stream and then becomes a trophozoite. It increases in size and forms a cyst. The mature spore develops here.

- **Control measures and Treatment:**

  As the spores of the parasite are very resistant to chemicals, the control measures are confined to prophylactic measures. These are as follows:

  - Spores are to be controlled from entering the aquarium water or pond water.
  - Fry and young fishes are more susceptible so they are separated from adults.
  - Aquaria and ponds are to be disinfected at regular intervals.

12.7.2 Sleeping sickness

- **Aetiology**:

  Blood flagellate, ***Trypanosoma*** sp. (Figure 12.6) is responsible for the disease sleeping sickness. Blood flagellates are transmitted mainly by leech.

- **Symptoms**:

  The affected fish lie in a sleeping posture on the water surface with little movement.

- **Control measures and Treatment:**

  - No treatment is yet available
  - Destruction or prevention of the disease transmitting organisms (leech) in the aquarium or pond may be adopted to control the disease.

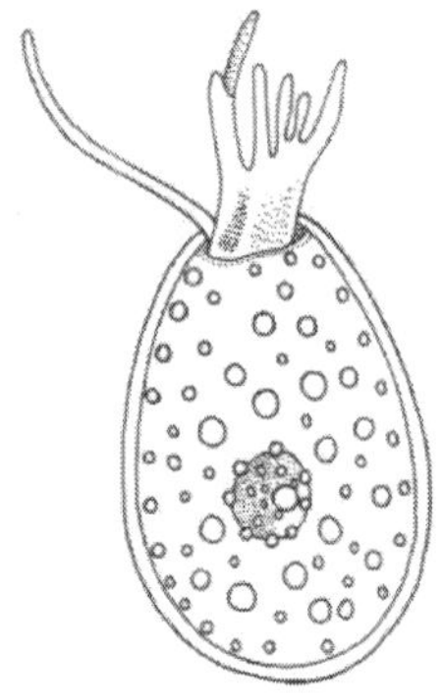

Figure 12.5 *Oodinium ocellatum*

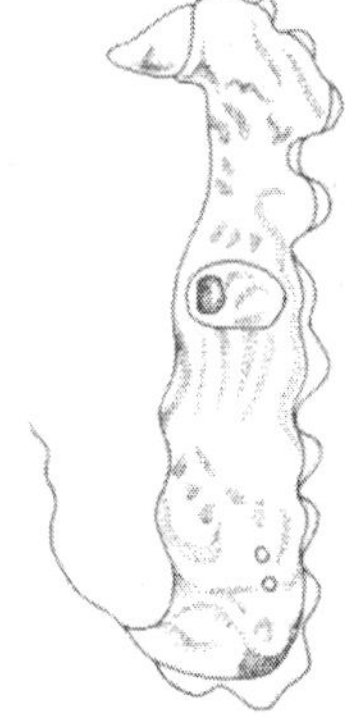

Figure 12.6 *Trypanosoma* sp.

12.7.3 Hexamitiasis

- **Aetiology:**

  Hexamita or *Spironucleus* is a flagellate protozoan inhabiting the intestinal tract of fishes (systemic infections can also occur).

- **Symptoms**:

  - Severe gastrointestinal disease in heavy infection.

- Head and lateral line erosion is caused as the disease is called "Hole in Head Disease".
- Common in Oscars and other cichlids.

- **Control measures and Treatment:**
  - Albendazole and metronidazole (trade name Flagyl) are most effective. During early infections, medicated fish food is used. In seriously affected fish, medication should be added to the water, at a dosage of 250 mg per US gallons, once per day for at least three days. Alongside the medication, improvements to diet and environment are absolutely essential.

12.8 Bacterial Diseases

Bacterial diseases are the most common problems encountered in ornamental fishes (Sanil and Vijayan, 2008).

Gram-negative bacteria such as *Aeromonas*, *Citobacter*, *Edwardsiella*, Flavo bacterium (*Flexibacter*), Mycobacterium, *Pseudomonas* and *Vibrio* cause major bacterial infections in ornamental fishes. *Streptococcus* the gram-positive bacteria cause infections in ornamental fishes. Bacterial infections may either be primary or secondary. Most bacterial fish pathogens are natural inhabitants of the marine or freshwater environment. Red and inflamed areas on the body and fins, raised scales, skin ulcers, exophthalmos, dropsy or swollen abdomen, fin rot etc., are common infections. Affected fish may be lethargic and anorexic. There may be lesions or haemorrhages in organs and/or a build-up of bloody scars. Bacterial pathogens are the causative agents of a large number of diseases in fishes. They are distributed worldwide and are found in nursery, rearing and grow-out ponds. They are serious threats to fish farms and sometimes wipe out the entire fish stock. Various branches of science such as symptomatology, clinical pathology, microbiology and molecular diagnosis are correlated in the diagnosis of the diseases caused by bacteria.

12.8.1 Fin rot and Tail rot

Fin Rot and **Tail Rot** is the most common bacterial disease occurring in aquaria, hatcheries, nurseries and grow-out ponds.

- **Aetiology:**

  The causative agents are ***Aeromonas*** spp. Particularly *A. hydrophila*, *Cytophaga* group and *Pseaudomonas flurescensis*.

- **Description**:

  Aeromonas bacteria are straight rods with non-sheathed flagella. *Pseudomonas fluroscensis* is common saprophytic bacteria that makes only a fluorescent pigment. *Cytophaga* spp. bacteria may appear as individual rods or filaments.

- **Symptoms**:
  - In affected fish, the tail and fin get necrosed, discoloured and sloughs off.

- Exposed fin rays, blood on edges of fins, reddened areas at base of fins, skin ulcers with grey and red margins.
- Cloudy eyes.

- **Control measures and Treatment:**

  Stressful environment and poor water quality enhance the disease.

  - Bath treatment in 1:2,000 $CuSO_4$ for 1-2 minutes or 1-2 ppm benzalkonium chloride for 1 hour.
  - Swabbing the wound with diluted (4:10) tincture iodine or concentrated $CuSO_4$ solution.
  - Maintain better water quality and avoiding over stocking is the part of good management practices.
  - An effective treatment is addition of antibiotic such as chloromycetin and tetracycline at the rate of 10 mg per litre of water.

12.8.2 Dropsy

Dropsy infection (Figure 12.7) is common in carps and gouramies, particularly in dwarf gourami.

- **Aetiology**:

  Causative agents are ***Aeromonas hydrophila*** and a few other *Aeromonas* species.

- **Symptoms:**

  - Bloating of the body due to renal failure. Bacteria- infected kidneys cause fluid accumulation in the body.
  - The epidermis and body cavities got filled with fluid, and scales protrude out. Eventually severe anaemia and heavy mortality occur.

- **Treatment and control measures:**

  - Terramycin is used in the feed
  - Treatment with chloromycetin (13 mg/litre of water) is effective and tetracycline at the rate of 10 mg per litre of water.

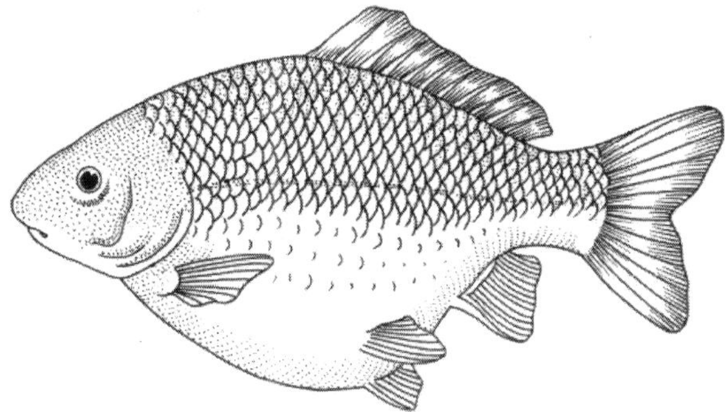

Figure 12.7 Dropsy infection in a fish.

12.8.3 Eye disease or Cloud eye

This disease is very common in Goldfish, Barbs, Angelfish, Mollies and Sword Tail.

- **Aetiology**:

  Causative agents are ***Aeromonas liquifasciens*** and ***Staphylococcus aureus***.

- **Symptoms**:
  - Cornea of the eyes are vascularised.
  - One or both the eyes become cloudy white (opaque).
  - Ultimately complete necrosis of the eyes and mass mortality of the fishes occur.
  - The optic nerves and brain are the primary sites of infection.

- **Control measures and Treatment:**
  - Bath treatment with Chloromycetin @ 8-10 mg/litre water is effective in early infection.
  - Use of $KMnO_4$ @ 0.1 ppm followed by liming @ 300 ppm can control the disease.

12.8.4 Columnaris or Cotton-wool disease or Mouth Fungus

It is a common bacterial disease of live-bearing ornamental fishes.

- **Aetiology**:

  The causative agent is ***Flexibacter columnaris***.

- **Symptoms:**
  - Raised whitish or greyish patches develop on the skin of head and back.
  - Skin around the mouth are particularly affected.
  - At the advanced stage of infestation, the body becomes ulcerated.
  - Fishes with severe mouth infection may die within 24 hours.

- **Control measures and Treatment:**
  - Bath treatment in 5 ppm of $KMnO_4$ for 2-5 minutes or 1: 2000 $CuSO_4$ solution (1 part $CuSO_4$: 2000 part water) for one or two minutes.
  - 5-10 ppm Chloramphenicol or 10-20 ppm Oxytetracycline for an hour.
  - Malachite green 1:15000 for 10 to 30 seconds.
  - Pond water with 3-5 ppm $KMnO_4$ or 1 mg of $CuSO_4$/ litre.

12.8.5 Ulcer disease or Ulcerative disease

Bacterial ulcer disease is one of the common fish disease problem particularly affecting goldfish and koi.

- **Aetiology**:

  Causative agents are mainly ***Aeromonas hydrophila*** and *Pseudomonas* bacteria. These organisms occur in freshwater and sewage. Some species are pathogenic for fish.

- **Symptoms:**

  Patches appear on the body and fins which later develops into open ulcers. Secondary infection may cause necrosis of muscle tissue and damage of internal organs like kidney, liver and spleen.

- **Control measures and Treatment:**

  External applications of antibiotics are found to be not very effective. So, antibiotics like Chloromycetin and Oxytetracycline are added to the food. The admissible dose is 50 mg/kg of fish/day. If severe infection occurs, fish should be destroyed and the water should be disinfected with 0.5 ppm solution of $KMnO_4$.

12.8.6 Edwardsiellosis

Brood fish population is affected by this septicemic disease.

- **Aetiology**:

  Edwardisiella tarda is the causative agent.

- **Symptoms:**

  Gas-filled abscesses may occur in the muscle. Minute haemorrhagic patches appear on skin and tail, and fishes become anaemic. Sometimes extreme morbidity occurs. Young fishes are more affected. Anatomical deformities happen initially and within 72 hours death occurs.

- **Control measures and Treatment:**

  - Water quality improvement in the hatchery is the most effective measure.
  - Application of Iodophor or Tetracycline is also effective.

12.8.7 Bacterial gill disease

This disease is wide spread and affect freshwater fishes.

- **Aetiology**:

  Flavobacterium branchiophila is the most important causative agent though there are other bacteria which cause the disease.

- **Symptoms:**

  Discolouration of the gills occur in the affected fishes and gill fringes become uneven. Gills torn with grey patches of necrosis. Morbidity and mortality occur.

- **Control measures and Treatment:**

  - Over-stocking of fish should be avoided.
  - Chloramine-T at the rate of 8-10 ppm for 1 hour is effective.
  - Bath treatment with 1-2 ppm $KMnO_4$ is also effective.

12.8.8 Epizootic Ulcerative Syndrome

Epizootic Ulcerative Syndrome or EUS, cause mass mortality in fishes and is a matter of great concern. Bottom-dwelling fishes are first to be affected then the disease spread to other fishes.

- **Aetiology:**

  Aeromonas hydrophila, A. punctata, Flavobacterium sp., *Pseudomonas* sp., *E. tarda, V. parahaemolyticus* and *Streptococcus* sp., are the bacteria and fungus like *Aphanomyces invadeus* have been isolated from affected fishes.

- **Symptoms:**

  Small haemorrhagic spots appear all over the body which then become circular deep ulcers with greyish slimy central necrotic areas. Surrounded by as zone of hyperaemia. Body eventually starts rotting when the fish is still alive and the fish ultimately die. Complete loss of epidermis in the ulcerative area of the skin is revealed through histopathological studies.

  The fungus penetrates into the muscles, bones and visceral organs producing characteristic mycotic granuloma.

  Deep ulcers, granulomatous lesions and high rate of mortality in several species of fishes are the diagnostic features of the disease.

- **Control measures and Treatment:**
  - Monoclonal antibody (Mab) based immunoblot has been developed for early detection of the disease (Das, 2011).
  - Lime @ 100-600kg/ha; KMnO4 @ 4ppm; bleaching @ 0.5ppm in pond water.
  - CIFAX developed by the Central Institute of Freshwater Aquaculture (CIFA), Bhubaneshwar is very effective in preventing and curing the disease.

12.9 Fish Viral Disease

It is advisable not to buy virus infected fish as fish viral disease are impossible to treat directly. It is better to quarantine all new fish prior to putting them in aquarium or fish pond.

The aquarist can recognize the symptoms of the viral infection of fish but positive identification by biomedical tests is generally beyond the reach of the average fish hobbyists. Transmission of virus between fish takes place by direct contact, by water, by wet objects.

KHV and DGIV are extremely contagious so quarantining procedures are essential. (Ganguly, 2017)

Some fish viral diseases appear to be prompted by poor aquarium conditions. In breeding also makes fish more vulnerable to fish viral infections. This is particularly relevant in case of fancy varieties of ornamental fishes.

12.9.1 Viral Encephalopathy and Retinopathy (VER) or Viral Nervous Necrosis (VNN)

This infection causes heavy mortality in several marine species.

- **Aetiology:**

  Causative agent is the Noda virus. Non-enveloped, icosahedral RNA virus belonging to the family Nodaviridae cause the disease. The virus spreads both horizontally and vertically.

- **Symptoms:**
  - Abnormal swimming behaviour.
  - Extensive cellular vacuolation.
  - Neuronal degeneration in the brain and retina are the main characteristic features of the disease.

- **Treatment and control measures**
  - Early detection of the virus
  - Pathogen-free brood stock.
  - Regulated movement of the fish are the measures to be implemented to control the disease.

12.9.2 Lymphocystic disease, Lymphocystis

Marine and brackish water fishes are more affected. Members of the genera *Scatophagus, Monodactylus* and *Chanda,* and freshwater fishes such as cichlids and gouramies. Heavy metals, poisons and consistently poor aquarium quality are the triggering factors of the disease.

- **Aetiology:**

  Caused by virus belonging to the Iridoviridae family.

- **Symptoms:**

  Nodular white swellings (cauliflower like) on fins or body. Infection occurs in the connective tissue cells of the fish. These cells then undergo extensive hypertrophy and the affected cells increasing thousand folds in size.

  Gross lesions appear white and granular and are seen on the skin and fins. Mouth and gills are also covered with the lesions.

- **Control measures:**
  - There is no proper chemotherapeutic treatment.
  - Proper water quality and nutrition is to be maintained.
  - In extreme cases, it is better to remove and destroy the infected fish as soon as possible.

12.9.3 Dwarf Gourami Iridovirus (DGIV)

In recent years, high incidences of dwarf gourami mortality have focussed attention on a virus known as Dwarf Gourami Iridovirus (DGIV).

This virus is probably specific to the dwarf gourami (*Colisa lalia*) and also other fancy varieties such as neon gourami and sunset gourami.

- **Symptoms:**
    - Loss of colour.
    - Decrease in activity and appetite.
    - Appearance of sores and lesions on the body.
    - Abdominal swelling.
    - Death results ultimately.
    - Highly contagious, completely untreatable and invariably fatal.
- **Control measures and Treatment:**
    - All new fish should be quarantined for at least six weeks prior to being placed in the main aquarium.

12.9.4 Megalocytiviruses

It consists of red sea bream iridoviruses (RSIV) and infectious spleen and kidney necrosis virus (ISKNV). Freshwater angelfish (*Pterophyllum scalare*) has been affected by megalocyti virus.

12.9.5 Fish pox/ Carp pox

- **Aetiology:**

  Fish Herpes virus that affects cyprinid fishes, including Koi and goldfish is **Cyprinid herpesvirus-1** (**CyHV-1**). It is not particularly infectious and is not life threatening.
- **Symptoms:**

  White, grey or pink growths on the skin and fins.
- **Control measures and Treatment:**
    - There are no treatments that can be used to clear the virus from carrier fish. But by manipulating temperatures losses can be minimized.

12.9.6 Herpesviral hematopoietic necrosis

- **Aetiology:**

  Causative agent is **Cyprinid herpesvirus 2** (**CyHV-2**) in goldfish.

  The clinically affected goldfishes, become anorectic and exhibit pale gills and ascites., spleen and kidneys become enlarged.

12.9.7 Koi Herpes Virus (KHV) Disease

This is one of the most lethal, emerging viral infections in carp ornamental fishes. It has been detected in ornamental Koi carp in Europe, Asia and the United States. Common carp (*Cyprinus carpio*), Koi and Mirror carp are affected by this highly infectious virus.

- **Aetiology:**

  Koi Herpes Virus or **Cyprinid herpesvirus-3** (**CyHV-3**) cause this infection.
- **Symptoms:**

  Bloody or discoloured gills, sunken eyes, pale or bloody patches on the skin and damage to the internal organs. KHV is highly contagious. Biomedical techniques such as testing the fish's blood serum for antibodies that the fish produces in response to the KHV infection.

- **Treatment and control measures:**
  - Increasing the water temperature to 30° C for some time can help.
  - Attenuated live vaccine appears to be the most appropriate for mass vaccination of carp.

12.10 Fungal Diseases in Ornamental Fishes

Fungal diseases are easy to identify under the microscope as a tight mat of fungal hyphae following a simple skin scraping. Fungi belonging to the genus *Saprolegnia* are the most commonly observed species affecting the tropical fishes. They are opportunistic pathogens/secondary invaders which typically colonize exposed damaged tissue. Normally if the fish is well supported with clean water and good food, the fungal tufts will slough off in time without the necessity of treatment but some severe conditions may demand chemotherapy.

12.10.1 Saprolegniasis

This fungal disease occurs generally in fry and fingerlings.

- **Aetiology:**

  Various species of ***Saprolegnia*** (Figure 12.8) cause the disease.
- **Symptoms:**

  White to brown cotton like growth appear (Figure 12.9) which consist of colonies of mycelium and filaments. These appear as small to large patches on various parts of the body like fins, gills, mouth, eyes or muscles. Netting injuries, over-crowded conditions or lesions caused by other infections may help to initiate the disease.
- **Control measures and Treatment:**
  - Bath treatment of infected fish in pond with 3-4% common salt or 1:2,000 $CuSO_4$ daily for 3-4 days.
  - Treatment with 1:1,000 malachite green for 30 seconds.
  - Eggs may be treated in hatcheries with 2-5 ppm good quality malachite green for 1-5 in or 1:500 diluted formalin for 10-15 min, followed by washing in flowing freshwater.

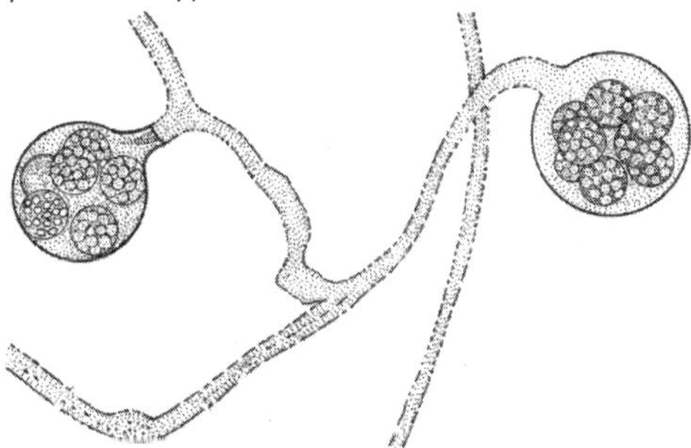

Figure 12.8 *Saprolegnia* sp.

Figure 12.9 Saprolegniasis infection in a fish.

12.10.2 Branchiomycosis/Fungal gill rot

This infection causes gill rot in fishes.

- **Aetiology:**

  Branchiomyces is the causative agent.

- **Symptoms:**

  Blood vessels of gills are invaded and necrosis of gills and its surrounding tissues occur. Gills become enlarged and are visible from outside.

- **Control measures and Treatment:**
  - Dipping of infected fish in 3-5% common salt solution.
  - Potassium permanganate of 5 ppm; 1:2,000 $CuSO_4$ per 1 min is effective.
  - Ponds can also be treated with 1ppm $CuSO_4$ followed by liming.
  - A flush treatment of malachite green (0.1 ppm) for 1 hr is found to be effective in controlling the fungal infection.

12.11 Helminth Diseases

12.11.1 Gyrodactylosis and Dactylogyrosis

These diseases affect fry, fingerlings and adult fishes causing huge losses. They are threat to the health of the fish maintained under crowded conditions such as aquaria and ponds.

- **Aetiology:**

  The causative agents of the disease are ***Gyrodactylus*** (skin fluke) (Figure 12.10 A) and ***Dactylogyrus*** (gill fluke) (Figure 12.10 B) the monogenetic trematodes (flukes).

- **Description:**

  In these parasites, the adhesive glands empty to the outside through ducts ending in bulbous terminals known as head organs. These monogenous have a direct life-cycle involving a single host. *Dactylogyrus* is oviparous laying eggs which liberate free swimming larvae. A new host is located by these larvae, become attached and then metamorphose into a mature worm. *Gyrodactylus* is viviparous and liberates young worms which attach to a new host. Eggs develop within the parental uterus.

- **Symptoms:**
  - In dactylogyrosis, the colour of the gills fades and there is excessive mucous secretion.
  - In gyrodactylosis, a fading of the body colour, dropping of scales and excessive mucous secretion.
  - There is growth reduction and morbidity in affected fishes.
- **Control measures and Treatment:**
  - Bath treatment with 3-5% NaCl for 10-15 min or 100 mg/litre formalin with aeration.
  - Pond treatment with 25 mg/litre formalin or 5mg/litre $KMnO_4$.

12.11.2 Black spot disease

Black spot disease is caused by the digenetic trematode.

- **Aetiology:**

  The parasite ***Diplostomum*** sp. (Figure 12.11) and other *Diplostomum* spp., are the causative agents.
- **Description:**

  The black spots represent encysted metacercaria (a stage in the lifecycle of the parasite) surrounded by reaction pigmentation. These digenetic trematodes are oviparous and release eggs in water. Miracidium, the ciliated larva hatches out and after swimming locates, the first intermediate host, a mollusc. Here, the parasite larva transforms into a cercaria larva which is locomotory. It swims out and again locates the secondary intermediate host, a fish. It penetrates and encysts into a metacercaria larval stage. This is seen as black spots on the body surface. The life-cycle is completed when the fish is infected with metacercaria is eaten by the final host, the bird.
- **Control measures and Treatment:**
  - Removal of the resident first intermediate host i.e., the molluscs and the final hosts from the waterbodies.

12.11.3 Ligulosis

Carps are affected by this disease.

- **Aetiology:**

  The causative agent is the plerocercoid stage of the parasite ***Ligula intestinalis*** (Figure 12.12)
- **Description:**

  It is a typical avian cestode. Life cycle of *Ligula* sp., begins with the release of its eggs along with faecal matter of bird. They hatch out into free-swimming coracidia larvae which are swallowed by cyclopoid copepods. Inside the parasite transform into procercoid larvae. When fishes eat these copepods, the larvae transform into the infective larval form, the plerocercoids. When the infected fish is eaten by a particular species of bird, the life cycle is completed.

- **Symptoms:**

  Infected fishes are dark coloured, anaemic. The abdomen is swollen due to the presence of large number of infective stages of the parasites.

- **Control measures and Treatment:**

  The definitive host, the ichthyophagous birds should be completely avoided and removed.

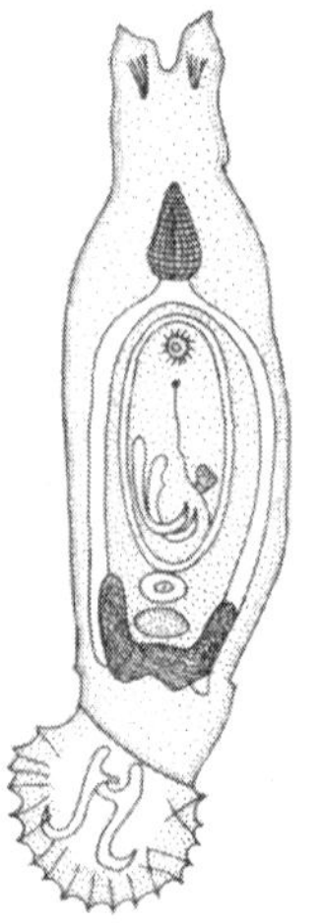

Figure 12.10 A *Gyrodactylus* sp.

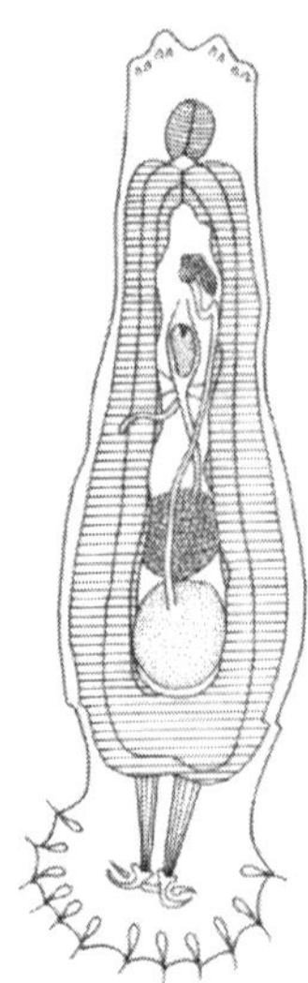

Figure 12.10 B *Dactylogyrus* sp.

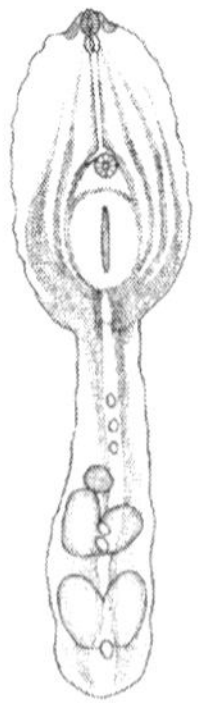

Figure 12.11 *Diplostomum* sp.

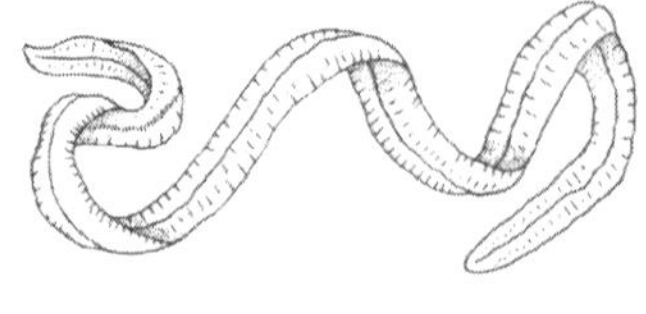

Figure 12.12 *Ligula intestinalis*

12.12 Acanthocephalan Diseases

These infest carps, catfishes and many other freshwater fishes.

- **Aetiology:**

  The causative agents are spiny -headed worms *Acanthogyrus acanthogyrus*, *Acanthosentis antespinus* and *Pallisentis* sp., the **Spiny-headed worms** (Figure 12.13).

- **Description:**

  The eggs of the parasites are released along with the faecal matter by the infected fish. The eggs are eaten by an invertebrate host (arthropod). The acanthor larva hatches out here and penetrates the host body cavity, hemocoel and develop into acanthella. Then it develops into the cystacanth larva. The cystacanth develops into an adult worm when the intermediate host is eaten by fish.

- **Symptoms:**
  - To determine infection, there is no visible external symptoms.
  - These parasites inhabit the alimentary tract of fishes attaching themselves by the hooks and spines of their proboscis.
  - The intestinal epithelia get damaged by these and become source of secondary infection.

- **Control measures and Treatment:**
  - No treatment is yet available.

12.13 Nematode Infection

- **Aetiology:**

  Causative agent is **Nematodes (Thread worm)**.

- **Symptoms**

  Worms hanging from the anal region of the fish. Heavy infestation causes hollow bellies.

- **Control measures and Treatment:**
- Thiabendazole as a nematode cure should be given with feed.

12.14 Leech Infection

- **Aetiology:**

  Causative agent is **External parasite leech.**

- **Symptoms:**

  Leeches are visible on the fish's skin

  These parasites suck and borrow into the surface of the fish.

- **Control measures and Treatment:**

  Bath treatment in 2.5 % solution of NaCl for 15 minutes.

- Adding Trichlorfon at 0.25 mg/l in the aquarium.
- Gently removing leeches from the body surface.
- Live plants should be removed, treated with $KMnO_4$ at 5 mg/l before replanting.

12.15 Crustacean Diseases

12.15.1 Argulosis

Carps are often affected by this parasitic disease. Many of the species are parasitic to freshwater and marine fishes.

- **Aetiology:**

  The causative agent is the brachyuran crustacea, *Argulus* spp. And *Argulosus foliaceus, A. bengalensis* and *A. siamensis* in Indian fishes. They are commonly referred to as **Fish lice** (Figure 12.14).

- **Description:**

  The ovoid body of parasite is dorsoventrally compressed. The cephalothorax is covered dorsally by carapace, and two conspicuous black spots are present. Both male and female are parasitic. Although usually attached to their hosts, these parasites leave their hosts periodically and are free swimming. It matures while inhabiting the host fish. After copulation, the female leaves the host and lays sticky eggs on the submerged vegetation, rocks, sticks etc. The free-living developmental stages are parasitic to fish. The time taken to complete life-cycle in *Argulus* spp. varies from three to six weeks.

- **Symptoms:**
  - The affected fishes are restless with entire swimming behaviour and loss of appetite.
  - With heavy infestation, growth retardation and death may occur.
  - Ulceration occurs on the body of fish at the site of the attachment of parasites.

- **Control measures and Treatment:**
  - Bath treatment with 3-5% NaCl or 100 ppm $KMnO_4$ or 2,000 ppm Lysol for 5-10 seconds.
  - Pond is treated with $KMnO_4$ @ 5 mg/litre.
  - Bamboo thats or corrugated sheets are placed in the water for the sticky eggs to get attached to. These are removed after a week and dried in the sun to kill the eggs of parasites.
  - Pesticides are sometimes used but these should never be used in tanks of the breeding stock.

12.15.2 Ergasilosis

Carps are generally infected by this disease.

- **Aetiology:**

  The causative agent is the female of the species of genus ***Ergasilus*** (Figure 12.15). Several freshwater species of this free-living copepod are parasitic on the gill filaments of the fish.

- **Description:**

  With the hatching of the eggs free swimming nauplii come out. After attaining sexual maturity, the females become parasitic and attach to their hosts by specialized second antennae, which have become modified as two muscular claws. The sexually mature males are free swimming in most cases, but parasitic males are not known. Infestation occurs in gills, buccal cavity, operculum and gills.

- **Symptoms:**
  - Affected fish shows surfacing, lethargy and restlessness.
  - With more infestation, gill tissues of the fish get damaged and there is retardation of growth.

- **Control measures and Treatment:**
  - Bath treatment with 1: 1000 glacial acetic acid for 5 minutes followed by dip in 1% NaCl for 1 hr or 2-3 % NaCl.
  - Pond treatment with $KMnO_4$ @ 5mg/litre.

12.15.3 Lernaeosis

This disease affects the gills and fins of freshwater fishes.

- **Aetiology:**

  The causative agents are ***Lernaea*** spp. Ectoparasitic copepod. Commonly known as Anchor worm. Two noted species are ***Lernaea chockoensis*** and ***Lernaea bengalensis*** (Figure 12.16).

- **Description:**

  The vermiform females, bearing horn- like processes on their cephalothorax, are parasitic. When attached to their hosts, the entire anterior end is deeply embedded in the host's flesh and permanently anchored by these horn-like processes. Two ovisacs (egg sacs) are attached to the posterior end. These females are 5-23 mm in length. The nauplii are free swimming and cling to the fish host. In the larval form, this parasite copulates and then the female leaves the temporary host and become free swimming. Then it finds a definitive host and becomes attached to it.

- **Symptoms:**
  - Infected locations are commonly haemorrhagic, spongy and necrotic.
  - As these parasites suck and feed on blood and tissue fluids of their hosts, severe damages are caused.
  - Heavy infections are known to kill the hosts.

- **Control measures and Treatment:**
  - Bath treatment with 1: 1000 glacial acetic acid for 5 minutes followed by dip in 1% NaCl for 1 hour.
  - Bath treatment with 2-3% NaCl.
  - Pond treatment with $KMnO_4$ @ 5 mg/ litre.

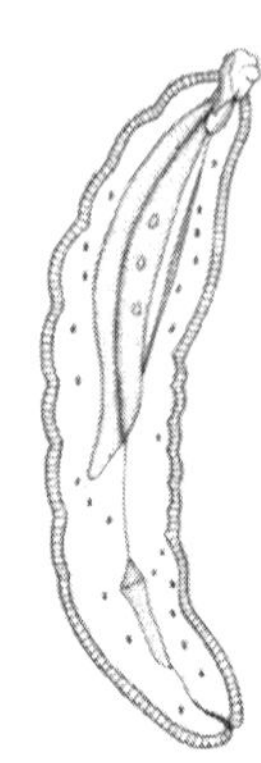

Figure 12.13 Spiny-headed worm

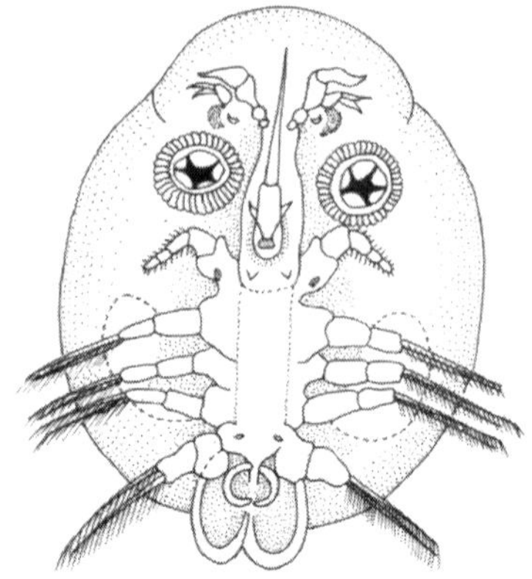

Figure 12.14 Fish lice

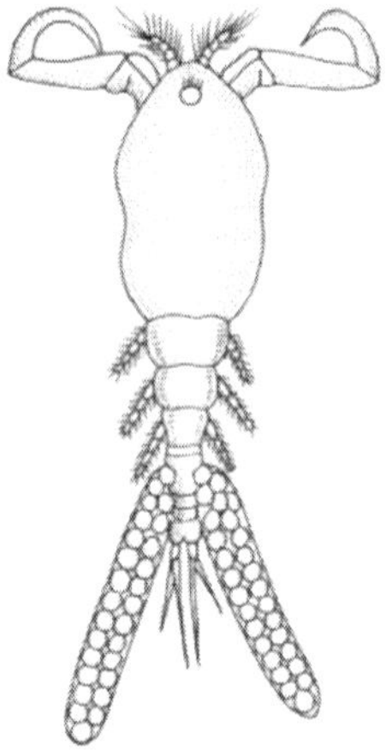

Figure 12.15 *Ergasilus* sp.

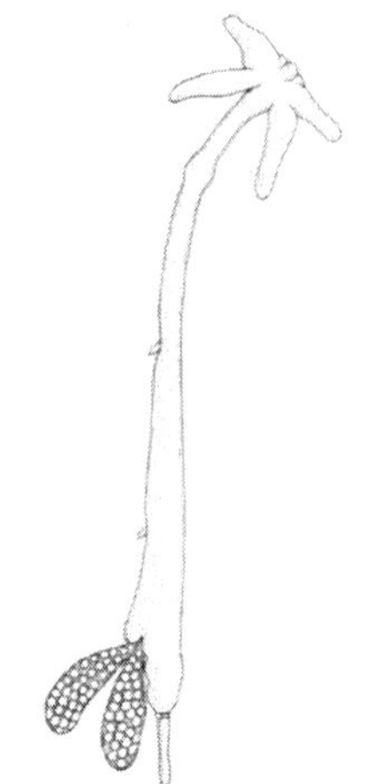

Figure 12.16 *Lernaea bengalensis*

12.16 Antibiotics used in Ornamental Fish Culture

Some antibiotics are used in ornamental fish culture either as a prophylactic or therapeutic manner (Table.12.1). Unfortunately, these antibiotics are often used carelessly leading to harmful effects. Some adverse effects observed are, reaction with the aquatic microflora, retention of harmful residues in aquatic organisms and development of drug-resistant bacteria and other pathogens (Das, 2011; Pandey, 2014).

Table 12.1: Antibiotics used in ornamental fish culture

Brand name	Antibiotics	Comments
Aurofac-20	Chlortetracycline	Used as feed supplement
AFS Forte Vetrodox	Line	Used as feed supplement
TM 50	Oxytetracycline	Used as feed supplement
Bifuran	Nitrofurazone+ Furazolidone	
Groviron	Furazolidone	Used as feed supplement
Bactrisol	Sulphadiazine+ Trimethoprim	Used as water-dispensable powder
Penicillin	Penicillin, amoxicillin, ampicillin	Feed supplement
Aquaflor	Florfenicol	Orally as feed supplement

12.17 Quarantine and Health Certification Measures to Secure Health Status of Fish

Invasive and **Exotic** ornamental fish species pose threat to the biological integrity of the native fish species. Moreover, new pathogens enter through the introduced fish and cause great harm to the aquatic biodiversity and aquaculture sector. An effective quarantine system is needed to assess the risks associated with movement of live aquatic animals, to avoid introduction of exotic pathogens. Development of standard disease surveillance and reporting system would assist the identification and distribution of pathogens affecting aquaculture sector. The National Bureau of Fish Genetic Resources (NBFGR) has prepared two important documents—1. National Strategic Plan for Aquatic Exotics and Quarantine and 2. Aquatic Exotics and Quarantine Guidelines. The objective is the safe introduction of exotics for aquaculture, while conserving biodiversity and preventing economic loss (Das, 2011). Health certification involves designation of competent authority, associating recognition of accredited laboratories across the country.

12.18 Immunodiagnostic Methods in Ornamental Fish Disease Detection

In ornamental fish disease diagnosis, the immunodiagnostic methods are very useful as these can specifically detect various microbial disease with extended sensitivity. Immunodiagnostic techniques in general are based on either the detection of an antigen with antibody of known specificity or detection of an antibody with a known antigen.

12.18.1 Immunodiagnostic methods in use are:

- Agglutination test—It is mostly used in case of diagnosis of bacterial diseases in fish. Modifications of this test are:
  - a) Quantitative agglutination test: Antiserum of various dilutions are taken to react with a fixed quantity of antigen or vice versa.
  - b) Latex agglutination test: Latex heads are used for antigen absorption to increase the size of clumps.

   c) Haemagglutination test: RBCs are used for absorption of antigen or antibody.

- Precipitation test—Precipitation bands are developed with the reaction of antigen and antibody. Modifications of this technique are:
   a) Agar gel precipitation test (AGPT): Here antigen and antibody are passed through agar until they form a precipitation line.
   b) Immunoelectrophoresis: The diffusion of antigen and antibody is enhanced by using an electric field. Modifications are, (i). Counter immunoeletrophoresis test; (ii). Rocket immunoelectrophoresis.
- Enzyme linked immuno sorbent assay (ELISA)—In this sensitive test, either the recognition of antigen by antibody— enzyme conjugate or recognition of antigen-antibody complex takes place by anti-species conjugate. Types of ELISA used in clinical diagnosis are:
   a) Competitive ELISA
   b) Sandwich ELISA
   c) Dot ELISA

12.18.2 Kits used for diagnosis

1. Spot agglutination kit developed by CIFA.
2. Dot-ELISA Kit
3. Antigen captured ELISA Kit
4. Competitive ELISA Kit
5. Monoclonal antibody (Mab) -based diagnostic kit.
6. An immunodot test based on Mab.
7. Fluorescent antibody technique—Fluorescent dye is used with antibody to detect specific pathogen inside the tissue itself.

12.18.3 Molecular diagnostic methods

1. Plasmid profiling, DNA restriction fragment analysis, randomly applied polymorphic DNA (RAPD), restriction fragment length polymorphism (RFLP), polymerase chain reaction (PCR), are effective diagnostic methods both in term of sensitivity and rapidity.
2. Polymerase chain reaction techniques:

   It is widely used in fish disease detection.
3. Reverse transcription polymerase chain reaction
4. Nested polymerase chain reaction
5. Multiple PCR
6. Real-time PCR
7. Nucleic acid probes and hybridization technique.
8. Genotyping technique.

9. Randomly amplified polymorphic DNA (RAPD) technique.
10. Ribotyping

12.18.4 Kits available

CIBA, Chennai has developed and commercialized DNA-based, Nested PCR diagnostic kit for WSSV. CMFRI, Kochi and Mangalore Fisheries college has developed PCR-based diagnostic kits for WSSV. CIFT developed improved Semi-nested PCR technique for early detection of WSSV. CIFT, CIBA and CIFA have standardized detection procedure for Noda virus in fish. CIBA IMMUNODOT developed an immune dot blot test for WSSV for early detection of WSSV in shrimps (WSSV: White Spot Syndrome Virus).

12.19 Immunoprophylaxis

Immunoprophylaxis is applied to get protection against diseases affecting ornamental fishes. It is of two different types, the active immunization (inoculation of antigen-vaccination) and the passive immunization (directly acquiring antibody through inoculation of antisera or from mother). In active immunization, fish can acquire immunity associated with production of humoral antibodies through oral administration of live or dead pathogens. A single type of antigen (monovalent) or a mixture of antigens (polyvalent) from more than one pathogen can be used. Monovalent, bivalent, trivalent or polyvalent vaccines, based on the number of serotypes of a pathogen from which the vaccine is prepared, can stimulate an immunity against one, two, three or several types of pathogens (Das, 2011). The passive immunization finds comparatively lesser application for practical purposes.

Vaccines, bacterins and immunostimulants are generally administered in four different ways (Lipton, 2006). Intraperitoneal, intramuscular or subcutaneous injections provide maximum level of protection. As these are stressful and labour-intensive, semi-automated devices are used. Vaccination by immersion, spray vaccination and immunization by oral route are the other methods. Instead of administration of antigen through injection in fish, oral vaccines are usually tried. Effective immunoprophylaxis in fish against natural infections by *Aromonas* and *Vibrio* has been achieved by oral vaccines. Vaccine concentrate or lyophilisate is added to the feed is one method. Another way is by administering antigen by hyperosmotic infiltration. Fish, in this case is dipped in hypertonic salt solutions and then transposed to a 2% antigen solution for about 2 minutes. The antigen penetrates the host then. The oral immunization depends on the nature of antigen, frequency of administration, species of fish and water temperature.

- Probiotics and immunostimulants

  Many microbial cell-wall products and plant extracts stimulate the immune system of fish and protects the animal against disease.

- Bacteriophage therapy

  Several phage viruses which are parasites on pathogenic bacteria have been tried to control fish disease.

- RNA interference

  RNA interference or post-transcriptional silencing of gene expression has proved its efficacy against certain pathogenic viruses.

12.20 Future Strategies for Health Management

Some of the strategies for fish health management to be adopted soon are mentioned here.

- Early detection of new pathogens, distribution and occurrence of diseases and quick reporting systems.
- Employment of skilled manpower and establishment of clinical laboratories.
- Development of molecular diagnostic techniques for early detection of carrier, sub-clinical and clinical conditions.
- Development of antibiotic and non-antibiotic drugs and also development of effective vaccines.
- Execution of proper quarantine system throughout the country.
- Specific research works to be continued on probiotics, vaccines, RNA and other molecular techniques of disease control.

12.21 References

Cheng T.C. *General Parasitology*. Second Edition. Harcourt Brace & Company Asia Pvt Ltd. 1999, pp. 1-827.

Das M.K. *Fish and Shrimp Health Management*. In Handbook of Fisheries and Aquaculture, ICAR, New Delhi. 2011, pp. 762-793.

Ganguly S. *Viral diseases infecting finfishes and ornamental fishes: A review of relevance to sustainable aquaculture*. Int. J. Pure App. Biosci. 5(1). 2017, pp. 282-284.

Lipton A.P. *Diseases in Ornamental Fishes and their control*. In: International Seminar on Ornamental Fish Breeding, Farming and Trade. 2006, pp. 109-114.

Pandey G. Ornamental fishes: health managements by disinfectants and antibiotics. Jigyasa- I 2014. 8 (1): 60-66.

Pelczar (JR) M.J., Chan E.C.S., Krieg N.R. *Microbiology*. Fifth Edition. Tata Mc Graw-Hill Publishing Company Limited. 1998, pp. 1-1918.

Sanil N.K. and Vijayan K.K. *Diseases in Ornamental Fishes*. In Ornamental Fish, Breeding, Farming and Trade (Eds. Kurup B.H., Boopendranath M.R., Ravindran K., Saira Banu, Nair A.G. Department of Fisheries. Government of Kerala, India. 2008, pp. 175-189.

13

Application of Genetics and Biotechnology in Ornamental Fish Keeping

13.1 Introduction

Recent advances in the research in genetics and biotechnology provided immense scope for their application in the ornamental fish keeping.

13.2 Cytogenetic Studies in Fishes

National Bureau of Fish Genetic Resources (NBFGR) had compiled a chromosome atlas on Indian fishes depicting karyotype of 128 teleost fishes found in Indian waters (Lal, 2011). Chromosomal information for more than 200 indigenous fishes has been documented. Application of restriction endonuclease banding and fluorescent in-situ hybridization (FISH) can be used for development of species/ stock specific molecular markers and identification of sex chromosomes.

13.2.1 Molecular Markers

Molecular markers are used in species discrimination, phylogeny and taxonomy validation, population genetics and stock identification and linkage mapping for selective breeding and functional genomics.

13.2.2 Methods for genetic improvement

Some approaches for genetic improvement in fishes are traditional, such as selective breeding and hybridization. A particular approach is chosen depending on the species, its biological traits, economic importance and the availability of the species.

13.2.2.1 Selective breeding

There are various approaches of selective breeding as mass selection, family selection, within family selection, combined selection etc. In case of fishes, most commonly followed methods are either family selection or mass selection. Mass selection can be used only for a single trait, family selection/combined selection can be used in selecting different traits such as growth, disease resistance, meat quality or reduced fat content etc. The selection methods act on the intraspecific variation available in the founder population and heritability of the desired trait. Improvement through selective breeding depends upon the extent of additive genetic variation in the selected population (Lal, 2011).

- **Marker used in selection**

  Development of molecular markers such as microsatellite DNA markers and single nucleotide polymorphisms (SNPs) have taken for developing linkage maps.

Marker assisted selection (MAS) is an important tool for genetic improvement and breeding programmes. In MAS, identified polymorphic markers are used to genotype families and progenies with desired trait data. The data is used to develop linkage maps to study linkage between markers and traits. Breeding are planned for families that show such linkages to achieve genetic improvement (Lal, 2011). In some fish species MAS have been applied.

13.2.2.2 Genome manipulation

Genome manipulation or chromosomal engineering is used to produce gynogenetic and androgenetic inbred lines and polyploid individuals. Highly inbred homozygous lines of mono-sex individuals are produced through the process of gynogenesis and androgenesis. The inbred individuals so produced when crossed with normal heterozygous ones may produce offspring heterosis, especially for growth.

13.2.2.3 Hybridization

In hybridization, crossbreeding between either member of different races or strains of the species(intraspecific) or between two species of the same genus (interspecific) and between species belonging to different genera (intergeneric) takes place. In this process, combination of positive traits of parent species in their hybrid offspring. Better growth, resistance to diseases, meat quality early or late maturity, better fecundity etc., are the positive traits.

- **Intraspecific hybridization**

  Offspring with heterosis are produced when different races or strains of the same species from different geographical regions are crossed. Stocks from different hatcheries, when crossed yielded progeny exhibiting heterosis for growth.

- **Interspecific hybridization**

  Intermediate traits, faster growth and improvement in body colours are observed in Indian major carps.

- **Intergeneric hybridization**

  Generally same characteristics are shown through intergeneric hybridization as through interspecific hybridization.

13.2.2.4 Hormonal sex reversal

Exogenous steroid hormones are used to control sex in fishes. Hormonal sex reversal techniques are applied in aquaculture for mass production of sterile monosex populations by interfering with genetic sex. Hormone administration is carried out either by mixing required quantity in the feed or dissolving in water medium (dip-treatment or immersion).

Early stages of spawn or fry are treated with androgens (male hormone), when all-male population is required and with oestrogens for all-female populations.

- **Androgens or male sex hormones**

  Male sex hormones, either natural or synthetic are used for production of all-male fish population. Among androgens, 17α methyl testosterone (MT) is the most widely used hormone and has been used for fishes belonging to the families, Salmonidae, Cyprinidae, Anabantidae, Poecilidae and Cyprinidontidae.

- **Oestrogen or female sex hormones**

  Estradiol 17β is the widely used female sex hormone to produce ovaries (feminization) in fishes. Two naturally occurring steroids are Oestradiol and Estrone. Ethyl oestradiol is an oestrogen preparation used for sex reversal and acts longer. For different fish species the dose requirements are different.

- **Disadvantages of hormonal sex reversal**

  Negative aspects are

  1. Residues of the administered steroid may be carcinogenic and may interfere with sex of consumers
  2. Sexual maturity is delayed in the sex reversed females.
  3. It is costly and time-consuming process.
  4. The process has to be repeated.

Therefore, use of hormonal manipulations should be carried out with necessary precautions.

13.2.2.5 Genetics in conservation of biodiversity

Decreased genetic variability may lead to increased levels of inbreeding and reduced fitness, so conservation of genetic variability is important.

This interdisciplinary stream of science of conservation genetics intend to apply genetic methods for the conservation and restoration of biodiversity. Subpopulations of the population of most of the species as called genetic stock, maintain genetic make-up or character-difference from other sub-population or genetic stock of the same species. This is the genetic variation within the species. Conservation and fishery management programmes should be stock specific. The variation is assessed within at intraspecies level, and between the populations using molecular markers such as allozyme, microsatellite DNA and mtDNA sequences. This is needed for planning rehabilitation of endangered fishes and for genetic improvement of cultivable fishes. An Outreach Programme to identify Fish Genetic Stocks of 10 cultivable fish and shell fishes of India have been carried out (initiated by ICAR) documenting molecular and biological descriptors of 10 species across the native distribution of wild population (Lal, 2011).

Genetic assessment of the aquatic resources is an utmost important step to conserve wild populations.

Replenishing and restructuring genetic status of the existing stocks/ populations by utilizing the available genetic information is important for sustainable utilization of the natural aquatic resources in future.

13.3 Application of Biotechnology

The knowledge of biotechnology has been utilized in aquaculture to enhance the quality of the cultivated aquatic species, conservation as well as management of genetic diversity.

13.3.1 DNA barcoding

DNA barcoding is the process of species identification by DNA sequence analysis of a uniform target gene.

The National Bureau of Fish Genetic Resources (NBFGR) is the centre to generate DNA barcodes of freshwater and marine finfishes of the region in collaboration with the global International Consortium for the Barcode of Life(cBOL)--FISH BOL (www.fishbol.org). The DNA barcodes (DNA sequence profile of 655 bp fragment of cytochrome c oxidase I) of more than 450 species has been completed in India (Gopalkrishnan et al.,2011). This has been used in the sustainable exploitation, management and conservation of Indian freshwater and marine water fish species.

13.4 Genetic Engineering and Transgenic Fishes

Genetically modified organism (**GMO**), **Transgenic** or **Living Modified Organism** (**LMO**) is an organism which has a foreign or modified gene integrated in its genome.

Genetic engineering is the tool which helps in connecting the gaps between an organism's natural characteristics and desirable characteristics. This is done by identifying and construction of the genes with desirable traits, cloning the genes in suitable vectors and transferring the gene construct into the fertilized egg before cleavage.

13.4.1 Applications of genetic engineering

- To increase growth and or efficiency of food conversion.
- To enhance the tolerance limit to environmental conditions.
- Production of new colour variants of sought- after ornamental fishes.
- Increasing commercially important flesh characteristics.
- Controlling reproductive activity and or sexual phenotype.
- Increasing resistance of species to pathogens or parasites.
- To control fertility.

13.4.2 Process of genetic engineering

- Identification of genes of interest—it can be chromosomal DNA or DNA copy of a mRNA or cDNA.
- Isolation of these specific genes.
- Amplifying the gene to produce many copies.
- Associating the gene with an appropriate promoter and poly A sequence and insertion into plasmids.

- Multiplying the plasmid in bacteria and recovering the cloned construct for injection.
- Transference of the construct into the recipient tissue, usually a fertilized egg.
- Integration of gene into recipient genome.
- Expression of gene in recipient genome.
- Inheritance of gene through further generations.

13.4.3 Transgenic fish research in India

The Central Institute of Fisheries Education (CIFE), Mumbai, Central Institute of Freshwater Aquaculture (CIFA), Bhubaneshwar, Centre for Cellular and Molecular Biology (CCMB), Hyderabad, and Madurai Kamaraj University, Madurai, are the organizations involved in fish transgenic research in India. The growth hormone gene constructs (both cDNA and genomic DNA) fully developed from these species are used to produce auto-transgenic *Catla catla, Labeo rohita* and *Clarias batrachus*. Isolation and characterization of anoxia and salt-tolerance genes from *Clarias batrachus*, mangrove plants, sea grasses and marine microbes have been going on.

13.4.4 Genetically Modified fish

GM fish are used in scientific research and kept as pets. They are produced as environment pollutant guard and for use in aquaculture.

The first transgenic fish were produced in China in 1985. This process of DNA manipulation is known as Cisgenesis when a gene is transferred between organisms that could be conventionally bred or transgenesis when a gene from one species is added to a different species. Gene transfer into the genome of the desired organism as in fish, requires a vector like lentivirus or mechanical/physical insertion of the altered genes into the nucleus of the host by means of micro-syringe or a gene gun (Genetically modified fish, n.d.).

13.4.4.1 Use in recreation

- **Glo Fish**

  A genetically engineered fluorescent fish which is patented and trade-marked is known as the Glo Fish. Zebra fish (*Brachydanio rerio*) is the first ornamental fish to be sold as Glo Fish in the pet stores. These are now available in colours like bright red, green, orange yellow, blue and purple fluorescent colours. Tetras are now sold in various colours such as "Electric Green", Sunburst Orange, Moonrise Pink, Starfire Red, Cosmic Blue, and Galactic Purple colours. Tiger barb is available in the "Electric Green" colour. Similarly, bright green fluorescent colour from jellyfish, red fluorescent from sea coral, orange-yellow fluorescent from a variant of the jelly fish gene are also available (Glo Fish, n.d.).

  It was in 1999, when Dr Zhiyuan Gong and his colleagues at the National University of Singapore initiated this while working with a gene that encodes the green fluorescent protein (GFP). A Texas Company Yorktown

Technologies obtained the worldwide rights to market the fluorescent zebrafish which Yorktown subsequently branded as Glo Fish. Taiwan became the first to authorize sales of a genetically modified organism as a pet. Zebrafish became the first genetically modified animal to became publicly available as a pet introduced for sale in 2003. They were quickly banned for sale in California. Glo Fish were sold only in the US in 2013.

After red fluorescent zebrafish which is trademarked as "Starfire Red", Yorktown Technologies subsequently developed and released a green fluorescent zebrafish and an orange-yellow fluorescent zebrafish in 2006. Blue and purple fluorescent zebrafish which are trademarked as "Electric Green", "Sunburst Orange", "Cosmic Blue" and "Galactic Purple" and incorporate genes from sea coral. In 2012, Yorktown Technologies introduced a new variety of "Electric Green" Glo Fish, derived from a different species of fish, the black tetra. Then, "Electric Green" barb, "Sunburst Orange" tetra and a "Moonrise Pink", "Starfire Red", and "Cosmic Blue" tetras were subsequently developed. Other transgenic lines of pet fish include Medaka which remain transparent throughout their lives and the pink body colour transgenic angelfish and lionhead goldfish expressing the *Acropora* coral's red fluorescent protein. In goldfish, the ocean pout type III antifreeze protein transgene has been successfully micro-injected and expressed. These transgenic goldfish showed higher cold tolerance (Genetically modified fish n.d.).

- **Source of colours**

  Sources of fluorescent protein genes include GFP (jellyfish), GFP (sea pansy), ds Red (mushroom coral), eqFP611(sea anemone), RTM S5(stony coral), dronpa (chalice coral), KFP (Venus bait anemone), eos FP (open brain coral), and dendra (octocoral). Scientist, in early 2014, discovered almost 200 species of naturally occurring fluorescent fish, and so it seems that the fluorescence trait is widespread in fish lives.

13.5 References

Gopalkrishnan A., Swaminathan T.R. and Lakra W.S. *Biotechnology in Fisheries and Aquaculture*. In Handbook of Fisheries and Aquaculture. ICAR, New Delhi. 2011, pp. 703-733.

Lal K.K. *Fish Genetics*. In Handbook of Fisheries and Aquaculture. ICAR, New Delhi. 2011, pp. 641-657.

Glo Fish. Wikipedia. The Free Encyclopaedia. en.wikipedia.org/wiki/Glo Fish. 8 September 2020. Access Date 28 September 2020.

Genetically modified fish. Wikipedia. The Free Encyclopaedia. en.wikipedia.org/wiki/Genetically_ modified_ fish. 6 June 2020. Access Date 28 September 2020.

14

Application of Best Management Practices in Ornamental Fish Farming

14.1 Introduction

Ornamental fish trade in India, though emerging, confront, the rapid rising of global markets, tough competition and increasing concern regarding health of the fish stock and degradation of environmental conditions.

In this Ornamental Fish Industry, several components are involved (Figure 14.1) and are to be dealt with. In view of that, scientific and technological research work, have established that the application of Best Management Practices (BMPs) is needed for the evolvement and improvement of most of the components of this trade.

Best Management Practices (BMPs) may be defined as the methods, measures, practices or techniques found to be the most effective and practical means in achieving an objective while making the optimum use of resources available.

BMPs may be exercised in case of site selection of the prospective farm, commercial production, water quality, feeding strategy, health management, transport and packaging, workers' and workplace safety, biosecurity and sustainable development of ornamental fish farming.

14.2 Adoption of BMP at the Ornamental Fish Farm Level

BMPs for ornamental fish farms have been adopted based on the following principles (Ponniah, 2011; Jain, 2016).

- Site selection and location of the farm based on, planned designing, water and soil quality, resource availability, quality construction and cost-efficiency.
- Commercial production with high-quality broodstock, efficient breeding methods and culture with increased rate of survival.
- Maintenance of water and soil quality.
- Feed management with maintenance of nutritional quality and avoidance of over-feeding and under-feeding.
- Health management through continuous surveillance.
- Proper transport and packaging practices.
- Maintenance of safety and security at the farm area.
- Sustainable development.

14.3 Appraisal of Adopting BMP

Some expected effects of application of BMPs in ornamental fish farms are as follows:

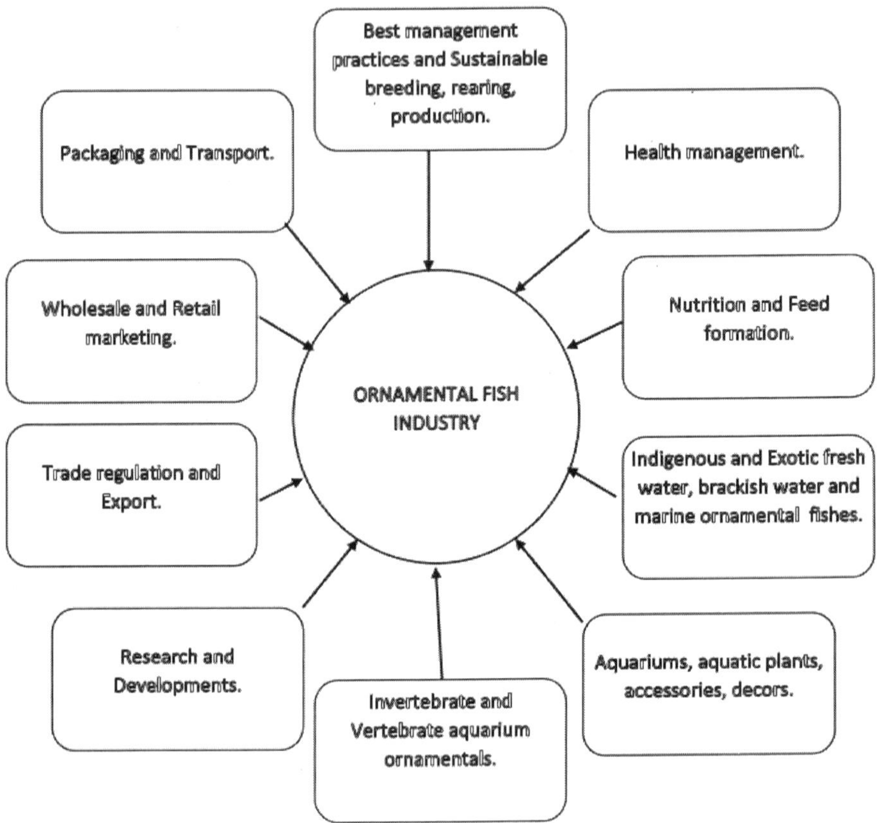

Figure 14.1 Components of Ornamental Fish Industry.

- Use of more locally available non-conventional products as fish feed ingredients and popularization of farm made feed preparation, feeding methods and feed storage techniques.
- Proper pond management with improved soil and water quality in ponds, reduced volume and pollution effect of pond effluents, improvement in environmental conditions, sustainability and profits.
- Maintenance of optimum environmental conditions.
- Use of enhancers and probiotics in place of antibiotics.
- Development of quarantine system in various parts of the country to prevent spread of pathogens from import/ consignments of fish and other ornamentals. Imported or newly acquired ornamental fish may carry disease and may infect valuable and healthy brood stock.
- Ensuring biosecurity, all the incoming stocks quarantined in the

designated area. All incoming stocks analysed, incoming water sources treated, equipment and material sterilized.

- For sustainable development, regular information flow among farming communities, technical and marketing resources and other relevant institutions for steady growth in the farm economy. Concerned government offices, institutions provide services through internet and mobile phones (e-government). Producers are provided with the knowledge and understanding with which location-specific problems may be solved.
- Ornamental fish industry being one of the fastest developing industries, there is an increasing need of proper information, technologies, farming techniques dissemination, such as Code of Good Management Conduct (Kumar & Pradhan, 2011). Ornamental fish farmers and entrepreneurs are in need of quality information regarding technological options in farming to and participate better in markets. They need to know not only the knowledge of the current prices of the commodity, but the trend of the market prices also. Modern information technology may thus be made a tool of sustainable development of the ornamental fish farming.

14.4 References

Jain A.K. *Need of best management practices for freshwater ornamental fish production. In: Best management practices for freshwater ornamental fish production*. (Eds: Jain A.K., Saini V.P. and Kaur V.I.). National Fisheries Development Board (Ministry of Agriculture and Farmer's Welfare, GOI), Hyderabad, India. 2016, pp. 2-6.

Kumar A.T. and Pradhan S. *Information and Communication Technologies in Fisheries*. In Handbook of Fisheries and Aquaculture. ICAR, New Delhi. 2011, pp. 1020-1040.

Ponniah A.G. *Shrimp Farming*. In Handbook of Fisheries and Aquaculture. ICAR, New Delhi. 2011, pp. 548-560.

15

Entrepreneurship Development and Economics of Fish Breeding and Culture Units

15.1 Introduction

Embracing entrepreneurship through ornamental fish farming has long been a popular practice in India. Many people from big cities, small towns and even from villages have opted this business of culturing and breeding of ornamental fishes by setting up farms. Now a days, not only traditional backyard type ornamental fish farming unit exist but major production facilities prevail, involving large investment. Unfortunately, some of these units are not performing successfully and economically. Studies and Research works suggested that there are several factors which may govern the fortune of these production units. In view of these, underlying factors are discussed.

15.2 Essential Approaches for Successful Breeding and Culture of Ornamental Fishes

- Proper training: Beginning of an ornamental fish unit should be preceded by proper training on the subject.
- Identification of a perfect location: An ideal location for the establishment of the farm is to be identified as an essential step. There are some aspects which are to be considered before purchasing the land. These are—
    - Permission from the competent authority.
    - Reviewing the activities of other ornamental fish farms in that area.
    - Recognising potential market, means of transportation and connectivity to the markets.
    - Estimation of demand and supply.
    - A view of the socio-cultural set up of the location.
    - Evaluation of availability and procurement of raw materials, human resource (labour, electrician, mechanic, plumber etc.), and electricity connection.
- Land requirement: For a small-scale unit, minimum land requirement is 500-1000 square feet and large-scale unit, 1 acre or more land is needed. Earthen ponds may be excavated in the land. Cool and shady place is suitable and so site selection is of prime importance.

- Suitability of the chosen site in terms water and soil: Source of water such as ground water, river streams, irrigation canals may be appraised. A constant supply of water is essential for breeding and rearing units. Location near stream is excellent for the supply of non-polluted water and flow-through rearing unit.
- Water quality of the location should better be tested in any water testing laboratory since on the water quality depends the selection of candidate species. Biofiltration unit is essential for maintaining suitable water quality.
- The quality of water in which the breeding and culture of ornamental fishes are to be carried out plays a very significant role. Parameters such as pH, alkalinity, hardness and salinity of the water available at the site should be within the range. Optimum range of essential physiochemical properties of the said water is shown (Table 15.1) and Water hardness level is also mentioned (Table 15.2).

Table 15.1: Some of the essential physicochemical properties of the facility water and their required optimum range

Physicochemical parameters of water	Optimum range
Temperature	24-28° C
pH	6.5-8.5
CO_2	< 5 ppm
Alkalinity	75-120 ppm as $CaCO_3$
Hardness	60-100 ppm as $CaCO_3$
Salinity	0-2.0g/ litre
Dissolved oxygen	6.0-8.0 ppm
Free ammonia	< 0.05 ppm
Ionized ammonia	< 0.1-0.4 ppm

Table 15.2: Water hardness levels and characteristics of the water

Water hardness level (ppm of $CaCO_3$)	Characteristics of water
0-70	Very soft
70-140	Soft
140-210	Medium hard
210-320	Fairly hard
320-540	Hard
>540	Very hard

- Soil quality: The quality of the soil at the farm site should be estimated. The parameters such as pH, clay percentage, organic carbon and nitrogen should be within range (Table 15.3). It should not be highly acidic (pH < 6.5) and highly alkaline (pH > 8.0) and should have good water retention capacity.

Table 15.3: Optimum range of the soil parameters indicating quality of soil

Soil parameters	Optimum range
pH	6.5-8.0
Clay percentage	20%
Organic carbon	1.5-2.5%
Nitrogen	50-75/ 100 gm of soil

- Quality of brood stock: For producing good quality fish seed, superior quality of brood stock should be selected. Rearing of desired quantity of younger fishes which can be developed into brood stock may be done in case of unavailability of brood stock. To improve breeding efficiency and for healthy offspring, brood stocks are to be pure line and of very good quality.
- Approach to the markets: To reach markets either domestic or international, the units better be located near airport, railway station or bus stand. Marketing network better be established by the owners of the breeding and rearing units.
- Quarantine: New incoming fishes are to be quarantined from resident stock. Quarantine tanks are to be set up to keep watch on the newcomers.

15.3 Culturable Ornamental Fish Species

First of all, it should be decided which ornamental fish species are to be cultured in the farm. Whether the fish are for domestic market or for export purpose. It is also to be decided whether it is of low cost or high cost. The facility should be designed accordingly.

- Varieties should be selected based on market demand.
- Easy to breed varieties may be chosen in the beginning.
- Stock of good quality and pure line brood stock should be developed.

15.4 Production Facilities

- Design and drawing of the construction facility: A detailed drawing, first a hand sketch then a final drawing by an architect/ civil engineers, using auto-cad software, is needed.
- Type of production facilities: Before the construction is started, it is to be decided about the level of production system, i.e., traditional, extensive, semi-intensive or intensive. The type of production facilities is to be constructed following the system. These are earthen ponds, poly-lined ponds, cement tanks, FRP tanks and glass tanks.
    - **Earthen ponds:**
        - A number of ponds (0.01-0.05 ha) with water depth of 3-4 ft are suitable. Koi carps, Oscars and young fishes can be reared here.
        - A site which retains water, without seepage or loss of water and do not get flooded during rainy season is suitable for pond construction.

- Rearing of ornamental fishes is done inside circular or rectangular net enclosures or hapa of suitable mesh size. These are installed in the ponds to avoid predators.

- **Poly-lined ponds:**
  - These are earthen ponds, covered with a Low Density Poly-Ethylene (LDPE) film of 150-500 μ (Jain, 2016). This cost effective and long- lasting facility are highly successful in rocky and sandy soils, in controlling seepage loss. The cost of poly-lining including poly-film of 250 μ and labour cost is about Rs 10 per square feet only.

- **Cement Tanks:**
  - For ornamental fish farming, cement tanks of different sizes and water holding capacity for both indoor (Figure 15.1) and outdoor (Figure 15.2) are very useful. Depending on the requirement cemented tanks are made up of ferro-cement, reinforced cement concrete (RCC) or bricks.
  - The ferro-cement tanks, circular or rectangular of different sizes can be used for rearing and breeding of livebearers, barbs, tetras and brood stocks.
  - The reinforced Cement Concrete (RCC) tanks are made up of steel bars and cement-concrete mortar. When these RCC tanks are constructed with the instructions from engineers.

- **Brick tanks:**
  - These are the most common, economical, easy and fast to build tanks. The average width of brick tanks should be about 5 ft. Here only one person can use a hand net and net out the fish.
  - All these cemented tanks must be dealt with water proofing material to avoid water leakage problems.
  - Brick tanks can be used for holding 3000 litres of water. If the water volume is more, RCC tanks should be used.

- **Fibre-Reinforced Plastic (FRP) tanks:**

  Fibre-reinforced plastic is a composite material made up of a polymer (epoxy, vinyl ester etc.) matrix reinforced with fibres (glass, carbon etc.). The FRP tanks are used in ornamental fish farming in hatcheries and intensive culture system. FRP tanks are available in circular (Figure 15.3), rectangular and square shapes. Though cost is more than the brick tanks, but are movable, easy to clean and hygienic. The minimum thickness of FRP tanks must be 3mm, up to 1000 litre capacity and increase proportionately (Jain, 2016).

- **Glass Tanks:**

  Glass tanks are used in ornamental fish for breeding (Figure 15.4) and culture purpose. Glass tanks are available in various sizes.

15.5 Types of Production Units in Ornamental Fish Farming Business

- Production through breeding and larval rearing up to 2-3 months.
- Culture/farming to produce fish from larvae to marketable size.

  Ornamental fish farmer or entrepreneur may opt for any one or both depending on the skilled technical manpower available (Swain, et al., 2010).

15.6 Use of Rain Water Harvesting System and Water-Recirculation Facility

- For rain water harvesting system, proper pipelines are to be fitted and channels are to be constructed so to direct all the rain water into collection chambers.
- All the harvested rain water should be stored properly.
- To re-use water at an ornamental fish breeding and culture farm, there should be a common-bio-filtration. In a bio-filtration system all the incoming water is passed through a filter bed (coarse gravel, activated charcoal, granular zeolite and layer of foam) which allow growth of few useful groups of bacteria (*Nitrosomonas* spp., *Nitrobacter* spp.). These bacteria decompose all the complex compounds into simpler gaseous forms that are released out from the water (Jain, 2016).

Figure 15.1 Indoor cement tanks. Figure 15.2 Outdoor cement tanks.

Figure 15.3 Circular plastic tanks. Figure 15.4 Glass tanks for breeding purpose.

15.7 Eco-Friendly Management of the Farm

The effluents of farm rich in suspended organic matters and nutrients should never be thrown directly in open water systems. So, there is a need of installation of a re-circulatory system. There should be series of drainage receiving tanks of recirculatory system so that all the suspended solids get easily settled in these tanks only before entering into bio-filter chambers. The settled waste should be removed and can be used as manure or for hydroponic cultivation.

15.8 Profit Analysis

It has been estimated that a beginner may earn about Rupees 14,137 per month by breeding, rearing and marketing popular ornamental fishes. Income may be increased with experience and by adopting strategies suitable for decreasing expenditure and by increasing number of breeding cycles per year. Proper breeding and rearing of high value live bearers and rare exotic ornamental fishes may fetch more profit. Economics of some ornamental fish culture units are presented here (Tables 15.4, 15.5 and 15.6) though these may vary with places and other factors governing the units.

Table 15.4: Economics of small-scale breeding and rearing facility for popular ornamental fish species

Fixed capital	**Amount in Rupees**
Low-cost shed of 300 sq. ft.	16,000
Breeding tank (concrete) (6′x 3′x 1′6″) @ Rs 4,000/tank (4 nos.)	16,000
Rearing tanks (concrete) (6′x 4′x 2′) @ Rs 6,000/tank (2 nos.)	12,000
Brood stock tanks (concrete) (6′x 4′x 2′) @ Rs 6,000/tank (2 nos.)	12,000
Larval tanks (4′x 1′6″x 1′10″) @ Rs 2,000/tank (10 nos.)	20,000
Bore-well facility with pump	10,000
Oxygen cylinder with accessories	10,000
Sub total	**96,000**
Recurring Expenditure (for one year)	
Brood fish (800 females+ 200 males) @ Rs 3/pc	3,000
Feed 150 kg/year @ 40/kg	6,000
Nets	3,000
Electricity and fuel	5,000
Wages 6,000/labour	72,000
Miscellaneous	5,000
Sub total	**94,000**

Fixed capital	Amount in Rupees
Total expenditure	
Recurring cost	94,000
Interest on fixed cost (12% per annum)	11,520
Interest on recurring cost (12% half yearly)	5,640
Depreciation (20% of fixed cost)	19,200
Total	**1,30,360**
Gross income	
Sale of 100,000 fish @Rs 3/pc, reared for one month	3,00,000
Net income (Gross income -Total expenditure) =Rs 3,00,000-130360	**169,640**
Monthly income	**14,137**

Table 15.5: Economics of large-scale breeding and rearing unit for live-bearer ornamental fishes

Investment Fixed capital	Amount in Rupees
(Land 1 ha) Shed with electrical connection (4,000 sq. ft.)	10,00,000
Breeding tanks (concrete, 6'x 3'x 2') @ Rs 4,000/tank, 25 nos.	1,00,000
Rearing tanks (concrete, 8'x 4'x 2') @ Rs 5000/tank, 30 nos.	1,50,000
Brood stock tanks (concrete, 6'x 3'x 2') @ Rs 4,000/tank, 30 nos.	1,20,000
Larval tanks (concrete, 3'x 2'x 1') @ Rs 2,500/tank, 20 nos.	50,000
Pump, overhead tank and water supply facility	2,00,000
Aeration system	2,00,000
Sub-total	**18,20,000**
Recurring expenditure	
Brood fish (5,000 females+ 2,000 males @ Rs 10.00/pc	70,000
Feed (500 kg/year @ Rs 80/kg)	40,000
Nets	30,000
Wages @ Rs 6,000/month for 3 labourers	216,000
Miscellaneous	50,000
Sub-total	**406,000**
Total expenditure	
Recurring expenditure	406,000

Investment Fixed capital	Amount in Rupees
Interest on capital cost (12%)	218400
Interest on recurring expenditure (12% for half yearly)	24360
Depreciation (10% of capital cost)	182000
Total	**830,760**
Gross income	
Fish sale (for 7,00,000 @ Rs 4.00/pc)	28,00,000
Net income (Gross income—total expenditure) =28,00,000—830,760	19,69,240
Monthly income	1,64,103

Table 15.6: Economics of large-scale breeding and rearing facility for popular exotic ornamental fishes in India.

Investment Fixed capital (Land 1.5 ha)	Amount in Rupees
Shed with electrical supply (4,000 sq. ft.)	10,00,000
Brood stock tank (concrete, 6′x 3′x 2′) @ Rs 4,000/tank, 30 nos.	1,20,000
Breeding tanks (concrete, 5′x 3′x 2′) @ Rs 3,000/tank, 30 nos.	90,000
Larval tank (concrete, 3′x 2′x 1′) @ Rs 2,000/tank, 50 nos.	1,00,000
Rearing tank (concrete, 8′x 4′x 2′) @ Rs 5,000/tank, 60 nos.	3,00,000
Earthen ponds (0.05 ha 2 nos./ large concrete tanks (50 sq. m. 6 nos.)	5,00,000
Pump, overhead tank, water supply facility	2,50,000
Aeration system	2,00,000
Generator, gas cylinder and others	2,00,000
Sub-total	**2,760,000**
Recurring expenditure	
Broodfish (3,000 females+ 4,000 males @ Rs 50.00/pc)	3,50,000
Feed (2,000 kg/year @ Rs 50/kg	1,00,000
Nets etc.	50,000
Wages @ Rs 6,000x 12 months x 4 nos. labourers	2,88,000
Miscellaneous	2,00,000
Sub-total	**9,88,000**

Investment	**Amount in Rupees**
Fixed capital (Land 1.5 ha)	
Total expenditure	
Recurring expenditure	9,88,000
Interest on capital cost (12%)	3,31,200
Interest on recurring expenditure (12% for half yearly)	59,280
Depreciation (10% of capital cost)	2,76,000
Total	**1,654,480**
Gross income	
Fish sale (for 12,00,000 nos. @ Rs 5/pc)	6,000,000
Net income (Gross income—Total income) =60,00,000—16,54,480	4,345,520
Monthly income	362,127

15.9 References

Jain A.K. *Setting up of freshwater ornamental fish production facilities.* In: Best management practices for freshwater ornamental fish production (Eds. Jain A.K., Saini V.P. and Kaur V.I.). NFDB. 2016, pp. 44-56.

Swain S.K., Sarangi N., Ayyappan S. *Ornamental Fish Farming.* Published by Directorate of Information and Publications of Agriculture. ICAR, New Delhi. 2010, pp. 1-145.

16

The Potential Scope of Ornamental Fish Industry and the Trade

16.1 Introduction

There is a growing interest in ornamental fish keeping and this has resulted in the steady expansion of its trade to several countries in the world. More than 99% of the global market for ornamental fishes belongs to the hobbyists while the rest i.e., less than 1% belongs to the public aquaria sector. At present, there are more than 100 million hobbyists in the world.

These fishes in trade show wide taxonomic and ecological diversity and divulge wide geographic distribution. Diversity is in terms of colour, body shape, movement, behavioural pattern, reproductive behaviour and feeding pattern and other adaptive features.

More than 125 countries are involved in this trade and the trade at retail level is worth US $ 18-20 billion (PMMSY, 2020) with an average annual growth of over 10%. More than 2 billion live ornamental fishes are traded (Satam et al., 2018). Entire industry including plants, accessories, aquariums, feed and drugs is estimated to be worth more than US $ 18-20 billion (De, 2016). The annual growth rate of total ornamental fish trade is 14%. India is considered a sleeping giant with yet untapped potential sources, and contributes 1.06 million US $ (0.4% of the total export) ranking 31st in the global ornamental fish trade (Raja et al., 2019; PMMSY, 2020). The ornamental fish trade is a foreign exchange earner, besides being source of employment. And so, has a noteworthy position in the economy of developed and developing countries.

16.2 Present Panorama

Developing countries in the tropical and sub-tropical regions are the major suppliers of the ornamental fishes, contributing 60% of the export overall. Application of advanced science and technology in the breeding, maintenance and transport had made the addition of more and more ornamental fishes possible every coming year (De, 2016). In the global ornamental fish trade about 2500 fish species are involved of which more than 60% are of freshwater origin and 40% are marine. These freshwater fishes are either captive-bred or collected from the wild. More than 15% (by value) of the market are dominated by marine fish species. About 98% of these fishes are collected from the wild and rest are captive bred. The marine fishes are becoming more popular and this sector is expanding rapidly.

16.3 Ornamental Fishes in the Global Fish Industry

About 50 species of freshwater fish dominate the market. Among these, live bearers, neon tetra, angel fish, gold fish, zebra danio and discuss are very popular. Though only few ornamentals are captive-bred, more than 90% of freshwater ornamental fishes exported from India are wild caught indigenous species (Swain, et al., 2011). Marine fish aquariums are popular nowadays. As prices are becoming affordable, technical and economic supports have also contributed to the popularity of marine aquariums.

16.3.1 Global exports

Global export of ornamental fishes is worth US $ 347.5 million in 2014. Asian countries contributed for US $ 197.7 million (57 % of the trade). Europe accounted for US $ 95.8 million, South America US $ 13.6 million, North America US $ 13.8 million, African countries US $ 9.6 million, Oceania US $ 4.9 million and the Middle East US $ 1.76 million (De, 2016). The export value for the global ornamental fish industry then declined, and stood at US $ 337.70. Singapore the ornamental fish capital of the world, was with exports worth of US $ 69.32 million, contributing about 20% of the total supply. This main trading hub in Asia, exported more than 30 % of the fish. Japan (US $ 41.34 million) is in the second place and the third position is occupied by the Czech Republic (US $ 32.0 million) followed by Thailand (US $ 23.31 million), Malaysia (US $ 22.62 million), Indonesia (US $ 21.54 million), Israel (US $ 19.04 million), Brazil (US $ 18.22 million), Sri Lanka (US$ 13.1 million) and Colombia (US $ 12.3 million).

The position of the countries based on the percentage of export of ornamental fishes in 2017, are in the order of Singapore (11.2), Spain (10.8), Japan (9.5), Myanmar (8.8), Indonesia (7.9), Czech Republic (5.8), Thailand (4.5), Malaysia (4.5), Netherlands (4.3), Sri Lanka (4.3) and USA (3.4).

16.3.2 Global imports

The USA was the single largest importer of ornamental fish in 2014, with 14.3 % (US $ 42.9 million) of total imports. Next was UK with US $ 29.5 million or 9.87% of total imports. The other major importing countries were Germany (US $ 23.4 million), Singapore (US $ 21.3 million), Japan (US $ 19.5 million), China/ Hongkong (US $ 19.3 million), France (US $ 16.7 million), the Netherlands (US $ 13.3 million), Italy (US $ 9.7 million), Malaysia (US $ 9.1 million), Canada (US $ 8.6 million) and Belgium (US $ 8.3 million). Singapore, Germany (Frankfurt), Hongkong, Malaysia and the Netherlands (Amsterdam) were the important trading hubs, re-exporting a major portion of their imports. The percentage share of the global ornamental fish imports in countries, in 2017, are USA (19.5), UK (7.1), Netherlands (6.3), Germany (6), China (5.9), Japan (5), Singapore (4.5), France (4.2), Hong Kong (3.8), Italy (3.3), and Spain (2.9).

16.4 Recent Supply Trends of Ornamental Fishes

In recent years, Czech Republic, Israel, Belgium and the Netherlands have started supplying ornamental fishes to European countries. Philippines, Morocco and the Pacific countries on the other hand have surfaced as major suppliers of marine ornamental fishes. In Europe, Japan, the Republic of Korea, Norway, Sweden and Switzerland, the import trend has been increased continuously. Besides Singapore and China, UAE, Germany (Frankfurt) and the Netherlands (Amsterdam). Brazil, Colombia and Peru have now been ranked as major suppliers.

16.5 Constraints

Anthropogenic factors play a major role in the setback of markets. These are intensive breeding programmes leading to susceptibility to diseases, resistance to antibiotics, poor brood stock quality, introduction of exotic species, over-exploitation of particular species, destructive fishing methods, poor handling and quarantine procedures. Besides these, erratic climatic conditions and seasonal weather pattern impose problems in the availability of certain species during certain periods of the year.

16.6 Green Certification and Ecolabelling

Green Certification

The extent and impact of huge uncontrolled exploitation of natural ornamental fish resources has only recently been identified. Development of a set of guidelines/code of practices of marine ornamental fishes by the Marine Stewardship Council (MSC) has taken place (Swain et al., 2011). Consequently, the Marine Products Export Development Authority (MPEDA) Kochi in 2011 has brought out the "**Guidelines for Green Certification of Freshwater Ornamental Fishes**". This is to create an awareness among local communities and stakeholders to desist from unlawful and illegal practices of catching ornamental fishes from the wild. The trade should also encourage protection of the habitat for an eco-friendly approach. Presently, exotic species dominate our domestic aquarium trade, and some are being bred for export. Unregulated/ illegal introduction of exotic species and control or elimination of invasive species pose problems and these issues have to be addressed. In this context, the guidelines incorporate the following aspects such as collection from the wild, handling, transport, holding, breeding and culture facilities, conditioning for export, infrastructure and maintenance of records to confirm to a value chain system for delivering healthy ornamental fishes to the trade and hobbyists (Silas et al., 2011). Vision of the guidelines is "conservation and sustainable development of freshwater ornamental fish sector involving collection, farming and marketing through a value chain system enabling certification at all stages". And to establish Green Certification (GC) system is to achieve sustainability and social security, to reduce risk and cost, and to enhance income is the mission. For wild caught native ornamental fishes, Green Certification will be made mandatory with a transition period of three years from the date of implementation of the guidelines. It is expected that the stakeholders, breeders and exporters will be self-motivated to implement the **Green Certification Guidelines** and the aquarium hobbyists may prefer buying the certified specimens, as the certification process is aimed at producing quality fishes, reducing mortality at all stages, fetching better prices, avoiding unethical practices and making all links of the chain sustainable and eco-friendly (Ramachandran, 2013).

Ecolabelling

Ecolabelling got entry in 1977, into the mainstream environmental policy, when the German government established the Blue Angel programme. Ecolabels are the seals of approval given to products that are deemed to have fewer impacts on the environment than the functionally or competitively similar products (Deere, 1999). There is a growing environmental concern and the business units have realised the need of environmental protection and sustainable utilization of natural sources into their marketing strategies. Strict implementation of ecolabelling and green certification is the need of the hour to strengthen India's export industry of ornamental fishes (Ramachandran, 2010). Ecolabelling and green certification schemes are available at local, national and international levels and for aquaculture, fisheries and ornamental fish trade. Ecolabelling promotes consumer choice, improves economic efficiency and enhance market development (Ramachandran, 2013, Nair and Parappurathu, 2020).

16.7 Ornamental Fish Trade in India

India is suitable for ornamental fish culture due to the rich diversity of fish species, favourable climate and cheap labour. There are 374 indigenous freshwater fish species, 700 indigenous and more than 300 exotic marine fish species. Two categories of ornamental fishes are marketed from India, indigenous and exotic.

North-eastern states, West Bengal, Kerala and Tamil Nadu are blessed with highly potential indigenous ornamental fishes. Indigenous fish resource of India is one of the richest in the world. There is a great scope in exploiting ornamental fishes inhabiting the hill steams, the major river systems, reservoirs and lakes. Rivers of North-eastern states and their Himalayan streams support large number of ornamental fish species. The North east India is acclaimed as one of the hotspots of freshwater fish diversity in the world. There are about twenty trash fishes in West Bengal which are potential ornamental fishes. The Western Ghats of India is a goldmine of tropical ornamental fishes. This hotspot exhibits exceptional mega biodiversity and high degree of endemism with respect to freshwater fishes.

The exotic ornamental fishes are marketed domestically and are dominating with 99%. In Indian market 288 exotic varieties are recorded. In different parts of India, more than 200 species of these freshwater exotic ornamental fishes are bred.

Ornamental fish industry involves several components. These components or units depend on efficient modern technologies, scientific knowledge and efficient manpower. Kolkata, Mumbai and Chennai are major exporting centres. About 80% trade consists of freshwater species and 20% from brackish and marine species.

The major share of this export comprises the captive collections from the North eastern states embracing as much as 85% of the total trade. Rest of the 15% is composed of exotic varieties mainly from Southern India. The domestic market of indigenous fishes is limited due to preference of exotic varieties.

Kolkata, Mumbai and Chennai are the main hubs for ornamental fish breeding and culture. Andhra Pradesh, Orissa, Kerala, Assam and Bihar have also started producing freshwater ornamental fishes at a commercial scale.

In Mumbai many breeders breed only rare and expensive fishes. Kolkata is the gateway for export as well as import of popular aquarium fishes.

Ornamental fish keeping an interesting activity for many, is generating income for unemployed youth, women and farmers. This concept of entrepreneurship development is gaining popularity every day. Availability of modern materials, equipment and air transport have made it possible for hobbyists to obtain wide variety of fishes from all over the world.

According to Marine Products Exports Development Authority of India, there are more than one million ornamental fish hobbyists in India who procure several varieties of fishes, display them in aquaria and maintain them successfully.

16.7.1 Domestic trade

Domestic ornamental fish trade is of about Rs. 500 crores. The trade encompasses fish (20%), aquarium accessories (63%), aquarium servicing (11%), and feed (6%). There is a tremendous scope for growth as the urban population will rise to 42.5% (547 million people) in 2025-26.

There are total 5,000 production units in India, out of this, West Bengal have 55%, Tamil Nadu 30%, Kerala 5%, Maharashtra and others 7%, North East and Islands 5%.

16.7.2 Export trade

The export trade of ornamental fishes is of Rs. 8.40 crores, during 2017-18, with 11.6% growth per year. According to National Fisheries Development Board (NFDB), 2017, ornamental fish export from India was maximum (about 125 MT, quantity wise) during 2012-13, with a decline to about 40 MT during 2015-16.

16.7.3 Production technology

Production practices of the ornamental fishes in India, are characterized by

- Based on Low-input: Low-output concepts and has developed in a cluster.
- The production system relies on excess input in terms of labour but less in terms of money, material and management.
- The cost of labour is usually not calculated and given much importance in backyard production units hence the total cost of the marketable fish is less.
- The breeding and culture of only low cost (guppy, platy, molly, widow tetra, rosy barb, koi carp) and medium cost (goldfish, gouramies, angels etc.) fishes are practiced in India.
- The breeding of high and premium value fish is practiced by few breeders only and that too in limited quantity.
- Lack of technical expertise, small size of the production units and limited demand of high and premium value fishes.

16.8 Ornamental Fish Trade and Employment Opportunities in India

Due to advancement in breeding technologies, transport operations and aquarium developments, several new species are involved in the trade each year. In developing countries, the international trade in ornamental fish is actually providing scope in employment for rural unemployed youth.

In India, rural population is 67.25% of the total population (2015). About 600 million people are still dependent on agricultural sector. As 60-70% of Indian population (directly or indirectly) depends upon Agriculture sector, it contributes 16-17% of the GDP. Developing skills of rural youth and making them employable would solve unemployment problem.

The Government of India, launched Swarna Jayanti Gram Swarozgar Yojana on April, 1, in 1999, under Ministry of Rural Development. The aim is to bring the assisted poor families above the poverty line by organising them into Self Help Groups (SHGs) through the process of social mobilisation, their training and capacity building and provision of income generation assisted through a mix of bank credit and government subsidy. Another Govt. of India project, Deen Dayal Upadhyaya Grameen Kausalya Yojana, was launched in 2015, under the same Ministry. Here the provision is to engage rural youth specially BPL and SC/ST segment of population in gainful employment through skill training programmes.

Marine products development Authority (MPEDA) has launched a subsidy scheme for setting up of breeding units of aquarium fishes and marketing societies in various states to facilitate the export. MPEDA provides subsidy for the ornamental fishery to the registered SHGs and marketing societies through the Ornamental Fish Assistance Scheme. Financial assistance is provided at the rate of 25 % of the total investment subject to maximum of 10 lakhs.

The Government of India, supports the National Bank for Agriculture and Rural Development (NABARD) to take up activities for economic empowerment of the rural poor.

Among the multiple schemes promoted by the National Agriculture Bank for Rural Development (NABARD), ornamental fish rearing is supported by its fisheries division. The project can be taken up either as a community managed programme by Self Help Groups or by fish farmers and by individual unemployed youth. The project costing about 7 lakhs will be refinanced by NABARD apart from subsidies offered by MPEDA.

Moreover, the National Fishery Development Board (NFDB) in collaboration with Central Institute of Fisheries Education, has started skill development programmes, particularly for women so as to upgrade their technical knowledge and expose them to best practices in the sector.

Ornamental fish farming can generate income not only for rural women but also for young unemployed people. There is a big scope of employment for youths.

The major part of India's ornamental fish trade (less than 1% of the world trade) is based on wild collection. It has a domestic market based on locally bred exotic species. The overall annual domestic trade is about 100 crores. The growth is at the rate of 20% per annum. There is an attempt to meet the massive demand by implementing suitable technology. By adopting the relatively simple and appropriate technology this sector is now trying to create substantial employment opportunities besides earning foreign exchange (Jain, 2016).

NABARD is developing rural entrepreneurship by encouraging fish culture. Many enthusiastic entrepreneurs are coming forward to establish breeding centres both in urban and rural areas.

In order to sustain growth in the ornamental fish farming and trade, the Government of India with schemes under NABARD and MPEDA aimed at culture- based development instead of capture of wild species. Organised trade in ornamental fish depends on assured and adequate supply which is possible only through mass breeding. NABARD is offering training to the integrated entrepreneurs, farmers and Self- Help Groups who are interested in the culture as it is economically viable and highly profitable in nature.

With the inception of National Agriculture Innovation Projects (NAIP) in Chitra Durga of Karnataka ornamental fish culture was introduced with initial investment from the project. Gross return per annum to the fish growers is Rs 25,000 with an expenditure of Rs 10,000.

In Kolathur, north Chennai, fish breeding was started as cottage industry by several people and has become the hub for ornamental fish trading in the country. About 600 varieties of ornamental fish and aquarium accessories are sold by the entrepreneurs. Women are engaged in ornamental fish culture at Sakkmalpuram. They could earn Rs 5000 to Rs 7000 every month. It has improved the living standards of rural women.

In West Bengal, there are 2.490 units of ornamental fish culture, out of which 920 are breeding or rearing farms and 12 are exporters (Anon, 2001). Foreign exchange worth of 0.49 US $ was earned here, from export of ornamental fishes during 2007-2008 (MPEDA, 2007; De and Ramachandran, 2011). About 50% share alone of the export value of ornamental fishes is with West Bengal. The business of these fishes is concentrated mainly in three districts viz., Howrah, North 24-Parganas and South 24- Parganas. In Howrah, areas of Bally, Jagacha, Domjur, Panchla, Sankrail and Jagatballavpur contribute largely in ornamental fish production and culture. Bishnupur, Mograhat, Falta and Budge Budge in South 24 -Parganas, have high level of ornamental fish trade (OFT) operations (Banerjee et al., 2019). Amatala (Udairampur), and Dasnagar (Howrah) are considered as the ornamental production hub of West Bengal (Swain et al., 2010). Galif Street, a weekly market near Hatibagan Kolkata is an important centre for selling of ornamental fishes and plants for aquarium (Figures 16.1, 16.2, 16.3).

In India, there are 212 MPEDA authorized breeding centres, in Kerala-126, Tamil Nadu-44, Madhya Pradesh-11, Himachal Pradesh-10, West Bengal-8, Maharastra-7, Rajasthan- 4, and Karnataka-2. There are 35 MPEDA authorized ornamental fish exporters in India, in West Bengal-12, Kerala-8, Tamil Nadu-8, Karnataka-4, and Maharashtra-3. About 90% of ornamental fish is traded from Kolkata port followed by Mumbai and Chennai.

Marketing channel operating network in Domestic Market:

Channel I Breeder cum Rearer → Wholesaler → Retailer → Domestic consumer

Channel II Breeder → Rearer → Wholesaler → Retailer → Domestic consumer

Channel III Breeder → Retailer → Domestic consumer

Marketing channel operating network in Export Market:

Channel I Collector → Local Transporter → Wholesaler → Exporter → Importer → Consumer abroad

Figure 16.1 Selling of ornamental fishes. Figure 16.2 Goldfish in market for sell.

Figure 16.3 Aquarium plants at Galif Street market.

The ornamental fish industry provides jobs to more than 50,000 people through 5,000 aquarium retail outlets and an equal number of production units. All these put together amount to a domestic aquarium trade of around Rs. 500 crores. This is at a time when 1.25 percent of 7.88 crores urban households in the country have an aquarium home or workplace.

There is a cyclic relation of demand and supply in the domestic aquarium trade, the supply in the form of retail outlets and production units triggered the demand from hobbyists. This is in turn, again stimulates the supply.

The country exported ornamental fish at 5.65 crore in 2015-16. Ornamental fish grown in captivity in India are not in demand in the International market because of the limited knowledge base of breeders and unhygienic traditional production practices.

A pilot project on Ornamental Fisheries is going to be launched by centre with a total outlay of 61.89 crore. Eight potential states have been identified for the pilot project viz., Assam, West Bengal, Odisha, Maharashtra, Gujarat, Karnataka, Tamil Nadu and Kerala.

Recognizing the potential and scope of ornamental fisheries, the Department of Animal Husbandry, Dairying and Fisheries, the Ministry of Agriculture and Farmers Welfare, has envisaged a programme to unlock the country's ornamental fisheries sector through a special drive by launching a pilot scheme for the development of ornamental fisheries with a total outlay of 61.89 crore. Implementation of the pilot-scale Ornamental Fisheries Project focuses mainly on creating an enabling environment for a sustainable and holistic development of ornamental fisheries for the socio-economic development of the people involved in this activity as well as for exports. The thrust areas have been identified for enhancing ornamental fisheries production through cluster- based farming and conservation of natural resources, both inland and marine, through habitat restoration and creating awareness amongst the stakeholders.

The major objectives of the pilot project are (i) to promote ornamental fish culture with cluster -based approach (ii) to augment ornamental fisheries trade and export earnings (iii) to create employment opportunities for the rural and peri urban population and (iv) use of modern technology and innovation to make ornamental fisheries a thriving activity. All the activities under the pilot project are classified in to four major groups viz., (a) activities related to production of ornamental fish, i.e., setting up of backyard rearing units, medium scale units, integrated-breeding-cum-rearing units etc., (b) activities related to aquarium fabrication, trade, and marketing (c) activities for promotion of ornamental fisheries sector, and (d) activities related to skill development and capacity building.

The pilot project on ornamental fisheries shall be implemented by the National Fisheries Development Board (NFDB) through the Fisheries Departments of states & UTs. The funding pattern is like that of CSS Blue Revolution, Integrated Development and Management of Fisheries.

Though ornamental fisheries do not directly contribute to the food and nutritional security, ornamental fish industry in India is small but vibrant, with potential for tremendous growth. The low production cost and high returns within a short span of time and the ever-growing demand, both in domestic and international markets, etc., are the major attractions.

16.9 Recent Initiatives taken for the Development of Ornamental Fisheries in India

NFDA provided financial assistance towards development of ornamental fisheries in the country for establishment of backyard rearing units, medium scale rearing units, integrated (breeding-cum-rearing) units, aquarium fabrication units and ornamental fish markets.

CMFRI initiated preparing a document on green certification of marine ornamental fish. Breeding and seed production techniques standardized for variety of marine ornamental fishes and supply of fingerlings to farmers. Production of "Varsha" series of ornamental fish feeds in commercial extruders.

MPEDA organized Domestic Fairs at various locations over the country to disseminate the vast potential of the fisheries, aquaculture and ornamental fish sectors. Conducting skill development programmes in ornamental fish farming. Introduced a scheme to award incentives on export of ornamental fishes.

16.10 Strategies Adopted by the PMMSY, Government of India

- Establishment of Ornamental Fish Brood bank.
- Establishment of wholesale ornamental fish markets.
- Establishment of ornamental aquatic plant units.
- Promoting national/ international aquaria shows.
- "ATMNIRBHAR BHARAT" in aquarium fabrication and accessories.
- Public promotion of ornamental fisheries.
- Development of marine ornamental fish culture.
- Development of freshwater ornamental fish culture.
- Capacity building programme.

16.11 Action Plan for Development of Ornamental Fisheries in India

Aquarainbow Vision—2030

Mission:

- Entrepreneurship and livelihood generation through development of ornamental fisheries and improvement in quality of life through promoting the hobby of aquarium keeping.

Vision:

- The hobby of aquarium keeping becomes a highly popular hobby both in urban and rural households.

Objective:

- To support the income of farmers and generate livelihood earning opportunities in rural India through introduction of ornamental fisheries activity.
- To provide self-employment and entrepreneurship development opportunities to rural and urban youth in the field of ornamental fisheries and allied sectors.
- To empower and encourage women, SCs, STs, other economically weaker sections of the society to substantiate their earnings.
- To promote private sector investment, augment export of ornamental fish from country and achieve a leading position in global ornamental fish trade.

- To promote the hobby of aquarium keeping, help the people of country to live a stress-free healthy life and improve the quality of life.

16.12 Financial Resources

- Investment of Rs. 576 crores: under Pradhan Mantri Matsya Sampada Yojana (PMMSY).
- Additional investment of Rs. 500 crores proposed under World Bank Scheme, that would catalyse about Rs. 2500 crores.
- Marketing infrastructure under FIDF.
- Government investments catalyse the growth of private sector on its own to about Rs. 10,000 crores.

16.13 Recommendations for Further Improvement in Ornamental Fisheries

These activities are recommended for further improvement in the development of ornamental fisheries:

- Setting up of production units including renovation.
- Encouraging women SHGs and unemployed youth for establishment of backyard enterprise for farming ornamental fishes.
- Development of proper marketing facilities to encourage more and more people towards this enterprise.
- Establishment of Public Aquaria, Ornamental Fish Promotion Centres and Bio-parks.
- More skill development training and capacity building programmes to be arranged.
- Incorporation of ornamental fisheries in course curriculum of more educational institutes, from school level to research level.
- Development of scientific, reference and popular literature on ornamental fish keeping, as well as promotion of making documentary /video-films.
- Establishment of aqua-clinic and fish hospitals for disease identification and control.
- Continuous survey and documentation work to be carried out.

16.14 References

Anon. *Economic review of West Bengal*. Bureau of Economics and Statistics. Government of West Bengal. 2001.

Banerjee S., Mahapatra B.K. and Mondal B. *Ornamental fish trading pattern in West Bengal*. International Journal of Fisheries and Aquatic Studies. 7(6). 2019, pp. 130-134.

De S.S. and Ramachandran A. *Marketing Channels in Ornamental Fish Trade in West Bengal*. Fishery Technology. 48 (2). 2011, pp. 163-170.

De V.K. *The Global trade in ornamental fish*. INFOFISH International 4. 2016, pp. 52-55.

Deere C. *Eco-labelling and sustainable fisheries*. FAO. http:// www.fao -org. 1999. Access Date 28 September 2019.

Jain A.K. *An overview of freshwater ornamental fish industry of India*. In: Best management practices for freshwater ornamental fish production (Eds: Jain AK, Saini VP and Kaur VI). NFDB (Ministry of Agriculture and Farmer's Welfare, GOI), Hyderabad, India. 2016, pp. 8-20.

MPEDA. *Marine products export development authority statistics review*. Kochi, Kerala, India. 2007, pp. 131-132.

Nair C.R. and Parappurathu S. 2020. *Who should certify the sustainability of our fisheries?* A property rights perspective on ecolabelling. Current Science 118 (10). 2020, pp. 1496-1499.

Promotion of ornamental fishes under PMMSY. Government of India. Ministry of Fisheries, Animal Husbandry & Dairying, Department of Fisheries. 2020, pp. 1-21.

Raja K., Anand P., Padmavathy S. and Sampathkumar J.S. *Present and future market trends of Indian ornamental fish sector.* International Journal of Fisheries and Aqua Studies. 7(2). 2019, pp. 06-15.

Ramachandran A. *Ecolabelling and Green Certification for effective Fisheries Management-An Analysis*. Researchgate.net/publication/ 287330532_ Ecolabelling_ and_ Green_ Certification_ for_ Effective_ Fisheries_ Management_ An _ Analysis.2010. Access Date 28 September 2019.

Ramachandran A. *Significance of Green Certification for sustaining the Ornamental Fish Resources with special emphasis on marketing of wild caught fishes*. www. researchgate.net> publication> 301551916_ significance. 2013. Access Date 28 September 2020.

Satam S.B., Sawant N.H., Ghughuskar M.M., Sahastrabuddhe V.D., Naik V.V., Pagarkar, A.U. *Ornamental Fishes: A new Avenue to Supplement Farm Income.* Advanced Agricultural Research & Technology Journal. II (2). 2018, pp. 193-197.

Sahayak S. *Marine Aquarium Trade in India,* MPEDA Newsletter, XIIV. 2009, pp. 31-33.

Silas E.G., Gopalakrishnan A., Ramachandran A., Anna Mercy T.V., Sarkar K., Pushpangadan K.R, Anil Kumar P., Ram Mohan M.K., and Anikuttan K. K. *Guidelines for green certification of freshwater ornamental fish*. The Marine Products Export Development Authority, Kochi, India. 2011, pp. 1- 106.

Swain S.K., Madhu K., Madhu R. and Gopakumar G. Ornamental fish breeding and culture. In: *Handbook of Fisheries and Aquaculture*. Published by Directorate of Knowledge management in Agriculture. ICAR. New Delhi 2011, pp. 500-532.

Swain S.K., Sarangi N., Ayyappan S. *Ornamental Fish Farming*. Published by Directorate of Information and Publications of Agriculture. ICAR. New Delhi. 2010, pp. 1-145.

Index

A

Absolute conversion ratio 548

Abudefduf saxatilis 332-333

Acanthocephalan diseases 619

Acanthodians 1

Acanthogyrus acanthogyrus 619

Acanthosentis antespinus 619

Acanthurus achilles 360

Acanthurus coeruleus 362

Acanthurus lineatus 363

Acanthurus pyroferus 366

Acanthurus sohal 371

Acanthurus tennenti 372

Acanthurus xanthopterus 376

Achilles tang 360

Acorus 452-453, 491

Acorus calamus 452-453

Acropora cervicornis 524

Acropora millepora 521-522

Acrylic tanks 412

Addis butterflyfish 329

Additive 627

Adventitious shoot 491

Aeoliscus strigatus 291, 357

Aeration xiv, 392, 418, 429, 431-432, 444-445, 560, 569, 577-578, 585, 592, 604, 617

Aerator 418, 431-432, 434

Aeromonas hydrophila 609-610, 612

Aeromonas liquifasciens 610

Aeromonas spp. 608

Aeroponics 494

Air pump 392, 432, 438, 448

Air stone xiv, 429, 432, 440, 572

Algae filters 438

Algae scrubber xiv, 419-420

Algal toxicosis 601

Alisma 453

Alisma plantago-aquatica 453

All glass tanks 412

Alternanthera 454

Altum angelfish 177

Amatitlania nigrofasciata 167, 168

Ameiurus melas 154-155

Amino acids 441, 541, 547, 558

Ammania 454

Ammonia 397, 419-421, 435-436, 445-447, 494, 547, 593-595, 600

Ammonification 446

Amphiphyte 449

Amphiprion frenatus 317

Amphiprion ocellaris 313-316, 565

Amphiprion percula 319, 589

Amyloodinum sp. 605

Anabas testudineus 89

Andaman Islands 96, 236, 245, 247, 323

Andinoacara rivulatus 171

Andropodium 581

Angelfish 176-179, 181, 320, 322, 367, 465, 568-569, 610, 614, 632

Anguilla bengalensis 74-75

Anisotremus virginicus 285, 339, 340

Anthopleura xanthogrammica 514

Antibiotics used in Ornamental Fish Culture xvi, 622-623

Anubias 454-455, 492-493

Aphanomyces invadeus 612

Aplocheilus panchax 95, 568

Aponogcton 450, 455 456, 490

Aponogeton crispus 456

Aponogeton madagascariensis 456
Aponogeton ulvaceus 456
Application of Biotechnology xvii, 630
Applications of genetic engineering 630
Aqua-clinic 655
Aquahouse 394, 398-399, 409
Aquaponics 494
Aquarainbow Vision—2030 xviii, 654
Aquarium Care xiv, 329, 447
Aquarium filter 434, 442
Aquarium heater xiv, 429, 433
Aquarium lighting xiv, 430, 433
Aquascaping 394, 399, 414, 422, 467
Argulosis 620
Argulosus foliaceus 620
Argulus bengalensis 620
Argulus siamensis 620
Argulus spp. 620
Arothron nigropunctatus 327
Arowana 146-148, 150-151
Artemia 53, 56, 58, 179, 325, 350, 523, 544, 553-554, 558-559, 571, 577, 590
Artificial propagation 492
Arulius barb 55
Asian bonytongue 149
Asian stone catfish 70-71
Asterias rubens 531
Astronotus ocellatus 183
Aulostomus chinensis 380
Automatic demand feeders 550
Azolla 449, 457-458
Azolla filiculoides 457

B

Bacopa 450, 458-459
Bacopa monnieri 459
Bacterial Diseases xvi, 608, 623
Bacterial gill disease 611
Bacteriophage therapy 625
Baffle filters 437
Balistoides viridescens 379
Barbonymus schwanenfeldii 185
Barilius bakeri 62
Barramundi 241
Barred spiny eel 79
Base rock 422
Bata 63
Beluga whale 535
Berlin method xiv, 420-421, 440
Best Management Practices (BMPs) 633
Betta splendens 211-212, 577-578
Bicolor anthias 324
Bicoloured foxface 349
Biological filtration 420, 424, 431, 435-436, 442
Biotope 395, 400
Black arowana 148
Black bullhead 154-155
Blackchin Guitarfish 308
Black ocellaris 315-316
Black Pacu 159
Black spot disease 617
Blacktail angelfish 322
Blacktip reef shark 304-305
Blood parrot cichlid 173
Blue-dotted hill trout 62
Blue panchax 95
Blue poison dart frog 500
Blue tang surgeonfish 362
Bodianus rufus 341
Bog-Wood xiv, 428
Bony fishes 1-2, 8, 254, 306, 310, 313
Botia lohachata 65-66
Bowmouth guitarfish 309
Box filter 417, 438
Brachionus 555, 590
Brachionus plicatilis 555, 590

Branching vase sponge 510
Branchiomyces 616
Branchiurus orientalis 224, 231, 235
Breeder 60, 172, 204, 572-573, 652
Breeding months 579, 587
Breeding of black molly (*Poecilia* sp.) 587
Breeding of Guppy 584
Breeding of livebearers xvi, 579, 582, 587, 639
Breeding of nest builders 577
Breeding of Platy fish 586
Breeding of swordtail 586
Breeding tank 54, 569-574, 577-578, 583, 586
Breeding trap 569-570, 583
Brick tanks 639
Bronze featherback 75
Bubble eye goldfish 190
Butter catfish 72
Button polyps "zoas" 526

C

Cabomba 450, 459-460, 577-578, 585
Cabomba aquatica 459-460
Calcium Reactor xiv, 420, 441
Callyspongia vaginalis 510-511
Canister filters 437-438
Captive breeding xvi, 66-67, 74, 152-154, 160, 166, 209, 238, 338, 347, 356, 360, 369, 389, 568, 572-579, 587-589
Captive breeding of *Betta splendens* 578
Captive breeding of *Carassius auratus* 573
Captive breeding of *Danio rerio* 572
Captive breeding of killifishes 576
Captive breeding of marine ornamental fishes in India 589
Captive breeding of Rasbora 575
Captive breeding of Tilapia 576
Carassius auratus 186, 565, 573
Carcharhinus melanopterus 304
Carcharias taurus 305
Caribbean spiny lobster 527-528
Carotenoids 543-544, 552, 563-565
Carp 6, 192, 194, 391, 565, 614-615, 649
Cartilaginous fishes 1, 249, 304
Catfish 6, 68-73, 151-153, 155, 423, 590
Celestial eye goldfish 190
Cement tanks 418, 578, 638-639
Centropyge eibli 322, 323
Ceratophyllum 450, 460-461
Ceratopteris 450, 461
Chaetodon auriga 328
Chaetodon semilarvatus 329-330
Chaetomorpha 419, 487
Chanda nama 81
Channa aurantimaculata 49, 85
Channa gachua 49, 85, 86
Channa marulius 49, 86, 87
Channa striata 88
Chelate 496
Chemical filtration 434
Chilodonella 604-605
Chilodonellosis 604
Chironomus larvae 86-87, 544, 558
Chlorella 555, 561
Chlorine 443-444, 494
Chocolate chip sea star 529
Chrysaora colorata 506, 511
Chrysaora fuscescens 512
Chrysiptera parasema 334
Climbing perch 89
Cloud eye 610
Clown tang 363
CMFRI 389, 552, 589, 625, 654
Columnaris 610
Comet goldfish 188, 191
Common carp 192, 614
Common clownfish 313

Common goldfish 188-189, 191
Common molly 196-197
Common sea star 531
Common sea urchin 533
Components of Ornamental Fish Industry 634
Conditioning of Fish xvi, 591-592
Controlled environment agriculture (CEA) system 494
Convict cichlid 167
Corythoichthys haematopterus 347-348
Cotton-wool disease 610
Crassula aquatica 484
Cryptocaryoniosis 602-603
Cryptocaryon irritans 602
Cryptocoryne 450, 461-463, 490-491, 493
Cryptocoryne beckettii 462
Cryptocoryne ciliate 462-463
Cryptocoryne cordata 462
Cuchia 77, 388
Cultiwool 495
Curled-gill goldfish 190
Cutting 455, 458, 467, 470, 490, 492
Cyclops 325, 335, 348, 519, 561-562
Cymodocea nodosa 484
Cyphotilapia frontosa 120, 164
Cyprinid herpesvirus-1 (CyHV-1) 614
Cyprinid herpesvirus 2 (CyHV-2) 614
Cyprinid herpesvirus-3 (CyHV-3) 614
Cyprinus rubrofuscus 192
Cystoseira 486
Cytophaga group 608

D

Dactylogyrosis 616-617
Danio rerio 59, 572
Daphnia 53, 56, 58, 71, 163, 171, 179, 442, 551, 554, 559-561, 571-572, 576
Dawkinsia arulius 12, 55, 56, 126
Dawkinsia assimilis 12, 57
Deep sand bed xiv, 421, 424, 440
Delphinapterus leucas 535
Dendrobates tinctorius 500-501
Dendrobates tinctorius "azureus" 500-501
Denitrification 438, 442, 446-447
Desjardini tang 365
Devil's hand leather coral 525
Diamondback terrapin 504
Diatom filters 437
Dietary ingredients 546
Dinoflagellate infection 605
Diplostomum sp. 617
Discus 181-183, 416, 443, 568-569
DNA barcoding 630
Dogface pufferfish 327
Domestic consumer 652
Domestic market 638, 648, 650, 652
Domestic ornamental fish trade 649
Doubleband surgeonfish 372
Drift Wood xiv, 66, 90, 428
Dropsy 608-609
Dropsy infection 609
Dry feeds 545, 550
DTPA (Diethylenetriaminepentaacetic acid) 496
Duncanopsammia axifuga 520
Dwarf gourami 90-91, 568, 577, 588, 609, 613
Dwarf Gourami Iridovirus (DGIV) 613
Dwarf seahorse 353

E

Earthen ponds 636, 638-639
Earthworms 61, 72, 79, 83, 86-87, 89-90, 150, 152, 154, 161, 167, 170, 175, 209, 328, 551, 554, 557, 574, 582
Echinodorus 450, 463-465, 490-491
Echinodorus grisebachii 464, 465
Echinus esculentus 533
Ecolabelling xviii, 647-648, 656

Economics of large-scale breeding and rearing facility 643
Economics of large-scale breeding and rearing unit 642
Economics of small-scale breeding and rearing facility 641
Ectoparasitic copepod 621
EDTA (Ethylenediaminetetraacetic acid) 496
Edwardsiellosis 611
Edwardisiella tarda 611
Efficiency of fish feeds 548
Egg buriers 567-568, 570, 576
Egg depositors 567-569, 574, 576
Eggfish goldfish 190-191
Egg-layers (oviparous) 567
Egg scatterers 360, 365, 567-569, 572
Eichhornia 465-466, 577
Eichhornia crassipes 465-466
Eimeria sp. 606
Eisenia fetida 553, 557
Eleocharis 466, 491
Eleocharis acicularis 466
Elodeid 449
Elongate glass perchlet 81
Emergent plants 449, 453
Emperor angelfish 320, 322
Endemics 7
Energy metabolism 547
Epistylis (Heteropolaria) infection 605
Epizootic Ulcerative Syndrome 611
Ergasilosis 620
Ergasilus 621
Eschmeyer's scorpionfish 351
Esomus danrica 61
Estuary 70, 222, 242
Etroplus maculatus 237
Eudrilus eugeniae 557
Euphyllia glabrescens 519
European bass 242-243
Exotic x, 7, 449, 599, 623, 641, 643, 647-648, 650
Exotic ornamental fish species 623
Export Market vii, 652
Export trade of ornamental fishes 649
External box filter 438
External parasite leech 619
External protozoan parasites 601
Eye disease 600, 610

F

Fabrication of an Aquarium xiv, 410
Fantail goldfish 189-190
Farm-made feed formulation 549
Feed dispensing methods 550
Feed formulation 549, 552
Feeding Practices xv, 543
Feeding rate xv, 541, 543, 551-552
Feeding ring 417
Feeding schedule 551
Feed Management xv, 544, 633
Ferro-cement tanks 639
Fertilizer mixes 496
Fertilizers 495, 543, 554
Fibre-Reinforced Plastic (FRP) tanks 639
Filter media 436-438
Filters xiv, 414, 417, 434, 436-439, 445, 450
Fin Rot 608
Fire coral 517-518
Fish Cam xiv, 428
Fish feed xv, 202, 207, 216, 541, 544, 546, 550-551, 553, 562, 564-565, 634
Fish feed ingredients 544, 553, 564-565, 634
Fish hospitals 655
Fish lice 620
Fish pox (Carp pox) 614
Fish Stock Calculator 415
Fish Viral Disease xvi, 612

Flavobacterium branchiophila 611
Flavobacterium sp. 612
Flexibacter columnaris 610
Floating plants 80, 449, 491, 571
Florida gar 214-215
Flowerhorn cichlid 174
Fluidized Bed Filter (FBF) 438
Flying barb 61, 588
Food conversion ratio (FCR) 548
Formulation 544, 546-547, 549, 552
Foxface rabbitfish 350
Free Carbon dioxide 443-444
Freeze dried feed 545
Freshwater angelfish 179, 614
Freshwater aquarium plants 450
Freshwater garfish 92
Frontosa cichlid 164

G

Gambusia affinis 206
Gangetic leaf fish 83
Gangetic mystus 67
Genetically modified fish 631-632
Genetically modified organism (GMO) 630
Genetic engineering xvii, 630
Genetics in conservation of biodiversity 629
Genome manipulation 628
Gentoo penguins 534-535
Giant gourami 208, 416
Gill disease 600-601, 611
Gilthead bream 243
Glass catfish 152-153
Glass cover 413
Glass tanks 410, 412, 592, 638-639
Glaucostegus cemiculus 308
Global export of ornamental fishes 646
Global imports 646
Glo Fish 631-632
Glossogobius giuris 52, 96, 228
Golden cobra snakehead 84-85
Golden severum 169-170
Golden trevally 342-343
Goldfish 172, 186-192, 391, 393, 396, 402, 443, 498, 551, 557, 563, 565, 568, 573, 610, 614, 632, 649
Gonopodium 194, 197, 199-201, 205, 207, 581-583, 585-587
Goussia sp. 606
Gnathanodon speciosus 342
Gracilaria 487
Gracilaria parvispora 486-487
Granular feed 545
Great snakehead 86
Green Certification Guidelines 647
Green hair algae 323, 419, 487
Green moray eel 336
Green surf anemone 514
Green swordtail 204
Green terror 171-172
Green water 554, 571, 590
Guidelines for Green Certification of Freshwater Ornamental Fishes 647
Gulf of Mannar 245, 247, 390
Guppy 194-195, 416, 582, 584-586, 604, 649
Gymnomuraena zebra 337
Gymnothorax tile 236
Gymnothorax funebris 336
Gyrodactylosis 616-617

H

Hand feeding 550
Hand nets 417, 434
Hang on Back (HOB) 437
Hara jerdoni 70-71
Heater xiv, 397, 429-430, 433, 447, 569, 571-572, 574, 585
Helanthium tenellum 484
Helminth Diseases xvi, 616
Helophyte 449

Hemoprotozoans. 606
Hermaphrodite 242, 244, 326, 342, 453, 527, 567
Heros severus 125. 169
Herpesviral hematopoietic necrosis 614
Heteractis magnifica 314, 317, 320, 515-516
Heteroconger hassi 338
Hexamita sp. 606
Hexamitiasis 607
Hippocampus guttulatus 356
Hippocampus reidi 354, 389-390
Hippocampus zosterae 353
Hormonal sex reversal 628-629
Hot Spot 7
Hybridization 624, 627-628
Hydrophyte 449
Hydroponics 494
Hygrophila 467, 490
Hypanus americanus 312
Hypostomus 151

I

Ich 376, 602-603
Ichthyobodo 603
Ichthyobodosis 603
Ichthyobodo necator 603
Ichthyobodo pyriformis 603
Ichthyophthiriasis 602
Ichthyophthirius multifiliis 602
Immunodiagnostic Methods xvii, 623
Immunoprophylaxis xvii, 625
Indian glass fish 82
Indian mottled eel 74
Indian mud moray 236
Infectious Diseases xvi, 601
Infusoria 553-555, 571, 573, 576-579, 584
Intergeneric hybridization 628
Interspecific hybridization 628
Intraspecific hybridization 628
Ipomoea 467-468
Isoetid 449
Iteroparous 567

J

Jaguar cichlid 166-167
Javanese cownose ray 311
Jikin goldfish 189

K

Kenyi cichlid 162-163
King Kong Parrot 174
Kits used for diagnosis 624
Koi 150, 192-194, 205, 212, 396, 416, 443, 563, 565, 610, 614, 638, 649
Koi Herpes Virus (KHV) Disease 614
Kreisel 397, 404
Kryptopterus bicirrhis 110, 152
Kryptopterus macrocephalus 153-154

L

Labelling requirements 596
Labeo bata 63
Labeo rohita 64, 631
Lagenandra 468
Lagenandra ovata 468
Lakshadweep islands 245
Larvophile 571
Laterite 423, 495
Lates calcarifer 241
LDPE (Low Density Polyethylene) bags 593
Lemna 449, 468-469
Lemna minor 469
Leopold's angelfish 178
Lepisosteus oculatus 214
Lepisosteus platyrhincus 215
Lernaea bengalensis 621
Lernaea chockoensis 621
Lernaea spp. 621
Lernaeosis 621

Lesser spiny eel 78
Light requirement in aquarium 497
Ligula intestinalis 617
Ligulosis 617
Limnophila 450, 469-470
Limnophila sessiliflora 470
Linckia multifora 509, 530
Lionhead goldfish 190, 632
Lipids 541-542, 582
Livebearers xvi, 194, 197-199, 201, 206, 569, 579-580, 582-584, 587-588, 639
Live-bearers (ovoviviparous) 567
Live Fish Food xv, 553
Live rock xiv, 318, 320, 339-340, 349, 366-367, 370, 397, 420-422, 425, 516, 520
Live sand xiv, 421, 425, 520
Lobophytum 525
Long-snouted seahorse 356
Longspine Snipefish 358
Ludwigia 450, 470, 490, 492
Ludwigia palustris 470
Lymphocystic disease 613
Lymphocystis 613
Lyretail anthias 325

M

Macquarium 404
Macrognathus aculeatus 78
Macrognathus pancalus 48, 79, 80
Macrophyte 449
Macroramphosus scolopax 358
Maculabatis gerrardi 223, 231, 234
Magnificent sea anemone 320, 515
Maintenance feed 551
Malaclemys terrapin 504-505
Mangrove 222, 239, 347, 487-488, 505, 529, 631
Marble 183, 212, 391, 423, 427, 570, 572
Marginal plants 450, 490
Marine mammal parks 396
Marine products development Authority (MPEDA) 650
Marine public aquariums 396, 515, 535
Marker used in selection 627
Marketing channel 652
Marsilea 471
Marsilea quadrifolia 471
Marthasterias glacialis 532
Mascara barb 57
Mash meal 545
Maylandia lombardoi 119, 162
M. catlae 606
Mechanical filtration 434
Megalocytiviruses 614
Mesocyclops 562
Messmate pipefish 347-348
Methods for genetic improvement 627
Metynnis argenteus 157-158
Microcyclops 562
Micro landschaft 396, 402
Micropropagation 493
Millepora alcicornis 517
Millepora coral 521
Mimic tang 366
Modern Public Aquariums 405
Moina 442, 551, 553, 559, 561, 576
Moinadaphnia 561
Moist feeds 550
Molecular diagnostic methods 624
Molecular markers 627, 629
Monodactylus argenteus 346
Monopterus cuchia 47, 77
Montipora capricornis 523
Mosquitofish 206-207
Mosquito larvae 177, 179, 207, 458, 551, 553-554, 557, 572, 574
Mottled linckia 530
Mouth brooders 567-568, 571, 576

Mouth Fungus 610
M. sphericum 606
Myriophyllum 471-472, 492, 577-578
Mystus cavasius 67
Mystus vittatus 68
Myxobolus bengalensis 606
Myxobolus spp. 606
Myxosporean Disease 606

N

Najas 450, 472
Nandus nandus 83-84
Nano reef 397
Naso vlamingii 373-374
National Agriculture Innovation Projects (NAIP) 651
National Bank for Agriculture and Rural Development (NABARD) 650
National Fishery Development Board (NFDB) 650
Natural carotenoids 563-565
Nemacheilus corica 66
Nematode Infection 619
Nematodes (Thread worm) 619
Neon tetra 156, 606, 646
Nest builders 213, 567-568, 571, 577
Nitrification 422, 445-446
Nitrifier bacteria 446
Nitrobacteria 446
Nitrogen cycle 422, 424, 435-436, 446-447
Nitrosomonas 446, 640
Nitrospira 446
Nodular Disease 606
Non-dry feeds 545
Notopterus notopterus 46, 75, 76
Nutritional diseases 601
Nymphae 491
Nymphaea pubescens 473
Nymph goldfish 189
Nymphoides 450, 473-474
Nymphoides indica 474

O

Oceanarium 395-396, 401, 405
Odonus niger 298, 377
Offsets 491
Oodiniasis (Velvet disease) 605
Oodinium ocellatum 605
Oodinium sp. 605
Open brain coral 518, 632
Oranda goldfish 189
Orange chromide 237
Orange clownfish 319
Oriental sole 235
Ornamental Fish Promotion Centres and Bio-parks 655
Oscar 183-184, 443
Osphronemus goramy 137, 208
Osteoglossum bicirrhosum 103, 146
Osteoglossum ferreirai 103, 146, 148
Ottelia 450, 474
Oviparous 66, 68, 93, 308, 320, 328, 342, 352, 358-359, 567, 616-617
Ovophile 163, 571
Ovoviviparous 235, 311, 313, 349, 354-355, 357, 567, 579-581
Ovuliparity 567

P

Pacific sea nettle 512
Packing xvi, 591-597
Packing density 594-595
Palk Bay 245, 247
Pallisentis sp. 619
Paludarium 395, 401, 501
Pangas catfish 69
Pangasius pangasius 69-70
Panulirus argus 527-528
Paracanthurus hepatus 296, 368
Parachcirodon innesi 113, 156
Parachromis managuensis 166

Paracyclops 562
Paramecium sp. 553
Paste feeds 545
Pearlscale goldfish 189
Pennant coralfish 331
Perionyx excavatus 557
Pico reef 397, 403
Pistia 449, 475, 491, 577
Pistia stratiotes 475, 491
Placoderms 1
Plankton 163, 324, 329, 337, 347, 355, 442, 511, 523, 554, 557
Plating Montipora 397, 523
Platy 201-204, 416, 582, 584, 586, 649
Plecostomus 151
Pethia conchonius 8, 52, 53, 128
Pleuston 449
PMMSY, Government of India 654
Poecilia latipinna 136, 198
Poecilia reticulata 194
Poecilia sphenops 136, 196, 197
Poecilia velifera 200
Poly-lined ponds 638-639
Polystyrene boxes 593-594
Pomacanthus imperator 320-321
Pompom goldfish 190
Pontederia 465
Pool barb 54
Pork fish 339
Power air pump 432
Power filter 417, 432
Powerhead 397, 431
Pre-Conditioning Preparations xvi, 592
Probiotics and immunostimulants 625
Process of genetic engineering 630
Production technology 649
Propagation of Aquarium Plants xv, 489, 498
Protandrous 315, 567, 589
Protein Efficiency Ratio (PER) 548-549
Proteins 335, 440-441, 541-542, 551, 578, 582
Protein skimmer 397, 420, 429, 440-441, 589
Protogynous 326, 342, 567
Protoreaster nodosus 509, 529
Protozoan Diseases xvi, 601
Pseudo-aquarium plants 452
Pseudanthias squamipinnis 259, 325, 326
Pseudomonas sp. 612
Pseudophallus 581
Pterophyllum altum 125, 176, 177
Pterophyllum leopoldi 125, 176, 178
Pterophyllum scalare 125, 176, 179
Pterois volitans 345
Public aquarium 174, 186, 235, 244, 308, 392, 405, 517
Public aquariums in India 406
Public Aquariums of the world 405
Public promotion of ornamental fisheries 654
Puntius sophore 54
Purple-striped jellyfish 511-512
Putitor mahseer 58
Pygocentrus nattereri 160-161
Pygoscelis papua 534

R

Raja undulata 307
Ray 5, 68, 95, 203, 208, 234, 307, 309, 311, 324, 326, 331, 362, 370
Rearer 652
Rearing tank 571-572, 590
Red-bellied piranha 160
Red-eared slider 502
Red lionfish 345-346
Red toothed triggerfish 377
Reef aquarium 317, 338, 349, 396-397, 403, 422, 429-431, 522, 526
Refugium 397, 441-442, 487

Regal tang 368

Reinforced Cement Concrete (RCC) tanks 639

Relative Food Conversion Ratio (RFCR) 548

Retailer 652

Reticulate loach 65

Retinopathy (VER) 612

Rhina ancylostoma 251, 309, 310

Rhinopias eschmeyeri 290, 351

Rhinoptera javanica 311

Rhizome dividing 492

Rhizomes 452, 456, 463, 473, 484, 488, 491-492

Rhizophora 487-488

*Rhizophora mangl*e 488

Riccia 449, 475-476

Riccia fluitans 476

River rock 423

RNA interference 626

Rockwool 494-495

Roho labeo 64

Rosy barb 52, 416, 443, 588, 649

Runners 465-466, 476, 479-480, 490-491

Ryukin goldfish 189

S

Sagittaria 476

Sailfin molly 198, 200

Sailfin Tang 365, 369, 375

Salinity 222, 241, 243, 315, 337, 394, 397, 429, 443, 446, 449, 519-520, 524, 526-527, 559, 637

Salvinia 449, 477

Sand tiger shark 305, 307, 390

Saprolegnia 615

Saprolegniasis 615-616

Sargocentron xantherythrum 359

Scatophagus argus 239

Scleropages formosus 103, 146, 149

Scoliosis 601

Scraper 417

Sea nettle 511-513

Semelparous 567

Sergeant major damselfish 332

Setting up of an aquarium 414, 419

Severum 169-171

Sexual propagation 475, 490

Shark 5, 70, 304-305, 307, 309, 390, 416

Sharpnose stingray 234

Shubunkin 188-189

Siamese fighting fish 211, 404

Siganus uspi 290, 349

Siganus vulpinus 350

SILO Feed 552

Silver arowana 146, 148

Silver dollar 157, 159

Silver moony 346

Sleeping sickness 607

Slender seahorse 354

Snakehead 84-86, 88

Snakelocks anemone 513-514

Snubnose pompano 343

Sohal tang 371

Sound and vibration 443, 445

Southern platy 201, 203

Southern stingray 312-313

Spanish hogfish 313, 341-342

Sparus aurata 243

Spawning tank 569, 571, 585

Spiny -headed worms 619

Spiny starfish 532

Spironucleus sp. 606

Sponge filter 569, 571, 573-574, 576

Spotted archerfish 238

Spotted gar 214

Spotted garden eel 338, 390

Spotted Scat 239-240

Spray bar 432
Square method 547-549
Staghorn coral 524
Staphylococcus aureus 610
Starvation 601
Stolon 455, 465-466, 473, 475, 490-491
Stone loach 66
Storage of fish feeds 550
Strategies for fish health management 626
Streptococcus sp. 612
Stress-mediated diseases 600
Striped dwarf catfish 68
Striped glass catfish 153
Striped panchax 93, 95
Striped razorfish 357
Striped snakehead 88
Striped squirrelfish 359
Submerged plants 82, 450, 461, 577
Sucker 151, 430, 491
Suckermouth catfish 151
Sump 396, 417, 419, 429, 441-442
Super Red Dragon Flowerhorn 175
Symphysodon aequifasciatus 126, 181, 182
Symphysodon discus 181-182
Symphysodon tarzoo 181, 182
Synnema 478, 492
Synthetic carotenoids 563

T

Tablet form 545
Tail Rot 608
Tank goby 96, 102
Taxiphyllum 478
Telescope goldfish 190
Tetrahymena infection 604
Tetrahymena pyriformis 604
Thai Silk Flowerhorn 175
Thalassia 488
Thalassia hemprichii 486, 488
Thermometer xiv, 430, 433
Threadfin butterflyfish 328
Three-spot gourami 209
Tinfoil barb 185-186
Tissue culture 490, 492-493
Titan triggerfish 379
Tomato clownfish 317-318
Torch coral 519
Tor putitora 58-59
Tosakin goldfish 190
Total Hardness 443-444
Toxotes chatareus 227, 238
Trachemys scripta 502
Trachinotus blochii 343, 344
Trachyphyllia geoffroyi 518
Transgenic Fishes xvii, 630
Transgenic fish research in India 631
Transgenic or Living Modified Organism (LMO) 630
Transporting and Transferring xvi, 596
Transport of Ornamental Fish xvi, 591, 596, 598
Tray feeding 550
Trichodina 603
Trichodinella 603
Trichodiniasis 603
Trichogaster lalius 50, 90, 91
Trichopodus trichopterus 138, 210
Trickle filters 437
Tripartiella 603
Tropocyclops 562
Trumpet fish 358, 380
Trypanosoma sp. 606-607
Tubifex 65, 91, 153-154, 172, 175, 204-205, 238, 544, 551, 553-554, 556, 572
Type of production facilities 638
Typha 450, 478-479

U

Ulcer disease or Ulcerative disease 610

Undergravel (UG) filters 438
Undulate skate 307

V

Vallisneria 450, 479, 490
Vallisneria spiralis 479
Variable platyfish 203
Vegetative propagation 470, 490
Veiltail goldfish 190
Vermiculite 423
Vertical farming 494, 497
Viral Encephalopathy 612
Viral Nervous Necrosis (VNN) 612
Virtual aquarium 404-405
Vitamins and mineral 558, 582
Vivarium 395, 400, 478
Viviparous 207, 305, 311, 488, 567, 579, 617
Viviparous livebearer fishes 579
V. parahaemolyticus 612

W

Wakin goldfish 189-190
Wallago 73
Wallago attu 73
Western mosquitofish 206
Wet feeds 545, 550
Whisker coral 520
White spot disease 602-603
White worms 153, 172, 182, 211, 557, 571
Wholesaler 652

X

Xenentodon cancila 51, 92, 145
Xiphophorus helleri 136, 204, 205
Xiphophorus maculatus 201
Xiphophorus variatus 203

Y

Yellowfin tang 376
Yellowtail damselfish 334
Yellow Tang 375
Yucatan molly 200

Z

Zebra danio 59, 568, 572, 646
Zebra moray eel 337
Zebrasoma desjardinii 294, 365
Zebrasoma flavescens 296, 375
Zebrasoma veliferum 246, 296, 369
Zoanthus 526
Zostera noltii 485